DATE DUE

WITHDRAWN

CORDILLERAN METAMORPHIC
CORE COMPLEXES

Peter Misch

In recognition of his pioneer work in the Cordilleran hinterland, this volume is dedicated to Peter Misch.

Peter Misch

In recognition of his pioneer work in the Cordilleran hinterland, this volume is dedicated to Peter Misch.

The Geological Society of America, Inc.
Memoir 153

Cordilleran Metamorphic Core Complexes

Edited by

Max D. Crittenden, Jr.
Geologic Division
Branch of Western Environmental Geology
345 Middlefield Road
Menlo Park, California 94025

Peter J. Coney
Department of Geosciences
University of Arizona
Tucson, Arizona 85721

George H. Davis
Department of Geosciences
University of Arizona
Tucson, Arizona 85721

1980

Copyright © 1980 by The Geological Society of America, Inc.
Copyright is not claimed on any material prepared by
Government employees within the scope of their employment.
Library of Congress Catalog Card Number 80-67489
ISBN 0-8137-1153-3

Published by
THE GEOLOGICAL SOCIETY OF AMERICA, INC.
P.O. Box 9140, 3300 Penrose Place
Boulder, Colorado 80301

Printed in the United States of America

Contents

Dedication to Peter Misch ii

PART 1. INTRODUCTION AND OVERVIEW

Introduction .... *Peter J. Coney*   3

Cordilleran metamorphic core complexes: An overview .... *Peter J. Coney*   7

PART 2. SOUTHERN BASIN AND RANGE

Structural characteristics of metamorphic core complexes, southern Arizona .... *George H. Davis*   35

Mylonitization and detachment faulting in the Whipple-Buckskin-Rawhide Mountains terrane, southeastern California and western Arizona .... *Gregory A. Davis, J. Lawford Anderson, Eric G. Frost, and Terry J. Shackelford*   79

Geologic and geochronologic reconnaissance of a northwest-trending zone of metamorphic core complexes in southern and western Arizona .... *William A. Rehrig and Stephen J. Reynolds*   131

Mid-Tertiary plutonism and mylonitization, South Mountains, central Arizona .... *Stephen J. Reynolds and William A. Rehrig*   159

Geology of a zone of metamorphic core complexes in southeastern Arizona .... *Norman G. Banks*   177

Evidence for multiple intrusion and deformation within the Santa Catalina–Rincon–Tortolita crystalline complex, southeastern Arizona .... *Stanley B. Keith, Stephen J. Reynolds, Paul E. Damon, Muhammad Shafiqullah, Donald E. Livingston, and Paul D. Pushkar*   217

Distribution and U-Pb isotope ages of some lineated plutons, northwestern Mexico .... *T. H. Anderson, L. T. Silver, and G. A. Salas*   269

PART 3. EASTERN GREAT BASIN

Transition from infrastructure to suprastructure in the northern Ruby Mountains, Nevada .... *Arthur W. Snoke*   287

Metamorphic infrastructure in the northern Ruby Mountains, Nevada .... *Keith A. Howard*   335

Structure and petrology of a Tertiary gneiss complex in northwestern Utah .... *Victoria R. Todd*   349

Fabrics and strains in quartzites of a metamorphic core complex, Raft River Mountains, Utah .... *Robert R. Compton*   385

Structural geology of the northern Albion Mountains, south-central Idaho .... *David M. Miller*   399

PART 4. THE NORTHWEST

Bitterroot dome–Sapphire tectonic block, an example of a plutonic-core gneiss-dome complex with its detached suprastructure .... *Donald W. Hyndman* 427

Structural and metamorphic evolution of the northeast flank of the Shuswap complex, southern Canoe River area, British Columbia .... *P. S. Simony, E. D. Ghent, D. Craw, W. Mitchell, and D. B. Robbins* 445

Kettle dome and related structures of northeastern Washington .... *Eric S. Cheney* 463

Metamorphic core complexes of the North American Cordillera: Summary ... *Max D. Crittenden, Jr.* 485

PART 1

INTRODUCTION AND OVERVIEW

Introduction

PETER J. CONEY
Department of Geosciences
University of Arizona
Tucson, Arizona 85721

Cordilleran metamorphic core complexes, the subject of this volume, are a group of generally domal or archlike, isolated uplifts of anomalously deformed, metamorphic and plutonic rocks overlain by a tectonically detached and distended unmetamorphosed cover. The features are scattered in a sinuous string along the axis of the eastern two-thirds of the North American Cordillera from southern Canada to northwestern Mexico. To date, more than 25 of them are known, and it is significant that more than half of them have been recognized only since 1970. They are, without question, the newest and most controversial addition to the recognized architecture of the eastern two-thirds of the Cordillera since the discovery, in the early 1960s, of the Tertiary calderas and their associated vast ignimbrite sheets.

The first hint of the existence and potential significance of these complexes was certainly from the early work of Peter Misch and his army of students who invaded the Great Basin in the 1950s. Misch (1960) discovered and emphasized scattered occurrences of a major subhorizontal dislocation plane or "decollement," as it was called, separating the unmetamorphosed Paleozoic miogeoclinal cover from a generally metamorphosed substratum. He found this relation repeatedly in many ranges scattered along, and west of, the Utah-Nevada border almost 200 km west of the already known low-angle, east-verging Mesozoic thrust faults in central Utah. What is most puzzling about this regional context is that the thrust faults in central Utah are typical "older-on-younger" faults similar to those found in foreland thrust belts throughout the world. On the other hand, the now infamous "decollements" of the hinterland, as the region came to be called, are typical "younger-on-older" faults associated with listric normal faults. These relations suggested that the basal shearing-off plane of a major foreland thrust belt was fortuitously exposed, perhaps due to Tertiary block faulting and uplift. Misch and his students tried to prove a geometric connection between the low-angle decollements of the hinterland and the eastern thrusts, but they could not. The broken continuity of the basin-range structure made this impossible. The matter was thus left to interpretation and discussion. The important result, however, was the identification of a new and important terrane and new problems. Another result was the beginning of what might be called a consensus that the strange features of the Nevada hinterland, namely the metamorphosed substratum and the overlying decollements, were genetically linked to the Cordilleran thrust belt far to the east.

During the 1960s, more attention was paid to the metamorphic substratum beneath the decollements, and it soon became known as an "infrastructure" beneath an unmetamorphosed and brittle "suprastructure" above (Armstrong, 1968; Armstrong and Hansen, 1966). The models presented were reminiscent of those from Caledonian-Alpine orogens in Greenland and western Europe. This set the stage for the Price and Mountjoy (1970) model for the eastern Canadian Cordillera, which was one of the most beautifully conceived and articulated tectonic proposals in the history of Cordilleran tectonic thought. This model invoked a massive Infrastructural culmination in the Shuswap metamorphic core zone which rose buoyantly to propel the foreland thrust belt eastward, largely by gravity. This model had great influence on workers in the United States and firmly established in peoples' minds a genetic relationship between the features of the hinterland and features of the thrust belts to the east. This was the consensus reached by the early 1970s, and it is reflected in a number of papers published about that time (for example, see Roberts and Crittenden, 1973; Hose and Danes, 1973). In this context, the degree to which gravity-sliding mechanisms were invoked to explain the proposed relationships between the distended hinterland and the compressed foreland to the exclusion of regional compression as a driving force is interesting from a historical point of view. When we consider the pre–"Plate Tectonics" setting of the times and what we now know about the Nevada hinterland, this is not surprising. This, however, was not the interpretation of Misch. He thought the eastern thrusts and the decollements were formed as the cratonic basement drove westward into deep-seated thrust roots, peeling and shearing off the Phanerozoic cover as it moved. He was thus invoking a model remarkably prophetic in the context of more recent plate tectonics models of convergent and collisional plate margin orogens.

In any event, it is with the above-mentioned consensus that the contents of this volume are to be compared. Since 1970, the situation has become one of considerable controversy and heated debate. In the process, some discoveries have been made and there have been some new insights into Cordilleran tectonics. Considerable confusion yet prevails, and there is still substantial, justifiable disagreement. The editors of this volume are convinced that the contents of this volume present a new perception of Tertiary Cordilleran tectonic evolution and establish the importance of Cordilleran metamorphic core complexes in this Tertiary history and in the architecture of the Cordillera.

We owe our readers an explanation of the term "Cordilleran metamorphic core complex." There has been objection to the name and widespread wonderment about its meaning. The name was coined, quite accidentally, by me (Coney, 1973, p. 723) in reference to the Shuswap metamorphic complex and its relationship to the Canadian Rocky Mountain foreland thrust belt. I am sure it was an unconscious reference to the term "metamorphic core zone" used by Canadian workers (Wheeler and Gabrielse, 1972) to describe the eastern Omineca crystalline belt in the Canadian Cordillera (see Monger and Hutchison, 1972). Since then, the term has been applied by me and others to the many metamorphic terranes and related features found southward in the Cordillera as they were recognized or reevaluated; these features are the subject of this volume. The term has stuck because the complexes seemed so distinctively peculiar, and they needed a special name. I make no apology for the term. It has proved useful for many of us, and it indicates that, although variety exists, the objects so identified have a considerable commonality as to physical features and age.

The events that brought about this new insight into Tertiary Cordilleran tectonics are recounted in Coney (this volume) and are reflected in the contents of this memoir. The most important aspect of this new vision of the Cordilleran hinterland was certainly the suspicion and eventual realization that there had been a strongly manifested and enigmatic Tertiary overprint superimposed on the region. The overprint was most clearly recognized in young K-Ar apparent ages of basement rocks and in the listric normal faulting and decollement zones characteristic of the unmetamorphosed cover terranes of the complexes. This activity followed Mesozoic thrusting and preceded basin-range block faulting, was more or less contemporaneous with widespread Tertiary ignimbrite eruptions, and created a strange image of tectonic response not easily understood then or even today.

In retrospect, the following developments proved significant: (1) publication of Armstrong's (1972) views on Tertiary denudational faulting in the Nevada hinterland and subsequent tests which supported his views (Coney, 1974; Todd, 1973); (2) discovery of the large number and great extent of metamorphic complexes in Arizona and recognition of their Tertiary age (Davis, 1973)—also important was the realization that these metamorphic culminations in Arizona were not in a hinterland behind a thrust belt of any age, which cast suspicions on the presumed genetic link between the northern complexes and the thrust belts to the east; and (3) the realization that most of the known complexes throughout the Cordillera had very similar lithologies and structures suggested a common origin. It was a particularly significant event for me when I first visited the Catalina-Rincon complex near Tucson, Arizona, with George Davis in early 1975. Having just finished field work in the Snake Range of eastern Nevada (Coney, 1974), I was surprised to find features startlingly like those I had become so familiar with far to the north. Expression of this created a sense of urgency, which eventually led us to join with Max Crittenden in convening a Penrose Conference on Cordilleran Metamorphic Core Complexes at Tanque Verde Ranch near Tucson in May 1977. The conference subsequently led to a symposium session on the complexes at the Geological Society of America Annual Meeting at Seattle in November 1977, then to this volume. We are grateful to the Society for all these opportunities.

It is remarkable how serendipitously the present state of knowledge evolved, as reflected in the contents of this volume. It is also instructive to note how scattered early works, which were seeds of what was to come, went largely unnoticed and have only recently flowered to significance and appreciation, albeit still not completely understood. Such work includes Anderson's (1971) study of Tertiary listric normal faulting south of Lake Mead and a number of early studies in Arizona and in the Nevada hinterland (for example, see Damon and others, 1963; Drewes, 1964; Moores and others, 1968).

It will be obvious to all readers that considerable controversy still exists and there are major problems surrounding the complexes which still need to be resolved. This is only normal for a new and complicated large-scale tectonic feature. At the same time, this situation adds to the excitement and clearly indicates need for continued study. A number of novel and somewhat contrasting models recently have been proposed to explain the new data surrounding the complexes, and there certainly will be more to follow. Several differing concepts can be found in this volume. This is also only natural and need not be taken too seriously at this time. There is much work to do yet. In any event, without theoretical abstractions to guide us, or to find fault with, there is little progress. As one of the authors who has contributed to this volume recently said in the midst of a spirited discussion on proper and improper terminology surrounding the complexes, 'We are re-writing the textbooks. . . .'' This, of course, remains to be seen, but the point is well taken, and there will be much continued spirited discussion before the issue is finally laid to rest.

In conclusion, the major contribution of all the work represented here is to signal an anomalous petrotectonic assemblage, now recognized from southern Canada to northwestern Mexico, which has an almost certain early to middle Tertiary evolutionary history. The evidence seems to indicate that a significant part of this assemblage was produced by extensional tectonics and that, whatever the process was, it postdates Sevier-Laramide compressional thrusting and predates late Tertiary extensional block faulting. It has been superimposed on, and confused with, earlier and mainly Mesozoic tectonic events, but the contents of this volume generally choose to emphasize the importance of the Tertiary events.

The most important controversy still remaining is the origin and significance of the mylonitic gneiss fabrics so characteristic of the basement cores of the complexes. Without question, the dramatic resemblance of these fabrics to those produced along deep-seated thrust faults in other parts of the world is the last remaining obstacle to what might be called a new consensus. Some workers have concluded that even this aspect of the complexes was produced by Tertiary regional extension. This fact will demand and inspire much-needed intense work in the future. It is remarkable how little detailed petrography and petrology have been done in these terranes. It is extremely important to date the terranes better, and if preliminary results are any indication, the dating of these basement terranes will be a geochronological nightmare which only detailed, multiple-attack methods combined with very careful field control will resolve.

REFERENCES CITED

Anderson, R. E., 1971, Thin skin distension in Tertiary rocks of southeastern Nevada: Geological Society of America Bulletin, v. 82, p. 43–58.

Armstrong, R. L., 1968, Mantled gneiss domes in the Albion Range, southern Idaho: Geological Society of America Bulletin, v. 79, p. 1295–1314.

—— 1972, Low-angle (denudation) faults, hinterland of the Sevier orogenic belt, eastern Nevada and western Utah: Geological Society of America Bulletin, v. 83, p. 1729–1754.

Armstrong, R. L., and Hansen, E., 1966, Cordilleran infrastructure in the eastern Great Basin: American Journal of Science, v. 264, p. 112–127.

Coney, P. J., 1973, Non-collision tectogenesis in western North America, *in* Tarling, D. H., and Runcorn, S. H., eds., Implications of continental drift to the earth sciences: New York, Academic Press, p. 713–727.

—— 1974, Structural analysis of the Snake Range 'décollement,' east-central Nevada: Geological Society of America Bulletin, v. 85, . 973–978.

—— 1980, Cordilleran metamorphic core complexes: An overview: Geological Society of America Memoir 153 (this volume).

Damon, P. E., Erickson, R. C., and Livingston, D. E., 1963, K-Ar dating of Basin and Range uplift, Catalina Mountains, Arizona: Nuclear Geophysics, National Academy of Sciences/National Research Council Publication 1075.

Davis, G. H., 1973, Mid-Tertiary gravity-glide folding near Tucson, Arizona: Geological Society of America Abstracts with Programs, v. 5, p. 592.

Drewes, H., 1964, Diverse and recurrent movement along segments of a major thrust fault in Schell Creek Range near Ely, Nevada, *in* Geological Survey research, 1964: U.S. Geological Survey Professional Paper 501-B, p. B20–B24.

Hose, R. K., and Danes, Z. F., 1973, Development of late Mesozoic to early Cenozoic structures in the eastern Great Basin, *in* DeJong, K. A., and Scholten, R., eds., Gravity and tectonics: New York, John Wiley & Sons, p. 429–442.

Misch, P., 1960, Regional structural reconnaissance in central-northeast Nevada and some adjacent areas: Observations and interpretations: Intermountain Association of Petroleum Geologists Guidebook for 11th Annual Field Conference, p. 17–42.

Monger, J.W.H., and Hutchison, W. W., 1971, Metamorphic map of the Canadian Cordillera: Canada Geological Survey Paper 70-33, 61 p.

Moores, E. M., Scott, R. B., and Lumsden, W. W., 1968, Tertiary tectonics of the White Pine–Grant Range region, east-central Nevada, and some regional implications: Geological Society of America Bulletin, v. 79, p. 1703–1726.

Price, R. A., and Mountjoy, E. W., 1970, Geologic structure of the Canadian Rocky Mountains between Bow and Athabasca Rivers—A progress report: Geological Association of Canada Special Paper No. 6, p. 7–25.

Roberts, R. J., and Crittenden, M. D., Jr., 1973, Orogenic mechanisms, Sevier orogenic belt, Nevada and Utah, *in* DeJong, J. A., and Scholten, R., eds., Gravity and tectonics: New York, John Wiley & Sons, p. 409–428.

Todd, V. R., 1973, Tectonic mobilization of Precambrian gneiss during Tertiary metamorphism and thrusting, Grouse Creek Mountains, northwestern Utah: Geological Society of America Abstracts with Programs, v. 5, p. 116.

Wheeler, J. O., and Gabrielse, H., 1972, The Cordilleran structural province, *in* Prince, R. A., and Douglas, R.J.W., eds., Variations in tectonic styles in Canada: Geological Association of Canada Special Paper No. 11, p. 9–81.

Geological Society of America
Memoir 153
1980

Cordilleran metamorphic core complexes: An overview

PETER J. CONEY
Department of Geosciences
University of Arizona
Tucson, Arizona 85721

ABSTRACT

More than 25 distinctive, isolated metamorphic terranes extend in a narrow, sinuous belt from southern Canada into northwestern Mexico along the axis of the North American Cordillera. Appreciation of these terranes has evolved slowly, and more than half of them have been recognized only since 1970. Growing evidence shows that these metamorphic terranes and related features evolved in part during early to middle Tertiary time (55 to 15 m.y. B.P.), that is, after the Laramide orogeny but before basin-range faulting. These terranes have been termed "metamorphic core complexes."

The complexes are characterized by a generally heterogeneous, older metamorphic-plutonic basement terrane overprinted by low-dipping lineated and foliated mylonitic and gneissic fabrics. An unmetamorphosed cover terrane is typically attenuated and sliced by numerous subhorizontal younger-on-older faults. Between the basement and the cover terranes is a zone of "decollement" and/or steep metamorphic gradient with much brecciation and kinematic structural relationships indicative of sliding and detachment. Plutonic rocks as young as early to middle Tertiary age are deformed in the basement terranes of many of the complexes, and some of the deformed cover includes continental sedimentary and volcanic rocks of early to middle Tertiary age.

Some complexes exhibit evidence of prolonged deformation and metamorphism extending back into Mesozoic and even Paleozoic time. All the complexes, however, reveal an early to middle Tertiary deformational and metamorphic overprint that is interpreted to be mainly of extensional origin. The extension coincided with a vast plutonic-volcanic flare-up of magmatic arc affinity mainly during Eocene time in the Pacific Northwest and mainly during late Eocene–Oligocene to middle Miocene time south of the Snake River Plain. The exact tectonic significance of the complexes remains obscure. Their extensional aspect clearly postdates, and seems unrelated to, Cretaceous and early Tertiary Sevier and Laramide compressional tectonics, but predates the more obvious late Tertiary basin-range extension and rifting.

INTRODUCTION

The North American Cordillera has more than 25 distinctive, isolated metamorphic terranes scattered along a narrow, sinuous belt from southern Canada to northwestern Mexico (Fig. 1.). Appreciation of these terranes has come slowly, and more than half of them have been recognized only since 1970. During the past 10 yr, a considerable debate has centered on them, and they have been termed "metamorphic core complexes" (Coney, 1B73a, 1976, 1978, 1979; Davis and Coney, 1979; Crittenden and others, 1978).

Orogenic theory long ago acknowledged the importance of regional metamorphism in the evolution of mountain systems. The exact tectonic significance of this metamorphism, however, remained elusive. Largely as a result of ideas of Wegmann (1935) and Haller (1956), the concept of an axial core zone of a mountain system evolved (de Sitter, 1956). This axial zone was visualized as being made up of metamorphic rocks that displayed evidence of extreme ductile flow, gneiss domes (Eskola, 1949), and related plutonic rocks. The zone was termed an "infrastructure" and was seen as distinct from an overlying or flanking, brittle, superficial "suprastructure." The boundary between these two contrasting domains was seen as generally sharp and characterized by steep thermal gradients and structural disharmony. Most of this theory was based on circum–North Atlantic Caledonian and Hercynian examples and the Alpine belt of southern Europe.

Zwart (1969) attempted classification of orogenic metamorphic facies into two basic types. He recognized a Hercynian end-member characterized by granites and by high-temperature and low- to moderate-pressure facies and an Alpine end-member characterized by high-pressure and low-temperature facies. He did not specify any particular tectonic process to explain these types.

A similar distinction was made by Miyashiro (1961, 1973), but was cast in a completely different light. Miyashiro recognized the same two contrasting facies, but he paired them in a single orogen, thus laying a cornerstone of plate-tectonics theory. For him, the high-temperature type was linked directly to processes in magmatic arcs, whereas the high-pressure type was linked to the trench. Both of these contrasting metamorphic facies were interpreted as the result of subduction of oceanic lithosphere along active arc-trench systems in island arcs or along consuming continental margins. Enter plate tectonics and the experience of the Pacific Ocean. Mountain-system evolution suddenly became more comprehensible (Hamilton, 1969; Dewey and Bird, 1970; Coney, 1970), and all moderate- to high-temperature Cordilleran-type metamorphism suddenly appeared to be related to processes associated with magmatic arcs.

METAMORPHISM IN THE NORTH AMERICAN CORDILLERA

Of particular interest to the concerns of this volume were developments in the 1960s which focused attention on a belt of metamorphic rocks directly west of the east-verging foreland fold and thrust belt of the North American Cordillera. Recognition of these metamorphic rocks produced various genetic interpretations that attempted to link the metamorphic core zone with the fold and thrust belt in models reminiscent of Caledonian and Alpine orogens. The earliest and most noteworthy were the ideas of Misch (1960), Armstrong and Hansen (1966), and Price and Mountjoy (1970). Out of these studies rose a consensus that the metamorphic core zone was a deep-seated infrastructural culmination, or hinterland, which evolved contemporaneously with and just west of the superficial thin-skinned fold and thrust belt.

To Misch (1960), the most important element of the eastern Nevada structural province, or hinterland, was a regional structural discontinuity, termed a "decollement." This discontinuity separated Precambrian basement and metamorphosed upper Precambrian–lower Paleozoic sedimentary rocks from an overlying unmetamorphosed allochthon. The type area for this regional decollement was the Snake Range of east-central Nevada. Misch reasoned that the decollement and the associated shearing and metamorphism along it were produced during the "mid-Mesozoic orogeny." He thought that the discontinuity was formed as cratonic basement moved westward into deep-seated thrust roots, peeling and shearing off the Phanerozoic cover as it moved. The implication was that the decollement was structurally continuous with pre-Laramide low-angle break-out thrust faults to the east in central Utah, along and west of the Wasatch line.

Working in the same region, Armstrong (1968b) and Armstrong and Hansen (1966) emphasized remobilization of basement rocks and metamorphism of the lowest part of the Phanerozoic cover during the mid-Mesozoic orogeny. In an analogy to Caledonian systems, they termed the remobilized core zone an infrastructure. In contrast to Misch, they emphasized mobility *below* the decollement (or abschrung zone) and contrasted this domain of recumbent folds and planar fabrics to a less-deformed, brittle suprastructure above.

They did not make any specific genetic link between the infrastructure-suprastructure tectonics of the hinterland and the fold and thrust belt to the east.

The Price and Mountjoy (1970) model for the tectonic evolution of the eastern Cordillera of southern Canada was one of the best-formulated and persuasive tectonic syntheses in the history of Cordilleran geologic thought. They proposed that a hot mobile infrastructure rose buoyantly in the Shuswap axial metamorphic core zone, gravitationally spread eastward, and propelled the Rocky Mountain foreland fold and thrust belt to the east. The deformation was seen as continuously evolving upward and eastward from Late Jurassic to early Tertiary time. Price and Mountjoy did not hesitate to directly link the evolving infrastructure with the foreland folding and thrust faulting to the east in a single grand genetic model.

The low-angle thrust faults along the Wasatch line in central and southern Utah are "older-on-younger" faults (Misch, 1960; Armstrong, 1968a) typical of foreland thrust belts the world over (Coney, 1973b). In the hinterland to the west, however, above the metamorphic rocks, low-angle faults are "younger-on-older" (Armstrong, 1972). Nothing quite like these widespread younger-on-older faults has been emphasized in the Canadian Rocky Mountains. In spite of this, the Price and Mountjoy model had great influence on workers in the western United States. As a result, subsequent syntheses attempted to apply modifications of the Canadian example to the battleground of eastern Nevada and western Utah (Roberts and Crittenden, 1973; Hose and Danes, 1973). Hose and Danes, for example, interpreted the decollement and younger-on-older faults in cover rocks as the result of extensional gravity-driven movement of the cover off an uplifted hinterland in eastern Nevada (see Fig. 3B). This cover terrane slid eastward to become the superficially telescoped older-on-younger thrust faults of central Utah.

The preceding account sets the stage for developments that were to cast considerable confusion over the simplicity and beauty of the early models. What followed, mostly after 1970, is one of the most fascinating debates in Cordilleran tectonic history.

The debate was predicted. Early workers noted certain data and relationships that were either troublesome or inconsistent with existing models. Damon (Damon and others, 1963; Mauger and others, 1968) found very young (middle-Tertiary) K-Ar "cooling ages" from metamorphic rocks in southern Arizona. Armstrong and Hansen (1966) found similarly young (Tertiary) ages from metamorphic rocks in Nevada. Misch (1960) was aware of Tertiary gravitational gliding and superficial brecciation in the Snake Range of eastern Nevada, and Drewes (1964) discussed multiple thrusting and gravity faulting extending into Tertiary time in the Schell Creek Range of eastern Nevada. Moores and others (1968) recognized Tertiary deformation and metamorphism in the White Pine–Grant Range of eastern Nevada. For many workers, most of these inconsistencies seem to have been explained as the result of a minor Tertiary overprint of a basically Mesozoic tectonic regime of metamorphism and

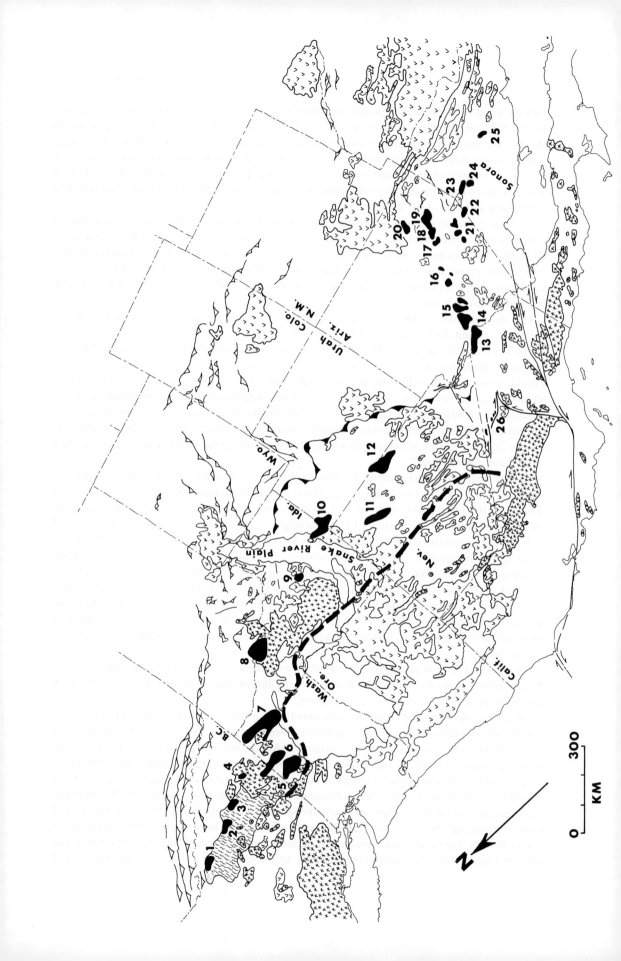

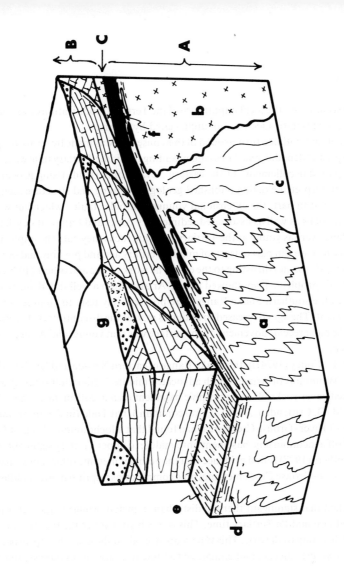

Figure 2. Schematic structural block diagram of typical domains of Cordilleran metamorphic core complexes; A, basement terrane; B, cover terrane; C, decollement zone; a, older metasedimentary rocks; b, older pluton; c, younger pluton (early to middle Tertiary); d, mylonitic foliation; e, mylonitic lineation; f, marble tectonite (black); g, lower to middle Tertiary sedimentary and volcanic rocks.

EXPLANATION

- Metamorphic Core Complex
- Major Batholith
- Early to Middle Tertiary Volcanic Rocks
- Shuswap Metamorphic Rocks
- Laramide Thrust Fault
- Sevier Thrust Fault
- Strontium 87/86 .706 Line

Figure 1 (facing page and explanation, above). Distribution and general regional tectonic setting of Cordilleran metamorphic core complexes, numbered as follows: 1, Frenchman's Cap; 2, Thor-Odin; 3, Pinnacles; 4, Valhalla; 5, Okanogan; 6, Kettle; 7, Selkirk; 8, Bitterroot (Idaho batholith); 9, Pioneer; 10, Albion–Raft River–Grouse Creek; 11, Ruby; 12, Snake Range; 13, Whipple; 14, Harcuvar; 15, Harquahalla; 16, South Mountains–White Tank; 17, Picacho; 18, Tortolita; 19, Catalina–Rincon; 20, Santa Teresa–Pinaleno; 21, Comobabi–Coyote; 22, Pozo Verde; 23, Magdalena; 24, Madera; 25, Mazatan; 26, Death Valley turtlebacks.

low-angle thrusting. In some cases, the local importance of the Tertiary modifications was adequately emphasized, but the regional significance was not appreciated by others.

The first clear statement that initiated turn-around of this consensus was made by Armstrong (1972). He proposed that widespread "denudational" low-angle faulting in middle Tertiary time was a possible explanation for the "regional decollement" of the eastern Nevada hinterland. Using geochronology and field relations, he reinterpreted existing published geologic mapping and structure sections. His results suggested that at least in part, the extensional younger-on-older faults, which cut well-dated middle Tertiary volcanic rocks as well as associated sedimentary rocks and flatten at depth to merge with the decollement plane, were as young as the Tertiary volcanic rocks they cut. The implication was clear. The decollement surface was, in part at least, of middle Tertiary age and possibly had only little or nothing to do with Sevier (Mesozoic) thrusting to the east. The work of Lee and others (1970) was very significant in regard to this problem. They showed that K-Ar ages from a well-dated Jurassic pluton below the decollement in the southern Snake Range were progressively reset to younger ages as one approached the decollement. The ages decreased to about 18 m.y. at the discontinuity. Lee and others (1970) concluded that the most recent movement on the surface was that young, and Armstrong (1972) entirely agreed with them.

Working independently, Coney (1974) in the Snake Range of eastern Nevada and Davis (1973, 1975) in the Catalina-Rincon Mountains of Arizona both advocated middle Tertiary low-angle gravity sliding of unmetamorphosed cover rocks off metamorphic basement on a decollement surface. At about the same time, I heard (M. D. Crittenden, Jr., 1972, oral commun.) that Todd (in Compton and others, 1977), working in the Raft River–Grouse Creek area, had found that allochthonous sheets of Paleozoic cover rocks had moved off a metamorphic basement onto middle to upper Tertiary sedimentary rocks. Finally, Compton and others (1977) found evidence that the younger metamorphic fabric characteristic of the Albion–Raft River–Grouse Creek metamorphic complex was imprinted on a middle Tertiary pluton.

All of this work implied that significant thermal disturbance, metamorphism, and deformation in the hinterland extended into middle Tertiary time. This is much younger than, and well clear of, the proven age of foreland thrusting to the east. This time sequence raised the disturbing prospect that we were dealing with a very young, special, and enigmatic tectonic response of obscure significance.

CHARACTERISTICS

The metamorphic complexes discussed in this symposium volume occur in a discontinuous belt extending from southern Canada south through the Cordillera into Sonora, Mexico (Fig. 1). In general, these complexes are characterized by distinctly similar rock types, structures, and fabric. These similarities are among the most remarkable aspects of the complexes and are noticed by anyone with more than a casual acquaintance with more than one of them. The striking similarities among the complexes form the thread that binds the issue before us.

Two distinct domains characterize the complexes (Fig. 2). These are a metamorphic-plutonic basement terrane and an overlying or adjacent unmetamorphosed cover. Separating the two is a sharp discontinuity, or zone, marking rapid or abrupt change in rock types and structure. Rarely do the rock types and structural fabric that are characteristic of either the basement or cover cross the discontinuity into the other domain.

The gross aspect of the complexes is domal or anticlinal, usually with an asymmetry such that one flank is slightly steeper than the other. The complexes usually form the highest mountains in their respective regions and may be recognized from afar by their distinctive low domal profile on the horizon.

Metamorphic-Plutonic Basement Terrane

The metamorphic basement is characterized by a low-dipping foliation whose attitude usually conforms to the overall domal or archlike shape of the complex. This foliation imparts a distinct gneissic aspect to the rock. Within the foliation plane is an inevitable mineral lineation. The bearing of the lineation (but not necessarily the plunge) is often remarkably constant within a given complex and sometimes in adjacent complexes as well. In more than 15 complexes located across a distance of 400 km in southern Arizona and Sonora, for example, the lineation bears about N60°E (Davis, this volume). To the north, the lineation is more variable, but generally trends either due west or northwest (Coney, 1974; Misch, 1960). In some cases, the lineation lies close to the axes of the domes or arches, but in other cases, it cuts across this trend. The dip of foliation on flanks of the domes rarely exceeds 20° to 30°. Both the foliation and lineation are usually described as "cataclastic," but do involve recrystallization, particularly in quartz, as well as cataclasis (Todd, this volume). The rocks are best described as mylonitic gneiss. Strain is extreme; the elongated minerals and stretched pebbles in conglomerate can attain axial ratios of 8:2:1 (Coney, 1974; Compton, this volume; Compton and others, 1977; Davis, this volume). The overall strain picture is one of maximum shortening and flattening perpendicular to the subhorizontal foliation plane and maximum extension parallel to the lineation (Compton, this volume). The direction of maximum elongation is frequently described as subparallel to recumbent, flattened, and attenuated minor fold axes. Davis has found small late-stage normal faults whose strike is perpendicular to the lineation (Davis and others, 1975; Davis, 1975, 1977, and this volume). These faults seem to be the result of progression from ductile behavior to brittle failure. On the fault surfaces are slickensides whose bearing is subparallel to the lineation.

In some complexes where erosion has cut deeply into the uplifts, the foliated and lineated fabric diminishes into either an earlier, usually steeper metamorphic fabric or a more homogeneous plutonic fabric (Coney, 1974; Todd, this volume; Reynolds and Rehrig, this volume). The deeper, earlier metamorphic fabrics are often quite complex and record polyphase deformation and complex history (Reesor, 1970; Hyndman, 1968; Miller, this volume). The distinctive mylonitic foliation and lineation are therefore superimposed on the earlier fabrics.

The protoliths of the metamorphic basement varied both in rock type and age. This is a point that cannot be overemphasized. In most cases, a single complex had several protoliths. In one complex or another, the protoliths include proven *older* Precambrian metasedimentary basement (Reynolds and Rehrig, this volume), *older* Precambrian plutons that intrude the metasedimentary rocks (Banks, this volume; Compton and others, 1977; Shakel and others, 1977), upper Precambrian sedimentary rocks, Paleozoic sedimentary rocks (Misch, 1960; Howard, 1971; Thorman, 1970), probable Mesozoic sedimentary rocks (Rehrig and Reynolds, this volume; Hyndman, 1968), Laramide Upper Cretaceous–lower Tertiary plutonic rocks (Anderson and others, this volume), and even lower to middle Tertiary plutonic rocks (Reynolds and Rehrig, this volume). All of the above protoliths have the distinctive late mylonitic foliation and lineation superimposed on them in one complex or another.

As one approaches the discontinuity separating the basement and cover terranes, all of the basement fabrics (including the late mylonitic fabric) are demonstrably brecciated. They are truncated at the discontinuity by a still later deformation apparently related to movement along the discontinuity (Coney, 1974). This latest deformation usually places unmetamorphosed cover rocks in direct contact with brecciated and locally truncated and disturbed basement rocks. Regionally, however, the discontinuity, or decollement, is subparallel or exactly parallel to the underlying mylonitic and gneissic fabric.

Granitic plutons are extremely common in the basement terrane. Of special interest are some described as the garnet-bearing two-mica type (Chappell and White, 1976; Best and others, 1974). Besides the plutons, pegmatitic and migmatitic rocks are common, as are other late-stage differentiates and leucocratic phases. These smaller bodies generally form sheetlike or lensoid masses and fine stringers that are subparallel to the foliation, but can also cross it. Some have the mylonitic fabric superimposed on them; others do not. At deeper levels, the larger plutons are commonly homogeneous, but as the decollement is approached, they progressively acquire the characteristic foliation and lineation. The pegmatitic and migmatitic bodies generally do not cross the decollement and are commonly terminated abruptly at it. In a few cases for which documentation is still emerging, even the largest plutons are apparently sill-like masses emplaced at or just below, and subparallel to, the decollement surface (Banks, this volume).

Metamorphic grade in the basement terrane is quite variable. In many complexes, the older and deeper metamorphic grade was quite high, and kyanite, sillimanite, and andulusite are not uncommon (Reesor, 1970; Armstrong, 1968b; Compton and others, 1977). The younger mylonitic fabric formed at more moderate conditions.

Unmetamorphosed Cover Terrane

The overlying cover terrane consists of unmetamorphosed rocks separated from the metamorphic-plutonic core by the decollement or by a zone of very steep metamorphic gradient. In many cases, little of the cover terrane remains because of erosion. Commonly, the only remnants are isolated "klippen," as they are usually termed, scattered around the margins of the complex.

Like the basement terrane, the types and ages of cover rocks are highly variable. In one complex or another, the cover rocks consist of slivers of original Precambrian basement (Davis, this volume; Drewes, 1977), upper Precambrian sedimentary rocks, Paleozoic sedimentary rocks (Coney, 1974; Compton and others, 1977; Thorman, 1970), probable Mesozoic sedimentary rocks (Rehrig and Reynolds, this volume), and Tertiary volcanic and sedimentary rocks (Davis and others, 1977 and this volume). All the cover rocks—including, it must be emphasized, lower to middle Tertiary volcanic and associated Tertiary sedimentary rocks (Davis and others, 1977 and this volume)—can be demonstrated to have moved along the decollement relative to the basement terrane.

Where sufficient cover terrane is preserved to make observations, structures within it are very complex. Workers are usually impressed by many low-angle, younger-on-older bedding-plane faults and by many extensional listric normal faults (Coney, 1974; Hose and Danes, 1973). The entire cover terrane can take on the character of a megabreccia. Faulting rarely penetrates into the basement terrane below the basal decollement. In other words, the decollement marks a discontinuity of extreme ductility contrast—brittle above, ductile below.

In some areas, detailed analysis of cover rocks reveals structures varying from minor folds to slickensides on faults, or on the decollement surface itself; such structures can be interpreted as reflecting movement of cover rocks down present dips of the decollement surface into adjacent basins (Coney, 1974; Davis, 1975). The movement directions inferred are in some cases at a high angle to the bearing of lineation in the underlying basement, whereas in other cases they are subparallel. Also, the earlier fabrics in the basement, including the mylonitic foliation and lineation, are redeformed (generally brittlely) into geometries consistent with the movement picture derived from the cover (Coney, 1974). This late movement seems to have been associated locally with intense brecciation in both cover rocks and in basement terrane just below the decollement. Water was abundant, from the evidence of clastic dikes and chloritic and hematitic fillings in the pervasive fractures.

The amount of extension in the cover terrane is typically dramatic. In several complexes, attenuation of the original stratigraphic sequence is extreme (Compton and others, 1977; Todd, this volume; Davis, 1975), and lithologic units once very high in the original cover sequence have been brought down into contact with the decollement surface. In some cases, the stratigraphic separation is greater than 2 km. It is not uncommon to find Tertiary rocks, the youngest of the original cover, brought down into tectonic contact with the basement terrane.

In many of the complexes, part of the cover sequence is an early to middle Tertiary continental sequence (usually red beds) composed of conglomerate, fanglomerate, sandstone, and siltstone, with some lake beds and evaporites (Pashley, 1966). These sedimentary rocks commonly are associated regionally with (usually overlain by) well-dated lower to middle Tertiary volcanic rocks (Armstrong, 1972) and are usually tilted to high angles. It is not insignificant that the direction of strike of these tilted sedimentary and associated volcanic sequences is perpendicular to the bearing of lineation in the basement terrace of the complexes over wide areas in Arizona. These relationships between the continental rocks and the complexes suggest a genetic link.

The sedimentary section is usually very thick, as much as several kilometres. In those few cases where investigations have been made (Pashley, 1966), current indicators and pebble counts suggest that the fluvial systems did not drain from the core complexes, but may have actually flowed toward them. Only in upper levels of the deposits do streams appear to have drained off the complexes, and it is only in these youngest horizons that clasts of foliated and lineated basement rocks occur (see also Todd, this volume).

The Decollement

The decollement or dislocation surface is the most distinctive aspect of the core complexes thus far recognized (Misch, 1960; Coney, 1974; Whitebread, 1968; Nelson, 1966, 1969; Davis and others, this volume; Miller, 1972). Although variations exist, the general characteristics of the decollement are so remarkably similar from one complex to another that the feature is instantly recognized.

In some complexes, particularly in central Nevada and Arizona, the decollement characteristically occurs either close to the Precambrian-Phanerozoic unconformity or in horizons directly above thick upper Precambrian–lower Paleozoic quartzite-siltstone sequences. Carbonates below the decollement are usually metamorphosed, attenuated, and intensely deformed into a marble tectonite rarely thicker than several tens of metres (Misch, 1960; Nelson, 1966; Coney, 1974; Whitebread, 1968). Davis (this volume) has referred to this tectonite band as a metamorphic carapace. Unmetamorphosed Paleozoic carbonates above the decollement can directly overlie their metamorphosed equivalents below the decollement. Where the basement is composed of plutonic rocks, the decollement usually lies directly above them, and the plutons never cross the decollement. A remarkable example in the southern Snake Range of Nevada (Whitebread, 1968, 1969; Lee and others, 1970) has a Jurassic pluton as basement. This pluton is overlain along a clearly tectonic low-dipping planar surface by as much as 30 m of marble tectonite. The decollement lies above the marble and is overlain by unmetamorphosed Cambrian limestone. Here, K-Ar ages on the Jurassic pluton decrease to 18 m.y. approaching the decollement (Lee and others, 1970; Armstrong, 1972; Coney, 1974).

The decollement surface typically is extremely sharp and very visible in topography. It is often highly polished and has slickensides; rock directly below can have the aspect of a fused paste or a breccia of welded small clasts.

Rocks near the decollement, both above and below, are usually brecciated and show extensive alteration (Reynolds and Rehrig, this volume). Retrograde chlorite is very common in brecciated basement rocks, and development of a distinctive hematitic red-stained fracture filling between fragments is ubiquitous. These breccias have the appearance of "exploded rocks." Even in thin section, so-called mylonitic zones are clearly without planar fabric, but rather show an exploded microbreccia aspect.

The decollement is generally best developed on one side of a particular complex, usually on the less steeply dipping flank of the commonly asymmetrical dome or arch (Reynolds and Rehrig, this volume). On the steeper flank, the discontinuity can be less tectonically abrupt with only very steep metamorphic gradient and little demonstrable evidence of major movement between cover and basement. In at least two complexes, several slices (or thin packets of discontinuity-bound rocks) extend across the dome or arch. Three separate slices are recognized in both the Raft River–Grouse Creek and Rincon complexes. In Raft River–Grouse Creek complex (Compton, 1972, 1975; Compton and others, 1977; Todd, this volume), the two lower slices are metamorphosed, but the upper is not. In the Rincon Mountains (Davis, this volume; Drewes, 1977), the lower slice is metamorphic Paleozoic rocks that are extremely attenuated, the middle slice is original Precambrian basement, and the upper slice is unmetamorphosed Paleozoic and Mesozoic rocks. The core in both examples is Precambrian granite, metasedimentary rocks, and lower to middle Tertiary plutons.

REGIONAL TECTONIC SETTING

The regional tectonic setting of Cordilleran metamorphic core complexes (Fig. 1) is in many ways the most puzzling aspect of the problem. This is so because no obvious regional tectonic relationship has been demonstrated to everyone's satisfaction. Nevertheless, it has become clear that the complexes are an important element in the overall architecture of the North American Cordillera.

Distribution of metamorphic rocks in western North America, excluding the cratonic Precambrian basement beneath the eastern margin of the orogen, grossly reveals two subparallel belts (King, 1969; Monger and Hutchison, 1971). The western belt is largely a metamorphic sheath below, within, and adjacent to the great belt of Cordilleran batholiths. This belt extends through the Canadian coastal plutonic complex, the Idaho batholith, and the Sierra Nevada–Peninsular Ranges batholith southward into western Mexico. An eastern belt (Coney, 1978) follows the Omineca crystalline complex of the eastern Cordillera in Canada and culminates in the Shuswap terrane of southern British Columbia to northwestern Washington (Cheney, 1977 and this volume; Fox and others, 1977). The western and eastern belts of metamorphic rocks appear to merge in the Idaho batholith (Miller and Engels, 1975), but to the south they again separate, and the eastern belt extends southward across eastern Nevada, southeastward across Arizona, then southward into Sonora. All the so-called Cordilleran metamorphic core complexes as defined here lie in the eastern belt, which extends at least from southern Canada to northern Mexico.

Most of the eastern belt developed either on or very close to the edge of the original North American Precambrian cratonic basement. In contrast, most of the western belt may have developed in magmatic arcs on oceanic crust or on crustal fragments that were subsequently accreted onto North America's continental margin (Jones and others, 1977; Monger, 1977; Dickinson, 1976; Coney, 1978). Finally, the apparent continuity of both belts does not imply historical continuity. The ages of batholiths in the western belt vary along strike, and much of the northern two-thirds of the eastern belt records a much more prolonged metamorphic history than the southern part. It is the most recent metamorphic-tectonic events of the eastern belt that we identify as characteristic of the core complexes, and these are superimposed on a diverse terrane, whose history and tectonic evolution predate the development of the core complexes.

Pre-Mesozoic Tectonic Trends

Of the 25 or so Cordilleran metamorphic core complexes currently identified, all evolved in terrane underlain by North American Precambrian continental cratonic basement, as defined by being inboard of the 0.706 $^{87}Sr/^{86}Sr$ contour line (see Fig. 1) (Armstrong and others, 1977; Kistler and Peterman, 1973). Two possible important exceptions are the Okanogan (Fox and others, 1977) and Kettle (Cheney, 1977 and this volume) complexes in Washington. In many complexes, the Precambrian age of the basement is either proved or implied by isotopic dating (Wanless and Ressor, 1975; Clark, 1973; Armstrong and Hills, 1967; Compton and others, 1977; Shakel and others, 1977), and regional relations seem to demand its presence in most of the remainder. Even when this Precambrian basement is looked at more closely, controls on evolution of metamorphic complexes are not obvious. Precambrian lithologic and structural trends, which are generally northeasterly, strike at high angles to the trend of the belt of complexes, and basement ages crossed by the belt range from greater than 2 b.y. in the north to about 1 b.y. in the south (King, 1969). The Arizona complexes may parallel a northwest trend of late Precambrian right-shear just southwest of the Colorado Plateau, but so do all other post-Paleozoic tectonic trends in that area.

From southern Nevada northward to the Selkirk complex in Idaho and Washington, the complexes lie within the thick Cordilleran miogeoclinal prism of latest Precambrian–Paleozoic age. On the other hand, in Arizona and Sonora, the complexes evolved well inboard of the Paleozoic miogeocline on what had been a thin cratonic shelf (Peirce, 1976). In contrast, the Okanogan and Kettle complexes seem to lie outboard of the Paleozoic miogeoclinal shelf edge. As a result, the complexes cannot be explained as being related to a zone of deep Paleozoic burial and thick Paleozoic sedimentary accumulation. The thickness of uppermost Precambrian–Paleozoic deposits over the eventual site of the core complexes is almost 15 km in southern Canada (Price and Mountjoy, 1970; Campbell, 1973), about 10 km in Nevada (Armstrong, 1968a), and only about 2 km in southern Arizona (Peirce, 1976).

Similarly, the distribution of the complexes varies relative to Paleozoic orogenic activity and metamorphism. The northern complexes are found in a region that was probably affected by middle and late Paleozoic Antler-Sonoma deformation and, in the case of the Shuswap complex, some Paleozoic metamorphism (Okulitch and others, 1975; Read and Okulitch, 1977; Brown and Tippett, 1978). Southward in Nevada and Arizona, the complexes lie well east or south of any profound Paleozoic thermal or tectonic events.

Relationship to Mesozoic–Early Cenozoic Trends

Latest Paleozoic–early Mesozoic time was a major transition in Cordilleran tectonic evolution (Coney, 1972, 1973a; Burchfiel and Davis, 1975). It marks inception of draping and accretion of magmatic arcs on the North American margin and the evolution of back-arc thrusting and folding inboard from the magmatic belts.

Genetic linking of the Cordilleran metamorphic core complexes to Cordilleran thrust belts has been the most persistent and persuasive argument put forward to explain the phenomena under study. The argument has manifested itself in several ways and has influenced discussion of Cordilleran tectonics in general as well as discussion of the complexes themselves. No aspect of the complexes has generated more controversy than this.

From Nevada northward, the metamorphic core complexes lie within a belt about 200 km west of the thin-skinned foreland folds and imbricate thrust faults so characteristic of the North American Cordillera and other orogens (Coney, 1976). Because this belt of metamorphic rocks lies behind the folds and thrust faults of the foreland, the belt has been termed a "hinterland." The tectonic events of the hinterland are supposed to have been deeper-seated than those of the foreland and included uplift, thermal disturbance, and remobilization. The assumption has been that the metamorphism in the hinterland accompanied the thrusting in the foreland; thus, the two responses were genetically linked. It was this assumption that so heavily influenced early models such as those of Misch (1960) and Price and Mountjoy (1970).

Paleozoic and lower Mesozoic rocks are clearly metamorphosed in Canada (Hyndman, 1968), and Paleozoic rocks are involved in Nevada (Misch, 1960). From regional relationships in Canada, some of the metamorphism is Paleozoic, and most of it is at least as old as Middle Jurassic (Wheeler and Gabrielse, 1972; Brown and Tippett, 1978). Ironically, recent work in the southern Rocky Mountains of Canada suggests that much of the polyphase deformation and metamorphism so characteristic of the Shuswap terrane is pre–Late Jurassic (Brown, 1978; Wheeler and Gabrielse, 1972) and thus *earlier* than the well-documented Late Jurassic to early Tertiary folding and thrusting to the east. Furthermore, the increasing evidence that the mylonitic metamorphism in United States core complexes is in part, at least, Tertiary (Compton and others, 1977; Cheney, this volume; Rehrig and Reynolds, 1977 and this volume; Anderson and others, this volume) and hence much *later* than Sevier or Laramide thrusting has clouded the issue of a genetic link between thrusting and metamorphism more than anything else.

In Arizona and Sonora, the complexes are not in a "hinterland" behind a thrust belt, but are in fact in the midst of a Laramide belt of deformation, although of a somewhat different character from the thin-skinned folds and thrusts of Sevier-Laramide age to the north (Davis, 1979). The deformation in the south was apparently deeper-seated and involved both basement and cover rock. Furthermore, Laramide tectonic features are clearly older than the core-complex metamorphism and deformation, which are superimposed on some postthrusting Laramide plutons and even on middle Tertiary plutons (Anderson and others, this volume; Reynolds and others, 1978; Reynolds and Rehrig, this volume). As is the case with so many other regional aspects of Cordilleran metamorphic complexes, the Arizona-Sonora examples show a major departure from relationships historically so suggestive to the north.

In Nevada—the original battleground of metamorphism and thrusting in the Cordillera—the relationships are less clear. Middle Mesozoic metamorphism has long been advocated in the hinterland (Misch, 1960) and certainly affected the Ruby Range (Howard, 1971 and this volume; Snoke, this volume) and probably the Snake Range as well. However, this is an older metamorphism and is observable only deep in the cores of some of the complexes. Superimposed on this earlier, generally steeper fabric is the later, shallow-dipping mylonitic fabric so characteristic of the complexes throughout their extent. The late metamorphism and associated deformation are at issue here. Their effects appear to be superimposed on plutons as young as middle Tertiary (Compton and others, 1977), thus much younger than the Late Jurassic to Late Cretaceous thrusting in the Sevier thrust belt of central Utah.

The assumption of a genetic link between the thrust belt and the core complexes has played another significant role in interpretations of the complexes. The argument has led the distinctive "decollement" surface and its associated mylonitization to be identified with the basal shear plane of the thrust belts (Misch, 1960). This interpretation has been variably invoked in Idaho (Hyndman and others, 1975), in Nevada and western Utah (Misch, 1960; Hose and Danes, 1973), and in Arizona (Drewes, 1976, 1978). Obviously, the thrusting would have to be of Sevier (Late Jurassic to Late Cretaceous) age in Nevada, of Sevier-Laramide age in Idaho, and of Laramide (Late Cretaceous to Eocene) age in Arizona. The fact that the decollement surface and the mylonitic deformation associated with it are now known to cut rocks as young as early to middle Tertiary in all these regions casts some doubt on this interpretation (see Fig. 5). As a result, some workers have more recently argued models of multiple thrusting (Drewes, 1976, 1978; Thorman, 1977) and/or middle Tertiary gravitational sliding on a decollement surface originally made during regional low-angle thrusting during Sevier or Laramide time (see discussion in Compton and Todd, 1979, and Crittenden, 1979). In some cases at least, these models seem geometrically difficult. Armstrong (1972), for example, has argued on geometric grounds that the decollement surface in the Snake Range of eastern Nevada is unrelated to the basal shearing plane of Sevier thrusts to the east (Fig. 3).

In any event, the debate concerns whether the characteristics so typical of the core complexes—namely, the decollement surface, the distinctive mylonitic foliated and lineated fabric, and younger-on-older "faults" in the cover—are genetically linked to regional thrusting of Mesozoic–early Cenozoic age. No one can deny deformation and even metamorphism of Sevier and/or Laramide age in some of the regions now occupied by core complexes. What is claimed is that the younger features typical of the complexes are superimposed on earlier fabrics; that the younger features are probably, for the most part, of Tertiary age; and that they are unrelated to the thrust belts at least in any direct way.

Because the core complexes are clearly, in part, of thermal origin, distribution of magmatic activity is of interest in discussing their evolution. Compilatons of Mesozoic–early Cenozoic magmatic activity reveal considerable complexity in both space and time. This activity encompasses the entire Cordillera and spreads over a far greater region than the rather narrow belt of complexes as defined here.

The main result of this magmatic activity was the eventual emplacement of a massive Cordilleran batholith system comprising the Canadian coastal plutonic complex, the Idaho batholith, and the Sierra Nevada–Peninsular Ranges batholith in California and western Mexico (King, 1969). These bodies are not all the same age. Furthermore, with the exception of the Idaho batholith, all are well west of the belt of metamorphic core complexes.

Lesser plutonic bodies extend eastward across the Cordillera into, and even east of, the core complex belt. In Canada, the Omineca crystalline belt has plutons of Late Jurassic and Cretaceous ages (Gabrielse and Reesor, 1974), most of which clearly crosscut the metamorphic fabrics typical of the Shuswap terrane (Hyndman, 1968).

In the United States, the Washington and Idaho complexes are superimposed on, or certainly co-existent with, the main batholith belt (Miller and Engels, 1975). Farther south, in Nevada, the complexes lie well east of the batholith belt, but well within a field of scattered Jurassic to Late Cretaceous plutons. These eastern plutons are perhaps best explained as scattered intrusions parasitic to the main batholith belt. They were emplaced where they are either by generation from deeper levels

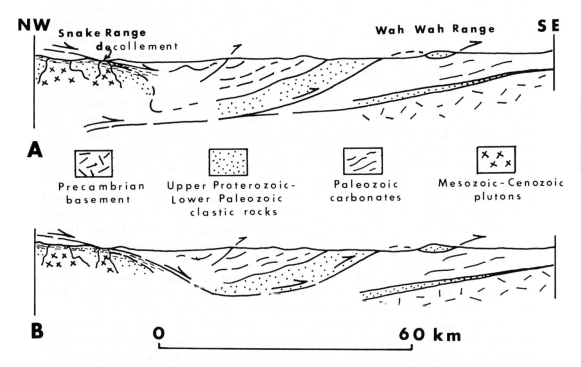

Figure 3. Two contrasting interpretations of structure from the Snake Range, Nevada, to the Wah Wah Range, Utah, based on a section by Armstrong (1972). (A) Deep-seated thrust-fault model showing basal Sevier thrust in Wah Wah Range rooting beneath the Snake Range. This interpretation was favored by Armstrong. The Snake Range decollement is not connected to the Sevier thrust faults, but is interpreted as a Tertiary denudation fault (Armstrong, 1971) off the Snake Range core complex. (B) Shallow thrust-fault model that connects the Snake Range decollement with Sevier thrust faults to the east. This interpretation was favored by Hose and Danes (1973).

of the subducted slabs or because of transient variable dips in these slabs. The Arizona and Sonora complexes are within a broad belt of mainly Late Cretaceous–ezrly Tertiary (Laramide) scattered magmatic activity that swept inboard from the Cretaceous Peninsular Ranges batholith along the coast after 80 m.y. B.P. (Silver and others, 1975; Coney and Reynolds, 1977).

In summary, Mesozoic–early Cenozoic magmatic arc activity spread over the entire Cordillera at one time or another, including the narrow belt of core complexes. Most of the volume of plutons was emplaced well west of the belt of core complexes. Finally, with the possible exception of the Omineca crystalline belt in Canada, no unique or distinct pre–middle Tertiary magmatic trend coincides with the belt of complexes, either in space or in time.

Relation to Middle Cenozoic Tectonic Trends

Between about 55 and 20 m.y. B.P., a very complex pattern of post-Laramide magmatic activity spread over the entire southern Cordillera (Armstrong, 1974; Coney and Reynolds, 1977; Noble, 1972; Lipman and others, 1971). The activity was characterized by enormous outbursts of caldera-associated ignimbrites and the emplacement of shallow plutons. This massive thermal disturbance reset radiometric ages over thousands of square kilometres and also caused much low-angle normal faulting (Rehrig and Heidrick, 1976; Anderson, 1971). The ignimbrite flare-up is very important because the core complexes seem to have evolved just prior to and during the ignimbrite outburst (Coney, 1974, 1978). What is puzzling, however, is that the magmatic activity covered a region far wider than that of the core-complex belt itself.

In Late Cretaceous time (Fig. 4A), a well-established magmatic arc terrane extended from Canada southward through the Idaho–Sierra Nevada–Peninsular Ranges batholiths into western Mexico (Coney, 1976, 1978; Armstrong, 1974). After 80 m.y. B.P., this Laramide magmatic activity swept rapidly eastward (Coney and Reynolds, 1977) across the southern Cordillera, then extinguished north of Arizona, except for very scattered activity (Fig. 4B). In Arizona and New Mexico (Coney and Reynolds, 1977) and in Mexico (Clark and others, 1978), the eastward sweep did not extinguish and reached nearly 800 km inboard by Eocene time (Fig. 4C). At about the same time, the entire Pacific Northwest erupted violently with Challis-Absaroka volcanism and shallow plutonism (Armstrong, 1974). This started a rapid return sweep of magmatic activity back across the western United States toward the continental margin. The return sweep was responsible for the vast ignimbrite flare-up (Fig. 4D) across Idaho, Washington, and Oregon; Utah and Nevada; New Mexico and Arizona; and all of central and northern Mexico (Coney and Reynolds, 1977). Only the Colorado Plateau was spared. By 15 to 20 m.y. B.P., the magmatic activity reached the coast and formed the Cascade magmatic arc trend in the Pacific Northwest, but transformed to a widespread bimodal basalt and rhyolite phase associated with basin-range faulting eastward and southward (Lipman and others, 1971) (Fig. 4E). There is considerable evidence, at least in the United States, that the core complexes developed during this massive return sweep of magmatic activity between 55 and 15 m.y. Furthermore, some evidence suggests that the complexes north of the Snake River Plain evolved during the Eocene Challis-Absaroka outburst (Reynolds, 1977, oral commun.; Cheney, this volume); likewise, those south of the Snake River Plain evolved during the Oligocene–early Miocene ignimbrite flare-up there (Coney, 1978).

The post-Laramide to middle Tertiary is a most puzzling time (Coney, 1978). The events clearly followed Laramide com -ressive deformation, and they seem to have begun by widespread erosion and beveling of Laramide landscapes (Epis and Chapin, 1975). Within the belt of core complexes, the relationship to the ignimbrites is often dramatic. In southern Arizona, some volcanic ranges are made up of vast ignimbrite sheets whose radiometric ages are essentially the same as cooling ages in metamorphic rocks within the adjacent core complex. Just why the core complexes should be restricted to such a narrow belt within this widespread panorama of ignimbrite eruption is not obvious.

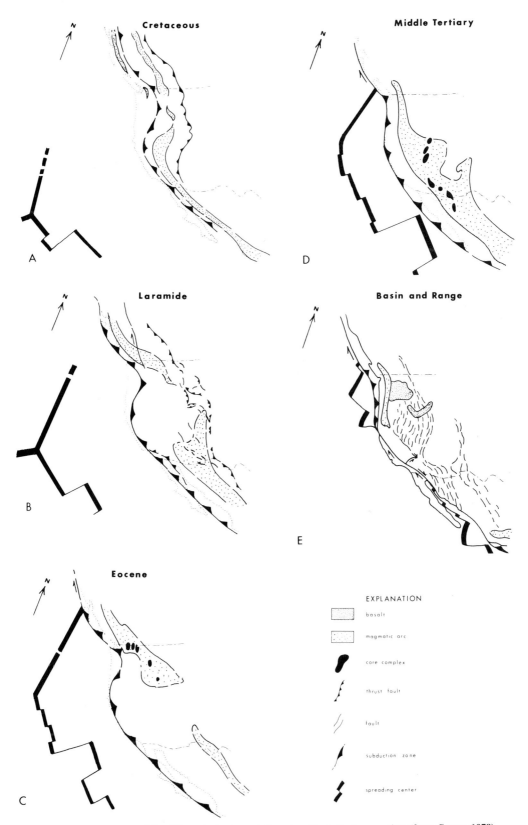

Figure 4. Major features of Cordilleran tectonic evolution since Early Cretaceous time (from Coney, 1978).

22   P. J. CONEY

Relationship to Late Cenozoic Tectonic Trends

One relationship on which almost all workers are agreed is that late Tertiary basin-range faulting seems to postdate most core-complex activity (Fig. 4E). In many areas, the steep block faulting clearly cuts metamorphic rocks, the decollement, or cover rocks (Coney, 1974; Eberly and Stanley, 1978). It is worth noting, however, that all the complexes south of the Snake River Plain occur within a region that was affected by this block faulting.

TECTONIC SIGNIFICANCE

The tectonic significance of Cordilleran metamorphic core complexes has been much debated during the past 20 yr. Before their significance can be fully understood it is necessary to recognize two distinct, and perhaps largely unrelated, aspects of their history. The first aspect is the earlier (mostly Mesozoic) history of many of the complexes, particularly those from western Arizona northward. The second aspect is the early to middle Tertiary history that is emerging as so important in all of the complexes, at least from northeastern Washington southward into Mexico. In my view, much of the confusion surrounding the complexes has been due to a lack of full appreciation of these younger Tertiary features and interpretation of these features to results of Mesozoic events. It was in relation to this that the full significance of the Arizona and Sonora examples emerged. This happened because they are not in the tectonic setting of the hinterland behind the Mesozoic thrust belts, which are so characteristic of the setting of those complexes to the north. The fact that the Arizona and Sonora complexes so remarkably resembled the younger aspects of the northern complexes lent support to the growing recognition of the importance of the Tertiary events throughout the belt.

In other words, there is considerable evidence that much of the metamorphism, deformation, and thermal disturbance so characteristic of Cordilleran metamorphic core complexes is of early to middle Tertiary age (Fig. 5). In many complexes, particularly those northward from Arizona, these mainly Tertiary features were superimposed on mainly Mesozoic metamorphic and deformational effects of the thrust-belt hinterland; however, most of the characteristic mylonitic fabrics, the decollement zones, and the chaotic structures in the cover rocks are, in part at least, the result of early to middle Tertiary tectonics.

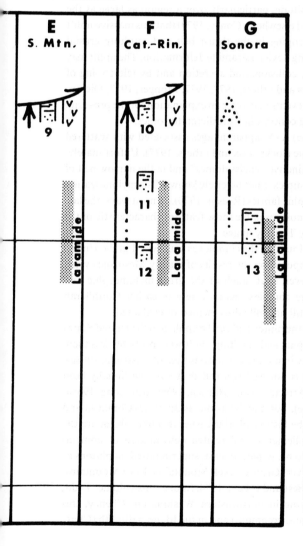

Figure 5 (facing pages). Age relationships in selected Cordilleran metamorphic core complexes. Columns as follows: A, northeastern Washington (Selkirk complex); B, Idaho batholith (Bitterroot complex); C, Albion–Raft River–Grouse Creek complex; D, western Arizona (Harcuvar complex); E, South Mountains complex south of Phoenix, Arizona: F, Catalina-Rincon complex near Tucson, Arizona; G, Sonora, Mexico. Stipple pattern is for age range of plutonic rocks with dash pattern at top signifying mylonitic gneissic fabric superimposed on the pluton: V pattern is age range of lower to middle Tertiary volcanic rocks in vicinity of each complex; vertical heavy arrow represents possible age limits for formation of overprinted mylonitic gneissic fabrics superimposed on the plutons and other basement terranes. Subhorizontal heavy barbed arrow is approximate older age limit on movement of cover rocks over basement terrane on "decollement" surfaces. In many places, this movement could be younger than indicated, but it can usually be shown to be older than latest Tertiary basin-range faulting. Vertical light stipple bands in each column are approximate durations of major compressional Sevier and/or Laramide deformation in thrust belts east of, or in vicinity of, respective core complexes. Numbered plutons as follows: 1, Silver Point quartz monzonite (Miller, 1972); 2, middle Cretaceous plutons (Miller and Engels, 1975); 3, Eocene plutons (S. J. Reynolds and W. A. Rehrig, 1979, oral communs.); 4, Bitterroot lobe of Idaho batholith (Chase and others, 1978); 5, Red Butte pluton (Compton and others, 1977); 6, Immigrant Pass pluton (Compton and others, 1977); 7, muscovite granite (Rehrig and Reynolds, this volume); 8, Tank Pass pluton (Rehrig and Reynolds, this volume); 9, South Mountains pluton (Reynolds and Rehrig, this volume); 10, Catalina granite (Shakel and others, 1977); 11, Wilderness granite (Shakel and others, 1977; Keith and others, this volume); 12, Leatherwood granite (Keith and others, this volume); 13, Sierra Mazatan (Anderson and others, this volume).

In the Shuswap complex of southern Canada, something on the order of 40,000 km^2 of metamorphic rock are exposed. It is the largest of all the Cordilleran metamorphic core complexes, and in the original conception of the problem, it was the type example and the source of the name (see my "Introduction" to this volume). As outlined earlier, studies indicate that some of the metamorphism is as old as Paleozoic and much of it is at least as old as Jurassic (Okulitch and others, 1975; Read and Okulitch, 1977; Brown and Tippett, 1978; Wheeler and Gabrielse, 1972; Hyndman, 1968). There are dated Upper Jurassic to middle Cretaceous plutons that crosscut metamorphic rocks (Gabrielse and Reesor, 1974).

In any event, the conclusion reached by some workers that much of the metamorphism in the Shuswap terrane is at least as old as Jurassic is of considerable importance for the Price and Mountjoy model relating the metamorphic hinterland to the Rocky Mountain thrust belt to the east. Their model invokes a mobile infrastructure that buoyantly rose and propelled the thrusts to the east, largely by gravity. This model has difficulties because if the metamorphism is of Jurassic age, then most of it was over before the thrusting began. Because the metamorphic hinterland exposes rocks once deeply buried and subjected to temperatures of 600°C and pressures approaching 4 kb, a postmetamorphic uplift of 10 km or more is demanded. An uplift of this magnitude, particularly over a region as large as the Shuswap complex, is not easy to explain. I have argued elsewhere (Coney, 1979) that this massive uplift is perhaps explained as being due to crustal telescoping in the thrust belt and resulting crustal thickening in the region of the hinterland mostly during Sevier-Laramide deformation. The uplift may have been also influenced by Mesozoic translation, collision, and accretion and by telescoping of exotic terranes against the Cordilleran margin (Jones and others, 1972, 1977; Monger, 1977; Dickinson, 1976; Coney, 1978). Furthermore, some of these events may offer an explanation for the pre–Late Jurassic metamorphism and deformation of growing concern in Cordilleran tectonics.

There are also many early Tertiary (mostly Eocene) K-Ar apparent ages associated with scattered shallow plutons and a widespread resetting of isotopic clocks (Fox and others, 1977). Unfortunately, the age of the apparent later arching in the three distinctive "gneiss domes" and in the narrow belt of late mylonitization along the eastern margin of the complex is not precisely known. The mylonitization is, however, apparently the youngest of the metamorphic fabrics (Reesor, 1970). In most ways, the late domes and the zone of east-dipping mylonitization most resemble those features characteristic of the complexes to the south considered here to be mainly Tertiary in age.

To what degree Tertiary features similar to those found in the complexes southward in the United States and northwestern Mexico have been superimposed on the results of mainly Mesozoic events described above is not yet known. It seems, however, that much of the gross metamorphic and structural character of the Canadian Shuswap core complex has an origin in earlier Cordilleran tectonic history. The later early Tertiary history is still not fully documented or evaluated.

A similar model of crustal telescoping and upift of the hinterland can be applied in the United States at least as far south as southern Nevada, but telescoping and resulting uplift were probably less there and mostly confined to the area of the Sevier orogeny in middle Cretaceous time (Armstrong, 1968a). The age of the older core-complex metamorphism is not well-controlled; it has traditionally been described as simply "mid-Mesozoic" (Misch, 1960; Armstrong and Hansen, 1966; Armstrong, 1968a; Howard, 1971 and this volume; Snoke, this volume), but may be as old as Jurassic (Compton and Todd, 1979). Superimposed on, or at least late in the history of, this earlier metamorphism are the shallow-dipping mylonitic fabrics, associated decollements, and related deformation of unmetamorphosed cover rocks. These late features appear to be, in part at least, superimposed on plutons as young as early to middle Tertiary age in the Raft River–Grouse Creek Mountains, Kern Mountains, and Ruby Mountains (Compton and others, 1977; Best and others, 1974; Snoke, this volume; Todd, this volume), and similar relationships are emerging in northeastern Washington (Cheney, this volume; Miller, 1972; Reynolds and Rehrig, 1978, oral commun.) and in the Idaho batholith (Chase and others, 1978). These events have reset K-Ar apparent ages to as young as 18 m.y. along the Snake Range decollement (Lee and others, 1970; Armstrong, 1972). They have produced field relationships such as "klippen" of Paleozoic rocks over middle to upper Tertiary sedimentary rocks in the Raft River–Grouse Creek Mountains (Todd, 1973 and this volume) and low-angle normal faults in middle Tertiary volcanic rocks (Armstrong, 1972). Coney (1974) inferred that minor structures in rocks associated with the Snake Range decollement were consistent with Armstrong's (1972) model of Tertiary denudational faulting in the Nevada hinterland. All of these features clearly predate late Tertiary basin-range faulting (Coney, 1974).

In Arizona and Sonora, the metamorphic core complexes are clearly of early to middle Tertiary age. In southwestern Arizona, an older Mesozoic metamorphism is overprinted by mylonitic gneissic fabrics in several complexes (Rehrig and Reynolds, 1977 and this volume), but several of the complexes of southeastern Arizona and Sonora are not complicated by widespread and complex pre-Tertiary metamorphism and deformation like those complexes to the north. More important, as already mentioned, the complexes here are not in a hinterland behind a thrust belt of any age but, instead, are partly in a belt of mainly brittle, deep-seated, basement-cored thrust uplifts of Laramide age accompanied by scattered Laramide plutons. Furthermore, the typical core complex fabrics and related decollements are superimposed on plutons ranging in age from about 55 m.y. in Sonora (Anderson and others, 1977 and this volume) to as young as 26 m.y. in the South Mountains near Phoenix (Reynolds and Rehrig, this volume; Reynolds and others, 1978). In the Catalina-Rincon complex near Tucson, a complicated history of plutonism, deformation, and metamorphism is recorded, but most workers now agree that features typical of the core complexes throughout the Cordillera are superimposed on plutonic rock as young as at least 50 m.y. (Shakel and others, 1977; Keith and others, this volume). From structural analysis, Davis (1975) inferred Tertiary movement of cover rocks down decollement surfaces in the Rincon Mountains.

It cannot be denied that the later, mainly Tertiary, overprint so characteristic of Cordilleran metamorphic core complexes was perhaps influenced by the preceding Mesozoic history. This is particularly true of the complexes in Nevada and northward into southern Canada. Exactly what the influence was, however, has been difficult to identify. Low-angle faults, some certainly of younger-on-older type, undoubtedly formed in the hinterland during Mesozoic time. They may have served to localize the Tertiary decollements. The Mesozoic metamorphism and deformation have already been mentioned. Perhaps one important influence was the Mesozoic uplift of the hinterland produced from crustal thickening behind the telescoping thrust belts to the east. This would permit the later thermal culminations and associated plutons of early to middle Tertiary age to likewise rise higher before being frozen in a reactive endothermic Phanerozoic cover. In any event, the typical location of the core complexes along the hinterland behind the thrust belt can hardly be a fortuitous accident. Even in Arizona and Sonora where the above relationships did not hold, the structurally and thermally battered ground inherited particularly from Laramide events could have prepared a weakened basement conducive to concentrating the Tertiary events.

Are the complexes basically classic gneiss domes (Eskola, 1949)? This question is often asked. Certainly, some show certain characteristics of classic gneiss domes, particularly the extreme stretching across the top, the overall domal or archlike geometry, and the steep metamorphic gradient. There are, however, certain difficulties. First, the actual doming is apparently a late feature superimposed on the metamorphic fabric of the basement. Second, the actual structural relief on the domes is not particularly great. For example, the amplitude of most of the domes is not more than 4 km in a wavelength of as much as 50 km; this gives an amplitude/wavelength ratio of 0.08—a fact reflected in the universal low-dipping foliation that rarely exceeds 20° to 30° (for example, see Drewes, 1977). This is considerably less structural relief than that usually depicted in classic gneiss domes or in experimental or theoretical modeling (Ramberg, 1972; Dixon, 1976) where amplitude/wavelength ratios are 0.25

to 0.5 or more. Third, it has already been emphasized in this paper that the mylonitic foliation and lineation so characteristic of the complexes in many cases actually seem to diminish and even disappear downward in the basement terrane. This argues for rigidity of this terrane in some cases since Precambrian time (Compton and others, 1977). It also argues against the deep-seated mobility that is characteristic of an "infrastructure" and usually cited as an essential ingredient of classic gneiss domes.

In any event, some workers in the Cordillera have found it useful to informally and tentatively reject the classic gneiss-dome concept if for no other reason than to aid objectivity and identification of the real characteristics of the Cordilleran complexes.

The evidence for extension in and adjacent to the complexes is extremely compelling. It was first acknowledged in the cover terranes (Armstrong, 1972; Anderson, 1979; Coney, 1974; Davis, 1973, 1975; see also Davis and others, this volume). Only more recently has it been proposed in the basement itself as manifested in the mylonitic foliation and lineation.

The work of George Davis (1973, 1975, 1977, Davis and others, 1975) in the Catalina-Rincon complex in Arizona first suggested this basement extension as a major aspect of core-complex evolution. Davis has subsequently (1977 and this volume; Davis and Coney, 1979) expanded these observations into the provocative concept that the complexes are in fact megaboudins. This concept is similar to recent interpretations of the Death Valley turtlebacks (Wright and others, 1974; Burchfiel and Stewart, 1966). Whatever they are, the evidence is that they were produced by extension and tectonic denudation. This is certainly the case with the cover terranes. Another early suggestion for regional extension was recognition of northwest-trending dikes cutting middle Tertiary plutons in southern Arizona (Rehrig and Heidrick, 1976). These dikes are oriented perpendicular to the lineation in the core complexes (Reynolds and Rehrig, this volume). This extension has also been recognized outside or adjacent to the main core-complex belt in areas of low-angle listric normal faults that cut middle Tertiary volcanic and sedimentary rocks but are cut by basin-range faults (Anderson, 1971; Eberly and Stanley, 1978; Rehrig and Heidrick, 1976; Davis and others, this volume).

It is remarkable how many of the complexes are characterized by lower to middle Tertiary granitic plutons. The late mylonitic fabrics are superimposed on many of these plutons, and evidence suggests that they were cooling and still partly mobile at the time of at least the earlier phases of the deformation recorded in the complexes. The granitic plutons precede, then become intimately associated with a massive thermal disturbance that is mainly of Eocene age (Fig. 4C) in the Pacific Northwest (Armstrong, 1974) and of late Eocene to middle Miocene age (Fig. 4D) in the south (Lipman and others, 1971). This vast ignimbrite flare-up has been interpreted as the result of the collapse and return (westward) sweep of a previously flattened Laramide Benioff zone during Eocene through Miocene time (Coney and Reynolds, 1977). A significant number of granitic plutons, particularly the earlier ones, are of the garnet-bearing two-mica type; this fact suggests a genetic association between the rock type and the evolution of the complexes. Perhaps the earlier two-mica plutons were generated during the late Laramide period of maximum flattening of the Benioff zone when Farallon lithosphere was essentially plated beneath North American lithosphere.

The belt of complexes has the character of an irregular, elongate and sinuous, large-scale pull-apart zone extending the length of the Cordillera at least from southern Canada into northwestern Mexico. The only large-scale phenomenon the zone seems to be associated with is the region that lies above the flattened or low-dipping Laramide (Late Cretaceous) Benioff zone which steepened and collapsed during early to middle Tertiary time. The evolution of the metamorphic core complexes as either separate distinct phases or as a continuum, probably endured 30 to 40 m.y. between about 55 and 15 m.y. B.P. in early to middle Tertiary time.

Just why the complexes formed when they did is not clear. The process began during Farallon–North America plate convergence at least 10 to 30 m.y. before even initial contact between the Pacific and North America plates and resulting growth of the San Andreas–Basin and Range transformtranspressive rifting (Fig. 4). They did not form behind a magmatic arc; they formed within a magmatic arc. But this magmatic arc was a very special one, perhaps the result of extreme flattening of the Laramide Benioff zone followed by its massive collapse and the resulting return sweep of arc activity across the Cordillera from the Pacific Northwest southward into Mexico. The thermal upwelling and instability implied in such a model gives at least an intuitive rationale for all that transpired.

The marriage of these petrologic, structural, and regional tectonic arguments has been productive, and the suggested relationships make the complexes appear more comprehensible and less confusing. There are, however, major issues still outstanding that need clarification and new insight as the contents of this volume point out. The relationships proposed here certainly provide a model to test by further detailed work in the individual complexes. I hope that such work is stimulated by the contents of this volume.

ACKNOWLEDGMENTS

Cordilleran metamorphic core complexes have been an interest of mine since 1970 when their enigmatic character and potential significance first emerged for me while making Cordilleran tectonic compilations. Early discussions with Peter Misch were most illuminating and helpful. R. Hose first kindly led me over the critical ground in eastern Nevada in 1971, and this led to my own field work in the Snake Range in 1972 and 1974. These field investigations were greatly facilitated by students at Middlebury College, R. Bowman, J. Grette, P. Harris, and R. Parrish, and by discussions with M. D. Crittenden, Jr., and R. Armstrong. Since then, I have benefited enormously from numerous field trips to most of the complexes in the company of R. E. Anderson, G. H. Davis, G. A. Davis, H. L. Foster, S. B. Keith, Gordon Haxel, Richard Nielsen, S. J. Reynolds, W. A. Rehrig, L. T. Silver, D. W. Shakel, B. W. Troxel, D. Templeman-Kluit, and F. R. Weber. Countless discussions with these individuals and others, particularly J. H. Stewart, have been stimulating and helpful. Hugh Gabrielse, J. E. Ressor, R. L. Brown, and R. B. Campbell have considerably improved my understanding of Canadian examples. The Penrose Conference on Cordilleran Metamorphic Core Complexes held at Tanque Verde Ranch near Tucson in 1977 was a turning point for all of us, and we are grateful to the Geological Society of America for making it possible. Over the past 8 yr, frequent discussions with W. R. Dickinson have been very valuable. I am particularly grateful to my colleagues G. H. Davis and P. E. Damon and students S. L. Reynolds and R. Rogers at the University of Arizona for detailed insight into the problem. Research Corporation funded my original field work and related projects in Cordilleran tectonics, and for this I am deeply indebted. R. L. Armstrong, Max D. Crittenden, Jr., and S. J. Reynolds reviewed early drafts of this paper and made many helpful suggestions; this does not imply that they necessarily agree with the interpretations or conclusions expressed.

REFERENCES CITED

Anderson, R. E., 1971, Thin skin distension in Tertiary rocks of southeastern Nevada: Geological Society of America Bulletin, v. 82, p. 43–58.

Anderson, T. H., Silver, L. T., and Salas, G. A., 1977, Metamorphic core complexes of the southern part of the North American Cordillera–northwestern Mexico: Geological Society of America Abstracts with Programs, v. 9, p. 881.

——1980, Distribution and U-Pb isotope ages of some lineated plutons, northwestern Mexico: Geological Society of America Memoir 153 (this volume).

Armstrong, R. L., 1968a, Sevier orogenic belt in Nevada and Utah: Geological Society of America Bulletin, v. 79, p. 429–458.

——1968b, Mantled gneiss domes in the Albion Range, southern Idaho: Geological Society of America Bulletin, v. 79, p. 1295–1314.

——1972, Low-angle (denudation) faults, hinterland of

the Sevier orogenic belt, eastern Nevada and western Utah: Geological Society of America Bulletin, v. 83, p. 1729-1754.

Armstrong, R. L., Geochronology of the Eocene volcanic-plutonic episode in Idaho: Northwest Geology, v. 3, p. 1-14.

Armstrong, R. L., and Hansen, E., 1966, Cordilleran infrastructure in the eastern Great Basin: American Journal of Science, v. 264, p. 112-127.

Armstrong, R. L., and Hills, F. A., 1967, Rubidium-strontium and potassium-argon geochronologic studies of mantled gneiss domes, Albion Range, southern Idaho, USA: Earth and Planetary Science Letters, v. 3, p. 114-124.

Armstrong, R. L., Taubeneck, R. W., and Hales, P. E., 1977, Rb-Sr and K-Ar geochronology of Mesozoic granitic rocks and their isotopic composition, Oregon, Washington, and Idaho: Geological Society of America Bulletin, v. 88, p. 397-411.

Banks, N. C., 1980, Geology of a zone of metamorphic core complexes in southeastern Arizona: Geological Society of America Memoir 153 (this volume).

Best, M. G., and others, 1974, Mica granite of the Kern Mountain pluton, eastern White Pine County, Nevada: Remobilized basement of the Cordilleran miogeosyncline?: Geological Society of America Bulletin, v. 85, 1277-1286.

Brown, R. L., 1978, Structural evolution of the southeast Canadian Cordillera: A new hypothesis: Tectonophysics, v. 48, p. 133-151.

Brown, R. L., and Tippett, C. R., 1978, The Selkirk fan structure of the southeastern Canadian Cordillera: Geological Society of America Bulletin, v. 89, p. 548-558.

Burchfiel, B. C., and Davis, G. A., 1975, Nature and controls of cordilleran orogenesis, western United States: Extensions of an earlier synthesis: American Journal of Science, v. 275-A, p. 363-396.

Burchfiel, B. C., and Stewart, J. H., 1966, "Pull-apart" origins of the central segment of Death Valley, California: Geological Society of America Bulletin, v. 77, p. 435-442.

Campbell, R. B., 1973, Structural cross-section and tectonic model of the southeastern Canadian Cordillera: Canadian Journal of Earth Sciences, v. 10, p. 1607-1620.

Chappell, B. W., and White, A.J.R., 1976, Two contrasting granite types: Pacific Geology, v. 8, p. 173-174.

Chase, R. B., Bickford, M. E., and Tripp, S. E., 1978, Rb-Sr and U-Pb isotopic studies of the northeastern Idaho batholith and border zone: Geological Society of America Bulletin, v. 89, p. 1325-1334.

Cheney, E. S., 1977, The Kettle dome: The southern extension of the Shuswap terrane into Washington: Geological Society of America Abstracts with Programs, v. 9, p. 926.

—— 1980, Kettle dome and related structures of northeastern Washington: Geological Society of America Memoir 153 (this volume).

Clark, K. F., and others, 1978, Continuity of magmatism in northern Mexico, 130 m.y. to present: Geological Society of America Abstracts with Programs, v. 10, p. 381.

Clark, S.H.B., 1973, Interpretation of a high-grade Precambrian terrane in northern Idaho: Geological Society of America Bulletin, v. 84, p. 1999-2004.

Compton, R. R., 1972, Geologic map of the Yost quadrangle, Box Elder County, Utah, and Cassia County, Idaho: U.S. Geological Survey Miscellaneous Geologic Investigations Map I-672, scale 1:31,680.

—— 1975, Geologic map of the Park Valley quadrangle, Box Elder County, Utah, and Cassia County, Idaho: U.S. Geological Survey Miscellaneous Geologic Investigations Map I-873, scale 1:3,680.

—— 1980, Fabrics and strains in quartzites of a metamorphic core complex, Raft River Mountains, Utah: Geological Society of America Memoir 153 (this volume).

Compton, R. R., and Todd, V. R., 1979, Oligocene and Miocene metamorphism, folding, and low-angle faulting in northwestern Utah: Discussion and Reply: Geological Society of America Bulletin, v. 90, p. 305-309.

Compton, R. R., and others, 1977, Oligocene and Miocene metamorphism, folding and low-angle faulting in northwestern Utah: Geological Society of America Bulletin, v. 88, p. 1237-1250.

Coney, P. J., 1970, The geotectonic cycle and the new global tectonics: Geological Society of America Bulletin, v. 81, p. 739-748.

—— 1972, Cordilleran tectonics and North America plate motion: American Journal of Science, v. 272, p. 603-628.

—— 1973a, Non-collision tectogenesis in western North America, in Tarling, D. H., and Runcorn, S. H., eds., Implications of continental drift to the earth sciences: New York, Academic Press, p. 713-717.

—— 1973b, Plate tectonics of marginal foreland thrust-fold belts: Geology, v. 1, p. 131-134.

—— 1974, Structural analysis of the Snake Range 'décollement,' east-central Nevada: Geological Society of America Bulletin, v. 85, p. 973-978.

—— 1976, Plate tectonics and the Laramide orogeny: New Mexico Geological Society Special Publication No. 6, p. 5-10.

—— 1978, Mesozoic-Cenozoic Cordilleran plate tectonics, in Smith, R. B., and Eaton, G. P., eds., Cenozoic tectonics and regional geophysics of the western Cordillera: Geological Society of America Memoir 152, p. 33-50.

—— 1979, Tertiary evolution of Cordilleran metamorphic core complexes, *in* Armentrout, J. W., and others, eds., Cenozoic paleogeography of western United States: Society of Economic Paleontologists and Mineralogists, Pacific Section Symposium III, p. 15–28.

Coney, P. J., and Reynolds, S. J., 1977, Cordilleran Benioff zones: Nature, v. 270, p. 403–406.

Crittenden, M. D., Jr., 1979, Oligocene and Miocene metamorphism, folding, and low-angle faulting in northwestern Utah: Discussion and Reply: Geological Society of America Bulletin, v. 90, p. 305–309.

Crittenden, M. D., Jr., Coney, P. J., and Davis, G. H., 1978, Penrose Conference report, Tectonic significance of metamorphic core complexes in the North American Cordillera: Geology, v. 6, p. 79–80.

Damon, P. E., Erickson, R. C., and Livingston, D. E., 1963, K-Ar dating of Basin and Range uplift, Catalina Mountains, Arizona: Nuclear Geophysics, National Academy of Sciences/National Research Council Publication 1075, p. 113–121.

Davis, G. A., and others, 1977, Enigmatic Miocene low-angle faulting, southeastern California and west-central Arizona—Suprastructural tectonics?: Geological Society of America Abstracts with Programs, v. 9, p. 943–944.

—— 1980, Mylonitization and detachment faulting in the Whipple-Buckskin-Rawhide Mountains terrane, southeastern California and western Arizona: Geological Society of America Memoir 153 (this volume).

Davis, G. H., 1973, Mid-Tertiary gravity-glide folding near Tucson, Arizona: Geological Society of America Abstracts with Programs, v. 5, p. 592.

—— 1977, Characteristics of metamorphic core complexes, southern Arizona: Geological Society of America Abstracts with Programs, v. 9, p. 944.

—— 1979, Laramide folding and faulting in southeastern Arizona: American Journal of Science, v. 279, p. 543–569.

—— 1980, Structural characteristics of metamorphic core complexes, southern Arizona: Geological Society of America Memoir 153 (this volume).

Davis, G. H., and Coney, P. J., 1979, Geologic development of Cordilleran metamorphic core complexes: Geology, v. 7, p. 120–124.

Davis, G. H., and others, 1975, Origin of lineation in the Catalina-Rincon-Tortolita gneiss complexes, Arizona: Geological Society of America Abstracts with Programs, v. 7, p. 602.

de Sitter, L. U., 1964, Structural geology: New York, McGraw-Hill, 551 p.

Dewey, J. F., and Bird, J. M., 1970, Mountain belts and the new global tectonics: Journal of Geophysical Research, v. 75, p. 2025–2647.

Dickinson, W. R., 1976, Sedimentary basins developed during evolution of Mesozoic-Cenozoic arc-trench systems in western North America: Canadian Journal of Earth Sciences, v. 13, p. 1268–1287.

Dixon, J. M., 1975, Finite strain and progressive deformation in models of diapiric structures: Tectonophysics, v. 28, p. 89–124.

Drewes, H., 1964, Diverse and recurrent movement along segments of a major thrust fault in Schell Creek Range near Ely, Nevada, *in* Geological Survey research, 1964: U.S. Geological Survey Professional Paper 501-B, p. B20–B24.

—— 1976, Laramide tectonics from Paradise to Hells Gate, southeastern Arizona: Arizona Geological Society Digest, v. 10, p. 151–168.

—— 1977, Geologic map and sections of the Rincon Valley quadrangle, Pima County, Arizona: U.S. Geological Survey Miscellaneous Investigations Map, scale 1:48,000.

—— 1978, The Cordilleran orogenic belt between Nevada and Chihuahua: Geological Society of America Bulletin, v. 89, p. 641–657.

Eberly, L. D., and Stanley, T. B., Jr., 1978, Cenozoic stratigraphy and geologic history of southwestern Arizona: Geological Society of America Bulletin, v. 89, p. 921–940.

Epis, R. C., and Chapin, C. E., 1975, Geomorphic and tectonic implications of the post-Laramide, late Eocene erosion surface in the Southern Rocky Mountains, *in* Curtis, B. F., ed., Cenozoic history of the Southern Rocky Mountains: Geological Society of America Memoir 144, p. 45–74.

Eskola, P., 1949, The problem of mantled gneiss domes: Geological Society of London Quarterly Journal, v. 104, p. 461–476.

Fox, K. R., Jr., Rinehart, C. D., and Engels, J. C., 1977, Plutonism and orogeny in north-central Washington—Timing and regional context: U.S. Geological Survey Professional Paper 989, 27 p.

Gabrielse, H., and Reesor, J. E., 1974, The nature and setting of granitic plutons in the central and eastern parts of the Canadian Cordillera: Pacific Geology, v. 8, p. 109–138.

Haller, J., 1956, Problems der Tiefentiktonic: Bauformen in migmatit-stockwork der ostgranlandischen Kaledoniden: Geologische Rundschau, v. 45, p. 159–167.

Hamilton, W., 1969, Mesozoic California and the underflow of Pacific Mantle: Geological Society of America Bulletin, v. 80, p. 2409–2430.

Hose, R. K., and Danes, Z. F., 1973, Development of late Mesozoic to early Cenozoic structures in the eastern Great Basin, *in* De Jong, K. A., and Scholten, R., eds., Gravity and tectonics: New York, John Wiley & Sons, p. 429–442.

Howard, K. A., 1971, Paleozoic metasediments in the northern Ruby Mountains, Nevada: Geological Society of America Bulletin, v. 82, p. 259–264.

Howard, K. A., 1980, Metamorphic infrastructure in the northern Ruby Mountains, Nevada: Geological Society of America Memoir 153 (this volume).

Hyndman, D. W., 1968, Mid-Mesozoic multiphase folding along the border of the Shuswap metamorphic complex: Geological Society of America Bulletin, v. 79, p. 575–587.

Hyndman, D. W., Talbot, J. L., and Chase, R. B., 1975, Boulder batholith, a result of emplacement of a block detached from the Idaho batholith infrastructure?: Geology, v. 3, p. 401–404.

Jones, D. L., Irwin, W. P., and Ovenshine, A. T., 1972, Southeastern Alaska—A displaced continental fragment?, in Geological Survey Research, 1972, Chapter B: U.S. Geological Survey Professional Paper 800-B, p. B211–B217.

Jones, D. L., Silberling, N. J., and Hillhouse, J., 1977, Wrangellia—A displaced terrane in northwestern North America: Canadian Journal of Earth Sciences, v. 14, p. 2565–2577.

Keith, S. B., and others, 1980, Evidence for multiple intrusion and deformation within the Santa Catalina–Rincon–Tortolita crystalline complex, southeastern Arizona: Geological Society of America Memoir 153 (this volume).

King, P. B., compiler, 1969, Tectonic map of North America: U.S. Geological Survey, scale 1:5,000,000.

Kistler, R. W., and Peterman, Z. E., 1973, Variations in Sr, Rb, K, Na, and initial Sr^{87}/Sr^{86} in Mesozoic granitic rocks and intruded wall rocks in central California: Geological Society of America Bulletin, v. 84, p. 3489–3511.

Lee, D. E., and others, 1970, Modification of potassium-argon ages by Tertiary thrusting in the Snake Range, White Pine County, Nevada: U.S. Geological Survey Professional Paper 700-D, p. D92–D102.

Lipman, P. W., Prostka, H. J., and Christiansen, R. L., 1971, Evolving subduction zones in the western United States, as interpreted from igneous rocks: Science, v. 174, p. 821–825.

Mauger, R. L., Damon, P. E., and Livingston, D. E., 1968, Cenozoic argon ages on metamorphic rocks from the Basin and Range province: American Journal of Science, v. 266, p. 579–589.

Miller, D. M., 1980, Structural geology of the northern Albion Mountains, south-central Idaho: Geological Society of America Memoir 153 (this volume).

Miller, F. K., 1972, The Newport fault and associated mylonites, northwestern Washington: U.S. Geological Survey Professional Paper 750-D, p. 77–79.

Miller, F. K., and Engels, J. C., 1975, Distribution and trends of discordant ages of the plutonic rocks of northeastern Washington and northern Idaho: Geological Society of America Bulletin, v. 86, p. 517–528.

Misch, P., 1960, Regional structural reconnaissance in central-northeast Nevada and some adjacent areas: Observations and interpretations: Intermountain Association of Petroleum Geologists Guidebook for 11th Annual Field Conference, p. 17–42.

Miyashiro, A., 1961, Evolution of metamorphic belts: Journal of Petrology, v. 2, p. 277–311.

——— 1973, Paired and unpaired metamorphic belts: Tectonophysics, v. 17, p. 241–254.

Monger, J.W.H., 1977, Upper Paleozoic rocks of the western Canadian Cordillera and their bearing on Cordilleran evolution: Canadian Journal of Earth Sciences, v. 14, p. 1832–1859.

Monger, J.W.H., and Hutchison, W. W., 1971, Metamorphic map of the Canadian Cordillera: Canada Geological Survey Paper 70-33, 61 p.

Moores, E. M., Scott, R. B., and Lumsden, W. W., 1968, Tertiary tectonics of the White Pine–Grant Range region, east-central Nevada, and some regional implications: Geological Society of America Bulletin, v. 79, p. 1703–1726.

Nelson, R. B., 1966, Structural development of northernmost Snake Range, Kern Mountains, and Deep Creek Range: Nevada-Utah: American Association of Petroleum Geologists, v. 50, p. 921–951.

——— 1969, Relation and history of structures in a sedimentary succession with deeper metamorphic structures, eastern Great Basin: American Association of Petroleum Geologists Bulletin, v. 52, p. 307–339.

Noble, D. C., 1972, Some observations on the Cenozoic volcano-tectonic evolution of the Great Basin, Western United States: Earth and Planetary Science Letters, v. 17, p. 142–150.

Okulitch, A. V., Wanless, R. K., and Loveridge, W. D., 1975, Devonia plutonism in south-central British Columbia: Canadian Journal of Earth Sciences, v. 12, p. 1760–1769.

Pashley, E. F., 1966, Structure and stratigraphy of the central, northern, and eastern parts of the Tucson basin, Arizona [Ph.D. dissert.]: Tucson, University of Arizona, 273 p.

Peirce, H. W., 1976, Elements of Paleozoic tectonics in Arizona: Arizona Geological Society Digest, v. 10, p. 37–58.

Price, R. A., and Mountjoy, E. W., 1970, Geologic structure of the Canadian Rocky Mountains between Bow and Athabasca Rivers—A progress report: Geological Association of Canada Special Paper No. 6, p. 7–25.

Ramberg, H., 1972, Theoretical models of density stratification and dispersion in the earth's crust: Journal of Geophysical Research, v. 77, p. 877–889.

Read, P. B., and Okulitch, A. V., 1977, The Triassic unconformity of south-central British Columbia: Canadian Journal of Earth Sciences, v. 14, p. 606–638.

Reesor, J. E., 1970, Some aspects of structural evolution

and regional setting in part of the Shuswap metamorphic complex: Geological Association of Canada Special Paper No. 6, p. 73–86.

Rehrig, W. A., and Heidrick, T. L., 1976, A northwest zone of tectonic stress during the Laramide and late Tertiary intrusive periods, Basin and Range province, Arizona: Arizona Geological Society Digest, v. 10, p. 205–228.

Rehrig, W. A., and Reynolds, S. J., 1977, A northwest zone of metamorphic core complexes in Arizona: Geological Society of America Abstracts with Programs, v. 9, p. 1139.

——1980, Geologic and geochronologic reconnaissance of a northwest-trending zone of metamorphic core complexes in southern and western Arizona: Geological Society of America Memoir 153 (this volume).

Reynolds, S. J., and Rehrig, W. A., 1980, Mid-Tertiary plutonism and mylonitization, South Mountains, central Arizona: Geological Society of America Memoir 153 (this volume).

Reynolds, S. J., Rehrig, W. A., and Damon, P. E., 1978, Metamorphic core complex terrain at South Mountain, near Phoenix, Arizona: Geological Society of America Abstracts with Programs, v. 10, p. 143–144.

Roberts, R. J., and Crittenden, M. D., Jr., 1973, Orogenic mechanisms, Sevier orogenic belt, Nevada and Utah, in De Jong, K. A., and Scholten, R., eds., Gravity and tectonics: New York, John Wiley & Sons, p. 409–428.

Shakel, D. W., Silver, L. T., Damon, P. E., 1977, Observations on the history of the gneissic core complex, Santa Catalina Mountains, southern Arizona: Geological Society of America Abstracts with Programs, v. 9, p. 1169.

Silver, L. T., Early, T. O., and Anderson, T. H., 1975, Petrological, geochemical, and geochronological assymmetries of the Peninsula Ranges batholith: Geological Society of America Abstracts with Programs, v. 7, p. 375.

Snoke, A. W., 1980, Transition from infrastructure to suprastructure in the northern Ruby Mountains, Nevada: Geological Society of America Memoir 153 (this volume).

Thorman, C. H., 1970, Metamorphosed and nonmetamorphosed Paleozoic rocks in Wood Hills and Pequop Mountains, northeast Nevada: Geological Society of America Bulletin, v. 81, p. 2412–2448.

——1977, Gravity-induced folding off a gneiss dome complex, Rincon Mountains, Arizona—A discussion: Geological Society of America Bulletin, v. 88, p. 1211–1212.

Todd, V. R., 1973, Tectonic mobilization of Precambrian gneiss during Tertiary metamorphism and thrusting, Grouse Creek Mountains, north-western Utah: Geological Society of America Abstracts with Programs, v. 5, p. 116.

——1980, Structure and petrology of a Tertiary gneiss complex in northwestern Utah: Geological Society of America Memoir 153 (this volume).

Wanless, R. K., and Reesor, J. E., 1975, Precambrian zircon age of orthogneiss in the Shuswap metamorphic complex, British Columbia: Canadian Journal of Earth Sciences, v. 12, p. 326–332.

Wegmann, C. E., 1935, Zur Deutung der Migmatite: Geologische Rundschau, v. 26, p. 303–350.

Wheeler, J. O., and Gabrielse, H., 1972, The Cordilleran structural province, in Prince, R. A., and Douglas, R.J.W., eds., Variations in tectonic styles in Canada: Geological Association of Canada Special Paper No. 11, p. 9–81.

Whitebread, D. H., 1968, Snake Range décollement and related structures in the southern Snake Range, eastern Nevada [abs.]: Geological Society of America Special Paper 101, p. 345–346.

——1969, Geologic map of the Wheeler Peak and Garrison quadrangles, Nevada and Utah: U.S. Geological Survey Miscellaneous Geologic Investigations Map I-578, scale 1:48,000.

Wright, L. A., Otton, J. K., and Troxel, B. W., 1974, Turtleback surfaces of Death Valley viewed as phenomena of extensional tectonics: Geology, v. 2, p. 79–80.

Zwart, H. J., 1969, Metamorphic facies series in the European orogenic belts and their bearing on the causes of orogeny, in Wynne-Edwards, H. R., ed., Age relations in high-grade metamorphic terrains: Geological Association of Canada Special Paper No. 5, p. 7–16.

MANUSCRIPT RECEIVED BY THE SOCIETY JUNE 21, 1979

MANUSCRIPT ACCEPTED AUGUST 7, 1979

Printed in U.S.A.

PART 2

SOUTHERN BASIN AND RANGE

Structural characteristics of metamorphic core complexes, southern Arizona

GEORGE H. DAVIS
Department of Geosciences
University of Arizona
Tucson, Arizona 85721

ABSTRACT

Metamorphic core complexes in southern Arizona may be subdivided into four elements: core, metamorphic carapace, decollement, and cover. Cores consist chiefly of mylonitic augen gneiss that is, for the most part, derived from Precambrian and Phanerozoic plutonic rocks of granitic composition. Foliation in the cores is characteristically low dipping and commonly defines large upright, doubly plunging foliation arches. The mylonitic augen gneisses everywhere display low-dipping mineral lineation of a cataclastic nature. Lineation within any given metamorphic core complex is generally remarkably systematic in orientation. In southeastern Arizona the lineation trends northeast to east-northeast; in south-central Arizona, it trends north-south. Ductile normal faults are locally abundant in the core rocks and always are oriented perpendicular to lineation. Core rocks in places are clearly transitional laterally or downward into undeformed protolith.

The metamorphic carapace consists of penetratively deformed younger Precambrian and Phanerozoic strata metamorphosed to greenschist-amphibolite grade. It forms a crudely tabular layer that locally overlies the crystalline rocks of the core. The contact is commonly so tight that the rocks of the carapace appear to be plated onto the crystalline rocks of the core. Within the metamorphic carapace, overturned to recumbent folds are ubiquitous, transposition is rampant, and boudins and pinch-and-swell features are commonplace. In spite of spectacular deformation, individual formations within the metamorphic carapace are arranged in normal stratigraphic order.

A decollement, marked by brittle low-angle faulting and shearing, separates carapace and cover or, where carapace is absent, core and cover. The surface thus separates rocks of remarkably contrasting deformational styles. Where the surface "caps" core rocks, a decollement zone is formed beneath the decollement and consists of a distinctive crudely tabular zone of crushed and granulated but strongly indurated mylonite, mylonitic gneiss, microbreccia, and chlorite breccia. Striking "younger-on-older" fault relations characterize the decollement. Decollements in this area typically separate orthogneisses, which were derived in part from Precambrian rocks, from unmetamorphosed upper Paleozoic, Mesozoic, or Tertiary strata. Overturned asymmetric folds, detached isoclinal folds, and unbroken cascades of recumbent folds are abundant in many of the cover sheets.

The metamorphic core complexes in southern Arizona are interpreted to be products of high-temperature extensional deformation, regional in extent, superseded by moderately ductile to moderately brittle tectonic denudation. Rocks in the augen gneissic core and metamorphic carapace were affected by profound ductile through brittle extension and flow in the direction of mineral lineation. Evolution of the decollement zones largely postdated the development of foliation and lineation.

Mechanics of strain are interpreted in the context of megaboudinage. Cores are pictured as parts of megaboudins imposed on heterogeneous crustal rocks. Profound thinning of younger Precambrian and Paleozoic sediments through flow during deformation had the effect of significantly decreasing the stratigraphic separation between individual Phanerozoic formations and the Precambrian basement. Intrafolial folds developed in the carapace as products of passive flow. The relatively brittle, massive crystalline basement responded to extension and flattening by ductile normal faulting and the development of penetrative foliation and lineation. The high-temperature extensional disturbance, which began in early Tertiary time and ended about 25 m.y. ago(?), was accompanied by moderately ductile to moderately brittle tectonic denudation and gravity-induced folding of cover rocks. Major listric normal faulting postdated the development of tectonite, shaped the internal fabric of the decollement zones, and effected final movements of the covers.

INTRODUCTION

Purpose

The purpose of this paper is to describe the characteristics of metamorphic core complexes as disclosed by attributes of core-complex terranes in southern Arizona. Emphasis is focused on structural elements and relationships that reveal important mechanical aspects of the kinematic and dynamic evolution of the core complex rocks.

Physical Model

Defining what constitutes characteristics of metamorphic core complexes is a scale-dependent process. Viewed as regional tectonic elements along the length of the western Cordillera from Sonora to southern Idaho, the metamorphic core complexes might be considered "small" outcrop areas of relatively high topographic relief that comprise arches of distinctively deformed and metamorphosed igneous and sedimentary rocks. They occur exclusively in the Basin and Range province. The deformed crystalline rocks are separated from unmetamorphosed country rock by decollement marking strikingly sharp thermal-strain gradients. Resting on the decollement are tiny thin plates of deformed but generally unmetamorphosed Phanerozoic strata. Commonly, the isotopic systems of core complex rocks disclose a mid-Tertiary thermal disturbance. The complexes are commonly co-spatial with Oligocene-Miocene continental sedimentary sequences and Miocene ignimbritic volcanic rocks.

Viewed macroscopically at the scale of several mountain ranges, individual metamorphic core complexes in southern Arizona may be subdivided into four elements: core, metamorphic carapace, decollement, and cover (left half of Fig. 1). Cores consist of either type A—mylonitic augen gneissic, granodioritic rocks with screens of mylonitic schist and metasedimentary metavolcanic rocks—or type B—cataclastically deformed metasedimentary and metavolcanic rocks intruded lit-par-lit by deformed granitic and pegmatitic layers. The core rocks in places are clearly transitional laterally or downward into nontectonite protoliths. Where cores are type A, a distinct metamorphic carapace can

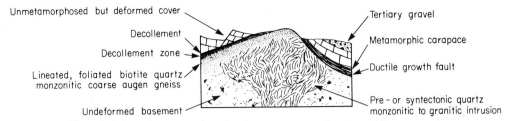

Figure 1. Schematic cross sections showing components of metamorphic core complex.

often be distinguished (center and rights parts of Fig. 1). The carapace consists of penetratively deformed younger Precambrian and Phanerozoic strata metamorphosed to greenschist or amphibolite grade. It forms a crudely tabular layer that overlies concordantly the crystalline rocks of the core. The contact itself is a tectonic contact, commonly so tight (that is, lacking a visible physical discontinuity) that the rocks of the metamorphic carapace appear to be "smeared," "welded," or "plated" onto the crystalline rocks. "Plated" will be used in this paper to describe such contact relationships, following the usage of Engel (1949) in the Adirondacks. In contrast to the tight contact between core and carapace, the contact between cover and carapace, or cover and core, is a decollement marked by brittle low-angle faulting and shearing. The decollement serves to separate two different distinctive strain styles. Typically, the decollement lies atop the core without significant intervening metamorphic carapace (left half of Fig. 1). Below such surfaces, mylonitic augen gneisses and metamorphosed strata of the cores are transformed into a low-dipping, resistant decollement zone of fine-grained cataclastically deformed rocks. The sharp, planar upper surface of this zone, the decollement itself, is overlain by semiconcordant thin sheets containing tectonic slices of unmetamorphosed Phanerozoic strata and Precambrian basement arranged in normal stratigraphic order (left half of Fig. 1).

Another way in which core, carapace, decollement, decollement zone(s), and cover can be bound together is shown in the right half of Figure 1. Where metamorphic carapace is preserved in significant thickness, the major detachment or decollement occurs at the top of the carapace. A resistant tabular decollement zone derived from intensive deformation of the metamorphosed carapace strata may locally develop directly below the detachment surface. But, additionally, a ductile fault zone may form just below the plated contact between core and carapace. The physical nature of the ductile fault zone in the crystalline rocks below the carapace is physically similar, although thinner, than the typical decollement zone that is derived from mylonitic gneiss.

At the mesoscopic scale, the array of structures is awesome in its diversity and pervasiveness: the cores contain low-dipping cataclastic foliation, systematically aligned low-plunging mineral lineation, slickensides, normal ductile faults, abundant boudins and pinch-and-swell features, and recumbent to overturned tight ptygmatic folds in aplites and pegmatites. Zones of contact between dissimilar augen gneisses are frequently marked by folded mylonitic schist. Rocks of the metamorphic carapace show extreme thinning and attenuation. The common array of structures includes tight isoclinal recumbent to overturned intrafolial folds, axial-plane cleavage, boudins, pinch-and-swell, flattened-stretched pebble metaconglomerate, lineation, and ductile faults. The metamorphic fabric is generally overprinted by pervasive fractures. The decollement zone, where it forms at the expense of mylonitic augen gneiss, is structurally a massive to moderately foliated resistant layer of microbreccia, chlorite breccia, mylonite, and mylonitic gneiss shattered by closely spaced fractures. It typically occurs at the upper surface of type A cores and is marked by a planar top. Although the mylonitic rocks are usually derived from augen gneiss, original lineation and foliation are rotated, overprinted, and masked by cataclastic granulation. Above the detachment surface atop the decollement zone, there may be some metamorphic carapace, but generally there are unmetamorphosed but deformed cover rocks characterized by overturned intraformational asymmetric folds, bedding-plane cleavage, and faults. Although in normal stratigraphic succession, formations represented in the cover rocks are markedly tectonically thinned or may be absent altogether.

Region of Study

The tectonic setting of southeastern Arizona is ideal for examining the structural geology of metamorphic core complexes and for interpreting their dynamic significance. Within this region, outside the metamorphic core complexes, granitic gneiss is insignificant volumetrically and, thus, not to be confused with gneisses in the core complexes themselves. Furthermore, in southeastern Arizona, no tectonites exist in the Precambrian and Phanerozoic rocks, except for 1.6- to 1.8-b.y.-old Pinal Schist, and, thus, the tectonites in the core complex terranes are unique to the region. This statement does not hold for the setting of core complexes west of Tucson in the Papago Indian Reservation. There, Mesozoic sedimentary and volcanic country rocks are metamorphosed to phyllite and schist. Finally, excellent exposures and great structural relief afford examination of a diversity of rocks at many different structural levels.

Southeastern Arizona may be an ideal setting in yet another, but more controversial, respect. Although rocks in southeastern Arizona have been affected by multiple, superposed deformations in the Phanerozoic, the deformations may not have produced at any time a regional low-angle overthrusting. If it is true that some core complexes of southeastern Arizona occur in a setting where major overthrusting *has not* occurred, then the descriptive characteristics of these complexes would have a significant bearing on dynamic interpretations of other complexes in the western Cordillera. The various models proposed for the nature of Sevier-Laramide deformation in southeastern Arizona will not be presented here. Papers that bear on the problem and that provide background for considering the metamorphic core complexes within their regional and historical framework include Cooper and Silver (1964), Davis (1977b, 1979), Drewes (1973, 1976a, 1978b), Gilluly (1956), Hansen and others (1978), Keith and Barrett (1976), Rehrig and Heidrick (1972), and Thorman (1977).

When compared with the normal country-rock terrane of southern Arizona, the rocks of the metamorphic core complexes are *striking* dynamo-thermal anomalies. The classic geologic terrane generally consists of (1) low-grade metasedimentary Pinal Schist (1.6 to 1.8 b.y. old) intruded by largely undeformed Precambrian (1.55 to 1.65 b.y. old) granite plutons (Ransome, 1903; Silver, 1955; Silver and Deutsch, 1961; Erickson, 1962; Cooper and Silver, 1964); (2) unfoliated, unlineated, 1.4- to 1.5-b.y.-old "anorogenic" Precambrian granitic rocks (Drewes, 1976b; Silver and others, 1977a, 1977b); (3) a thin (about 2,000 m) sequence of faulted, generally unfolded, commonly homoclinal, younger Precambrian and Paleozoic sedimentary rocks (Silver, 1955; Bryant, 1955, 1968; Butler, 1971); (4) a thick sequence of faulted, commonly wedge-shaped deposits of Triassic(?)-Jurassic volcanic and sedimentary rocks (Hayes and others, 1965; Simons and others, 1966; Hayes and Drewes, 1968; Hayes, 1970b; Cooper, 1971; Drewes, 1971); and (5) a thick sequence of gently to tightly folded Cretaceous sedimentary rocks (Hayes and Drewes, 1968; Drewes and Finnell, 1968; Finnell, 1970; Hayes, 1970a; Drewes, 1972). Transformation of these country rocks to tectonites in the sites of the metamorphic core complexes is believed to provide kinematic and dynamic clues to processes responsible for the evolution of core complexes in general.

The presently known metamorphic core-complex terranes in the Basin and Range province of southernmost Arizona are shown in Figure 2. Core-complex terranes are exposed in the Santa Catalina, Rincon, and Tanque Verde Mountains bounding Tucson on the north and east; in the Tortolita Mountains, Suizo Hills, Durham Hills, and southern Picacho Mountains in the corridor between Tucson and Phoenix; in the northern Pinaleno Mountains and eastern Santa Teresa Mountains (Jackson Mountain) west of Safford; and, to the southwest of Tucson, in the Coyote Mountains, the southern Babokivari Mountains (Pozo Verde Mountain), part of the Comobabi Mountains, Sierra

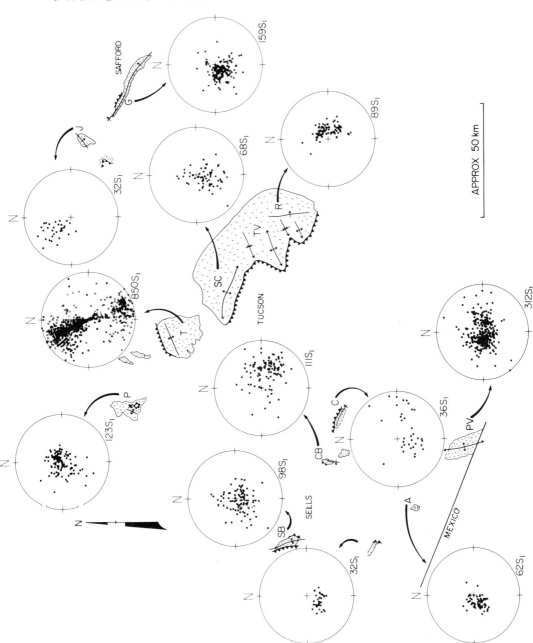

Figure 2. Map showing distribution of metamorphic core complexes in southern Arizona. Base used in preparation of figure is a palinspastic reconstruction of southern Arizona (prepared by Peter J. Coney, Richard Nielson, and me) that portrays the geology of southern Arizona after Laramide and before basin and range tectonism. The reconstruction assumes a conservative 10% to 15% extension in the last 16 m.y. J = Jackson Mountain in the eastern Santa Teresa Mountains; G = Graham Mountains (also known as Pinaleno Mountains); P = Picacho Mountains; T = Tortolita Mountains; SC = Santa Catalina Mountains; TV = Tanque Verde Mountains; R = Rincon Mountains; SB = Sierra Blanca; NC = northern Comobabi Mountains; C = Coyote Mountains; PV = Pozo Verde Mountains; A = Alvarez Mountains; K = Kupk Hills. Sawtooth symbol = decollement zone. Lower-hemisphere projections show poles to foliation in rocks of the core and metamorphic carapace rocks.

Blanca, Alvarez Mountain, and Kupk Hills. Noteworthy is the change in trend of the belt of metamorphic core complexes from N50°W in western and central Arizona to north-south from Tucson to Hermosilla, Sonora.

The Rincon–Santa Catalina–Tortolita metamorphic core complex contains well-exposed cataclastically deformed crystalline rocks and elegantly deformed metasedimentary and sedimentary rocks. Although the evolution of the rocks in these mountains is by no means completely understood, there exists a large and significant amount of data regarding the geologic and geochronologic relationships. Geologic maps of major parts of the Rincon–Santa Catalina–Tortolita complex have been made by Moore and others (1941), Pashley (1966), Drewes (1975, 1878a), Creasey and Theodore (1975), Davis and others (1975), Budden (1975), Banks (1976), and Banks and others (1977). Geochronologic data, including K-Ar, Rb-Sr, fission-track, and U-Pb, have been contributed by Damon and others (1963), Damon (1968a, 1968b), D. E. Livingston (unpub. Rb-Sr data, Laboratory of Geochronology, the University of Arizona), Shakel (1974), Creasey and others (1976), Shakel and others (1977), and Keith and others (this volume). The structural fabric of the rocks has been analyzed by Mayo (1964), Waag (1968), Peterson (1968), Davis (1973, 1975, 1977a), Davis and others (1974, 1975), and Davis and Frost (1976).

In contrast to the geology of the Rincon–Santa Catalina–Tortolita complex, the geology of the other ranges shown in Figure 2 is relatively little known. These terranes are dominated geologically by augen gneissic rocks and, unfortunately, are generally lacking in metasedimentary carapace rocks and unmetamorphosed but deformed cover rocks. Small-scale county geologic maps prepared by the Arizona Bureau of Mines provide some data (Wilson and others, 1957; Wilson and Moore, 1959; and Wilson and others, 1960). The Picacho Mountains have been mapped in reconnaissance fashion by Yeend (1976). Similarly, Haxel and others (1978) have mapped the Comobabi Mountains. Wargo and Kurtz (1956) mapped the geology of the Coyote Mountains. May and Haxel (1979) have recently mapped the Sells quadrangle. My own contributions to the geologic mapping include 1:24,000 mapping in the Tortolita Mountains (Davis and others, 1975; Banks and others, 1977), large-scale mapping in the Rincon Mountains (Davis and others, 1974; Davis and Frost, 1976), 1:62,500 mapping of the Coyote, Picacho, Sierra Blanca, Pozo Verde Mountains, Jackson Mountain, and the northeastern Pinaleno Mountains, and 1:20,000 mapping in Happy Valley (Davis and others, in prep.).

In this presentation, no attempt is made to outline the geology of the complexes, range by range. Rather the characteristics of the metamorphic core complexes are discussed from inside out, from augen gneissic core, through metasedimentary carapace, through decollement zone, to unmetamorphosed but folded and/or faulted cover rocks.

METAMORPHIC CORE

Augen Gneiss

Macroscopic Structures. By far the greatest volume (>90%) of rocks in the metamorphic core complexes of southern Arizona comprise the cores. These individual core zones underlie exposed surface areas of as much as 450 km^2 and, within single complexes, express an exposed vertical relief of as much as 1,800 m. These cores of mostly mylonitic augen gneiss undergird some of the highest summits in the Mountain subprovince of the Basin and Rane province (for example, Mount Lemmon in the Santa Catalina Mountains and Mica Mountain in the Rincon Mountains at about 2,700 m; Fig. 3).

Foliation in the individual mylonitic augen gneissic cores of the complexes is characteristically low dipping (see Fig. 2) and commonly defines large upright, doubly plunging foliation arches or antiforms,

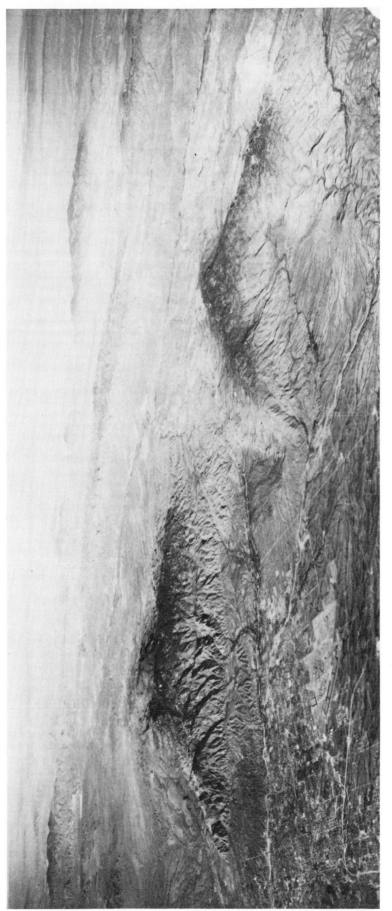

Figure 3. Northwest-directed U-2 oblique aerial photograph showing the Santa Catalina Mountains and Tanque Verde–Rincon Mountains, north and east of Tucson, respectively. Ranges are composed dominantly of foliated, lineated quartz monzonitic augen gneiss. The gneisses are separated from unmetamorphosed but deformed sedimentary cover rocks by a curviplanar decollement zone that marks the basin-mountain interface on the south side of the Santa Catalina Mountains and the western and southern sides of the Tanque Verde–Rincon Mountains. Rocks of the metamorphic carapace are preserved along the very summit of the Santa Catalina Mountains and on the eastern side of the Rincon Mountains. The northern part of the Pinaleno Mountains (dark range in right background), 95 km northeast of the Rincon Mountains, displays deformation of the metamorphic core complex as well.

half-arches, upright synforms or troughs, and irregular ameboid domical structures. Some of the doubly plunging arches, like the Tanque Verde antiform (Fig. 4A) and Sierra Blanca (Fig. 4B), have exceptional physiographic expression. Tanque Verde antiform merges with two other antiforms and a synform in the Rincon Mountains to define a crudely ameboid-shaped domical structure (Moore and others, 1941; Pashley, 1966; Drewes, 1978a). Other archlike structures are found in the forerange of the Santa Catalina Mountains (Pashley, 1966; Peterson, 1968) and the Pozo Verde Mountains (Fig. 4C). Quaquaversal dips characterize the low-dipping foliation in the Picacho Mountains (Fig. 4D), and these define a crude, irregular dome. Half-arches are the dominant macroscopic structural form of the Coyote Mountains (Fig. 4E), part of the Comobabi Mountains (Fig. 4F), Jackson Mountain (Fig. 4G), and the northeastern Pinaleno Mountains (Fig. 4I). The internal structure of foliation in the central Tortolita Mountains appears to be archlike (Fig. 4H), but mapping reveals that no actual closure exists.

In general, the penetrative mineral lineation within the arched augen gneissic rocks is coaxial with the axis (axes) of arching. The only significant exception is the forerange of the Santa Catalina Mountains (Peterson, 1968). There, formation of the east forerange fold quite clearly resulted in rotation of the original undirectional penetrative lineation.

Lithology

Coarse-Grained Quartz Monzonitic Augen Gneiss. The mylonitic gneisses in the cores of some of the terranes are derived in part from coarse-grained Precambrian quartz monzonite, approximately 1.4 to 1.5 b.y. old. This interpretation has been offered by U.S. Geological Survey geologists in the Santa Catalina–Rincon–Tortolita complex for coarse-grained, dark-colored, biotite augen gneiss and schist (Fig. 5). This rock has been mapped as such by Drewes (1975, 1978a) in the Rincon and Tanque Verde Mountains, by Creasey and Theodore (1975) in the eastern part of the Santa Catalina Mountains, and by Davis and others (1975) and Banks and others (1977) in the Tortolita Mountains. To the north of Tucson, Banks and others (1977) mapped lithologically similar rock in the Durham and Suizo Hills, and I mapped it in parts of the Picacho Mountains. The progressive transformation of the recognizable nonfoliated Precambrian porphyritic quartz monzonite to coarse-grained augen gneiss, and even to mylonitic schist, can be observed at a number of locations in southern Arizona (Davis and others, 1975; Banks and others, 1977; Banks, this volume). The most recently discovered locale in which this transition can be observed is in the northeastern Pinaleno Mountains (Fig. 4I). Isotopic data also have confirmed that at least part of the coarse-grained augen gneiss was derived from Precambrian porphyritic quartz monzonite. Shakel and others (1977) reported that zircons from coarse-grained biotite augen gneiss in the Santa Catalina forerange indeed yield an age of $1,440 \pm 10$ m.y.

The microscopic petrography of the coarse-grained quartz monzonitic augen gneiss has been described in detail by a number of workers, especially Peterson (1968) and Sherwonit (1974), and they are described in this volume by Banks. In outcrop, the augen gneiss is relatively dark colored and cut by the penetrative flat to moderately dipping cataclastic foliation (Fig. 6A). Concordant aplite and pegmatite veins serve to enhance the foliation, and such sills and veins commonly display abundant boudinage and pinch-and-swell. Aplites and pegmatites that cut the foliation at high angles are almost always ptygmatically folded (Fig. 6B). These folds are axial planar with respect to the low-dipping penetrative foliation and, thus, are typically strongly overturned to recumbent.

Medium-Grained Quartz Monzonitic Augen Gneiss

In addition to the mylonitic gneisses derived from the Precambrian quartz monzonite, all of the Arizona core complexes expose abundant mylonitic augen gneisses that are derived from quartz

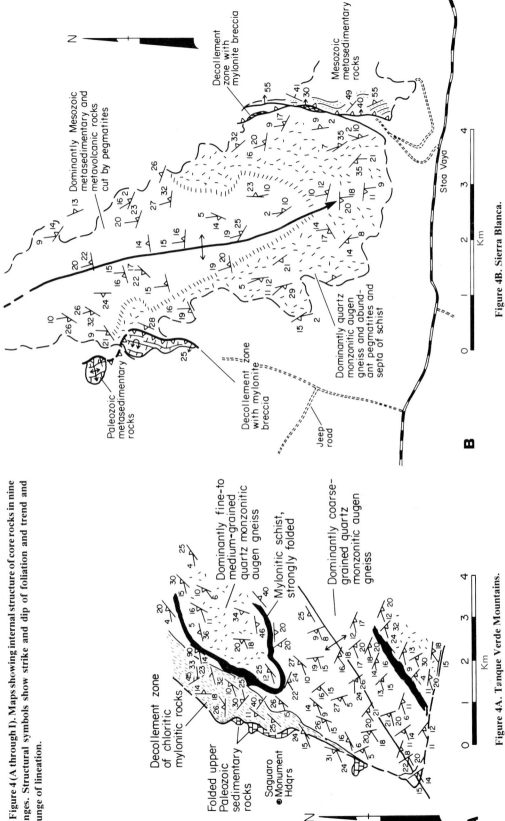

Figure 4 (A through I). Maps showing internal structure of core rocks in nine ranges. Structural symbols show strike and dip of foliation and trend and plunge of lineation.

Figure 4A. Tanque Verde Mountains.

Figure 4B. Sierra Blanca.

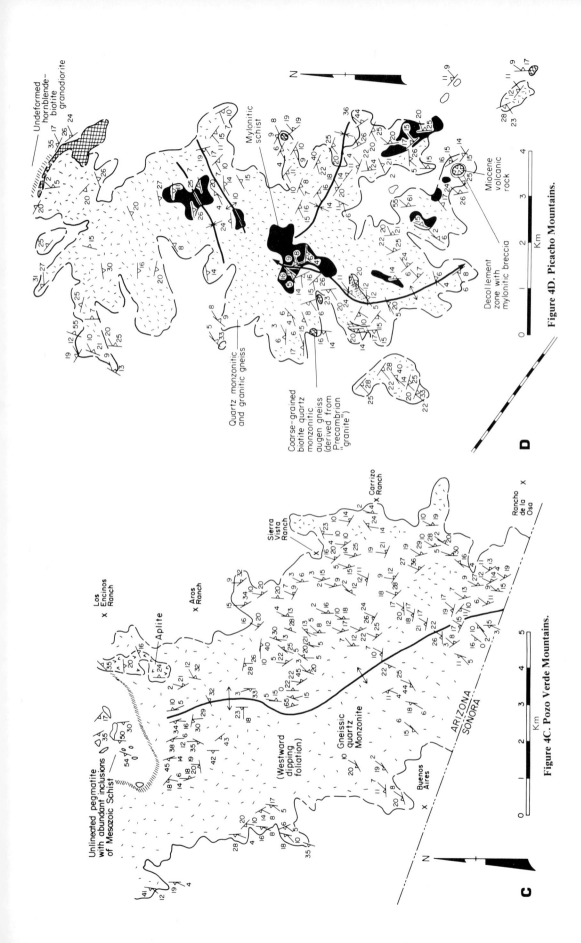

Figure 4C. Pozo Verde Mountains.

Figure 4D. Picacho Mountains.

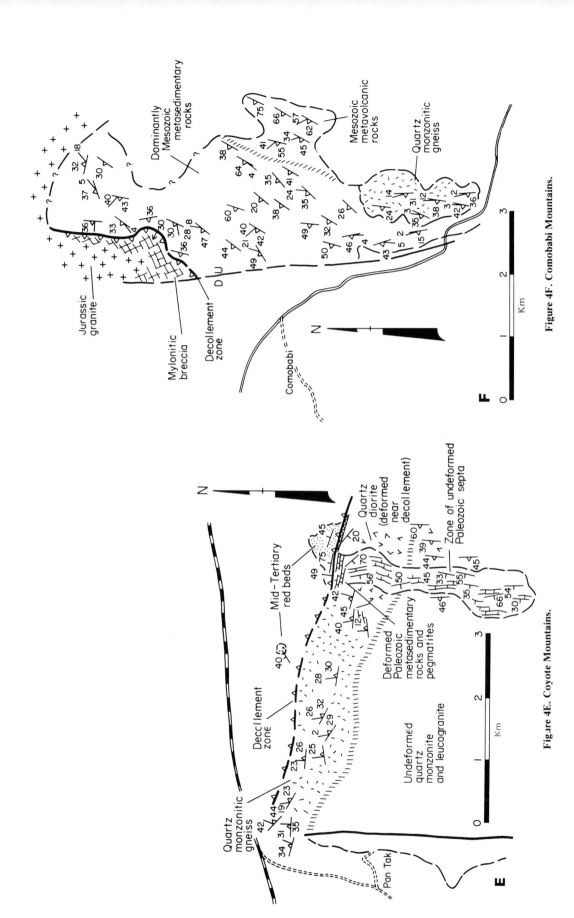

Figure 4E. Coyote Mountains.

Figure 4F. Comobabi Mountains.

Figure 4G. Jackson Mountains.

Figure 4H. Tortolita Mountains.

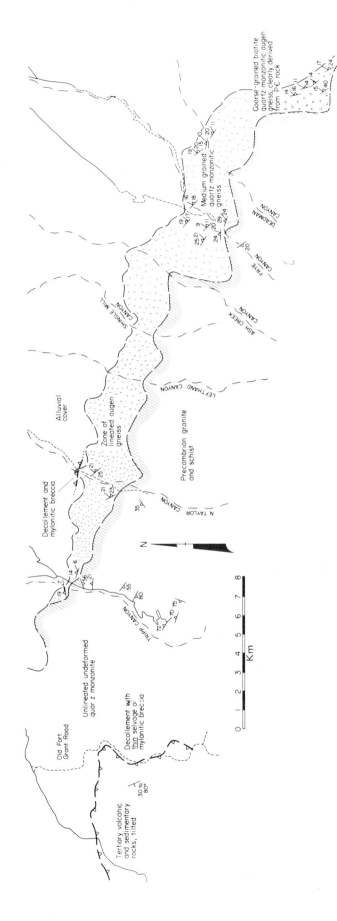

Figure 4I. Pinaleno Mountains.

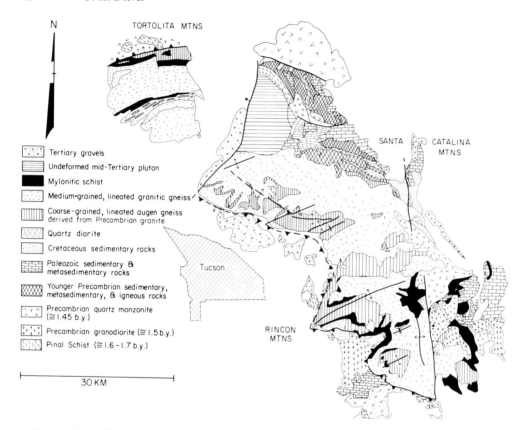

Figure 5. Generalized structure map of the Rincon–Santa Catalina–Tortolita complex. Adapted and interpreted from the maps of Banks (1974), Creasey and Theodore (1975), Drewes (1975, 1978a), Davis and others (1975), and Banks and others (1977).

monzonite, granodiorite, granite, and alaskite of Phanerozoic age. Many textural and compositional varieties of these rocks exist, but most tend to be medium grained and quartz monzonitic in composition. Garnet-bearing, two-mica granites are not uncommon. As in the case of the coarse-grained augen gneiss, a distinguishing structural characteristic of these rocks is low-dipping cataclastic foliation and penetrative lineation.

In the Rincon Mountains, the medium-grained augen gneisses have been mapped by Drewes (1975, 1978a) as belonging to the Precambrian(?) Wrong Mountain quartz monzonite. In the Santa Catalina Mountains, Creasey and Theodore (1975), Banks (1976), and Creasey and others (1976) have included the medium-grained augen gneisses as components of a mid-Tertiary composite batholith (quartz monzonite of Sanmaniego Ridge). In the Tortolita Mountains, Davis and others (1975) and Budden (1976) mapped three main phases of medium-grained augen gneiss that range from biotite granodiorite to granite in composition.

The dominant rocks in the Picacho Mountains, Pinaleno, and Jackson Mountain core-complex terranes are fine- to medium-grained quartz monzonitic augen gneisses. These display a striking variety of textures, with deformational fabric ranging from barely recognizable to mylonitic and schistose. Similar rocks dominate the terrane in the southern Pozo Verde Mountains, Alvarez Mountains, and Kupk Hills. To the north in the Coyote Mountains, alaskite, biotite quartz diorite with septa of Paleozoic rocks, and quartz monzonite were mapped by Wargo and Kurz (1956). In the northern Coyote Mountains these phases are augen gneissic and marked by penetrative lineation and low- to moderate-dipping cataclastic foliation. Pegmatites are abundant. In the Sierra Blanca and part of the northern Comobabi Mountains, lineated, foliated fine- to medium-grained quartz monzonite augen gneiss, alaskite, pegmatite, and minor biotite granodiorite invade a terrane of metamorphosed Mesozoic sedimentary and volcanic rocks. Mapping, dating, and correlating all of the fine- to medium-grained augen gneiss at a 1:24,000 scale constitutes a major regional geologic problem!

The petrography of the medium-grained augen gneisses has been examined most carefully in the Santa Catalina Mountains by Pilkington (1962), Peterson (1968), Sherwonit (1974), Shakel (1975), and Banks (this volume). In outcrop the rocks are light colored, hypidiomorphic, equigranular, and generally homogeneous. As in the case of the coarse-grained augen gneiss, foliation is typically defined by thin laminae of quartz, parallelism of micas, and aligned augen of cataclastically deformed feldspar. The rocks are cut by abundant pegmatite, aplite, and flat joints. The striking physiographic expression of the penetrative low-dipping foliation and jointing calls attention to these core-complex gneisses from afar.

The general distribution of the medium-grained augen gneisses in the Rincon–Santa Catalina–Tortolita complex is shown in Figure 5. Establishing the exact age of these gneisses has been difficult, for the rocks are characterized by profoundly disturbed Rb-Sr isotopic systems. K-Ar and fission-track ages (from the work of Damon and his associates of the University of Arizona, and Creasey and others of the U.S. Geological Survey) typically range from 30 m.y. to 24 m.y. for the quartz monzonitic gneisses and associated pegmatites and aplites. Through the years, these dates have been interpreted variously as cooling ages by some and as emplacement ages by others. For example: (1) Creasey and others (1976) and Banks (this volume) regard the entire body of fine- to medium-grained augen gneisses as a composite batholith 25 m.y. old that was penetratively deformed at the time of emplacement. (2) Shakel and others (1977) reported that part of the so-called composite batholith is indeed 27 m.y. old, but that monzonites from a two-mica, garnet-bearing part of the mass known locally as the Wilderness granite gives U-Th-Pb ages of 44 to 47 m.y. They interpreted the Wilderness granite to be undeformed and concluded that the last major gneiss-forming event was no younger than 44 m.y. (3) Drewes (1978a) considered the mass of fine- to medium-grained quartz monzonite augen gneiss in the Rincon Mountains to be Precambrian in age or to have been derived from Precambrian rocks.

The article by Keith and others (this volume) provides a useful summary of the geochronological problems and offers some new interpretations. It is becoming increasingly well documented that the core-complex terranes contain both mid-Tertiary *and* pre–mid-Tertiary Phanerozoic plutons. Recent isotopic data suggest that some of the mid-Tertiary bodies have been affected by low-dipping foliation and penetrative lineation (Rehrig and Reynolds, 1978, and this volume). The Tertiary event, whatever its dynamic nature, imposed its thermal-tectonic signature on all core rocks, regardless of whether they existed before the event or were emplaced during the event.

Mylonitic Schist

The contact zones between the coarse- and the medium-grained augen gneissic phases are commonly marked by strongly foliated, folded rocks that herein are referred to as mylonitic schist. Such highly deformed, cataclastic rocks are well exposed in Tanque Verde Mountain, the forerange of the Santa Catalina Mountains, and in the Little Rincon, Tortolita, Picacho, and Sierra Blanca Mountains. These rocks are fine to very fine grained and range in color from brown to steel gray to black (Fig. 6C). White broken feldspar chips, irregular in size and shape, commonly "float" in the fine-grained, mylonitic matrix. Deformed concordant aplite and pegmatite layers and veins accentuate the foliation. Penetrative folding is a characteristic of these rocks and presumably reflects movements that helped to shape the mylonitic, foliate fabric. The folds are intrafolial, tight to isoclinal, overturned to recumbent structures whose axes lie in the low-dipping foliation. Orientations of folds in mylonitic schist in a

Figure 6 (facing pages). Tracings of photographs of structures in coarse-grained quartz monzonitic augen gneiss. (A) Foliation as seen in outcrop. (B) Ptygmatically folded aplite dikes cutting augen gneiss. (C) Folded mylonitic schist. (D) Transposed foliation in coarse-grained quartz monzonitic augen gneiss.

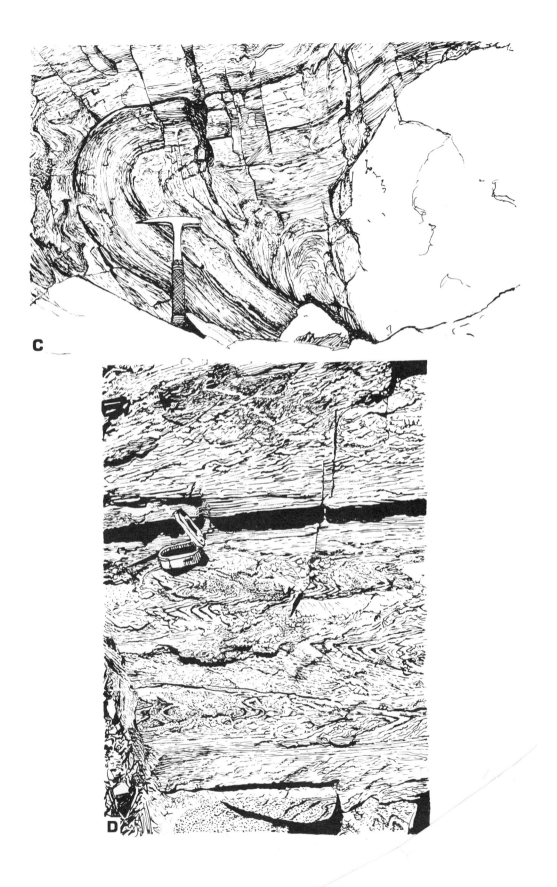

number of ranges in southern Arizona are portrayed stereographically in Figure 12. Foliation is essentially concordant with that of the adjacent augen gneissic phases. Foliation surfaces in the mylonitic schist are not always marked by mesoscopically recognizable mineral lineation, but where lineation occurs, it too is parallel to that of the augen gneissic phases.

It is not clear in all cases whether the mylonitic schist has evolved cataclastically from the coarse-grained quartz monzonitic augen gneiss, the fine- to medium-grained quartz monzonitic augen gneiss, or both. Adjacent coarse-grained augen gneiss is commonly marked by transposition and folding of the penetrative low-dipping foliation (S_1), producing zones of moderately dipping crenulation foliation (S_2; Fig 6D). The fine- to medium-grained quartz monzonitic augen gneisses and associated pegmatites in contact with the mylonitic schist commonly display rather remarkable overturned to recumbent folds.

A significant field problem is distinguishing the mylonitic schist derived by cataclasis of augen gneiss from actual Pinal Schist. Mayo's (1964) investigation of folds in the schist in the forerange of the Santa Catalina Mountains is, in my judgment, an analysis of structures in mylonitic schists cataclastically derived from orthogneiss. The folds there occur in a deformed zone between the coarse- and the medium-grained augen gneissic phases. Davis and others (1975), in mapping the geology of the Tortolita Mountains, noted the transformation of undeformed porphyritic quartz monzonite to coarse-grained quartz monzonitic augen gneiss. Banks demonstrated to us that even the very fine grained, folded schistose rocks that we had mapped as Pinal Schist may be in large part derived from quartz monzonite (Banks and others, 1978). Drewes's (1978a) geologic map of the Rincon and Tanque Verde Mountains shows Pinal Schist in narrow bands, generally occurring in the contact zone between the coarse-grained and the fine- to medium-grained augen gneissic phases. Some of this rock is indeed Pinal Schist but much of it was derived from orthogneiss (Fig. 4A). Wright (1978) has prepared a large-scale map of part of the mylonitic schist zone on the northwest flank of Tanque Verde Mountain showing its relation to augen gneissic phases (Fig. 7). Yeend (1976) mapped undifferentiated schist in the Picacho Mountains and suggested (in the explanation of his map) that these may have been derived from sedimentary rocks. These schists, for the most part, cap the mountain and closely resemble mylonitic schists found in other ranges. Some of the schists that occur at midslope on the western flank are located at the contact between the coarse-grained and the fine- to medium-grained augen gneiss as are those in other ranges.

At the scale of individual outcrops, mylonitic schist forms thin bands and layers at contacts between the medium-grained augen gneiss and concordant aplite and pegmatite. Such mylonitic schists are especially well developed in the Sierra Blanca. Black mylonites form at the margins of the abundant pegmatite-aplite sills. Such relations draw attention to the role of ductility contrast in controlling development of the mylonitic schist.

Lineation

The augen gneisses everywhere display low-plunging mineral lineation of a cataclastic nature. The lineation generally is penetrative on the scale of hand specimen, with a delicacy of development dependent in large part on grain size of the host rock. Figure 8A reveals the physical nature of the lineation in coarse-grained augen gneiss in outcrop. The lineation has a variety of forms, but generally is expressed in the plane of foliation of the augen gneisses by the alignment of long directions of inequant feldspar and quartz-feldspar augen, striae and slickensides, elongate aggregates and streaks of crushed minerals, and crenulations on quartz ribbons. In the Santa Catalina Mountains alone, the lineation pervades approximately 2,000 m of augen gneiss. Lineation is by no means restricted to the augen gneiss layers, mylonite, but occurs penetratively on concordant or discordant *low-dipping* aplite and pegmatite zones, transposition foliation surfaces, and low-angle normal faults. Even in some

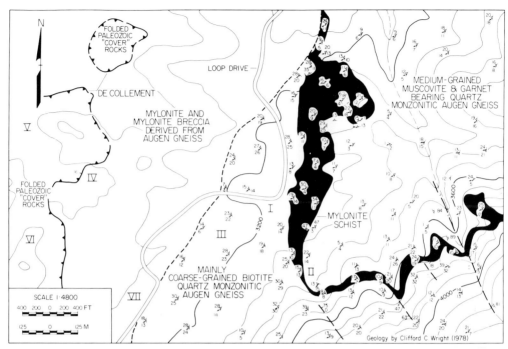

Figure 7. Structural geologic map of part of Saguaro National Monument showing relationship of mylonitic schist to augen gneisses and the decollement zone (from Wright, 1978).

equigranular hypidiomorphic quartz monzonitic bodies lacking penetrative lineation, such as those in the central Tortolita Mountains or the southern part of Jackson Mountain, the lineation locally occurs on shallow-dipping aplite and pegmatite layers that have served to localize tectonic movements.

Within any given metamorphic core complex the lineation is generally remarkably systematic in orientation (Fig. 8B). Lineation in the Picacho, Tortolita, and most of the Santa Catalina and Rincon Mountains trends N50° E to N70° E. Departure from the modal N60° E trend is evident between the Santa Catalina and Rincon Mountains, where lineation is approximately north-south, and near the crest of the Santa Catalina Mountains where it trends north-northeast. In Jackson Mountain and the northern Pinaleno Mountains, the trend of lineation is on the average N30° E and N40° E, respectively. Southwest of Tucson, lineation trends about N10° W in the Sierra Blanca, Comobabi Mountains, and Kupk Hills; north-south in the Coyote and Alvarez Mountains; and N30° E in the Pozo Verde Mountains. This systematic swing in lineation trend occurs in the inner arc of the bend in the belt of metamorphic core complexes in southern Arizona. In essence, three regional structural domains can be distinguished in southern Arizona on the basis of homogeneity of penetrative lineation (Fig. 8C): a western, "Papago" domain characterized by north-trending lineation, a central, "Catalina" domain of east-northeast–trending lineation, and an eastern, "Pinaleno" domain characterized by northeast-trending lineation. Axes of the folds in mylonitic schists in the core rocks are generally oriented parallel to lineation. Axes of boudinage and pinch-and-swell structures, abundant in the lineated core rocks, are more diffuse but tend to lie at right angles to lineation.

Ductile Normal Faults

Ductile normal faults (Fig. 9A), first discovered by Davis and others (1975) in the Tortolita Mountains, are associated with mylonitic, lineated rocks and typically result in impressive local thinning of

Figure 8A. Penetrative lineation in coarse-grained quartz monzonite augen gneiss.

Figure 8B. Map showing distribution of metamorphic core complexes in southern Arizona (see Fig. 2 caption for explanation of symbols); lower-hemisphere projections disclose orientations of lineation in core rocks.

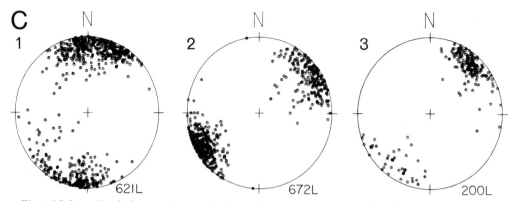

Figure 8C. Lower-hemisphere equal-area projections of lineation orientations in (1) the "Papago" domain (Sierra Blanca, Comobabi, Coyote, Pozo Verde, Alvarez, and Kupk Mountains), (2) the "Catalina" domain (Rincon, Tanque Verde, Santa Catalina, Tortolita, and Picacho Mountains), and (3) the "Pinaleno" domain (Jackson and Pinaleno Mountains).

these rocks within the core. The ductile normal faults are always oriented at right angles to the lineation, regardless of its absolute orientation. I have observed them in all of the ranges shown in Figure 2 except the Pozo Verde Mountains, Alvarez Mountains, and Kupk Hills. Resultant folds that naturally arise from the ductile faulting are superimposed on folds in mylonitic schist. Where examined by Davis and others (1975) in the Tortolita Mountains, the antiforms and synforms are gentle to open, characterized by wavelengths and amplitudes of approximately 6 to 3 m, respectively. Axial surfaces are moderately steep to steeply dipping and strike at right angles to the lineation. In profile, the associated folds are seen to produce marked thinning in the zones of maximum inflection. In some cases, the hinge zones of the folds lack visible offset, and ductile flexing appears to be superseded by actual normal faulting (Fig. 9B).

METASEDIMENTARY CARAPACE

General Relations

In southern Arizona metamorphic core complexes, the metasedimentary carapace, where present, consists of younger Precambrian and lower Paleozoic metasedimentary rocks that rest concordantly atop type A core rocks. Strata in the carapace are commonly metamorphosed to upper greenschist and amphibolite grade and form a relatively thin, tabular sheet. The rocks appear to be concordantly welded or *plated* to underlying crystalline rocks. For example, in Happy Valley east of the Rincon Mountains, there are many outcrops where not even a crack separates younger Precambrian quartzite or marble of the metasedimentary carapace from underlying, cataclastically foliated crystalline basement. Along the crest of the Santa Catalina Mountains (Fig. 5), in the vicinity of Mount Lemmon and Mount Bigelow (Waag, 1968), weakly foliated and lineated medium-grained augen gneiss concordantly intrudes the basal part of the metasedimentary carapace that consists of younger Precambrian Apache Group rocks, there converted to phyllite, amphibolite schist, quartz-sericite schist, quartzite, marble, and metaconglomerate. These are overlain (still within the metamorphic carapace) by marble, calc-silicate, and phyllitic rocks derived from Paleozoic formations. The carapace generally rests in low-angle contact on either (1) medium-grained quartz monzonitic augen gneiss of Tertiary(?) age or (2) moderately deformed coarse-grained augen gneiss derived from 1.4- to 1.5-b.y.-old porphyritic quartz monzonite (Banks, 1977).

In Happy Valley east of the Rincon Mountains (Fig. 5), the rocks of the metasedimentary carapace

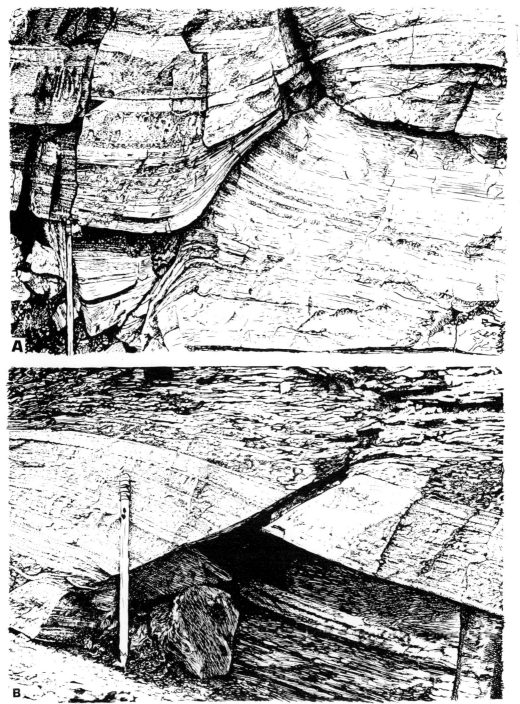

Figure 9. Tracings of photographs. (A) Ductile normal fault zone. (B) Normal-slip fault.

are exposed over large areas (Drewes, 1975). They include unequivocal metamorphosed Apache Group and lower Paleozoic formations. Upper Paleozoic formations also appear to be present in great volume but are harder to distinguish. Dominant and/or distinctive rock types include quartzite, marble, calc-silicate and marble sequences, phyllite, and minor flattened pebble metaconglomerate. These rocks rest in concordant, low- to moderate-dipping contact on Precambrian granitic basement rocks thought to be deformed 1.4- to 1.5-b.y.-old granodioritic quartz monzonitic rocks, and 1.55- to 1.65-b.y.-old granite (Drewes, 1975). Although the metamorphic carapace of the Santa Catalina Mountains rests directly on a vast thickness of penetratively lineated and foliated augen gneiss, in Happy Valley 35 km to the southeast the basement is generally unlineated and gently to moderately foliated. The rocks of the metamorphic carapace are separated from granitic basement by a thin (<20 m) zone of microbreccia, mylonite, and mylonitic gneiss. Thin tectonic slices (<10 m) of the mylonitic gneiss occur from place to place within the lowermost part of the metasedimentary carapace. On the mesoscopic scale, the base of the metasedimentary carapace is planar and strictly concordant with the foliation in the gneiss immediately below. Viewed macroscopically, the surface is smooth and gently warped into systematic upright antiforms and synforms. Locally, the surface is deformed into surprisingly tight antiforms that project upward in almost tent-like fashion. The vertical relief on such sharp, tight structures is at least 10 to 20 m. At many locations in Happy Valley, granite gneiss below the contact displays tight to isoclinal recumbent folds.

Except for the Rincon–Santa Catalina–Tortolita complex, the characteristic features of a metamorphic carapace are seldom exposed in southern Arizona. In the Tortolita Mountains, metasedimentary carapace rocks derived from Apache Group and lower Paleozoic rocks are found in the west-central part of the range (Fig. 5), where they rest in low-dipping contact on lineated, foliated, fine- to medium-grained quartz monzonitic augen gneiss. Homoclinal but internally deformed quartzite of the Pinal Schist is plated onto fine- to medium-grained quartz monzonitic augen gneiss in Jackson Mountain (Fig. 2). In the Coyote Mountains, highly foliated and attenuated quartzite, possibly Cambrian Bolsa quartzite, rests concordantly on thick, lineated, pegmatitic intrusions, which in turn cap foliated and lineated leucocratic augen gneiss. The foliated, folded, and lineated metasedimentary and metavolcanic rocks (Mesozoic?) in the Sierra Blanca are not considered strictly "metamorphic carapace," for they do not clearly rest on a discrete concordant augen gneiss substructure. Rather, they are invaded concordantly and discordantly by pegmatite, alaskite, quartz monzonite, and biotite granodiorite, which are foliated and lineated. Anderson and others (this volume) describe similar relations in northern Sonora.

Mesoscopic Structures

Rocks of the metasedimentary carapace, wherever found, display very distinct structural characteristics (Davis, 1975; Schloderer, 1974; Waag, 1968; Frost and Davis, 1976; Frost, 1977). Strongly overturned to recumbent folds are ubiquitous. Protoliths for the phyllite, schist, marble, and quartzite have been deformed by transposition into tectonites characterized by intrafolial, commonly rootless, tight to isoclinal folds (Figs. 10A, 10B, 10C). Depending on mechanical characteristics of the original rocks, the folds may form by passive flow, passive slip, and/or flexural flow. The passive folds typify homogeneous domains of quartzite or marble. The most outstanding flexural forms occur in interlayered calc-silicate and marble. Where the rocks are dominantly marble and contain only thin brittle calc-silicate or quartzite struts, the "competent" layers are typically distended, attenuated, and transposed (Figs. 10C, 10D, 10E). Fragments of the originally continuous layers form boudins, isolated fold hinges, and assorted tectonic inclusions in the marble matrix (Fig. 10F). In the plane of foliation and lithologic layering of calc-silicate and marble, gash fracturing and orthogonally disposed mineral lineation are locally strongly developed (Fig. 10G).

Conglomerates within the metamorphic carapace are transformed into subhorizontally flattened quartzite-pebble units. The flattened pebbles are typically profoundly elongated parallel to the lineation in underlying augen gneiss. Where the pebbles are flattened but not elongate, underlying cataclastically deformed augen gneiss is foliated but not lineated. In the Tortolita Mountains, Davis and others (1975) measured the orientation and dimensions of 86 flattened elongate quartzite pebbles within Barnes metaconglomerate and compared these to undeformed clasts at a nearby locality in the Santa Catalina Mountains (Fig. 11). Axial ratios were computed to be 9:2:1, the plane of flattening is subhorizontal, and the direction of elongation (N58°E) is identical to that of the adjoining augen gneiss.

Within the metasedimentary carapace, foliation and lithologic layering are generally shallow dipping and strongly developed. Axial planes of tight to isoclinal, overturned to recumbent folds are generally parallel to foliation and layering. The folds are commonly reclined relative to layering and foliation. Seen on the scale of large single outcrops, the foliation and layering are marked by pinch-and-swell and boudinage, with graceful gentle changes in attitude. Predictably, boudinage and pinch-and-swell are best developed where rocks of contrasting ductilities are juxtaposed.

Stratigraphic Relationships

Individual formations within the metamorphic carapace are arranged in normal stratigraphic order, but they are generally tectonically thinned or locally thickened. Specific estimates of the change in thickness are very difficult to make because of the uncertainties inherent in correlating the strongly deformed lithotectonic units with southeastern Arizona stratigraphy. In many places in Happy Valley, the thinning appears to exceed 75%.

Thinning and thickening within the younger Precambrian and Paleozoic sequences have been achieved by passive flowing, including transposition (Frost, 1977) of units rendered ductile during the deformational process. On the mesoscopic scale, the mode of thinning is explicitly displayed in the stretched-flattened pebble metaconglomerates, passively folded marble and calc-silicate rocks, transposed schists, and dismembered, attenuated quartzite layers in ductile matrix. Detailed mapping reveals interesting macroscopic adjustments to the "thinning process," mainly a heterogeneous distribution of lithologic units of contrasting mechanical properties. Mapping by Plut (1968), Drewes (1975), and G. H. Davis and others (in prep.) in Happy Valley has revealed that the massive quartzites of the Apache Group and Cambrian Bolsa Quartzite only locally rest (in parautochthonous contact) on basement. More commonly, these quartzites have been faulted out entirely, and marbles inferred to be of early to mid-Paleozoic age rest directly on basement. The map pattern of such relations demonstrates lateral movement and concentration of ductile materials as a response to thinning, but carried out so that young rocks are always moved over older rocks.

Kinematic Significance

The structural fabric of the metasedimentary rocks is intimately coordinated with that of the underlying augen gneiss. Fabrics of both structural units are marked by profound flattening perpendicular to subhorizontal layering and foliation (see Fig. 2) and by extension parallel to lineation (see Fig. 8B) as denoted by boudinage, deformed pebbles, and the ductile normal fault zones. Davis and others (1975) analyzed structures in the Tortolita Mountains and showed that the lineated surfaces within fine- to medium-grained quartz monzonite augen gneiss accommodated vertically directed flattening and profound east-northeast extension; they also showed that the fold and stretched-pebble fabric in the immediately adjacent metamorphic carapace formed as a response to the same deforming process. Critical to that analysis was recognition of the coordinated nature of the augen gneiss fabric to

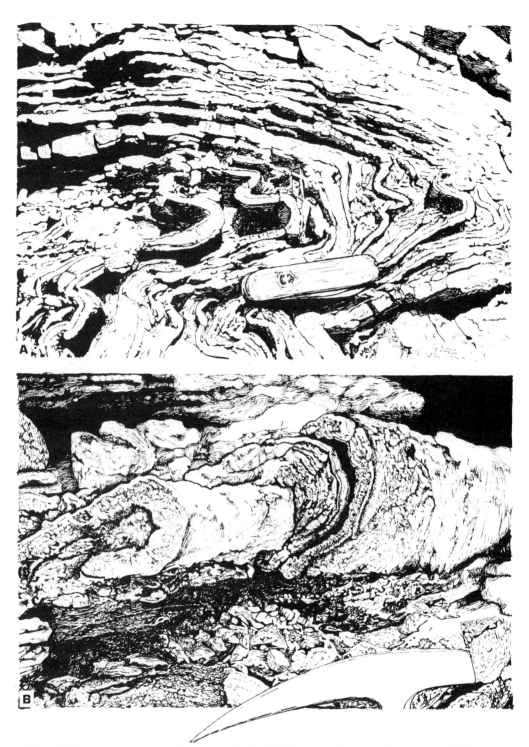

Figure 10. Structures in metamorphic carapace. (A, B, C) Folds in calc-silicate–marble sequences. (D) Lozenge-shaped boudins (traced from photo taken by Eric G. Frost). (E) Tectonic inclusions derived from folded, attenuated calc-slicate and quartzite layers. (F) Boudin of calc-silicate rock in marble. (G) Penetrative gash fractures. (Continued on following three pages.)

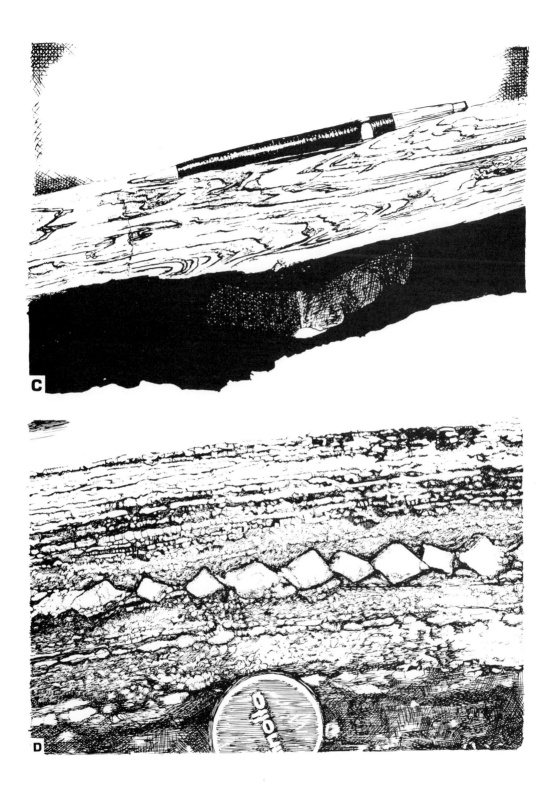

Figure 10 (continued).

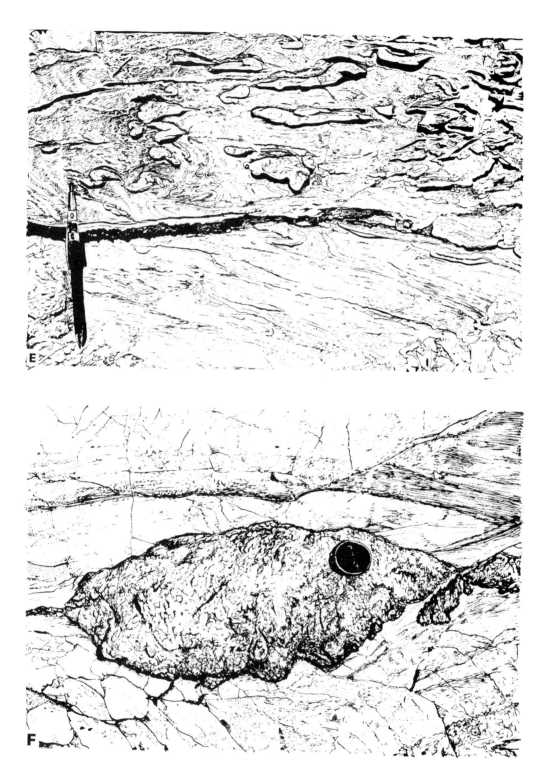

Figure 10 (continued).

Figure 10 (continued).

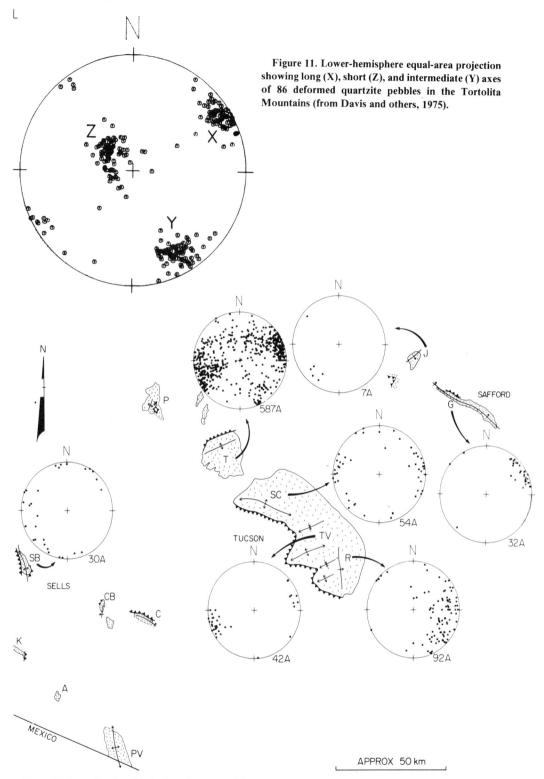

Figure 11. Lower-hemisphere equal-area projection showing long (X), short (Z), and intermediate (Y) axes of 86 deformed quartzite pebbles in the Tortolita Mountains (from Davis and others, 1975).

Figure 12. Map showing distribution of metamorphic core complexes in southern Arizona and lower-hemisphere projections of orientations of axes in metamorphic carapace (R, SC, T, J, and SB) and mylonitic schist (G, P, T, and TV).

that of the metamorphic carapace: (1) the modal long axis of pebbles is strictly parallel to lineation in adjacent augen gneiss, (2) the fold axes in the metamorphic carapace parallel the lineation orientation, and (3) ductile normal faults are orthogonal to lineation. On the basis of the strict compatibility of the normal-slip ductile and brittle faults to inferred principal strain directions inferred from the stretched-pebble data, the "stretching" of the quartzite pebbles and the development of normal-slip faults were interpreted as the early and late stages, respectively, of a ductile-to-brittle continuum of extensional deformation.

Fold axes in rocks in the metasedimentary carapace are generally difficult to evaluate in regard to slip-line direction. The fundamental problem is establishing for certain that specific asymmetric folds are indeed first-order folds within the transposed sequences. Furthermore, the fold axes in the metasedimentary rocks generally display a broad range of orientation within the plane of slip or flow (Fig. 12). Although details of contrasting interpretations of kinematics and dynamics could be cited, I believe it is most important to emphasize that (1) within the array of variably oriented overturned to recumbent folds, reclined folds are commonly the preferred mode, (2) the axes of reclined folds tend to be coaxial with lineation in ductile rocks like marble, (3) mineral lineation in marble is essentially orthogonal to penetrative gash fracturing in mechanically suitable lithologies in the metamorphic carapace, and (4) lineation in underlying augen gneiss is parallel to that in marble within the metamorphic carapace.

The above observations and inferences are consistent with the interpretation that both the augen gneiss and metamorphic carapace were affected by flattening and extension and that these processes resulted in profound thinning through flow of the ductile, originally bedded, carapace strata. Flow in the metamorphic carapace was parallel to mineral lineation, and during progressive deformation fold hinges rotated partly or wholly into alignment with the flow direction. The degree of rotation was partly related to the mechanical properties of the deforming sequence.

DECOLLEMENT ZONES

A decollement separates unmetamorphosed cover rocks above from tectonites of the core and/or carapace below. The surface separates rocks of remarkably contrasting deformational styles. Most often in southern Arizona examples, the decollement marks the top of augen gneissic quartz monzonitic core rocks and the base of unmetamorphosed cover; metamorphic carapace rocks are generally absent (Figs. 13, 14). Below the decollement surface, a "decollement zone" is usually conspicuous and consists of a distinctive crudely tabular zone of crushed and granulated but strongly indurated fine-grained mylonitic rocks (mylonite, mylonitic gneiss, microbreccia, and chlorite breccia). The decollement zones are distinctive in their physiographic and structural expression, for the mylonitic rocks form resistant benches or cliffs that are capped by planar upper surfaces (Figs. 13, 14). These so-called decollement zones occur on one or two flanks of the gneissic cores of individual complexes (Fig. 2). Where metamorphic carapace is present, the decollement separates carapace from cover, and a resistant decollement zone of intensely deformed metasedimentary rocks may cap the top of the carapace.

The decollement zones commonly display striking "younger on older" fault relations involving tens to hundreds of metres of stratigraphic separation. They typically separate orthogneisses, derived in part from Precambrian rock or from unmetamorphosed upper Paleozoic or Mesozoic or Tertiary deformed cover rocks. This array of structural and petrologic characteristics has prompted many workers to interpret them as thrust faults (Thorman, 1977; Drewes, 1978b).

The decollement zone in the Santa Catalina and Rincon Mountains, known locally as the Catalina fault, crops out along a sinuous trace, tens of kilometres in length, on the south and west flanks of the

Figure 13. Decollement zone (resistant ledge) in the southern Rincon Mountains. Upper planar surface of decollement zone is the so-called Catalina fault. Overlying the surface is the Precambrian Rincon Valley granodiorite and unmetamorphosed Permian strata.

complex (Figs. 2, 5). The decollement zone separates augen gneiss (below) from a variety of deformed but generally unmetamorphosed cover rocks, including Paleozoic limestone, sandstone, and shale, Mesozoic shale, dolomite, and conglomerate, and Oligocene-Miocene red beds (Fig. 14). The dip of the decollement zone is generally less than 15° or 20°. The only exposures are commonly confined to the pediment-mountain interface (Fig. 3); at no place is the zone known to crop out at a high level on the mountain flank.

Viewed at the scale of the complex, the decollement zone forms a smoothly arcuate surface conforming to the macroscopic structural geometry of the augen gneiss. Viewed at the mesoscopic scale, the zone may be concordant or discordant to foliation in underlying augen gneiss (Fig. 4A).

Although the contact of the decollement zone with the overlying deformed cover rocks is sharp and smooth, the contact with the underlying augen gneiss is generally gradational and ill defined. The thickness of the zone varies from several metres to several tens of metres.

In outcrop, rocks of the decollement zone are brown to brownish-green, chloritic, fine-grained mylonites. They are pervasively overprinted by shattering along closely spaced fractures. Where strongly foliated, rocks of the decollement zone are deformed by kink folding. Within the highly deformed rock suite, lineation and augen gneissic foliation fabrics can be recognized, but these are generally masked by the mylonitization. It seems evident that the rocks of the decollement zone were produced at the expense of already lineated coarse-grained augen gneiss, fine- to medium-grained augen gneiss, and pegmatites and aplites.

The internal structure of the decollement zone in the Santa Catalina–Rincon complex is highly variable. In places, the zone shows homoclinal foliation, with or without northeast-trending lineation. In other places, for example, on the northwest flank of Tanque Verde Mountain, the relict lineation and foliation are systematically rotated.

Decollement zones of this type are by no means restricted to the Santa Catalina and Rincon Mountains. Such zones crop out along the north end of the Tortolita Mountains, at the southeastern

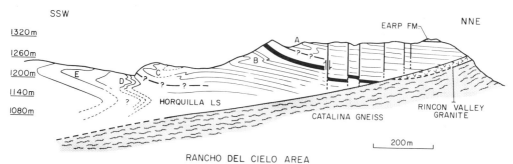

Figure 14. Cross section of decollement zone and upper plate rocks on the southern flank of the Rincon Mountains (from Davis and others, 1974).

end of the Picacho Mountains, along the north end of the Coyote Mountains, and on the west flank of Sierra Blanca (Fig. 2). Also, one exposure of the otherwise concealed decollement zone on the east flank of the Pinaleno Mountains was discovered (Fig. 4I).

At the north end of the Tortolita Mountains, the decollement zone strikes east-west, dips gently north, and separates undeformed (1.4- to 1.5-b.y.-old) porphyritic biotite quartz monzonite and normal Pinal Schist on the north from cataclastically deformed porphyritic biotite quartz monzonite of Tertiary(?) age on the south. Cataclasis has transformed the quartz monzonite to augen gneiss and schist.

A low-dipping chloritic, mylonitic layer defines a decollement zone at the southeastern end of the Picacho Mountains. The zone marks the deformed upper surface of exposures of medium-grained quartz monzonitic augen gneiss. The upper surface of the zone is planar; it separates the cataclastically deformed augen gneisses from overlying Tertiary (23-m.y.-old) basalt. The interface of contact is itself a fault contact, marked by slickenslided surfaces that overprint the mylonitic fabric of the decollement zone. Slickensides plunge gently southeastward. The decollement zone grades downward over a 10- to 20-m interval into typical lineated, foliated augen gneiss.

In the Coyote Mountains southwest of Tucson (Fig. 2), a decollement zone expressing physical properties identical to the Catalina fault crops out along the northern pediment of the mountain. Its strike is east-west, and it dips northerly from 20° to 35°. Brown-green cataclastically deformed, chloritic, resistant rocks occur on the "footwall" of the zone, including biotite quartz diorite, marble and calc-silicate rocks of a metasedimentary carapace, and leucocratic granitic and quartz monzonite augen gneiss. Overlying the decollement zone, in sharp planar contact, are red beds inferred to be Oligocene-Miocene in age.

On the west flank of Sierra Blanca, unmetamorphosed to slightly metamorphosed Permian sedimentary strata are separated from the crystalline core rocks by a pronounced, gently west-dipping decollement zone. Again, the rock of the decollement zone is chloritic, highly fractured but highly indurated mylonite, mylonitic gneiss, and microbreccia. Its thickness is about 10 m, although the base of the zone grades into underlying Mesozoic schists and biotite granodiorite.

UNMETAMORPHOSED BUT DEFORMED COVER ROCKS

The so-called cover rocks that overlie the augen gneissic core and metamorphic carapace of the core-complex terranes are interesting and instructive in their gross configuration and internal structure. For the most part, the cover rocks directly overlie well-developed decollement zones of the type already described. Cover rocks above the decollement zone (Catalina fault) in the Santa Catalina and

Rincon Mountains form thin plates that include slices of Paleozoic, Cretaceous, and Tertiary formations (Fig. 5). They include shattered Precambrian granite as well. Along the front of the Santa Catalina Mountains, the zone separates augen gneiss from the Oligocene-Miocene continental red beds, notably mudstone, siltstone, sandstone, and conglomerate. Clasts within the lower part of the sequence bear no resemblance to rocks within the core complex itself, except in the upper beds where the "influx" of clasts of limestone and lineated augen gneiss document an unroofing of the complex (Pashley, 1968).

Miocene beds locally rest on the decollement zone on the south and west flanks of the Rincon Mountains; additionally, Precambrian, Paleozoic, and Mesozoic strata lie atop the zone as well (Drewes, 1978a). Completeness (incompleteness?) of stratigraphy in these sections is highly variable. Sedimentary rocks on the west side of the Rincon Mountains within the Saguaro National Monument (East) consist of limestone, dolomite, and shale of Permian age, as well as remnants of Mississippian(?), Pennsylvanian, and Cretaceous formations. Along the southeast flank of the Tanque Verde antiform, Cretaceous shale and limestone, with interbedded siltstone, dolomite, and limestone conglomerate, lie directly on the decollement. The thickness of the sheet is less than 90 m. Sedimentary rocks near Colossal Cave on the south side of the Rincon Mountains form a sheet approximately 150 m thick that rests on the decollement zone. The rocks consist of limestone interbedded shale and include formations of Cambrian through Permian age. At the southeasternmost corner of the Rincon Mountains, a 75-m sheet of Paleozoic rocks rests on the decollement zone. Although formations from Cambrian to Permian time are represented, the thickness of the sequence is less than 10% of the full Paleozoic section exposed in the Whetstone Mountains only 20 km south.

Structures in such deformed cover rocks are described in some detail in Davis (1975), Liming (1974), and Davis and Frost (1976). The individual sheets of Phanerozoic strata range from about 40 to 120 m in thickness. Strata within each sheet are generally unmetamorphosed, except near the base where limestones are commonly marbleized over thicknesses of 10 m or so. However, in the Martinez Ranch area, low-grade marbles can be found 60 m above the decollement zone. The sheets of deformed cover rock are grossly concordant with the underlying decollement zone, but in detail it can be demonstrated that discordance exists between layering in the sheets and foliation in the underlying gneisses.

Overturned asymmetric folds, detached isoclinal folds, and unbroken cascades of recumbent folds characterize the sheets of the Paleozoic and Mesozoic cover rocks (Fig. 15). Such folds by no means characterize normal Paleozoic-Mesozoic sedimentary country rock in southeastern Arizona (Davis, 1979). Most of the folds are transitional between ideal parallel and ideal similar folds and, thus, are characterized by some hinge-zone thickening (Davis, 1975). The scarcity of axial-plane cleavage, the abundance of bedding-plane cleavage, and the obvious influence of layering on the morphology of folds indicate that the folds evolved through slippage between layers and flow within layers.

Cover rocks in the Coyote Mountains include Mesozoic volcanic rocks and Oligocene-Miocene red beds. The small patch of cover at the southeast end of the Picacho Mountains is Miocene basalt. In the Sierra Blanca, cover rocks are folded upper Paleozoic strata.

INTERPRETIVE REMARKS

The metamorphic core complexes in southern Arizona are fundamentally products of high-temperature extensional deformation, regional in extent, superseded by moderately ductile to moderately brittle tectonic denudation. During the disturbance, rocks in the augen gneissic core and metamorphic carapace were produced by profound ductile through brittle extension and flow in the direction of mineral lineation, accompanied by vertical flattening and transposition. Tectonites developed at the expense of Precambrian crystalline basement; Phanerozoic quartz monzonite, granodiorite, and granite bodies that were emplaced both before and during the disturbance; and younger Precambrian, Paleozoic, and Mesozoic sedimentary rocks that were metamorphosed during the disturbance. Folds rotated in the plane of flow during deformation; this produced a broad range of axial trends in general, except in zones of mylonitic schist where movement was localized and fold hinges rotated into strict parallelism with lineation. Decreasing ductility through time is recorded by features such as folding of lineation and refolding of folds along ductile normal fault zones and superimposition of brittle normal-slip faults upon the ductile faults.

Evolution of the decollement zones largely postdated the development of foliation and lineation fabric in the augen gneiss and metamorphic carapace. Progressive cataclastic granulation of augen gneiss or metasedimentary rocks in the decollement zones was accompanied by rotation of the original

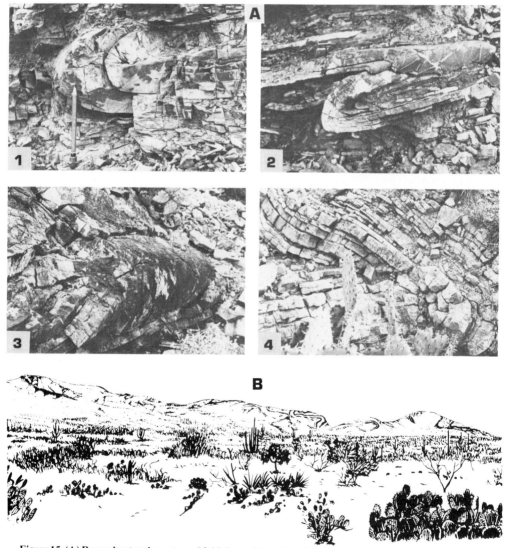

Figure 15. (A) Recumbent and overturned folds in mudstone: 1, dolomite; 2, 3, limestone; 4, Cretaceous strata on the south flank of Tanque Verde Mountain. (B) Overturned to recumbent large-scale folds in Paleozoic strata on the south flank of the Rincon Mountains (tracing of photograph from Pashley, 1966). Cliff in dark shadow at left is that portrayed in Figure 13.

foliation. The foliation in turn controlled the formation of late-stage kink folds. Brittle structures in the plates of unmetamorphosed but deformed cover rocks mostly postdate development of the fabric of the augen gneiss core and metamorphic carapace. Folds and faults formed in the cover rocks during gravity-induced gliding and listric normal faulting.

Ductility contrast influenced the structural deformation as viewed at the outcrop scale (Fig. 16). Likewise, ductility contrast was the major influence in predetermining the nature, relative position, and configuration of the major structural units within the complexes (Fig. 2), that is, core, carapace, decollement zone, and cover.

The array and distribution of structures in the metamorphic core complexes have proved difficult to interpret. Structures in the cover rocks have been traditionally interpreted as products of regional thrusting, but recent work suggests that the structures actually evolved through denudation in middle to late Tertiary time (Davis, 1973, 1975; Coney, 1974; Compton and others, 1977; Davis and Frost, 1976). Structures in the augen gneiss of the core rocks and the metamorphic carapace have been interpreted variously as products of overthrusting (Drewes, 1976a, 1978b; Thorman, 1977), diapiric batholithic emplacement (Creasey and others, 1976), and mantled-gneiss doming (Peterson, 1968; Waag, 1968). The presence of metamorphic core complexes in southeastern Arizona, in a region where the presence of regional low-angle overthrusts has not been proved, weakens the possibility of a dynamic linkage between thrusting and the penetrative structures found in core and carapace rocks. Although metamorphic core complexes bear a superficial resemblance to mantled gneiss domes and igneous batholithic domes, the application of pure granite tectonic and/or diapiric models fails to explain important details of the structural geology of the complexes.

Structures in the augen gneissic core and carapace rocks can be best understood when the cores are pictured as components of megaboudins imposed on heterogeneous crustal rocks (Davis, 1977a; Davis

Figure 16. Tracing of photograph of boudined and folded schist and aplite that illustrates the influence of ductility contrast on deformational style.

and Coney, 1979). By way of explanation, it is instructive to consider Figure 17, a tracing of a photograph from Ramberg's (1955) paper on boudinage. It is useful to consider the scale of this figure not as 0 to 5 cm, but as 0 to 5 km; to consider the enveloping mica schist as a small-scale analog of the metamorphic carapace; and to consider the upper part of the pegmatite boudin as an analog of the augen gneissic core. The interface between the contrasting rock types should be viewed in this discussion as the surface of great unconformity. If this image and Figures 18 and 19 are used as guides, the mechanical interpretation of core-complex evolution may be more easily understood.

Specifically, the megaboudinage concept may provide a basis for understanding the macroscopic and mesoscopic structures as defined here: (1) *augen gneissic core*—upper part of crystalline basement rocks and pretectonic plutons extended in a way to accommodate syntectonic intrusions; (2) *low-dipping foliation*—formed as a result of moderately brittle to moderately ductile response to extensional deformation and flattening in a zone of finite thickness at outer margin of megaboudined "basement"; (3) *penetrative lineation*—response to stretching of moderately brittle to moderately ductile augen gneissic core; (4) *ptygmatic folds*—expression of flattening of augen gneissic core during extension; (5) *mylonitic schist*—produced by strain concentration localized by shearing of the lithologic contact between moderately brittle crystalline basement and moderately ductile to ductile syntectonic instrusion; (6) *metamorphic carapace*—ductile Phanerozoic layered rocks that were stretched, attenuated, and flattened during the passive-flow accommodation to ever-increasing surface area of contact with underlying augen gneissic core; (7) *intrafolial folds*—perturbations in flow regime at surfaces of interface of layers of contrasting ductilities; (8) *lineation, including stretched pebbles*—reflection of profound plastic extension of highly ductile materials; (9) *gash fractures*—late-stage brittle deformational response to extension; (10) *cascades of recumbent folds*—infilling of necked zones by plastic enveloping matrix; (11) *decollement zone*—granulation and rotation of augen gneissic and metamorphic carapace rocks; viewed as "necking" of the moderately brittle megaboudin; or cataclastic degradation of corner of basement block that before mid-Tertiary time, owing to early Mesozoic and Laramide tectonic movements, was in high-angle contact with layered strata; (12) *cover rocks*—Phanerozoic strata or Precambrian crystalline rocks that, for the most part, were above the level of thermal metamorphism but were nonetheless denuded (Davis and Coney, 1979), both during and after the formation of lineated tectonite.

Although still in development, the megaboudinage concept may provide a new way for assessing relations that have proved to be impossible to explain by more traditional structural mechanisms. For example, decollement zones commonly occur only on one or two flanks of specific complexes. In the Rincon Mountains, the decollement zone occurs on the south and west and caps penetratively foliated and lineated augen gneissic core rocks. On the east side of the Rincon Mountains in Happy Valley, the crystalline igneous rocks are overlain directly by quartzite, calc-silicate rocks, and marble of the metamorphic core complex, *without* an intervening strongly developed decollement zone. Furthermore, the crystalline rocks are unlineated and moderately foliated at best. One interpretation lies in

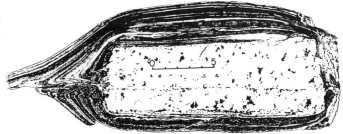

Figure 17. Tracing of photograph in Ramberg's (1955) paper on boudinage. See text for explanation.

Figure 18. Boudinage characteristics (from Ramsay, 1967). Stippling added to denote regions of likely cataclastic deformation. Decollement zones would preferentially develop at corners of blocks in upper diagram and within necked ends of lensoids in lower diagram.

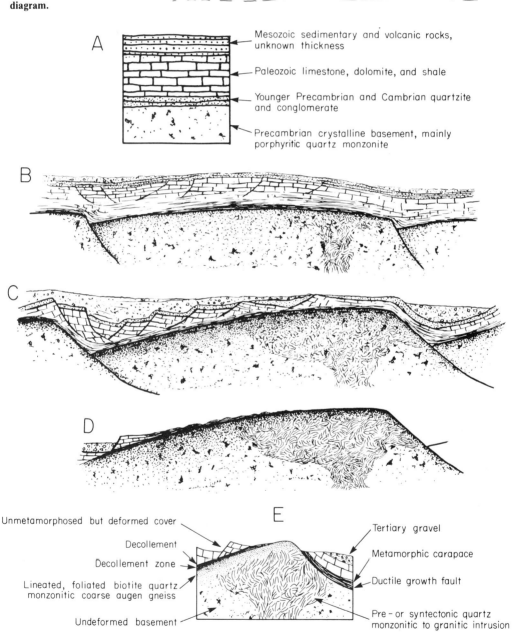

Figure 19. Generalized tectonic reconstruction of the geologic evolution of metamorphic core complexes.

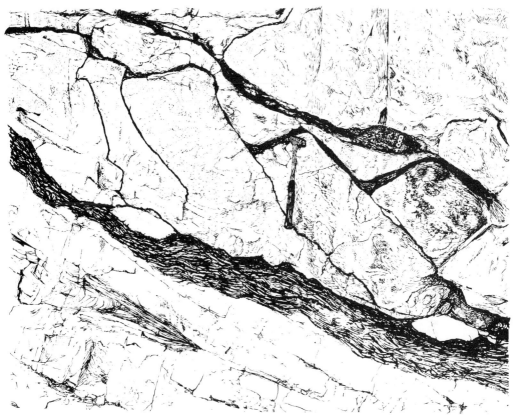

Figure 20. Outcrop-scale example of the ductile plating of a relatively thin lithological layer (white, pinched-and-swelled unit above hammer) onto the footwall of underlying, normal faulted "basement" layer. Quartzofeldspathic gneiss and schist in zone of lineated augen gneiss in the Pinaleno Mountains (see Fig. 4I).

considering the so-called Rincon dome as akin to part of a lozenge-shaped megaboudin (Rast, 1956). When considered in this way, an important distinction comes to light: the decollement may coincide with the great unconformity; on the other hand, rocks of the metamorphic carapace (in this example) may *never* have been in depositional contact with the underlying crystalline rocks, but progressively flowed into contact by spreading over the surface of developing normal faults that define the eastern edge of the lozenge-shaped megaboudin (Figs. 19C, 19D, 19E). The unlineated, moderately to poorly foliated crystalline rocks reflect a deeper (and hotter) structural level than those augen gneisses beneath the decollement zone. The normal faulting portrayed in Figure 19 constitutes a new concept in large-scale faulting, one that is termed herein "ductile growth faulting." The relationship shown in Figure 20 is a wonderful outcrop example of this type of "disharmonic" faulting. The term "growth" implies that the area of the surface on which metamorphic carapace materials are stretched and welded to the crystalline basement continually increases during the life of the fault. The faulting reflects a mode by which the surface area of the crystalline basement can be increased during extension. Such faults and associated folds were recognized at the outcrop scale by Davis and others (1975) and termed "ductile normal faults."

This emphasis on megaboudinage is not intended to suggest that each metamorphic core complex is an individual boudin. Rather, the core complexes are partial exposures of a regional system of extensional deformation where properties might be best understood in the context of megaboudinage. Countless styles of boudins have been reported, both by field geologists and experimentalists. The properties of boudin systems depend on many factors, most important of which are ductility contrast and percent strain. The challenge in further investigations of core complexes will be (1) documenting the great variety of macroscopic strain styles that have evolved diachronously in early to mid-Tertiary time, (2) attempting to establish relations between structure relief and ancient or present topography (Max Crittenden, Jr., 1979, personal commun.), and (3) assessing the superimposition of brittle, mid-Miocene deformation on the pre-existing, ductile extensional systems.

Stretching of basement in a manner akin to megaboudinage took place after Laramide and before basin and range tectonism. Early to middle Tertiary granitic plutons which were emplaced during the crustal stretching were affected by movements that produced penetrative foliation and lineation. Topographic and structural basins created by regional pinch-and-swell were synchronously filled by early to mid-Tertiary continental sediments. Brittle faulting of necked megaboudins prompted the outpouring of mid-Tertiary ignimbrites through rifted basement. High pore-fluid pressure concentrated in the vicinity of decollement zones during the dynamo-thermal event favored near-wholesale denudation of the metamorphic carapace and cover from the culminations and flanks of the megaboudin. Late-stage mid-Miocene listric normal faulting further denuded the cover and heightened the physical distinctiveness of the decollement zones. The stretching and thinning of lithosphere that resulted in metamorphic core complexes led to conditions within the western Cordillera which favored the *collapse* that formed the basin and range structure.

ACKNOWLEDGMENTS

Foremost among those that I would like to thank are present and former students, especially Eric Frost, Brett Liming, John Schloderer, Monte Swan, Terry Budden, Stan Keith, Ken Brook, Chuck Kiven, Bob Varga, Steve Reynolds, Gene Suemnicht, Chuck Kluth, and Stevey Lingrey. I would acknowledge the helpfulness of provocative discussions with geologists who have worked in the Rincon-Catalina-Tortolita terrane, especially Norm Banks, Paul Damon, Harald Drewes, Evans Mayo, and Doug Shakel. I will be forever grateful for the endless hours of discussion with my friends and colleagues Tom Anderson, Peter Coney, and Greg Davis.

Structural investigations in the augen gneissic core zones and metamorphic carapace were supported through National Science Foundation Grant EAR 76-84167. Also, Continental Oil Company, through the help and cooperation of William Rehrig, provided financial aid to support Steve Lingrey as a field associate during the second summer of the project. The Department of Geosciences at the University of Arizona has provided continued financial and logistical support of my structural investigations over the years. I am grateful to the University of Arizona's Bureau of Audiovisual Services for drafting of maps and figures and to David O'Day for his artistic rendering of the illustrations. I extend my appreciation to Cliff Wright and Jim Hardy who served as willing field assistants.

Early drafts of the manuscripts were read and edited by Tom Anderson, Bob Compton, Peter Coney, and Max Crittenden, Jr. To them I am very grateful.

REFERENCES CITED

Anderson, T. H., Silver, L. T., and Salas, G. A., 1980, Distribution and U-Pb isotope ages of some lineated plutons, northwestern Mexico: Geological Society of America Memoir 153 (this volume).

Banks, N. G., 1976, Reconnaissance geologic map of the Mount Lemmon quadrangle, Arizona: U.S. Geological Survey Miscellaneous Field Studies Map MF-747, scale 1:62,500.

——1980, Geology of a zone of metamorphic core complexes in southeastern Arizona: Geological Society of America Memoir 153 (this volume).

Banks, N. G., and others, 1977, Reconnaissance geologic map of the Tortolita Mountains quadrangle, Arizona: U.S. Geological Survey Miscellaneous Field Studies Map MF-864, with text.

Bryant, D. L., 1955, Stratigraphy of the Permian System in southern Arizona [Ph.D. dissert.]: Tucson, University of Arizona, 209 p.

——1968, Diagnostic characteristics of the Paleozoic formations of southeastern Arizona: Arizona Geological Society Guidebook III, p. 33–47.

Budden, R. T., 1975, The Tortolita and Santa Catalina Mountains—A spatially continuous gneissic complex [M.S. thesis]: Tucson, University of Arizona, 133 p.

Butler, W. C., 1971, Permian sedimentary environments in southeastern Arizona: Arizona Geological Society Digest, v. 9, p. 71–94.

Compton, R. R., and others, 1977, Oligocene and Miocene metamorphism, folding, and low-angle faulting in northwestern Utah: Geological Society of America Bulletin, v. 88, p. 1237–1250.

Coney, P. J., 1974, Structural analysis of the Snake Range "decollement," east-central Nevada: Geological Society of America Bulletin, v. 85, p. 973–978.

Cooper, J. R., 1971, Mesozoic stratigraphy of the Sierrita Mountains, Pima County, Arizona: U.S. Geological Survey Professional Paper 658D, 42 p.

Cooper, J. R., and Silver, L. T., 1964, Geology and ore deposits of the Dragoon quadrangle, Cochise County, Arizona: U.S. Geological Survey Professional Paper 416, 196 p.

Creasey, S. C., and Theodore, T. G., 1975, Preliminary reconnaissance geologic map of the Bellota Ranch 15-minute quadrangle, Pima County, Arizona: U.S. Geological Survey Open-File Report 75-295, scale 1:31,680.

Creasey, S. C., and others, 1976, Middle Tertiary plutonism in the Santa Catalina and Tortolita Mountains, Arizona: U.S. Geological Survey Open-File Report 76-262, 20 p.

Damon, P. E., 1968a, Application of the potassium-argon method to the dating of igneous and metamorphic rocks within the Basin Ranges of the Southwest: Arizona Geological Society Guidebook III, p. 7–20.

——1968b, Potassium-argon dating of igneous and metamorphic rocks with applications to the Basin Ranges of Arizona and Sonora, in Hamilton, E. I., and Farquhar, R. M., eds., Radiometric dating for geologists: New York, Interscience Publications, p. 1–71.

Damon, P. E., Erickson, R. C., and Livingston, D. E., 1963, K-Ar dating of Basin and Range uplift, Catalina Mountains, Arizona: Nuclear Geophysics-Nuclear Science Series, no. 38, p. 113–121.

Davis, G. H., 1973, Mid-Tertiary gravity-glide folding near Tucson, Arizona [abs.]: Geological Society of America Abstracts with Programs, v. 5, p. 592.

——1975, Gravity-induced folding off a gneiss dome complex, Rincon Mountains, Arizona: Geological Society of America Bulletin, v. 86, p. 979–990.

——1977a, Characteristics of metamorphic core complexes, southern Arizona [abs.]: Geological Society of America Abstracts with Programs, v. 9, p. 944.

——1977b, Gravity-induced folding off a gneiss dome complex, Rincon Mountains, Arizona: A reply: Geological Society of America Bulletin, v. 88, p. 1212–1216.

——1979, Laramide folding and faulting in southeastern Arizona: American Journal of Science, v. 279, p. 543–569.

Davis, G. H., and Coney, P. J., 1979, Geological development of the Cordilleran metamorphic core complexes: Geology, v. 7, p. 120–124.

Davis, G. H., and Frost, E. G., 1976, Internal structure and mechanism of emplacement of a small gravity-slide sheet, Saguaro National Monument (East), Tucson, Arizona: Geological Society Digest, v. 10, p. 287–304.

Davis, G. H., and others, 1974, Recumbent folds—Focus of an investigative workshop in tectonics: Journal of Geological Education, v. 22, p. 204–208.

Davis, G. H., and others, 1975, Origin of lineation in the Catalina-Rincon-Tortolita gneiss complex, Arizona: Geological Society of America Abstracts with Programs, v. 7, p. 602.

Drewes, Harald, 1971, Mesozoic stratigraphy of the Santa Rita Mountains, southeast of Tucson, Arizona: U.S. Geological Survey Professional Paper 658-C, 81 p.

——1972, Structural geology of the Santa Rita Mountains, southeast of Tucson, Arizona: U.S. Geological Survey Professional Paper 748, 35 p.

——1973, Large-scale thrust faulting in southeastern Arizona: Geological Society of America Abstracts with Programs, v. 5, p. 35.

——1975, Geologic map and sections of the Happy Valley quadrangle, Cochise County, Arizona: U.S. Geological Survey Miscellaneous Investigations Map I-832.

——1976a, Laramide tectonics from Paradise to Hell's Gate, southeastern Arizona: Arizona Geological Society Digest, v. 10, p. 151–167.

——1976b, Plutonic rocks of the Santa Rita Mountains, southeast of Tucson, Arizona: U.S. Geological Survey Professional Paper 915, 75 p.

——1978a, Geologic map and sections of the Rincon Valley quadrangle, Pima County, Arizona: U.S. Geological Survey Miscellaneous Investigations Map.

——1978b, The Cordilleran orogenic belt between Nevada and Chihuahua: Geological Society of America Bulletin, v. 89, p. 641–657.

Drewes, Harald, and Finnell, T. L., 1968, Mesozoic stratigraphy and Laramide tectonics of the Santa Rita and Empire Mountains southeast of Tucson, Arizona, in Titley, S. R., ed., Southern Arizona Guidebook III: Tucson, Arizona Geological Society, p. 315–324.

Engle, A.E.J., 1949, Studies of cleavage in the metasedimentary rocks of the northwest Adirondack Mountains, New York: American Geophysical Union Transactions, v. 30, p. 767–784.

Erickson, R. C., 1962, Petrology and structure of an exposure of the Pinal Schist, Santa Catalina Mountains, Arizona [M.S. thesis]: Tucson, University of Arizona, 71 p.

Finnell, T. L., 1970, Formations of the Bisbee Group, Empire Mountains quadrangle, Pima County, Arizona, in Cohee, G. V., Bates, R. G., and Wright, W. B., Changes in stratigraphic nomenclature by U.S. Geological Survey, 1968: U.S. Geological Survey Bulletin, 1294-A, p. A28–A35.

Frost, E. G., 1977, Mid-Tertiary, gravity-induced deformation in Happy Valley, Pima and Cochise Counties, Arizona [M.S. thesis]: Tucson, University of Arizona, 86 p.

Frost, E. G., and Davis, G. H., Mid-Tertiary, gravity-induced folding and transposition in Happy Valley, Pima and Cochise Counties, Arizona: Geological Society of America Abstracts with Programs, v. 8, p. 876–877.

Gilluly, James, 1956, General geology of central Cochise County, Arizona: U.S. Geological Survey Professional Paper 281, 169 p.

Hansen, A. R., Moulton, F. C., and Owings, B. F., 1978, The Utah-Arizona hingeline-thrust belt, a potential new hydrocarbon province [abs.]: New Mexico Geological Society 29th Field Conference.

Haxel, G., and others, 1978, Reconnaissance geological map of the Comobabi Mountains quadrangle, Pima County, Arizona: U.S. Geological Survey Miscellaneous Field Studies Map MF-964, scale 1:62,500.

Hayes, P. T., 1970a, Cretaceous paleogeography of southeastern Arizona and adjacent area: U.S. Geological Survey Professional Paper 658-B, 42 p.

—— 1970b, Mesozoic stratigraphy of the Mule and Huachuca Mountains, Arizona: U.S. Geological Survey Professional Paper 658-A, 28 p.

Hayes, P. T., and Drewes, Harald, 1968, Mesozoic sedimentary and volcanic rocks of southeastern Arizona, in Titley, S. R., ed., Southern Arizona Guidebook III: Tucson, Arizona Geological Society, p. 49–58.

Hayes, P. T., Simons, F. S., and Raup, R. B., 1965, Lower Mesozoic extrusive rocks in southeastern Arizona—The Canelo Hills Volcanics: U.S. Geological Survey Bulletin, 1194-M, p. M1–M9.

Keith, S. B., and Barrett, L. F., 1976, Tectonics of the central Dragoon Mountains: A new look: Arizona Geological Society Digest, v. 10, p. 169–204.

Keith, S. B., and others, 1980, Evidence for multiple intrusion and deformation within the Santa Catalina–Rincon–Tortolita crystalline complex, southeastern Arizona; Geological Society of America Memoir 153 (this volume).

Liming, R. B., 1974, Geology and kinematic analysis of deformation in the Martinez Ranch area, Pima County, Arizona [M.S. thesis]: Tucson, University of Arizona, 86 p.

May, D. J., and Haxel, G., 1979, Reconnaissance bedrock geologic map of the Sells quadrangle, Pima County, Arizona: U.S. Geological Survey Miscellaneous Field Studies Map, scale 1:62,500 (in press).

Mayo, E. B., 1964, Folds in gneiss beyond North Campbell Avenue, Tucson, Arizona: Arizona Geological Society Digest, v. 7, p. 123–145.

Moore, B. N., and others, 1941, Geology of the Tucson quadrangle, Arizona: U.S. Geological Survey Open-File Report, 20 p.

Pashley, E. F., 1966, Structure and stratigraphy of the central, northern, and eastern parts of the Tucson basin, Pima County, Arizona [Ph.D. dissert.]: Tucson, University of Arizona, 273 p.

Peterson, R. C., 1968, A structural study of the east end of the Catalina forerange, Pima County, Arizona [Ph.D. dissert.]: Tucson, University of Arizona, 105 p.

Pilkington, H. D., 1962, Structure and petrology of a part of the east flank of the Santa Catalina Mountains, Pima County, Arizona [Ph.D. dissert.]: Tucson, University of Arizona, 120 p.

Plut, F. W., 1968, Geology of the Eagle Peak–Hell's Gate area, Happy Valley quadrangle, Cochise County, Arizona [M.S. thesis]: Tucson, University of Arizona, 78 p.

Ramberg, H., 1955, Natural and experimental boudinage and pinch-and-swell structures: Journal of Geology, v. 63, p. 512–526.

Ramsay, J. G., 1967, Folding and fracturing of rocks: New York, McGraw-Hill Book Company, 568 p.

Ransome, F. L., 1903, Geology of the Globe copper district, Arizona: U.S. Geological Survey professional Paper 12, 168 p.

Rast, N., 1956, The origin and significance of boudinage: Geology Magazine, v. 93, p. 401–408.

Rehrig, W. A., and Heidrick, T. L., 1972, Regional fracturing in Laramide stocks of Arizona and its relationship to porphyry copper mineralization: Economic Geology, v. 67, p. 198–213.

Rehrig, W. A., and Reynolds, S. J., 1980, Geologic and geochronologic reconnaissance of a northwest-trending zone of metamorphic core complexes in southern and western Arizona: Geological Society of America Memoir 153 (this volume).

Reynolds, S. J., Rehrig, W. A., and Damon, P. E., 1978, Metamorphic core complex terrain at South Mountain, near Phoenix, Arizona: Geological Society of America Abstracts with Programs, v. 10, p. 143–144.

Schlodercr, J. P., 1974, Geology and kinematic analysis of deformation in the Redington Pass area, Pima County, Arizona [M.S. thesis]: Tucson, University of Arizona, 60 p.

Shakel, D. W., 1974, The geology of layered gneisses in part of the Santa Catalina forerange, Pima County, Arizona [M.S. thesis]: Tucson, University of Arizona, 233 p.

Shakel, D. W., Silver, L. T., and Damon, P. E., 1977, Observations on the history of the gneiss core complex, Santa Catalina Mountains, southern Arizona: Geological Society of America Abstracts with Programs, v. 9, p. 1169.

Sherwonit, W. E., 1974, A petrographic study of the Catalina gneiss in the forerange of the Santa Catalina Mountains, Arizona [M.S. thesis]: Tucson, University of Arizona, 165 p.

Silver, L. T., 1955, The structure and petrology of the Johnny Lyon Hills area, Cochise County, Arizona [Ph.D. dissert.]: Pasadena, California Institute of Technology, 407 p.

Silver, L. T., and Deutsch, A., 1961, Uranium-lead method on zircons: California Institute of Technology, Division of Geological Sciences Publication 1012.

Silver, L. T., and others, 1977a, Chronostratigraphic elements of the Precambrian rocks of the southwestern and far western United States: Geological Society of America Abstracts with Programs, v. 9, p. 1176.

Silver, L. T., and others, 1977b, The 1.4–1.5 b.y. transcontinental anorogenic plutonic perforation of North America: Geological Society of America Abstracts with Programs, v. 9, p. 1176–1177.

Simons, F. S., and others, 1966, Exotic blocks and coarse breccias in Mesozoic volcanic rocks of southeastern Arizona, *in* Geological Survey Research 1966: U.S. Geological Survey Professional Paper 550-D, p. D12–D22.

Thorman, C. H., 1977, Gravity-induced folding off a gneiss dome complex, Rincon Mountains, Arizona: Discussion: Geological Society of America Bulletin, v. 88, p. 1211–1212.

Waag, C. J., 1968, Structural geology of the Mt. Bigelow, Bear Wallow, Mt. Lemmon area, Santa Catalina Mtns., Arizona [Ph.D. dissert.]: Tucson, University of Arizona, 133 p.

Wargo, J. G., and Kurtz, W. L., 1956, Geologic and tectonic features of the Coyote Mountains, Arizona: Ohio Journal of Sciences, v. 56, p. 10–16.

Wilson, E. D., Moore, R. T., and Peirce, H. W., 1957, of Pinal County, Arizona: Arizona Bureau of Mines, scale 1:375,000.

Wilson, E.D., Moore, R. T., and Peirce, H. W., 1957, Geologic map of Maricopa County, Arizona: Arizona Bureau of Mines, scale 1:357,000.

Wilson, E. D., Moore, R. T., and O'Haire, R. T., 1960, Geologic map of Pima and Santa Cruz Counties, Arizona: Arizona Bureau of Mines, scale 1:375,000.

Wright, C. C., 1978, Folds in mylonite schist in the Rincon Mountains, Tucson, Arizona [senior thesis]: Northfield, Minnesota, Carleton College, 26 p.

Yeend, W. E., 1976, Reconnaissance geologic map of the Picacho Mountains quadrangle, Arizona: U.S. Geological Survey Field Studies Map MF-778.

MANUSCRIPT RECEIVED BY THE SOCIETY JUNE 1, 1979
MANUSCRIPT ACCEPTED AUGUST 7, 1979

Mylonitization and detachment faulting in the Whipple-Buckskin-Rawhide Mountains terrane, southeastern California and western Arizona

GREGORY A. DAVIS
J. LAWFORD ANDERSON
Eric G. FROST
Department of Geological Sciences
University of Southern California
Los Angeles, California 90007

TERRY J. SHACKELFORD
Research Center
Marathon Oil Company
Littleton, Colorado 80120

ABSTRACT

Field studies in the Whipple Mountains, southeastern California, and in the Buckskin and Rawhide Mountains, western Arizona, have defined the existence of an Oligocene(?) to middle Miocene gravity slide complex that is at least 100 km across in the direction of its transport (N50° ± 10°E). The regionally developed complex is underlain by a subhorizontal detachment fault, named the Whipple detachment fault in western areas and the Rawhide detachment fault in eastern areas. The fault, which was warped and domed after its formation, separates a lower-plate assemblage of Precambrian to Mesozoic or lower Cenozoic igneous and metamorphic rocks and their deeper, mylonitic equivalents from an allochthonous, lithologically varied upper plate.

Most lower-plate crystalline rocks were subjected to regional Late Cretaceous and/or early Tertiary mylonitization and metamorphism. The abrupt (3- to 30 m wide) upper limit of mylonitization, the Whipple "mylonitic front," is a mappable zone of high strain and, presumably, high thermal gradient. In parts of the Whipple Mountains, mylonitization was accompanied by the intrusion of subhorizontal sheets or sills of adamellite to tonalite up to a few hundreds of metres thick, although elsewhere thick sections of mylonitic rocks are devoid of such sills. The sills include both peraluminous and metaluminous varieties and are compositionally distinct relative to plutons in the overlying upper plate, being richer in Al, Mg, Ca, Na, and Sr and depleted in K and Rb. The compositions of most of the minerals in the mylonitized sills and their country rock gneisses did not reequilibrate during metamorphism. However, reequilibrated phases do occur in the ultrafine-grained mylonitic matrix and in tension gashes developed perpendicularly to mylonitic lineation. As a result of incomplete reequilibration, bimodal compositional ranges exist for plagioclase, epidote, celadonitic muscovite, and biotite. The minimum depth for intrusion and mylonitization is estimated to be 9.6 km from consideration of the interaction of compositionally corrected curves of muscovite stability and the adamellite solidus. Metamorphic mineral assemblages and feldspar thermometry indicate that mylonitization occurred from solidus temperatures of the plutonic sills down to middle greenschist grade.

Allochthonous (upper-plate) units in the detachment complex include Precambrian to Mesozoic crystalline rocks, Paleozoic and Mesozoic metasedimentary rocks, Mesozoic metavolcanic rocks, and Tertiary volcanic and sedimentary rocks. The oldest Tertiary rocks are debris flows (some containing mylonitic rocks), fanglomerates, lacustrine sediments, and volcanic rocks, all of the Oligocene(?) to lower Miocene Gene Canyon Formation. Red beds and volcanic rocks of the Copper Basin Formation overlie Gene Canyon rocks unconformably and are tilted less steeply than the older Tertiary rocks along northeast-dipping listric normal faults that occur widely within the upper plate. In the Whipple Wash area of the eastern Whipple Mountains, volcanic rocks of the Copper Basin Formation sit unconformably on brecciated lower-plate mylonitic rocks in a channel cut ~70 m below the Whipple detachment fault. These volcanic rocks were themselves involved in renewed detachment faulting along that fault. Collectively, these stratigraphic-structural relations indicate that detachment faulting occurred during Tertiary sedimentation over a significant period of time, and was therefore of growth-fault rather than catastrophic nature. Upper Miocene valley-fill sediments and alkali basalts unconformably overlie upper-plate structures and tilted strata, thus providing an upper age limit for the detachment faulting.

Northeastward movement of the thin (<5 km) upper plate is believed to have occurred under the influence of gravity, although the Whipple-Rawhide detachment fault could not have originally dipped more than a few degrees. The head, or breakaway zone, of the crustal slide is apparently defined by northeast-dipping normal faults in the Mopah Range, just west of the Whipple Mountains. Central areas of the slide complex in the vicinity of the Colorado River (Whipple and Buckskin Mountains) are characterized by extreme distension of the detached slab along northwest-striking, northeast-dipping, listric normal faults. There is telescoping of allochthonous units in distal, or toe, portions of the slide complex in the Rawhide and Artillery Mountains of western Arizona, where thrust faulting of older rocks over rocks as young as middle Miocene is common. Northeastward displacements of allochthonous units in excess of several tens of kilometres are indicated by field relations in the Buckskin and Rawhide Mountains.

INTRODUCTION

Detachment Faults, Regional Setting

Nearly two decades ago, Peter Misch (1960) described an extensive terrane in the border area between northern Nevada and Utah that he thought was underlain by a low-angle regional fault (Fig. 1). He named this fault "Snake Range decollement" and the area in which it occurs "Northeastern Nevada structural province." The decollement typically separates complexly faulted, unmetamorphosed, upper-plate rocks from stratigraphically older, lower-plate crystalline complexes, the latter now known to include metamorphic rocks of both Mesozoic and Tertiary age. Misch noted that some mountain ranges in the structural province were fault blocks of basin-range type; others were "elongate-domal uplifts" with only subordinate block faulting. In recent years "decollement" structures

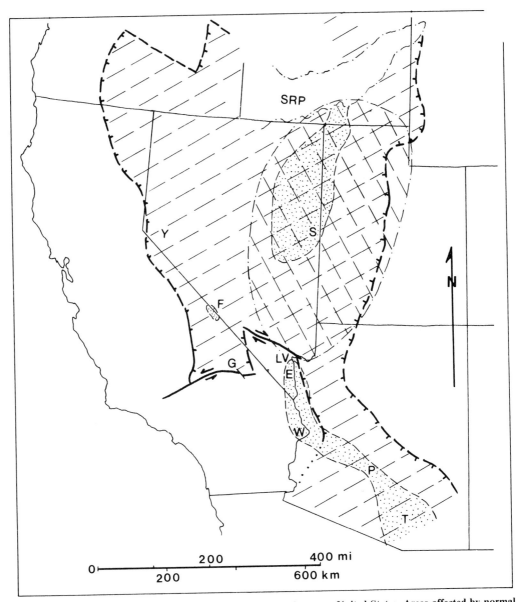

Figure 1. Distribution of extensional tectonics in the southwestern United States. Areas affected by normal faulting of basin-range type are indicated by northeast-southwest ruled pattern. Areas of low-angle detachment faulting are stippled. The inferred distribution of a crustal low-velocity layer (Smith, 1978) is shown by northwest-southeast ruled pattern. Geographic localities from north to south: SRP = Snake River Plain; Y = Yerington; S = Snake Range; F = Funeral Range; G = Garlock fault; LV = Las Vegas; E = Eldorado Mountains; W = Whipple Mountains; P = Phoenix; T = Tucson.

and domal ranges of the type recognized by Misch have also been discovered in more southerly areas of the Cordillera (Fig. 1), in the Death Valley area of California (Keene Wonder fault and Funeral Range; Reynolds, 1974) and, more extensively, in southernmost Nevada, southeastern California, and Arizona—the area discussed here.

Many of the low-angle faults in the Nevada-Utah region studied by Misch are in areas that lack

Tertiary rocks (or direct evidence for the involvement in faulting of Tertiary units) and are complicated by the superposed disruptive effects of basin-range faulting. As a consequence, the age and origin of these low-angle faults are highly controversial. Nevertheless, Armstrong (1972) presented a strong case that the Snake Range decollement and higher, related faults (1) were Tertiary in age or, at the least, showed Tertiary reactivation; (2) were formed in an extensional regime; and (3) resulted from thinning of supracrustal rocks by normal faulting above a basal detachment surface (denudational tectonics). In contrast, Misch (1971) considered the Snake Range decollement to be Mesozoic in age and compressional in origin. Hose and Danes (1973) offered still another interpretation: the decollement resulted from regional eastward gravity sliding during Mesozoic time.

The case for a Tertiary origin for the low-angle faults of the northern Nevada-Utah area is greatly enhanced by geologic relations in the closely similar (and comparably enigmatic) terrane of low-angle faults that begins near Hoover Dam in the Las Vegas area, follows the Colorado River trough as far south as Parker, then swings southeastward and extends across southern Arizona into areas near Tucson (Fig. 1). Here, on the flanks of a score of ranges that characteristically owe their elevation to doming or arching rather than to block faulting, are basal detachment faults that separate allochthonous upper-plate units from crystalline rocks in the cores of the ranges. Tertiary strata as young as middle Miocene age are widely involved in the low-angle faulting, as are Precambrian gneisses, Paleozoic and Mesozoic metasedimentary rocks, and Precambrian and Mesozoic plutons. Lower-plate rocks in the terrane include Precambrian gneisses and plutonic rocks, Mesozoic and Tertiary plutonic rocks, and strongly lineated mylonitic gneisses of Cretaceous(?) to Tertiary age.

This report treats the northwestern third of the Nevada-California-Arizona detachment terrane (Fig. 1), and summarizes data and conclusions drawn from an ongoing University of Southern California research program in the Colorado River area, primarily in the Whipple Mountains of southeastern California and in the conterminous Buckskin and Rawhide Mountains of west-central Arizona (Figs. 2, 3). We emphasize the fact that conclusions drawn from our studies may not be applicable to detachment terranes farther east, for example, in the Tucson area (G. H. Davis, 1975, 1977, and this volume), or in the northern Nevada-Utah terrane first described by Misch (1960). Ours are not the only investigations in the Whipple-Buckskin-Rawhide terrane. Carr and Dickey (1977), Lucchitta and Suneson (1977), and Rehrig and Reynolds (1977 and this volume) have also studied areas within or adjacent to the terrane.

Previous Studies, Colorado River Trough

F. L. Ransome (1931) may have been the first geologist to observe and report low-angle faults in the Colorado River trough. In a consulting report submitted to the Metropolitan Water District of Southern California, he described the existence in the upper Bowmans Wash area, southeastern Whipple Mountains, of a low-angle thrust fault, along which Tertiary rocks had been displaced across crystalline basement rocks (Figs. 3, 4).

Curiously, Kemnitzer in a later study (1937), although aware of Ransome's findings, concluded that the dominant structures in the range consisted of northwest-striking normal faults that cut both Tertiary and underlying basement rocks. Although he recognized evidence for movement along some contacts between Tertiary rocks and basement rocks, he regarded all such contacts as unconformities. "Thrust faulting," he reported (p. 126), "which might have involved the Tertiary mantle as an over-riding block on the Basement Complex has been considered and is regarded as untenable."

Terry (1972) later recognized low-angle faulting in the Whipple Mountains, but not with the relations described by Ransome. She mapped an extensive shallow-dipping fault around the northern, eastern, and southern periphery of the range between contrasting upper- and lower-plate crystalline assemblages. She, like Kemnitzer, interpreted all contacts between Tertiary strata and crystalline

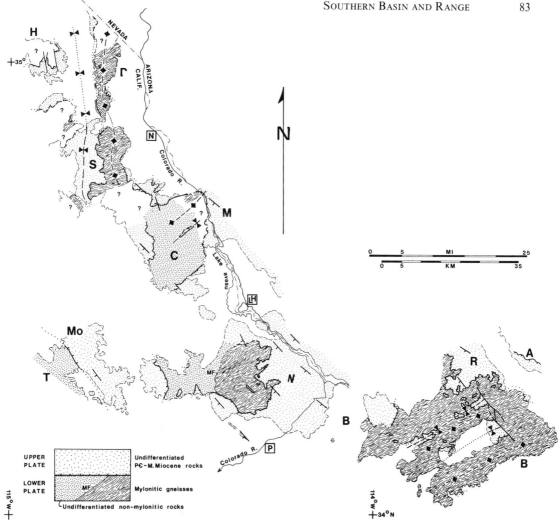

Figure 2. Tertiary low-angle detachment fault complex, southeastern California and western Arizona. Upper- and lower-plate units are separated by a regional low-angle fault(s) (hachured contact), except between the Mopah (Mo) and Turtle (T) ranges where the contact is an east-dipping normal fault. Mylonitic and nonmylonitic lower-plate assemblages are separated by a "mylonitic front" (MF) in the Whipple Mountains. Heavy dashed lines represent the axial traces of broad antiformal (triangles point outward) and synformal (triangles point inward) folds that warp the regional detachment fault(s). Strike and dip symbols show orientation of bedding in Tertiary units contained within upper-plate tilted fault blocks. Communities (letters in squares): N = Needles; LH = Lake Havasu City; P = Parker. Mountain ranges from northwest to southeast: H = Homer; D = Dead; S = Sacramento; M = Mohave; C = Chemehuevi; T = Turtle; Mo = Mopah; W = Whipple; B = Buckskin; R = Rawhide; A = Artillery. Geologic relations north of the Whipple Mountains first ascertained by geologists of the Southern Pacific Land Company (unpub. studies), and confirmed and supplemented by E. G. Frost and G. A. Davis. Geologic relations in the Rawhide Mountains from Shackelford (1976).

basement rocks as unconformities. Because the involvement of Tertiary rocks in low-angle faulting was not recognized, Terry regarded the low-angle fault contact which she had mapped between crystalline rocks as a thrust fault (the Whipple Mountains thrust fault) of Mesozoic age.

In neighboring ranges, "thrust faults" had also been recognized by other workers. Wilson and Moore (1959) and Wilson (1960) mapped low-angle faults in the Buckskin and Rawhide Mountains, Arizona,

Figure 3. View to north of Whipple detachment surface as exposed in area west of Bowmans Wash. Note steep orientation of southwest-dipping Tertiary strata into the fault surface. Lower-plate rocks are mylonitic gneisses.

Figure 4. View due south of southwest-tilted Tertiary strata cut by detachment fault, south-central Whipple Mountains. The characteristic ledge seen directly below the fault surface is composed of microbreccia overlying sheared, shattered, and altered mylonitic gneisses.

which Wilson (1962) regarded as thrust faults of Laramide or younger age. At the same time, geologists of the Southern Pacific Company were mapping several mountain ranges in California near Needles (Fig. 2). Typical of their findings is the report by Coonrad (1960) for an area in the eastern Chemehuevi Mountains. He described extensive "thrust faults" involving crystalline and upper Tertiary rocks, and perceptively stated (p. 20) that the "thrust faults in this area seem to be part of an active zone of thrust adjustment that extends at least from the Harquahala Mountains in Western Arizona to the Homer Mountain region northwest of Needles, a distance of 120 miles." Referring specifically to the Whipple Mountains, Coonrad stated (p. 21) that "most, if not all, of the Tertiary rocks described by Kemnitzer (1937) in the Parker Dam area are probably in thrust fault relationships."

Anderson's study (1971) of low-angle faulting in the Eldorado Mountains, Nevada, 100 km (63 mi) north of Needles, was the first to attribute the low-angle faults of the Colorado River area to noncompressional tectonics. He described the Eldorado Mountains as an area of major imbricate normal faulting accompanied by pronounced eastward rotation of Tertiary strata. Miocene normal faults, which dip westward, were interpreted as flattening downward and merging with a subhorizontal basal fault surface below which extension by normal faulting had not occurred. Displacement of Tertiary rocks above the inferred basal surface was, as a consequence of fault geometries, westward (S70°W) relative to lower-plate autochthonous units.

PRE–MIDDLE MIOCENE ROCK UNITS, WHIPPLE-BUCKSKIN-RAWHIDE MOUNTAINS

General Statement

Remapping of the Whipple Mountains fault recognized by Ransome (1931) and Terry (1972) demonstrated that it is of Tertiary age and that it underlies an extensive (>3,000 km^2) normal-faulted terrane that includes the Whipple, Buckskin, and Rawhide Mountains (Shackelford, 1976, 1977; Lingrey and others, 1977; Davis and others, 1977). The fault may be coextensive with similar low-angle faults originally mapped by Southern Pacific Company geologists on the flanks of the Chemehuevi, Sacramento, Homer, and Dead Mountains to the north (Fig. 2).

The rock units of the Whipple, Buckskin, and Rawhide Mountains can be divided into three assemblages, of which the older two were involved in pre–middle Miocene deformations. The youngest assemblage consists of sedimentary and volcanic units that postdate most Miocene low-angle faulting; they are treated in a subsequent section. The two deformed assemblages are separated by a prominent detachment surface, or fault, that is regionally subhorizontal in orientation. Rocks above this detachment surface are allochthonous to varying degrees. These upper-plate rocks include representatives of the entire stratigraphic sequence in this region (Precambrian to middle Miocene), as well as igneous intrusive rocks of Precambrian, Mesozoic, and Cenozoic age. Crystalline rocks of the lower plate probably include most upper-plate units, with the striking exception of Tertiary sedimentary and volcanic strata. With only one extremely important exception, areally limited Tertiary strata in the Whipple Wash area, Tertiary sedimentary or volcanic rocks have not yet been recognized in an autochthonous (lower-plate) position. This exception is discussed elsewhere. Coonrad (1960, p. 20) reported a similar relation in the distribution of Tertiary rocks in the extensive region studied by Southern Pacific Company geologists: "At no place within the region studied have the thrusted rocks of Tertiary age been found in unequivocal original depositional position." Lower-plate rocks in areas east and southeast of the central Whipple Mountains characteristically exhibit the effects of a regional metamorphic and mylonitic event of probably Late Cretaceous and/or early Tertiary age.

The descriptions of upper- and lower-plate rock units that follow are drawn primarily from recent and ongoing studies in the Whipple and western Buckskin Mountains. Shackelford (1976; 1980, in prep.) has described the petrology of faulted units in the Rawhide Mountains that generally resemble the units treated here.

Upper-Plate Units (Allochthonous)

Crystalline Basement Rocks. Crystalline rocks of the upper plate primarily form the eastern third of the Whipple Mountains (Fig. 5). Preliminary aspects of their petrology have been reported by Podruski (1979) and Anderson and others (1979).

The oldest units in the upper-plate crystalline complex (Fig. 6) are a middle- to upper-amphibolite–grade, Precambrian metamorphic package consisting of quartzo-feldspathic metasedimentary rocks, amphibolite, and a granitic metaplutonic suite. The metasedimentary rocks consist of meta-arkose, quartz arenite, and graywacke. Where these rocks have not been migmatized, laminar and nonlaminar (cross-bedding, channels) sedimentary structures are well preserved. Layers of felsic metavolcanic rock are present, but are rare. A near-vertical tectonic foliation is generally parallel to bedding; small-scale, isoclinal folds with fold axes parallel to foliation are common. Mineralogically, the metasedimentary rocks are fine to medium grained, granoblastic to weakly foliated, and consist of quartz, alkali feldspar, plagioclase (An_{20-34}), biotite, Fe-Ti oxides ± muscovite, garnet, sillimanite, and retrograde chlorite. Key metamorphic assemblages include quartz + muscovite + plagioclase + alkali feldspar + garnet + biotite + sillimanite (alkali-feldspar–sillimanite zone of Evans and Guidotti, 1966), and quartz + plagioclase + garnet + biotite + sillimanite (upper sillimanite zone to alkali-feldspar–sillimanite zone). Migmatitic rocks contain abundant leucosomal veins of quartz, alkali feldspar, and albite (An_{06-10}) and postkinematic garnet porphyroblasts (to 2 cm in diameter) in medium-grained quartz-feldspar-biotite gneiss.

Layers of amphibolite (0.1 to 350 m thick) are common throughout the metasedimentary section, and are concordant to locally discordant with the relict bedding of the metasedimentary rocks. The amphibolite possesses the same foliation as the metasedimentary rocks and contains the primary metamorphic assemblage of andesine-labradorite (An_{33-55}) + hornblende ± cummingtonite, and retrograde chlorite, epidote, and actinolite. Interpreted as sills, the chemistry of the amphibolites (assuming isochemical metamorphism) indicates a quartz tholeiitic composition (SiO_2 = 51.3% to 51.7% by weight with normative hypersthene and quartz). A metaplutonic suite consisting of biotite-hornblende adamellite and granodiorite, and lesser biotite granite (Fig. 7; Table 1), intrudes the metasedimentary rocks and amphibolite. All three plutonic units occur as elliptical to lensoidal bodies with contacts and foliation concordant with the foliation of the older units (with local exceptions). The granodiorite is dark gray, medium grained, and consists of plagioclase (An_{32-35}) and alkali feldspar (Or_{87-89}) in a foliated matrix of hastingsitic hornblende, biotite, and quartz. The foliated adamellite is characterized by medium- to coarse-grained alkali-feldspar augen (Or_{85}; 0.3 to 1.2 cm) enclosed by interstitial plagioclase (An_{27-28}), quartz, and biotite ± hornblende. The weakly foliated granite phase is red, coarse grained, and inequigranular. It contains quartz, alkali feldspar, and interstitial plagioclase and biotite. Accessories in all granitic units are allanite (nonmetamict), sphene, magnetite (with little or no ilmenite), apatite, and zircon. The composition of the three rock types is subalkalic, and uniformly too high in Fe/Mg to be calc-alkaline (Fig. 8). With respect to K, Rb, Ba, and Sr (Fig. 9), the units are compositionally distinct, which indicates that they do not form a fractionation sequence, but that they represent three separate intrusive events.

The oldest postmetamorphic-plutonic unit is a dark gray, porphyritic quartz monzodiorite (SiO_2 = 60.9% to 63.8%) restricted to large inclusions (0.2 to 1.2 km in diameter) in a granite porphyry described below. The rock contains scattered phenocrysts (1% to 5%) of euhedral to subhedral plagioclase (An_{29-37}) and lesser amounts of alkali feldspar (Or_{79-83}) and idiomorphic quartz set in a fine-grained

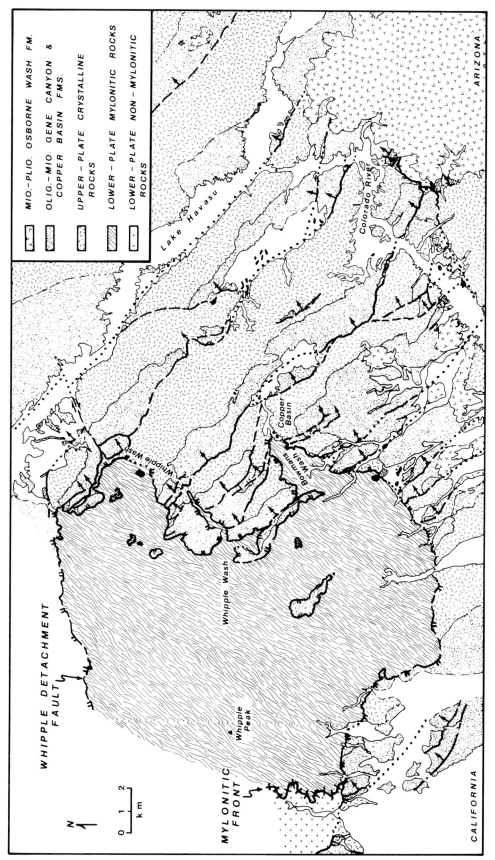

Figure 5. Geologic map of Whipple Mountains, San Bernardino County, southeastern California.

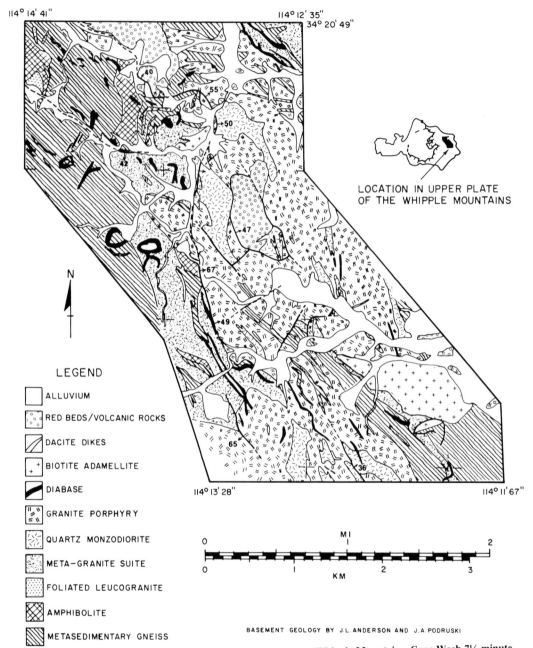

Figure 6. Geologic map of upper-plate crystalline rocks, eastern Whipple Mountains, Gene Wash 7½-minute quadrangle.

matrix of quartz, two feldspars, biotite, ferroedenitic to hastingsitic hornblende, and accessories of allanite, apatite, zircon, and magnetite ± ilmenite. The restricted occurrence and hypabyssal texture suggest that the inclusions are stope blocks from the roof of the granite porphyry pluton. The rock exhibits various degrees of recrystallization and incipient foliation related to proximity of the contacts with the porphyry. Compositionally, the quartz monzodiorite is tholeiitic (Fig. 8) and similar to the

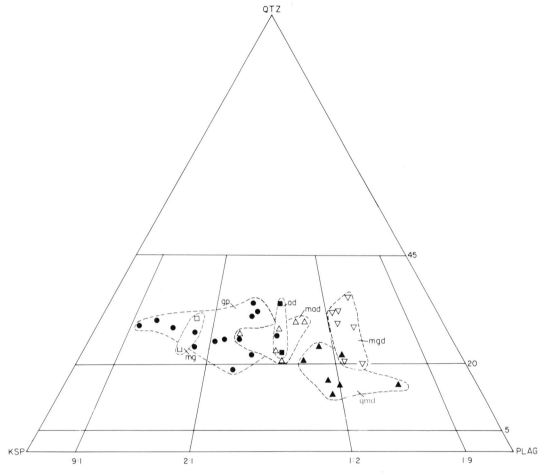

Figure 7. Modal composition of upper-plate granitic plutons in terms of quartz, plagioclase, alkali feldspar. Pluton symbols are mgd = metagranodiorite; mad = meta-adamellite; mg = metagranite (all of metaplutonic suite); qmd = quartz monzodiorite; gp = granite porphyry; and ad = biotite adamellite.

granodiorite phase of the metaplutonic suite in most major elements. It likewise has a high level of Ba (1,300 to 1,650 ppm) and a moderate level of Sr (278 to 325 ppm), although it is uniformly lower in Al_2O_3 (13.2% to 13.8% versus 14.6% to 15.6% by weight).

A granite porphyry is the major crystalline unit of the eastern Whipple Mountains upper-plate basement complex. This unit exhibits sharp intrusive contacts (generally concordant) with all the above lithologies. The rock is distinctive in having abundant (34% to 70%), coarse-grained, alkali-feldspar phenocrysts (1 to 4 cm; Or_{82-88}) that are aligned in a well-defined planar flow fabric. Interstitial to the large alkali feldspars are anhedral quartz, plagioclase (An_{24-29}), biotite $\pm$ hastingsite to hastingsitic hornblende, and accessories of allanite, sphene, magnetite $\pm$ ilmenite, apatite, and zircon. The pronounced flow fabric and texture of the porphyry indicate that it was forcefully intruded as a crystal mush. Like the metaplutonic suite and the quartz monzodiorite, the porphyry is subalkalic and is too iron-rich and aluminum-poor to be calc-alkaline. The granite porphyry does have a wide range of modal and chemical composition (SiO_2 = 64.7% to 69.9% by weight; K_2O = 4.26% to 6.61% by weight). Less silicic members have less quartz and alkali feldspar, more plagioclase and total mafic minerals (to 18.9% by volume), and increased hornblende over biotite. A fractional crystallization model (Podrus-

TABLE 1. AVERAGE MAJOR AND TRACE ELEMENT COMPOSITION OF UPPER-PLATE CRYSTALLINE UNITS

	am*	mgrd	mad	mgr	qmd	gp	d_1	ad	dac	d_2
No[+]	2	4	4	2	5	9	4	1	2	2
SiO_2	51.51	61.11	71.82	70.71	62.17	67.00	47.13	72.25	69.38	45.05
TiO_2	1.40	1.46	0.45	0.31	1.40	0.90	1.76	0.15	0.20	1.63
Al_2O_3	15.69	15.08	13.23	14.10	13.53	13.94	16.79	14.98	15.98	16.25
FeO[§]	12.18	7.82	3.97	2.06	8.13	5.15	11.77	1.18	1.74	10.13
MgO	4.50	2.02	0.79	0.32	1.55	1.00	7.13	0.34	0.38	8.67
MnO	0.230	0.120	0.061	0.040	0.130	0.080	0.184	0.081	0.066	0.184
CaO	8.29	4.12	1.77	1.32	4.07	2.44	9.55	2.01	3.29	11.77
Na_2O	3.30	2.93	2.66	2.59	2.75	2.73	2.77	3.48	3.79	2.38
K_2O	1.23	4.03	4.74	6.35	4.13	5.28	1.23	4.68	3.96	1.40
Total	98.33	98.69	99.49	97.80	97.86	98.52	98.30	99.15	98.80	98.20
K_2O+Na_2O	4.53	6.81	7.40	8.94	6.88	8.01	3.98	8.15	7.73	3.78
$\frac{FeO}{(FeO+MgO)}$	0.73	0.795	0.834	0.866	0.840	0.837	0.622	0.776	0.822	0.539
"Al"[#]	0.687	0.919	1.069	1.043	0.826	0.953	0.723	1.038	0.970	0.609
Li	4.70	14.80	8.70	22.50	22.70	15.30	9.30	5.60	5.91	42.1
Rb	29.5	159.0	139.0	348.0	121.0	230.0	40.1	82.0	92.7	40.8
Sr	125	272	123	119	291	188	322	424	293	429
Ba	318	1,390	918	567	1,458	1,111	304	530	625	240
Ba/Sr	2.55	5.10	7.43	4.76	5.01	5.89	0.940	1.25	2.13	0.56
Ba/Rb	10.5	8.71	6.59	1.63	12.1	4.84	7.58	6.46	6.74	5.87
Rb/Sr	0.237	0.588	1.13	2.92	0.414	1.22	0.125	0.193	0.317	0.095
K/Rb	46.1	210	283	151	284	191	254	474	344	286

Note: Analyses performed at U.S.C. Petrochemistry Laboratory by J. A. Podruski, Gregory A. Benson, and J. Lawford Anderson. Major elements as wt %. Li, Rb, Sr, Ba as ppm.

*Unit symbols: am = amphibolite, mgrd = metagranodiorite, mad = meta-adamellite, mgr = metagranite, qmd = quartz monzodiorite, gp = granite porphyry, d_1 = early diabase, ad = biotite adamellite, dac = dacite dikes, d_2 = late alkalic diabase and lamprophyre.

[+]Number of analyses averaged.

[§]All Fe as FeO.

[#]"Al" = molecular proportions of Al/(K + Na + Ca); peraluminous when >1, metaluminous <1.

ki, 1979) based on trace-element data of nine samples indicates about 58% fractionation of hornblende and lesser amounts of plagioclase, alkali feldspar, and biotite. The absolute age of the porphyry is presently unknown. It has textural, mineralogical, and compositional affinities to unique, upper Precambrian, potassic, iron-rich granites of 1.4- to 1.5-b.y. age (Silver and others, 1977; Anderson and Cullers, 1978), but may be as young as Jurassic age as indicated by a 156-m.y. K-Ar date on biotite (Terry, 1972).

Numerous dikes and sills (1 to 10 m wide and up to 1.1 km long) of diabase that range in composition from olivine tholeiite to alkali-olivine basalt (SiO_2 = 45.2% to 47.7% by weight) intrude the granite porphyry and all older units. The diabase ranges in grain size from fine (at chilled contacts) to coarse and has a distinctive ophiticlike texture of euhedral, normally zoned plagioclase (An_{33-62}) enclosed, or

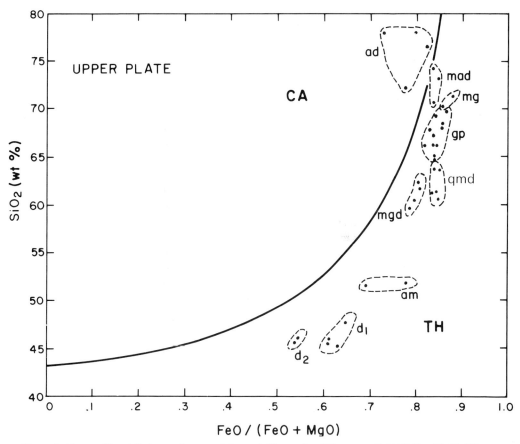

Figure 8. Compositional features of upper-plate crystalline rocks: SiO_2 versus $FeO/(FeO + MgO)$. Diagram modified after Miyashiro (1974). All Fe as FeO. Pluton symbols are as in Figure 7 and as follows: am = amphibolite; d_1 = early diabase; d_2 = late diabase and lamprophyre.

partially enclosed, within late magmatic(?), interstitial, pale-green, actinolitic hornblende. Secondary deuteric minerals are common and include fine-grained, dark-green, more aluminous, magnesio-hornblende (Al_2O_3 increases from 4.1% to 10.4% by weight), biotite, chlorite, epidote, and calcite. Accessory minerals include apatite, magnetite, and ilmenite.

The youngest granitic event in the crystalline basement of the upper plate is represented by several leucocratic biotite and biotite–muscovite adamellite plutons that intrude the granite porphyry, diabase, and the older rocks. Biotite K-Ar dates of 81 ± 2 m.y. (Terry, 1972, southern Whipples) and 74 ± 3.2 m.y. (Bull and Ku, 1975) suggest a Late Cretaceous emplacement age. These pluglike bodies typically measure 0.81 to 1.6 km in diameter and are generally massive and unmetamorphosed. The rock is even-grained consisting of quartz, alkali feldspar (Or_{88-93}), oligoclase (An_{20-27}), and lesser amounts of biotite (≤3%) ± muscovite; accessories include allanite, zircon, apatite, and magnetite. Compositionally, this is the only calc-alkaline granitic unit that has been found in the upper plate and, as such, is distinct from all lithologies described above. It is also compositionally peraluminous (based on molecular proportions of K, Na, Ca, and Al) as substantiated by the biotite ± muscovite mineralogy.

Subsequent igneous activity is represented by dikes of Tertiary(?) dacite (volcanic feeders?) followed by late dikes and sills of pyroxene lamprophyre and hornblende-bearing alkalic diabase. This diabase

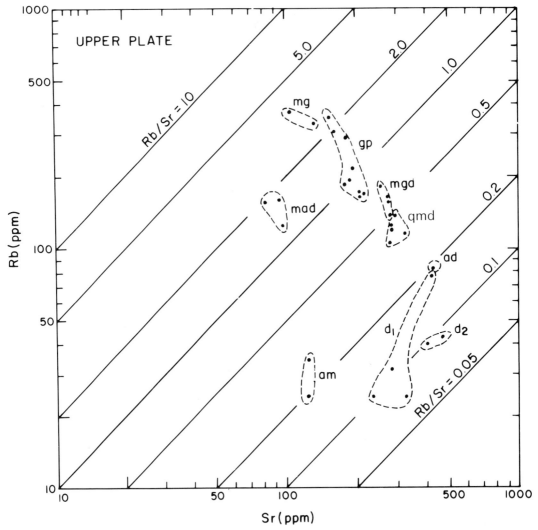

Figure 9. Compositional features of upper-plate crystalline rocks: Rb (ppm) versus Sr (ppm). Pluton symbols same as in Figures 7 and 8.

constitutes the third generation of mafic dike and sill emplacement into rocks of the upper plate. They are compositionally similar to the earlier diabase, except for slightly higher levels of CaO and MgO. Texturally, they differ in having prismatic, euhedral hornblende (not interstitial) and interstitial (not euhedral) plagioclase.

Paleozoic Metasedimentary Rocks. Paleozoic rocks are present within the western Buckskin and Rawhide Mountains but appear to be absent from the Whipple Mountains. Within the western Buckskin Mountains, these Paleozoic metasedimentary rocks occur in three separate blocks, each ~2 km² in extent. Numerous, isolated patches of metasedimentary rocks crop out in the Rawhide Mountains, most as klippen resting on the major detachment surface (Shackelford, 1976).

All the Paleozoic rocks in the Buckskin-Rawhide terrane consist of highly deformed marble, quartzite, and phyllite. Original bedding has been almost completely destroyed, leaving a pronounced transposition foliation in these greenschist-grade metasedimentary rocks. In the western Buckskin

Mountains, this foliation defines several large, northeast-trending folds. Individual formations are tectonically intermixed within these folds and are characterized by pronounced tectonic thinning, or attenuation, on the limbs of some folds. A similar deformational style has been documented to the southwest in the Big Maria Mountains by Hamilton (1971).

The age of these rocks and their correlation with other Paleozoic sections are as yet unsolved problems. A prominent white marble that contains abundant brown chert as transposed layers is probably correlative with the Permian Kaibab Limestone of the Grand Canyon section. A thick quartzite adjacent to the Kaibab(?) may be the Permian Coconino Sandstone, although Carr and Dickey (1977) have suggested that it may be more closely related to the Queantoweap Formation of McNair (1951). The identity of other units is even more tentative, but all are probably upper Paleozoic or Mesozoic. Rocks that appear to be correlative with the Cambrian Bright Angel Shale and Tapeats Sandstone are present in the Rawhide Mountains (Shackelford, 1976).

Similar though considerably more complete sections are to the southwest in the Big Maria Mountains (Hamilton, 1964; 1979, written commun.). Another section similar to that in the Buckskin Mountains is in the Arica Mountains, 60 km west of Parker, Arizona (Edward Steiner, work in progress). Other documented Paleozoic sections that appear to be similar to the rocks in the Rawhide and Buckskin Mountains are present in the Plomosa Mountains (Miller, 1970) and Harquahala Mountains (Varga, 1977) of west-central Arizona.

Mesozoic(?) Rocks. There are Mesozoic(?) rocks within the southwestern Buckskin Mountains near Osborne Wash and in the central Buckskin Mountains in the Mineral Wash and Planet Ranch areas. Similar, primarily metavolcanic rocks are present in the Rawhide Mountains, but could also be correlative with Precambrian metavolcanic rocks (Shackelford, 1976). In the Buckskin Mountains, these rocks consist primarily of phyllites composed of quartz, feldspar, sericite, talc, and epidote. These phyllites were originally a thick sequence of sandstones, siltstones, arkoses, and mudstones. A rhyolite flow is also present within the section in the southwestern Buckskin Mountains. These rocks are exposed in a section that is ~1,300 m thick. The actual stratigraphic thickness is unknown.

Several conglomeratic layers near the base of this section are each only a few metres thick. Clasts consist mostly of quartzite, although there are some carbonate and adamellite clasts. Pebbles within the lowermost conglomerate are elongated in a northeast-southwest direction and are contained within phyllite that has locally developed a northeast-trending lineation. Metasedimentary and metavolcanic rocks in the Rawhide Mountains also contain a northeast-trending lineation (Shackelford, 1976). This orientation is somewhat variable where the phyllitic rocks have been highly folded. Some quartzites within the Paleozoic section exhibit a faint, northeast-trending lineation, and tremolite crystals within some marbles are locally elongate in a northeast-southwest direction. Metamorphism of upper-plate Paleozoic and Mesozoic(?) rocks probably occurred during the same event responsible for metamorphism and the development of northeast-trending mineral (mostly quartz) lineations in widespread mylonitic gneisses of the lower plate.

Gene Canyon Formation. The oldest Tertiary units in the Whipple-Buckskin-Rawhide terrane are the Gene Canyon Formation and the probably correlative Artillery Formation of Lasky and Webber (1949). Named by Ransome (1931) in his geologic reconnaissance for the California aqueduct project, the Gene Canyon Formation consists of volcanic and clastic rocks that were deposited in local basins whose locations were significantly different from present-day basins. In its type area, the Gene Canyon Formation is composed of 585 m of interbedded red-to-buff sandstone, amygdaloidal andesitic(?) volcanic rocks, coarse debris flows of crystalline rocks, and siltstone (Kemnitzer, 1937); limestone is an important component of the formation in other areas. Individual units are discontinuous along strike, extending for only a few kilometres. Gene Canyon sedimentary rocks mantled a low topography that was developed atop the crystalline basement rocks of the upper plate. The upper surface of the formation is a marked angular unconformity in the type area; the unconformity records a break in

Tertiary sedimentation and a time when erosion produced a relief of 50 to 100 m. Lacustrine beds thicken to the southwest of the type area; this suggests that the basin of deposition deepened in that direction. Northeast of the Whipple Mountains in the Aubrey Hills, Arizona, the Gene Canyon section consists of a thick sequence of volcaniclastic rocks that appear to interfinger with lacustrine and fanglomerate deposits to the south.

Monolithologic breccias, or debris flows, are widespread within the Gene Canyon Formation and consist of fragments of quartzite derived from Mesozoic(?) metasedimentary rocks and mylonitic (Fig. 10) and nonmylonitic adamellites and granodiorites. Some clasts within the debris flows are more than 5 m in diameter. One section in the eastern Whipples contains a coherent block of metaigneous rock ~0.5 km long and 50 m thick. Coarse fanglomerate deposits with large, angular clasts of Mesozoic(?) metasedimentary rocks as well as crystalline rocks are also widespread. These phyllitic, very fissile, metasedimentary clasts strongly suggest that the fanglomerates were transported only short distances and across an area of high relief. The abundance of such coarse clasts and debris flows suggests that the Gene Canyon Formation is in part derived from erosion of active fault scarps that formed at the same time as the Gene Canyon Formation. The overall facies environment may be caldera related, as has been suggested for similar rocks to the east of the Rawhide Mountains at the Anderson Uranium Mine (Sherborne and others, 1979).

The age of the Gene Canyon Formation is poorly known. Volcanic rocks from the overlying Copper Basin Formation have been dated by Kuniyoshi and Freeman (1974) at 19 to 21 m.y. (K-Ar), or early Miocene. An artiodactyl (giant pig) track found by Kemnitzer (1937) in shale beds of the Gene Canyon Formation implies a late Oligocene to early Miocene age for the unit. A similar Oligocene to middle

Figure 10. Large clast of mylonitic biotite adamellite in monolithologic breccia of Gene Canyon Formation, south of Gene Wash Reservoir. Note strong quartz grain–defined lineation. Width of clast ~20 cm.

Miocene age appears to be applicable to the Artillery Formation in the Rawhide Mountains area (Eberly and Stanley, 1978).

Copper Basin Formation. Overlying the Gene Canyon Formation is the Copper Basin Formation (Ransome, 1931; Kemnitzer, 1937), which consists of red sandstones, conglomerates, and siltstones with interbedded andesitic(?) volcanic rocks. The Copper Basin Formation rests unconformably on Gene Canyon rocks around the eastern and southern margins of the present Whipple Mountains, but was deposited directly on upper-plate crystalline rocks in most areas near the interior of the range (Copper Basin, Bowmans Wash, Whipple Wash areas). Southwest of Copper Basin and in the Aubrey Hills, Arizona, the two formations appear to grade into each other across a section of volcanic rocks whose affinity to either formation is as yet unknown.

Kemnitzer (1937) suggested that the formation is at least 750 m thick, although all sections of the formation have truncated upper boundaries. Fault-bounded sections contain clastic units at least that thick in the southern Whipples. In the central Whipples, volcanic sections 500 to 600 m thick underlie these clastic rocks. The absence of a thick sandstone section in the central Whipples appears to result from erosion rather than nondeposition. Primary sedimentary features such as mudcracks, ripple marks, and cross-bedding are abundant within the formation and suggest a fluvial or fanglomerate environment of deposition for much of the unit. Mylonitic clasts are sporadic within the Copper Basin Formation in the southern and eastern Whipples near the contact with the Gene Canyon Formation. In the area near the Colorado River around Whipple Wash, clasts of all types of mylonitic gneisses are abundant in tilted fanglomerats above the basal Copper Basin(?) volcanic rocks (Lopez, 1980). Mylonitic clasts are also abundant in complexly faulted Copper Basin fanglomerates on the northern edge of the Whipples.

The age of the Copper Basin Formation is not precisely known. Volcanic rocks that are probably interbedded within the formation in the southern Whipples have been dated by Kuniyoshi and Freeman (1974) as 19 to 21 m.y. old. Animal tracks found by both Kemnitzer (1937) and Gassaway (1972, 1977) also suggest a Miocene age for the Copper Basin Formation and the correlative Chapin Wash Formation in the Rawhide Mountains area (Lasky and Webber, 1949).

Volcanic and sedimentary rocks that are probably of Copper Basin age occupy a uniquely autochthonous and allochthonous setting in the Whipple Wash area. Here (Figs. 11, 12) erosion has carved a channel across allochthonous upper-plate rocks, the Whipple detachment surface, and sheared autochthonous rocks below that fault (Figs. 13, 14). Copper Basin(?) strata were then deposited unconformably atop both upper- and lower-plate units. In this area, the young Tertiary sequence is thus autochthonous, but in areas east of the main channel filling in Whipple Wash, the young sequence is itself displaced by listric normal faults that flatten into the original detachment surface (Fig. 13).

Lower-Plate Units

General Statement. Because of the domical form of the Whiple Mountains, autochthonous crystalline rock units below the Whipple detachment surface occupy the core of the range and underlie its highest peaks. Crystalline rocks in the central and eastern core of the range generally resemble the lower-plate rocks of the eastern Buckskin and Rawhide Mountains (Fig. 2) and have the foliated and lineated mylonitic fabric so typical of the Arizona "core complexes" (G. H. Davis, this volume; Rehrig and Reynolds, this volume). Of particular interest, however, is the fact that mylonitic gneisses and related rocks do not compose *all* of the Whipple Mountains autochthon. The crystalline, lower-plate rocks in the western half of the range (west of ~114°27'W long) are a heterogeneous, nonmylonitic assemblage of older gneisses, plutonic rocks (both foliated and nonfoliated), and hypabyssal dikes. The mylonitic and nonmylonitic assemblages are separated in the field by a narrow, west-dipping zone of structural transition. Because it can be demonstrated that the zone represents the mappable upper limit

Figure 11. View to east-northeast down Whipple Wash, northeastern Whipple Mounains, showing autochthonous canyon fill of Miocene volcanic and sedimentary rocks. The canyon was eroded into lower-plate rocks below the Whipple detachment surface, which can be seen as the slightly east-dipping planar contact in the upper right-hand quarter of the photograph.

of a pervasive mylonitic overprinting upon older rocks, it is here termed a "mylonitic front" and is described below.

Older Metamorphic Rocks. Lower-plate rocks that are older than the mylonitic event include a layered gneiss unit, foliated calc-alkaline plutons, and some hypabyssal dikes (to be described in a separate section).

The layered gneiss unit consists of interlayered biotite quartzo-feldspathic gneiss, granitic augen gneiss, and amphibolite. This package is similar to the metamorphic package of the upper plate (pre–quartz monzodiorite and granite porphyry). The quartzo-feldspathic gneiss is fine to medium grained and contains the metamorphic assemblage quartz + plagioclase (oligoclase to andesine) + alkali feldspar + biotite ± epidote, hornblende, and retrograde chlorite. Pelitic interlayers contain quartz + oligoclase + muscovite + biotite + garnet + sillimanite. As shown below, these assemblages persisted through the mylonitic event but are considered relict. At deeper structural levels, the gneiss becomes migmatitic with alternating dark, melanosomal bands of quartz + plagioclase (An_{28-31}) + alkali feldspar + biotite ± hornblende and granitic (leucosomal) veins separated by a mafic selvage rich in biotite and hornblende. Interlayered within the quartzo-feldspathic gneiss are layers of amphibolite and granitic augen gneiss, both commonly 2 to 10 m thick. The augen gneiss contains large (to 5 cm) and abundant (34% to 65%) alkali-feldspar augen set in a medium-grained matrix of two feldspars, biotite, and hornblende. Compositionally (Table 2), the augen gneiss is iron rich (relative to magnesium) and potassic, and thus has affinities to the metagranitic suite of the upper plate.

Plutons of hornblende-biotite quartz diorite, garnet-biotite granodiorite, and two-mica adamellite

Figure 12. Closer view of unconformity seen in Figure 11. Lower-plate rocks are shattered and altered mylonitic gneisses.

intrude these gneisses and share the same foliation. The quartz diorite occurs as lensoidal bodies aligned parallel to the northeast-striking foliation. It is light to dark gray, medium grained, and not well foliated because of low amounts of modal quartz (<10% by volume). A weak foliation is defined by hornblende, biotite, and quartz which flow around the larger crystals (2 to 5 mm) of plagioclase (An_{34-36}). Epidote is abundant and occurs both as small porphyroblasts and as fine-grained granules in the matrix.

The garnet-biotite granodiorite also occurs as a lensoidal body (700 by 2,000 m) in sharp contact with the layered gneiss unit. Dark gray in color, the rock is medium grained with a foliation defined by biotite, feldspar, and quartz which wrap around distinctive crystals (2 to 5 mm) of garnet. Locally, particularly near inclusions of amphibolite, the granodiorite lacks garnet and contains hornblende.

The two-mica adamellite occurs as light-colored, irregular plugs greater than 600 m in diameter. The rock is fine to medium grained and is uniform in composition (71.9% to 73.4% by weight of SiO_2). It consists of quartz, two feldspars (plagioclase and alkali feldspar) and subequal amounts of biotite and muscovite. Although metamorphosed, the plagioclase has vague outlines of relict normal and oscillatory zoning (An_{19-30}). Compositionally, the pluton is low in K_2O (2.43% to 3.66% by weight) and is peraluminous.

Mylonitic Front. The assemblage of gneisses and the two-mica adamellite just described also lie structurally above a thick (>2 km) sequence of mylonitic gneisses and related rocks. Rocks termed "mylonitic" possess a pervasive foliation and lineation. They are characterized by the contrasting brittle and ductile behavior of their mineral components. Whereas quartz, micas, and some amphiboles flowed and/or recrystallized during deformation, other minerals (feldspar, some amphiboles,

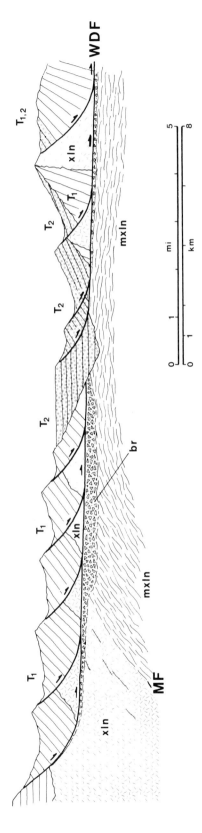

Figure 13. Diagrammatic cross-section across the Whipple Mountains, southeastern California, illustrating simplified middle Miocene geologic relations prior to domal uplift and warping of the Whipple detachment fault (WDF). The cross-section illustrates evidence for two phases of rotational normal fault displacement along the Whipple detachment surface. T_1 = older Tertiary (Gene Canyon) sedimentary and volcanic rocks; T_2 = younger Tertiary (Copper Basin?) sedimentary and volcanic rocks deposited across the Whipple fault prior to their involvement in renewed detachment faulting.

MF = mylonitic front, the mappable contact between undifferentiated lower-plate metamorphic and intrusive rocks (xln) and their largely mylonitic equivalents (mxln); br = breccias developed below the Whipple fault. Horizontal scale is approximate, but break-away zone for detachment faulting on left side of section probably lies in Mopah Range, 25 km to west (see Fig. 2). Vertical dimensions are less accurate and are intended to be diagrammatic only.

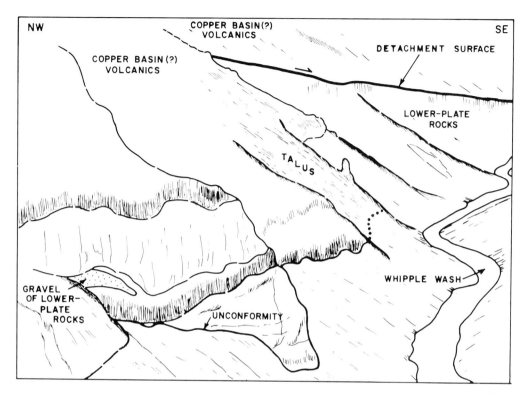

Figure 14. Line drawing of geologic relations shown in Figure 11.

epidote, and sphene) behaved brittlely. The semicataclastic foliation that developed during the mylonitic event generally dips at angles less than 30°. The lineation is defined by flattened and strongly elongated quartz grains and by trains of biotite and broken feldspar porphyroclasts. Quartz-poor rocks such as amphibolite, hornblendite, and quartz diorite do not develop a good mylonitic foliation or lineation.

As discussed below, this mylonitic assemblage includes rock types found above the mylonitic front as well as sill-like granitic plutons and hypabyssal bodies that have not been recognized above the front. The front is a well-defined and mappable structural boundary between the two assemblages. It is not, however, a discrete fault surface. It is a zone of high strain gradient, locally as narrow as 3 m, but more typically as wide as 10 to 30 m. Except for the shattered and altered zone beneath the younger Whipple detachment fault, where its trace is obscured, the front is readily visible in the field as the boundary between inclined, strongly foliated and lineated mylonitic rocks and overlying rocks that are more heterogeneous in appearance and that lack a topography-controlling foliation (Fig. 15).

In secs. 9 and 10, T. 2 N., R. 24 E. (Whipple Mountains S.W., 7½-minute quadrangle), west of and above the front, quartzo-feldspathic gneisses (with subordinate amphibolite) and an intrusive two-mica adamellite pluton share a common foliation that strikes N60° to 85°E and dips 50° ± 15°SE (Fig. 16). This orientation persists eastward to the mylonitic front and is preserved locally in outcrop-sized domains beneath it. In contrast, rocks below the mylonitic front in this area have a penetrative, southwest-plunging lineation and a foliation that strikes northwestward and dips to the southwest at variable angles (15° to 65°; Fig. 16) These penetrative fabric elements are not restricted to rocks below the front, but occur discontinuously and with decreasing frequency upward in all rock types above and within 100 m of the front. More puzzlingly, they are also found in some isolated, fine-grained dikes of

TABLE 2. AVERAGE MAJOR AND TRACE ELEMENT COMPOSITION OF LOWER-PLATE CRYSTALLINE UNITS

	Older Metamorphic Rocks				Synkinematic Sills and Associated Rocks					Dike Rocks	
	agn*	sag	hbd	fad	pgd	gtm	tmt	fgd	ad	and	od
No[+]	1	1	1	3	3	2	2	1	2	5	2
SiO_2	72.65	63.87	60.82	72.65	67.50	71.33	66.63	69.62	73.47	64.32	48.56
TiO_2	0.55	0.97	0.66	0.15	0.43	0.16	0.35	0.35	0.21	0.64	1.37
Al_2O_3	13.19	15.38	16.87	14.74	16.57	15.85	16.98	15.57	14.70	15.42	15.33
FeO[§]	3.15	5.48	5.25	1.29	2.51	1.16	2.87	1.63	0.77	4.01	9.51
MgO	0.70	1.61	2.77	0.44	0.89	0.42	1.25	0.73	0.20	2.31	9.26
MnO	0.018	0.032	0.117	0.029	0.020	0.078	0.065	0.021	0.016	0.064	0.151
CaO	2.11	3.28	5.46	2.34	2.93	2.47	4.05	2.53	1.58	3.84	9.41
Na_2O	2.64	2.80	4.01	3.84	5.00	4.29	4.32	4.78	4.08	3.79	2.95
K_2O	4.55	4.53	2.11	2.86	2.95	3.57	2.04	3.03	4.26	3.47	0.92
Total	99.56	97.95	98.07	98.34	98.80	99.32	98.56	98.26	99.29	97.86	97.46
K_2O+Na_2O	7.19	7.03	6.12	6.70	7.95	7.85	6.36	7.81	8.34	7.26	3.87
$\frac{FeO}{(FeO+MgO)}$	0.818	0.773	0.655	0.746	0.708	0.734	0.698	0.691	0.794	0.635	0.507
"Al"[#]	1.007	0.994	0.897	1.079	0.990	1.029	1.019	0.989	1.036	0.922	0.668
Rb	92.4	121.0	60.0	65.2	57.6	67.2	42.9	66.5	111.0	77.5	17.5
Sr	256	457	919	594	814	540	883	867	360	452	406
Ba	991	2,139	1,065	843	1,210	1,037	1,406	781	689	1,164	408
Ba/Sr	3.87	4.68	1.16	1.42	1.49	1.91	1.62	0.901	1.91	2.57	1.00
Ba/Rb	10.7	17.7	17.8	2.93	21.0	17.9	32.6	11.7	6.21	15.0	23.4
Rb/Sr	0.361	0.265	0.065	0.1098	0.071	0.128	0.049	0.077	0.308	0.172	0.043
K/Rb	409	301	282	364	425	447	380	378	319	372	438

Note: Analyses performed at U.S.C. Petrochemistry Laboratory by Gregory S. Benson, Mark C. Rowley, Whitney Moore, and J. Lawford Anderson. Major elements as wt %. Rb, Sr, Ba as ppm.

*Unit symbols: agn = biotite augen gneiss (above front), sag = mylonitized augen gneiss from Swansea area, hbd = mylonitized hornblende biotite quartz diorite, fad = foliated two-mica adamellite (above front), pgd = porphyritic granodiorite sill, gtm = garnet, two-mica adamellite sill, tmt = two-mica tonalite sill, fgd = mylonitized fine-grained granodiorite dike rock in sills, ad = mylonitized biotite adamellite (clasts from Gene Canyon Formation), and = andesite-dacite dike swarm, and od = olivine tholeiite diabase dike.

[+]Number of analyses averaged.

[§]All Fe as FeO.

[#]"Al" = molecular proportions of $Al_2O_3/(CaO+Na_2O+K_2O)$.

silicic to intermediate composition at structural levels up to 600 or 700 m above the front. The mylonitic foliation and lineation are not seen in the gneissic wallrocks of the dikes, and appear to develop only in those dikes that have shallow dips. G. H. Davis (this volume) reports similar field relations in southeastern Arizona, where a comparable lineation appears locally in shallow-dipping aplite and pegmatite layers enclosed in nonlineated plutonic rocks. In the Whipple Mountains, it seems that factors that inhibit development of the mylonitic fabric in rocks above the front include (1) the existence of an earlier (gneissic) foliation, and (2) an intermediate to steep orientation of that foliation with respect to the front.

The physical means by which the front was formed have not yet been completely resolved. The front is

Figure 15. View to west in south-central Whipple Mountains. Homogeneous-appearing, dark, foliated rocks which dip to left (WSW) in central part of photo are mylonitic gneisses that lie below the mylonitic front (Figs. 5, 16). Rocks above the mylonitic front are heterogeneous-appearing, lighter-shaded units (upper left-hand quarter of photo) that are nonmylonitic, but largely compositionally equivalent to rocks immediately below the front. Light-shaded rocks in lower right foreground are mylonitic porphyritic granodiorite in the upper part of a sill-like pluton (Fig. 16).

the uppermost level of a penetrative mylonitic fabric impressed or superimposed upon pre-existing foliated and nonfoliated rocks (Fig. 17). On the basis of preliminary studies, the development of the mylonitic front seems to be associated with the transposition of older foliation in some rocks (perhaps within zones of high flattening strain), but is more characteristically due to the rotation of pre-existing foliation into parallelism with the front and to pronounced thinning by flattening of the rotated rocks. In some outcrops of a foliated, quartz-rich, two-mica adamellite above the front, the older gneissic foliation is crosscut by the younger mylonitic foliation. As might be expected, these outcrops exhibit a strongly lineated appearance due to intersection of the two S-surfaces (as well as to the closely parallel, quartz-defined, mineral lineation). Transposition of the earlier foliation in gneissic units near the mylonitic front does not, however, seem to be a pervasive process. Rather, the mylonitic lineation in numerous outcrops is overprinted on the older gneissic foliation, but characteristically only when that foliation has been rotated into parallelism with the nearby mylonitic front. This orientation-controlled overprinting can appear or disappear within a few centimetres as the attitude of the older foliation changes out of or back into its original northeast strike and steep to intermediate dips.

Synkinematic Sills and Late Kinematic Plutons. Mylonitization was accompanied in the central, northern, and eastern Whipple Mountains by the intrusion of near-horizontal sheets, or sills, or adamellite to tonalite (Fig. 18) measuring less than a metre to several hundred metres thick. In the southeastern Whipple Mountains, however, such sills are missing in the thick section of mylonitic gneisses. The sills are interpreted as synkinematic because (1) they lie parallel to mylonitic foliation in

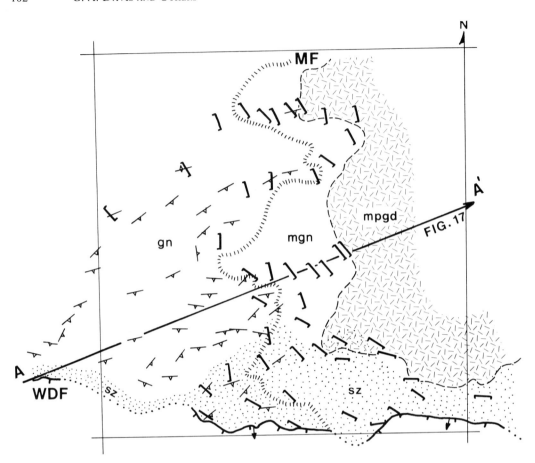

Figure 16. Simplified geologic map of secs. 2, 3, 10, 11, T. 2 N., R. 24 E., Whipple Mountains S.W. 7½-minute quadrangle. Area outlined equals 4 mi². Map shows major geologic relations across west-dipping mylonitic front (MF) in the south-central Whipple Mountains. Units: gn = undifferentiated gneisses and intrusive igneous rocks above mylonitic front; mgn = mylonitized equivalents of gn below mylonitic front; mpgd = upper part of mylonitized porphyritic granodiorite "sill." Pre-mylonitic foliation shown by attitudes with open barbs; younger mylonitic foliation shown by heavier lined attitudes with double dip symbols. Stippled pattern (SZ) represents zone of shearing, shattering, and alteration below south-dipping Whipple detachment fault (WDF). This zone is overprinted across lower-plate rocks both above and below the mylonitic front.

country-rock gneisses, (2) they, and gneisses in contact with them, are more strongly mylonitized than gneisses more distant from the sills, (3) they are cut by mylonitized aplite sheets that have contacts and a parallel mylonitic foliation that is discordant with the mylonitic foliation of the sills, and (4) the pegmatitic interiors of some adamellite sills are only weakly mylonitized.

The mylonitization of pre-sill country rocks is most intense within or on strike with the zone of synkinematic sills. In areas on strike, but lacking the sills, mylonitization is pervasive throughout sections over a kilometre thick. Conversely, in areas containing sills, intense mylonitization is restricted to rocks between the mylonitic front and the top of the highest sill, in the sills themselves, and in immediately adjacent gneisses. Between the sills, and below the lowest sill, the mylonitization lessens and the foliation of the older gneisses has a steeper attitude. Here, the mylonitization occurs only as a moderate to mild overprint.

Each of the major sills is lithologically distinct from the others. The highest sills include two biotite granodiorites, one porphyritic and one not porphyritic. The porphyritic granodiorite is a thick sill

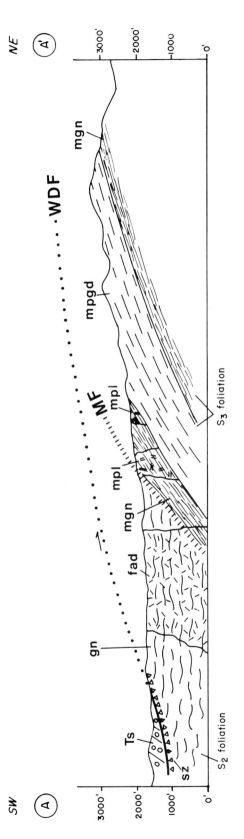

Figure 17. East-northeast–west-southwest cross-section through area of Figure 16 showing geologic relations across Whipple detachment fault (WDF) and mylonitic front (MF) in the south-central Whipple Mountains. Tertiary sedimentary rocks (Ts) lie above the fault. A shear/shatter zone (SZ) lies below it. Lower-plate rocks below the fault, but above the mylonitic front: gn = older layered gneiss; fad = foliated two-mica adamellite (not distinguished on Fig. 16). Rock units below the front: mgn = mylonitic gneisses; mpgd = mylonitic porphyritic granodiorite; mpl = mylonitic plutonic rocks (not distinguished on Fig. 16).

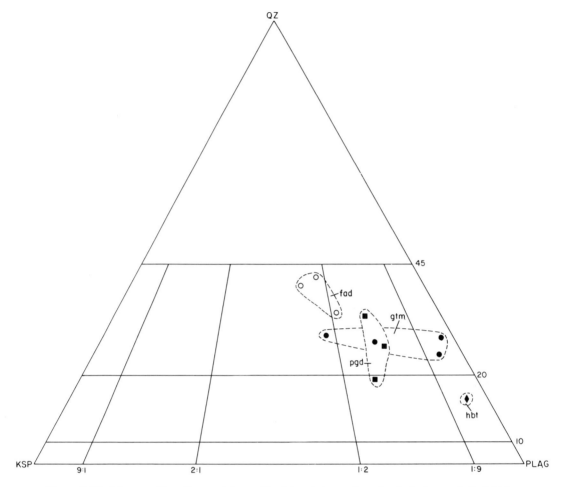

Figure 18. Modal composition of lower-plate granitic plutons in terms of quartz, plagioclase, and alkali feldspar. Pluton symbols are fad = foliated two-mica adamellite (above mylonitic front); hbt = hornblende-biotite tonalite; pgd = porphyritic granodiorite; gtm = garnet, two mica tonalite-adamellite series (last three all below mylonitic front).

(maximum thickness of 500 m) that occurs 250 m below the mylonitic front in the central part of the range (Fig. 17). Compositionally (Table 2), the intrusion is calc-alkaline (Fig. 19), as are all the sills, with intermediate levels of SiO_2 (66.7% to 68.8% by weight) and K_2O (2.73% to 3.25% by weight). The rock is light gray, medium grained, and contains scattered alkali-feldspar augen (0.3 to 1.2 cm, 12% to 18% by volume) set in a foliated matrix of biotite, quartz, plagioclase ($An_{15.8}$ to $An_{25.3}$), and accessory minerals include sphene, allanite, and magnetite (the Fe-Ti oxide, ilmenite, has yet to be found in any of the sills).

The intensity of mylonitization is variable. Where highly mylonitized, feldspars in the foliated groundmass (exclusive of alkali-feldspar augen) display two grain-size distributions. Fine-grained, granoblastic feldspar (mostly plagioclase) coexists with epidote and green-brown biotite. It has an average grain size of 0.01 to 0.02 mm and occurs as tails on larger feldspars and in layers separated by undulatory quartz lenses (0.1 to 1.0 mm in diameter). Crystals of plagioclase (0.5 to 2.0 mm; bent albite twins are common), sphene, brown biotite, and epidote are surrounded by the fine-grained quartz-feldspar matrix. The greener, matrix biotite is compositionally distinct from the coarser, brown biotite

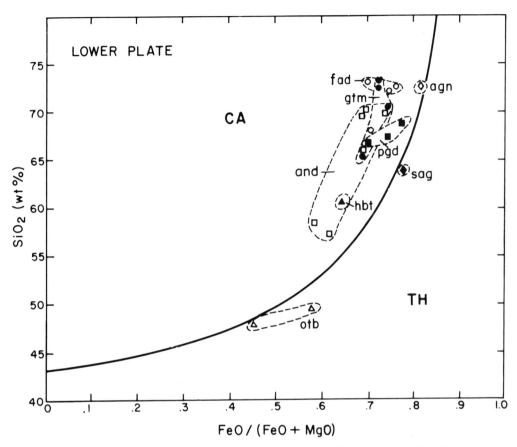

Figure 19. Compositional features of lower-plate crystalline rocks: SiO_2 (wt%) versus $FeO/(FeO + MgO)$. Pluton symbols are the same as in Figure 18 in addition to the following: agn = biotite-hornblende augen gneiss; and = andesite-dacite dike swarm; otb = olivine tholeiite dike swarm (all above mylonitic front); sag = augen gneiss of Swansea area (below mylonitic front).

in being depleted in Ti, Al, and Fe and enriched in Mg and Si. The large epidote crystals have dark-brown, pleochroic, allanitic cores rich in Ce_2O_3 (to 8.1% by weight) and La_2O_3 (to 4.6% by weight). Epidote rims and matrix epidotes are colorless and have a standard epidote composition. Crosscutting granodiorite dikes (also with the same mylonitic fabric) contain textures that are similar but consistent with their hypabyssal occurrence. Large plagioclase porphyroclsts in these dikes are evidently relict phenocrysts, as they contain rather delicate oscillatory zoning (Fig. 20). The zoning reflects changes in composition from $An_{26.6}Ab_{72.1}Or_{1.3}$ to $An_{22.3}Ab_{76.2}Or_{1.5}$; rims have the compositions $An_{25.1}Ab_{73.2}Or_{1.6}$. In contrast, the very fine-grained plagioclase coexisting with epidote in the mylonitized matrix is rather uniform in composition, ranging from $An_{19.6}Ab_{79.1}Or_{1.3}$ to $An_{19.2}Ab_{79.6}Or_{1.2}$. Matrix alkali feldspar has the composition $Or_{92.0}Ab_{7.2}An_{0.1}Cn_{0.7}$.

The biotite granodiorite (nonporphyritic) sill is also thick (600 m), but due to its high structural position, it is widely truncated by the Whipple detachment fault in the central and eastern parts of the range. The rock is whitish to light gray in color, medium grained, and, with a quartz + two-feldspar (plagioclase>alkali feldspar) + biotite mineralogy, it is compositionally similar to the granodiorite described above.

Three structurally lower sills in the eastern Whipple Mountains are all peraluminous and have two micas, biotite and celadonitic muscovite. In descending structural position, these are a two-mica

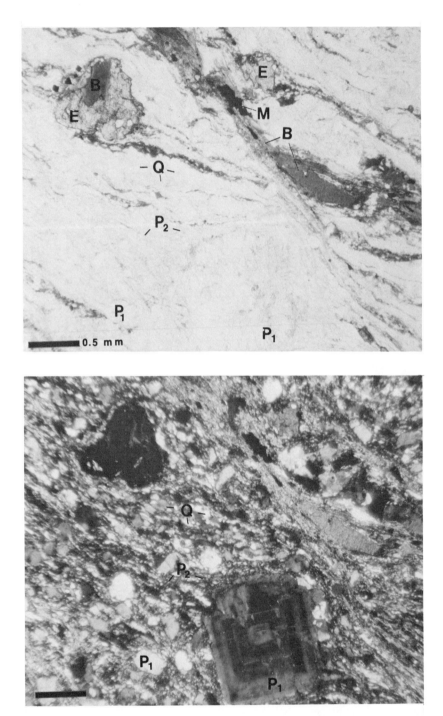

Figure 20. Photomicrograph of specimen from mylonitized granodiorite dike in porphyritic granodiorite sill below mylonitic front. Top, plane light; bottom, same view in cross-polarized light. Note oscillatory zoned plagioclase floating in mylonitic matrix. Compositional data are given in the text. Minerals are P_1 = relict plagioclase, P_2 = metamorphic plagioclase, E = epidote, M = magnetite, Q = quartz, B = biotite. Scale bar represents 0.5 mm.

adamellite, a garnet two-mica adamellite, and a two-mica tonalite. The interior portions of the two adamellite sills (both of which have a maximum thickness of 75 m) are commonly pegmatitic and only weakly foliated, a relation that attests to the coincident timing of intrusion and mylonitization. The adamellites are both light gray, medium grained, and compositionally similar, except that the lower adamellite ranges to granodiorite in composition and contains garnet (to 1.6% by volume). The garnet is principally a mixture of almandine, spessartine, and grossular components. A typical composition is $Al_{43.9}Sp_{22.1}Gr_{23.9}Py_{5.9}An_{4.2}$. The proportion of the two micas is subequal. The muscovite is fairly celadonitic and is discussed in more detail in the following section. As found in all sills analyzed to date, the biotites are uniformly intermediate in composition with typical $Mg/(Mg+Fe)$ ranging from 0.51 to 0.58. A compositional gap was found between porphyroclastic and matrix (mylonitic) plagioclase in one sample, from $An_{21.9}Ab_{77.1}Or_{1.0}$ to $An_{4.1}Ab_{95.1}Or_{0.8}$. Other samples had no compositional gaps, or less well-defined ones.

The lowest sill in the core is a 240-m-thick, dark-gray, two-mica tonalite. It is fine to medium grained and contains porphyroclasts of plagioclase (An_{19} to An_{33}), epidote with dark-brown allanitic cores, two micas (biotite > muscovite), and apatite, set in a fine-grained, foliated matrix of quartz, feldspar, two micas, and epidote. The modal alkali-feldspar content is low (1.8 to 4.0%). Although the tonalite is compositionally distinct from the garnet, two-mica adamellite (SiO_2 = 65.3% to 68.0% and 70.2% to 72.5% by weight tonalite and adamellite, respectively; K_2O = 1.65% to 2.43% and 2.82% to 4.31% by weight; Sr = 833 to 932 and 498 to 582 ppm), some compositional continuity suggests that these intrusive rocks may be comagmatic.

Small plugs of biotite adamellite occur within the lower plate, especially in proximity to the Whipple fault in the southeastern part of the Whipple Mountains. These plutonic rocks are compositionally and mineralogically equivalent to the 74- to 81-m.y.-old biotite adamellite that forms small stocks in the upper plate of the Whipple Mountains. Emplacement of the lower-plate biotite adamellite is considered to be a late kinematic event with respect to mylonitization. These plugs crosscut mylonitic gneisses and, in the southern Whipple Mountains, a mylonitized biotite adamellite sill; nevertheless, they commonly possess a weak to moderately developed foliation and quartz lineation of mylonitic type.

Dike Rocks. A major north-northwest-trending dike swarm of calc-alkaline andesite and dacite and olivine tholeiite diabase intrudes lower-plate rocks above the mylonitic front in the central Whipple Mountains. The andesite and dacite represent a fractionation series (Fig. 21) in which SiO_2 ranges from 56.4% to 70.0% by weight. The rocks range in color from dark gray to light tan and contain 10% to 15% by volume of plagioclase phenocrysts (with oscillatory zoning from An_{34} to An_{41}) and lesser amounts of biotite and/or hornblende, magnetite, and ilmenite in a fine-grained to aphanitic matrix. The more dacitic members contain biotite as the sole mafic mineral. They also have sparse phenocrysts of alkali feldspar and quartz and irregular zones of quartz-feldspar granophyric intergrowths in the matrix. Intermediate rock types have both biotite and hornblende, whereas hornblende is the sole mafic in the andesitic members.

The diabase dikes are olivine tholeiite in composition. They contain a fine-grained, equigranular assemblage of randomly oriented plagioclase laths (An_{34} to An_{64}) and interstitial olivine and brown hornblende. Alkali feldspar and biotite are rare interstitial phases. Clinopyroxene occurs as relict cores in some of the hornblende. Deuteric alteration effects include pale-green actinolite and colorless cummingtonite on the hornblende and olivine. Field studies in eastern parts of the dike swarm above the mylonitic front suggest that at least some of the calc-alkaline members of the swarm predate the mylonitic event. Some dikes of dacitic to andesitic composition in proximity to the front exhibit mylonitic fabric elements although others do not. Although the andesite-dacite series is grossly similar to the synkinematic sills below the front, samples analyzed to date are compositionally different and do not represent the same magmatic episode. At a given silica content, the dike swarm is more potassic (Fig. 21) and poorer in Al_2O_3 and Sr than the sill rocks. The possibility of more than one dike-forming

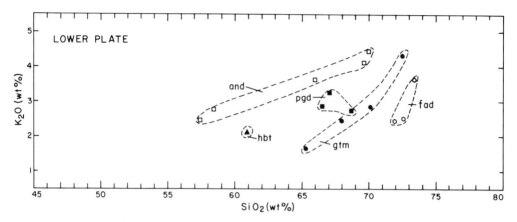

Figure 21. Compositional features of lower-plate crystalline rocks: SiO$_2$ versus K$_2$O (wt%). Pluton symbols same as in Figures 18 and 19.

event is undeniable, yet we hesitate to correlate any of these dike rocks with the Oligocene-Miocene volcanic rocks of the upper-plate Gene Canyon and Copper Basin Formations because of the likelihood of large transport of the upper plate along the Whipple detachment fault.

Depth and Metamorphic Grade of Mylonitization. The occurrence of primary muscovite in the peraluminous sill-like plutons allows an estimate of the minimum pressure and depth of crystallization, and perhaps of mylonitization, by consideration of the interaction of muscovite stability and granitic melt solidus curves in P-T-X space. These relations have recently been discussed by Thompson and Algor (1977). Primary (magmatic) muscovite is rare in epizonal granites because it is not stable at solidus temperatures at low pressure. One estimate of this low-pressure limit is the crossing of the reaction curves defining the wet granite solidus and the upper stability of muscovite + quartz (muscovite [Mu] + quartz [Qz] = alkali feldspar [Ksp] + aluminosilicate [Als] + H$_2$O) (Thompson, 1974; Chatterjee and Johannes, 1974; Skippen, 1977). This occurs at ~3.3 kb (or ~11.8 km). However, correct application of this experimental work requires that one consider the composition of the silicate liquid and solid phases involved. In this case, we are considering melts of tonalite to adamellite composition and a phengitic muscovite with a significant celadonite (K(Mg, Fe^{2+})$_2$Si$_4$O$_{10}$(OH)$_2$) component. The tonalite solidus intersects the muscovite curve at higher pressure (4.2 kb). In contrast, celadonite solution in muscovite increases its stability to higher temperature, which, in turn, causes the intersection with either solidus curve to occur at lower pressure. Moreover, the true lower pressure limit of muscovite + silicate melt occurs at the intersection of the solidus and the discontinuous reaction Mu + Ab + Qz = Ksp + Als + H$_2$O (Evans and Guidotti, 1966; Thompson, 1974), which occurs at about 20 °C lower temperature and 600 bars higher pressure. Electron microprobe analyses of several muscovites in the two-mica, adamellite sill yielded $X_{(KAl_2(Si_3AlO_{10})(OH)_2}$ (mole fraction of end-member muscovite) ranging from 0.700 to 0.807, the remainder being the ferriceladonite component K(Mg,Fe^{2+})(Al,Fe^{3+})Si$_4$O$_{10}$(OH)$_2$.

Although phengitic muscovite is commonly regarded as a product of low-temperature and/or high-pressure conditions (Ernst, 1963; Velde, 1965, 1967, 1972), primary muscovite in sillimanite-free, two-mica granites is commonly phengitic with mole fraction celadonite ranging from 0.11 to 0.29 (Best and others, 1974; Guidotti, 1978; Anderson, unpub. data). To test whether this phengitic composition is primary or due to re-equilibration accompanying mylonitization, we have attempted to fully characterize the compositional variability of muscovite in the mylonitized two-mica sills. Muscovite displays a range of textural habits including coarse-grained flakes that are oblique to the foliation, fine-grained muscovite aligned in the mylonitic matrix, and sericitic muscovite inclusions in feldspar.

Although the mole fraction of celadonite (X_{cel}) in these separate textural habits is remarkably uniform (ranging from 0.21 to 0.28), the major variability is in Ti which systematically varies with texture. Weight percent TiO_2 ranges from 0.63% to 1.19% in coarse muscovite, 0.13% to 0.39% in fine-grained matrix muscovite, and 0.05% to 0.10% in sericite. Guidotti and others (1976) have demonstrated that the Ti-saturation limit in muscovite decreases with lower temperature. In accord with this, we interpret the composition of the matrix muscovite to have formed during mylonitization; the coarse muscovite is regarded as relict and consequently useful for determining minimum depth of emplacement. The fact that the mole fraction of celadonite remained uniform may be indicative of a low-pressure environment during both intrusion and mylonitization (Guidotti and Sassi, 1976).

There is some uncertainty in the calculation of the displacement of the muscovite + quartz terminal-stability curve because of present uncertainty of activity-composition expressions for muscovite-celadonite solutions. The activity of $KAlSi_3O_8$ in alkali feldspar was calculated using Margules parameters of Thompson and Waldbaum (1969). The activity of $KAl_2(Si_3Al)O_{10}(OH)_2$ in muscovite was determined using an ideal multi-site mixing model. For the composition of these phases in the two-mica adamellite, the muscovite + quartz stability curve (using the free energy expression of Skippen, 1977) is shifted to 50 to 55 °C higher temperature at constant pressure. This curve intersects the wet adamellite solidus (Piwinskii, 1968; Boetcher and Wyllie, 1968) at 2.0 kb. A similar shift in the reaction Mu + Ab + Qz = Ksp + Als + H_2O yields a minimum pressure estimate of 2.6 kb or 9.6 km. This is indicative of an upper crustal origin and is currently our best estimate of depth. However, this minimum depth estimate will increase if P_{H_2O} is less than P_{total} and if muscovite in the two-mica tonalite sills is also primary. If the muscovite (X_{cel} = 0.75 to 0.80) in the tonalites is also primary, as in the adamellites, then our preliminary depth estimates increase to 11.8 km. These considerations are being reevaluated in our continuing investigation.

Estimation of metamorphic grade is made difficult by the general lack of thorough reequilibration of minerals during mylonitization. Our data indicate that most of the individual minerals in the mylonitic gneisses are relict from the preceding igneous or metamorphic history (see Fig. 22). The incomplete reequilibration has resulted in compositional dissimilarities or gaps between porphyroclasts and matrix occurrences for plagioclase, epidote, biotite, and muscovite. This important aspect is qualitatively an indication of (1) low temperature, (2) low f_{H_2O}, and/or (3) a restricted time interval for the metamorphic-mylonitic event. To get at the true re-equilibrated or metamorphic phases, we have directed our attention (1) to the very fine-grained (0.01 to 0.1 mm) mylonitic matrix in these rocks (in contrast to the larger crystals floating in this matrix) and (2) to mineralogy within tension gashes that developed perpendicularly to the mylonitic lineation. Our estimates to date exhibit a clear and real range in metamorphic grade from middle greenschist to lower amphibolite. One tension gash specimen contained coarse, prismatic magnesio-hornblendes oriented parallel to the lineation in enclosing rocks and coexisting with quartz + oligoclase ($An_{21.9}Ab_{75.9}Or_{2.2}$) + epidote + magnetite (essentially pure Fe_3O_4) + sphene + late-stage, untwinned orthoclase ($Or_{82.9}Ab_{12.1}An_{0.01}Cn_{4.0}$-structure confirmed by X-ray diffraction). This assemblage is consistent with metamorphic grades from uppermost greenschist to lower amphibolite facies. Feldspar thermometry calculations (Stormer, 1975; Whitney and Stormer, 1977) yield temperature estimates of 527 to 538 °C at an assumed total pressure of 2 to 4 kb. This is our highest estimate of grade.

As described earlier, there are gaps in plagioclase compositions between the grains in the very fine-grained mylonitic matrix and the larger plagioclase crystals enclosed in this matrix. In one sample this gap was from An_{26-22} to An_{18-19}, whereas in another sample the gap was from An_{22} to An_{04}. The gaps are a good indication that the matrix plagioclase has reequilibrated. The matrix mineral assemblage for both samples is the same, being quartz + plagioclase (oligoclase and albite, respectively) + alkali feldspar (Or_{92} and Or_{95}, respectively) + biotite + chlorite + epidote. Epidote, which occurs

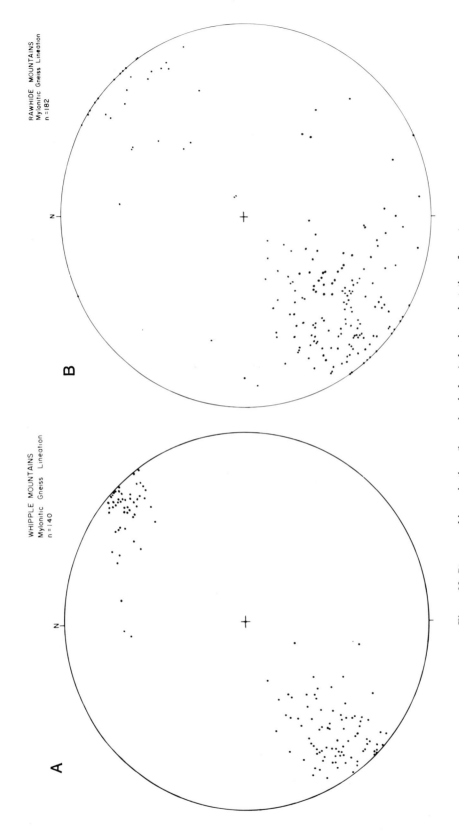

Figure 22. Stereographic projections (lower hemisphere) showing orientation of quartz-defined lineation in mylonitic rocks below detachment surface. (A) Whipple Mountains; (B) Rawhide Mountains.

in the matrix and as overgrowths on relict, rare-earth–enriched allanite, is evidently the by-product of plagioclase retrogression. In the first sample, this occurred at metamorphic grades above the "plagioclase-jump" (uppermost greenschist grade) of Winkler (1976), whereas in the second sample (albite forming), it occurred at lower grades. Feldspar thermometry confirms these conclusions, yielding estimates of 439 to 458 °C and 410 to 430 °C, respectively.

In conclusion, there appears to be a real variation in metamorphic grade during mylonitization. The spatial distribution of grade with structural depth is now being studied. Although there is some uncertainty in the absolute temperature calculations, the differences should be reliable and are consistent with differences in mineral assemblages.

Origin of the Mylonitic Rocks. The mylonitic rocks of the Whipple-Bucksin-Rawhide Mountains are the consequence of a regional metamorphic and deformational event that is as yet poorly dated and incompletely understood. Although the mylonitic rocks of our study area extend southeastward into the Harcuvar and Harquahala ranges of western Arizona, it is not demonstrable (and may never be) that the mylonitic rocks of these areas are continuous at depth with the strikingly similar rocks of south-central and southeastern Arizona (G. H. Davis, this volume). Age data presented here and in papers by G. H. Davis, and Rehrig and Reynolds (this volume) tentatively suggest that the mylonitic rocks of south-central and southeastern Arizona areas may be younger than those of western Arizona and southeastern California; no stronger statement is warranted owing to the sparseness of geochronologic data available from the western areas.

The Whipple Mountains may be unique among the ranges in which mylonitic rocks are known to occur because of the preservation in the Whipples of an original upper boundary to the regional mylonitic assemblage. Mylonitic rocks in the cores of the Arizona ranges are typically truncated upward by low-angle detachment faults or covered by younger deposits. The Whipple mylonitic front (Fig. 15) is startling because the transition between the thick ($>$2.5 km) assemblage of mylonitic rocks and structurally higher, nonmylonitic units is so abrupt (3 to 30 m).

In some parts of the range there is a close spatial association of the mylonitic front with plutonic and hypabyssal sills below the front. Intrusion probably occurred in Late Cretaceous to earliest Tertiary time on the basis of limited age controls and the Laramide age of similar, garnetiferous, two-mica plutons in southwestern Arizona (J. Wright, 1979, personal commun.) In light of field evidence for synkinematic intrusion of the sills during mylonitization, we initially considered the possibility that the location of the front, and therefore the upper limit of pervasive mylonitization, was thermally controlled by the highest level of sill emplacement. Subsequent studies, however, reveal that the sills of the central and eastern Whipple Mountains thin southward and interfinger out into well-developed mylonitic gneisses that are devoid of sills. Therefore, in southern parts of the Whipple Mountains, the apparently requisite thermal control on the upper limit of regional mylonitization cannot be related to sheetlike igneous intrusion.

Nevertheless, as discussed above, sills, where present, did influence the geometry and degree of development of mylonitization in a vertical section. The two strain components of mylonitization, flattening perpendicular to foliation and northeast-southwest extension parallel to the quartz-defined lineation, were enhanced in the hot sills and their thermally weakened gneissic wall rocks. Between such sills the steeply to moderately dipping, pre-existing foliation of Precambrian(?) gneisses and metaplutonic rocks either incompletely rotated into parallelism with the shallowly dipping front or, as above the front, escaped rotation altogether. However, along strike to the south where the sills are absent, mylonitic strain and rotational shallowing of pre-existing foliation developed uniformly throughout a thick ($>$1-km) section of gneisses below the front.

As amplified below, we believe that the Whipple-Buckskin-Rawhide assemblage of mylonitic rocks developed significantly before the deposition of Oligocene(?) and Miocene strata. Thus, we are not inclined to relate the extensional faulting of such strata to deeper crustal extension by semiductile mylonitic flow. If northeast-southwest extension occurred within the crystalline rocks above the mylonitic front in the central Whipple Mountains while it was occurring below by semiductile flow, its mode in the structurally higher rocks is not yet clear to us. Nevertheless, north-northwest-striking dikes (described in a previous section) occur above the mylonitic front and may prove to be temporally and kinematically related to extension beneath it.

PRE–MIDDLE MIOCENE STRUCTURES RELATED TO DETACHMENT FAULTING

Whipple-Buckskin-Rawhide Detachment Surface

The most impressive single structural element of the Whipple-Buckskin-Rawhide Mountains terrane is the detachment surface, a gently dipping and sharply defined fault (or faults) that separates the terrane into upper- and lower-plate assemblages. The surface is spectacularly exposed at numerous localities (Fig. 23). It is clearly visible in Landsat and Skylab photographs as an irregular contact between light-colored, lower-plate rocks (typically pale-green) and darker, upper-plate rocks (typically reddish-brown). It physically resembles the "decollement" of other workers in other areas (for example, the Rincon Mountains near Tucson, G. H. Davis, this volume), although its formation was not stratigraphically controlled. The Whipple fault is the fault recognized by Ransome (1931) and Terry (1972) as the "Bowmans Wash thrust" and the "Whipple Mountains thrust," respectively. The term "Rawhide fault" was introduced by Shackelford (1976) to refer to the master detachment surface in the Rawhide Mountains. We believe it likely that the Whipple and Rawhide faults are simple separate portions of the same regional detachment surface.

The fault is most obvious in the field where Tertiary strata dip discordantly into it (Figs. 3, 4), although it is generally overlain by crystalline rocks. Aphanitic and in some places flinty microbreccias underlie the surface along much of its trace. These finely comminuted rocks form a layer that is typically 2 to 25 cm thick and resistant to weathering and erosion. As a consequence, the fault surface is commonly left exhumed when sheared and brecciated upper-plate rocks are eroded (Fig. 23). The resistant cap of microbreccia (Fig. 24) produces a distinctive ledge immediately below the fault. Recently exhumed exposures of the detachment surface are commonly remarkably planar and typically display either a dark-orange to reddish-brown polish or dull patina. Slickenside striae on the surface are not as common as might be expected, but can generally be found if areas of the planar fault surface are scrutinized closely. Striae on the surface, or within a few centimetres of it, consistently trend northeast—usually within 10° of N50°E.

Although the actual contact between upper and lower plates is knife-sharp, movement along the detachment surface has had a profound effect on rocks below most segments of the fault. Enigmatically, shearing and brecciation of allochthonous rocks above the fault are generally confined to within several metres of the fault surface. Below the fault, however, the effects of faulting can extend downward as far as 250 to 300 m. These effects include the comminution of lower-plate crystalline rocks to fine-grained, structureless cataclasites (for which the protoliths are generally unrecognizable), the shearing and rotation of lower-plate rocks, and in situ shattering. An intense and pervasive alteration occurs throughout the entire structurally disturbed zone. Chlorite and epidote are major secondary minerals that give the rocks of the altered and disturbed zone a dark- to pale-green color. The disturbed zone in the lower plate was also a preferential site for iron and copper sulfide mineralization (at least some of it after detachment faulting), and apparently, at least locally, for the intrusion of mafic dikes (now altered). In some areas (for example, the southeastern corner of the Whipple Mountains S.W. 7½-minute quadrangle) the lower limit of extremely comminuted and altered rocks is

Figure 23. Exhumed detachment surface, extreme southeast corner of Whipple Mountains S.W. 7½-minute quadrangle.

Figure 24. Microbreccia developed directly below detachment surface, north side of Bowmans Wash. Note clasts of lower-plate crystalline rocks in the fine-grained black breccia, and irregular seams of breccia in the underlying crystalline rocks.

a low-angle detachment fault that seems to have formed at the contact between mylonitic gneisses below the Whipple fault and the top of a structurally lower granitic sill. The thickness of the disturbed and altered zone decreases markedly to the west in sec. 10, T. 2 N., R. 24 E. (Figs. 13, 16), where the Whipple fault overlies lower-plate rocks above the mylonitic front. This relation may be explained by a westward decrease (eastward increase) in displacement along the detachment surface. Alternatively, it may be that the penetratively deformed and comminuted mylonitic rocks were more susceptible to brittle deformation beneath the overriding plate than were their more steeply dipping, coarser-grained counterparts above the front. For whatever reason, the disturbed zone beneath the Whipple fault decreases in thickness westward and is locally absent or, at best, only a few metres thick in lower-plate areas west of the mylonitic front.

Suggestions by Davis and Coney (1979) that detachment faults similar to our detachment surface form catastrophically or explosively seem inapplicable to the terrane described here. As discussed elsewhere in this paper, at least two discrete phases of movement can be documented for the Whipple fault in the Whipple Wash area (Fig. 13). Here and at other localities, older Tertiary strata dip more steeply toward the detachment surface than do younger strata. This relation argues for continued sedimentation during the normal faulting of upper-plate strata, since normal faulting is responsible for their southwestward rotation. If, as seems probable, such growth faults flatten into the detachment surface (see next section), the gently dipping surface must also be a growth fault and one with a longevity of displacement equal to that of the upper-plate, listric normal faults (Rehrig and Reynolds, this volume, reach a similar conclusion).

We infer from geologic reconnaissance in the Colorado River trough (ours and that of Southern Pacific Company geologists nearly two decades ago) that the detachment surface is a regional surface that underlies not only parts of the Whipple-Buckskin-Rawhide Mountains but parts of all other ranges to the north shown on Figure 2. Posttectonic warping and doming of the allochthonous terrane have led to the exposure of the detachment surface around the margin of the uplifted ranges. It seems likely that the surface must lie at depth between the areas of uplift, for example, beneath Chemehuevi Wash between the Whipple and Chemehuevi Mountains. Different parts of the surface may have been active at different times, as evidenced by the rejuvenation of the Whipple fault in areas northeast of the central reaches of Whipple Wash (Fig. 13).

Kinematic relations described in detail below indicate that tectonic transport of upper-plate units in the area of Figure 2 was consistently northeastward, irrespective of their position on the various domes or arches. The domes should not be mistaken for relatively minor culminations on an initially curviplanar detachment surface. Structural relief on the Whipple dome is in excess of 2 km, and outward dips of the warped Whipple fault range from 1° to 40°. No evidence has been found that allochthonous units moved radially away from uplifted areas, in contrast to the conclusions drawn by Coney (1974) for units above the Snake Range decollement in the Snake Range, Nevada, and by G. H. Davis (1975) for folded rocks above the Catalina fault in the Rincon Mountains near Tucson.

If our inference of the regionality of the detachment surface in the Colorado River trough is correct, then the problems of how it formed and what localized it are compounded by the geologic variability of the various lower-plate terranes. Reasons for that localization have not been adequately resolved. In the Whipple Mountains the detachment fault developed primarily within crystalline rocks, although local detachment along the unconformity between the basement and its Tertiary cover appears to have occurred during at least two discrete phases of low-angle faulting. The possibility exists that the detachment surface was in part localized in the range by the mylonitic front, which, as the upper limit of the strongly foliated mylonitic assemblage, may have provided a planar anisotropy favorable to detachment. Detachment of crystalline rocks along the mylonitic front may explain why their structurally lower mylonitic equivalents are essentially restricted to the lower plate; to date, mylonitic rocks have been found at only one locality above the Whipple fault.

Other relationships argue against simple detachment of upper-plate units along the mylonitic front—among them the widespread discordance between the Whipple fault and the mylonitic foliation in the lower plate, and the occurrence of upper-plate rocks with a mylonitic overprinting in the Rawhide Mountains. Furthermore, mylonitic boulders in conglomerate units of the Oligocene(?) Gene Canyon Formation indicate that the mylonitic front had been breached locally by erosion prior to Tertiary sedimentation and detachment faulting. Finally, the location of the detachment surface in the western Whipple Mountains and in the Chemehuevi and Eldorado Mountains to the north could not have been controlled by a mylonitic front because mylonitic rocks are absent or uncommon in those areas.

The alteration and mineralization of rocks below the detachment surface (both above and below the mylonitic front) is suggestive of other possible controls on detachment—most obviously, thermal weakening and/or abnormal fluid pressures. Although the concept of high fluid pressures may seem attractive as a means of facilitating upper-plate displacement along the subhorizontal detachment surface (by reducing frictional resistance), the presence of the thick, sheared, and shattered zone beneath the surface is instructive. This zone indicates that stresses engendered by the overriding plate produced deformation far below the detachment surface, and that the fault between upper and lower plates cannot, therefore, be considered as a surface along which frictional resistance was either negligible or absent. Frankly, we do not understand the mechanics of faulting along the Whipple-Buckskin-Rawhide detachment faults.

Internal Structure of Upper and Lower Plates

Although areal differences exist in the Tertiary structure of the Whipple-Buckskin-Rawhide Mountains terrane, some generalizations regarding that structure are possible. High- and low-angle faults are present in both upper and lower plates, although as a rule the former is more faulted than the latter. Normal faults are widespread above the Whipple-Rawhide detachment fault and, surprisingly, below that fault in some areas (for example, in the southeastern Whipple Mountains; Rawhide Mountains).

Most of the steeper faults in both plates strike northwest (N45° ± 15°W) and dip northeastward, but southwest-dipping normal faults are present (more in the Rawhides than the Whipples), as are moderately dipping to low-angle reverse faults (again, more in the Rawhides). Normal faults that extend downward to the detachment surface are in some places truncated by the detachment surface; in other places, the normal faults are flattened to merge with the detachment surface. Other minor normal faults have similar geometric relations to low-angle faults in both the upper and lower plates.

Low-angle faults that are parallel or subparallel to the Whipple-Rawhide fault have been mapped in both plates but are more common in the Rawhide Mountains than in the Whipples. Viewed collectively, faults above and below the detachment fault have developed independently of each other. In other words, the upper- and lower-plate faults are confined to their respective plates and do not cross and offset the detachment fault between those plates. The existence of low-angle and normal faults in the lower plate of the Whipple-Buckskin-Rawhide terrane is significant because it indicates that the regionally extensive Whipple-Rawhide detachment surface is not everywhere the downward limit of Tertiary detachment faulting and upper-crustal distension.

Whipple and Western Buckskin Mountains

Kemnitzer (1937) correctly noted that northwest-striking (N50° ± 10°W), northeast-dipping normal faults control both the structure and topograpy of the eastern Whipple Mountains (Fig. 5). However, he did not clearly recognize his own map relations: namely, that while cutting Tertiary strata and underlying crystalline basement rocks, the normal faults are confined to structural levels above the Whipple fault.

The pronounced structural grain is produced by at least a dozen, major, northwest-trending normal faults. Nearly all these faults dip northeast (65° to 5°, most from 50° to 30°) and have progressively extended the upper plate in a northeast-southwest direction. Some faults exhibit shallower dips at lower structural (and topographic) levels. Small faults that are antithetic to the larger east-dipping structures are fairly widespread. Strike valleys along and between the major faults control the drainage pattern of the area and further accentuate the northwest-trending structural grain. Individual faults, such as the Copper fault (Fig. 25), have exposed lengths in excess of 20 km. If faults that are exposed east of the Osborne Wash Mesa are correlative with faults in the western Buckskin-Whipple Mountains, then these faults have lengths of over 40 km. Compared to their lengths, the widths of individual blocks are relatively thin, averaging only 3 to 4 km. The regularity of spacing between each of the major faults produces a corrugated pattern of alternating Tertiary and crystalline rocks (Fig. 5). Long, rather narrow (0.5- to 2-km) ridges of Tertiary rock are separated from each other by somewhat wider bands of upper-plate crystalline rocks. The major faults are somewhat sinuous in outcrop pattern and are marked by small tectonic slices of both Tertiary and crystalline rocks within re-entrants along the faults. Although the faults branch and interconnect in some places, most fault bocks maintain a fairly uniform width and individual character for their entire exposed lengths. The pattern of faulting, therefore, resembles a tilted deck of cards, each higher card having moved progressively farther away from a starting point (Fig. 13).

The structural style is very consistent between each of the fault blocks in the eastern Whipple Mountains. The eastern edge of most of the Tertiary sections is an unconformity that mantles old topography and contains channels filled with detritus from the underlying crystalline rocks. Beds are

Figure 25. Copper fault south of Colorado River, Arizona. Fault dips northeastward at approximately 30° to 35°.

vertical or steeply overturned eastward within some blocks but, most commonly, dip 30° to 60° to the southwest. These southwest-dipping Tertiary beds are truncated by the northeast-dipping normal faults. Displacement along many of these faults has been great enough to juxtapose the Tertiary rocks of the hanging wall against crystalline rocks in the footwall. Southwestward rotation of the Tertiary rocks during faulting is thus characteristic of most of the fault blocks. Because motion along the northwest-striking normal faults is almost exclusively downdip (Fig. 26), the orientation of tilted Tertiary rocks is another kinematic indicator for the northeast-directed extension in the faulted allochthon (compare Fig. 13). Except for the complicating effects of drag folds, the dip, or amount of rotation, of a Tertiary section seems to be a rough indicator of the amount of displacement of the hanging wall. Gently tilted rocks appear to have moved only short distances, whereas more steeply tilted rocks have moved greater distances. A clear progression of decreasing dip with decreasing age is seen in the Tertiary rocks. Rocks in the Gene Canyon Formation uniformly dip more steeply than rocks in the overlying Copper Basin Formation. This angular relationship can be easily seen in the type areas of the two formations, and was described by Ransome (1931) when he named the two units. Such a systematically variable dip seems to require that faulting and deposition were occurring synchronously. The large, northwest-trending faults should then be regarded as growth faults.

Figure 26. Prominent downdip striae on surface of Spring fault. Exposure seen on north side of road to Havasu Palms Resort.

Although Tertiary strata dip directly into the Whipple fault at many localities, crystalline rocks (largely Precambrian) lie above that fault at most localities (Fig. 5), particularly in the upper reaches of Bowmans Wash and in areas to the west-southwest. Low-angle faults typically separate these allochthonous crystalline rocks from higher, tilted, Tertiary strata. This relationship is so widespread as to suggest that the upper-plate unconformity between Tertiary units and older crystalline basement rocks was itself an important, preferential surface of detachment above the deeper, more throughgoing Whipple fault. Low-angle faults ($>30°$) are also present within the crystalline rocks of both the upper and lower plates, but are most common in the lower plate. They do not always dip in the direction of the detachment surface, but are presumed to reflect broadly synchronous and related deformations.

The principal example of strike-slip faulting found to date in the Whipple Mountains is an upper-plate fault exposed along the road to Havasu Palms, 3.2 km (2 mi) south of the resort (Fig. 5). This subvertical fault strikes N54°E, and its strike-slip displacement is confirmed by fault striae. It appears to be confined to Tertiary and basement units in the hanging wall of a normal-fault block (J. Lopez, 1978, personal commun. and M.S. thesis in progress) and is interpreted as a tear fault developed during differential upper-plate distension. This right-lateral tear fault also seems to be responsible for producing a major flexure in Tertiary strata in the Aubrey Hills, on the Arizona side of Lake Havasu. Because this fault is younger than the northwest-trending faults (several are offset along it), it may have formed during northeastward gravity gliding of rocks on the flanks of the Whipple Mountains dome.

Northwest-trending open folds have been mapped at a number of upper- and lower-plate localities. Some (synclines) are hanging-wall drag folds that developed locally along upper-plate normal faults, and thus formed during upper-plate distension and displacement along the detachment surface. However, one major asymmetric northwest-trending antiform (steep limb to northeast) in lower-plate mylonitic gneisses (southeastern corner of Whipple Mountains S.W. 7½-minute quadrangle) is truncated upward by a flat fault that separates the gneisses from comminuted and altered mylonitic rocks developed below the detachment surface. Here, then, folding of lower-plate rocks predates a phase of low-angle faulting related to the detachment surface.

Collectively, kinematic indicators in the Whipple Mountains (strike of normal faults, sense of rotation of hanging-wall strata, trends of drag folds, orientation of fault striae, and the one tear fault described above) show consistently that displacements along the Whipple fault and synchronous extension in upper-plate units were northeast-directed with trends varying from N40° to 60°E.

Rawhide Mountains

Tertiary fault development was more intensive and geometrically more complicated in the Rawhide Mountains (Fig. 27) than in the Whipple Mountains to the west. This relation is not particularly surprising because the Rawhide portion of the detached terrane is more distal with respect to probable source areas of allochthonous units than the Whipple Mountains. In the Rawhide Mountains a great number of high- and low-angle faults disrupt rocks above and below the Rawhide fault. These faults now dip southwestward at moderate angles and commonly separate major lithologic units of Precambrian(?), Paleozoic, Mesozoic(?), and Tertiary age. Minor structures associated with these early faults indicate northeast-directed movement along the faults. Although some appear to be reverse faults (based on associated drag folds), others are best interpreted as early formed, northeast-dipping, listric normal faults that have been back-rotated to their present position along younger, northeast-dipping listric faults (see Proffett, 1977, for amplification of this phenomenon). Faults with northwest strikes and northeast dips are well represented in the Rawhide Mountains, but they are only one component of a more complex structural assemblage, and they do not control the topographic grain of the area as they do in the Whipple Mountains.

Allochthonous units directly above the Rawhide detachment fault are bounded by a myriad of variably dipping low-angle faults that imbricate and shuffle stratigraphic units into both older-over-younger and younger-over-older configurations. This "chaos" of fault-bounded slices is overlain in areas to the north by a higher plate that contains two major southwest-tilted fault blocks (Figs. 27, 28). In each, Tertiary strata (Artillery and Chapin Wash Formations) overlie crystalline basement rocks of probably Precambrian age. The crystalline units have escaped the regional mylonitization seen in some lower rocks (metavolcanic, metasedimentary) above the Rawhide fault and in almost all rocks below it. The existence of this higher, relatively undeformed plate is anomalous. It appears to represent an overriding allochthon that has traveled farther from a southwestern source area than the chaotically faulted assemblage of, in part, younger upper-plate rocks on which it now rests. The implications of this relationship are discussed in the concluding sections of this paper.

A feature of the structure of the Rawhide Mountains worth emphasizing is the surprising similarity of low- and high-angle faulting in rocks both above and below the Rawhide fault. Low-angle faults with northeast-trending slickenside striae occur in both settings (Fig. 27), not uncommonly with higher normal faults that flatten into them. S-surfaces in the various upper-and lower-plate rock units seem to have been consistently rotated as the consequence of similar patterns of internal fault displacements. For example, the mean attitude of foliation in lower-plate mylonitic gneisses is N46°W, 26°SW. In Paleozoic and Tertiary (Artillery Formation) rocks above the Rawhide fault, the mean bedding orientations are N42°W, 36°SW and N41°W, 40°SW, respectively. The mean orientation for all S-surfaces in the Rawhide Mountains is N40°W, 36°SW, an attitude strikingly compatible with the prevailing N50°E direction of rotational tilting and tectonic transport throughout the area. This direction is indicated by the strike of most faults, by the consistent northeast-southwest orientation of slickenside striae, and by other mesoscopic Tertiary structural elements—including boudinage, small folds, and the intersection direction of conjugate normal faults. The orientation of minor fold axes and the sense of vergence on these folds define two slip planes (N36°W, 20°NE, and N32°W, 30°W) with slip lines plunging N54°E, 20° and S58°W, 30°, respectively. The N54°E, 20° slip line is the more strongly developed. The presence of two slip planes is probably largely the result of conjugate normal faulting during distension and the complex, often chaotic, imbrication of rock units in the upper plate.

Whatever the cause of Tertiary detachment and extensional tectonics in the Whipple-Buckskin-Rawhide Mountains terrane, it is abundantly clear and certainly important that in the Rawhide Mountains the effects of this deformation were not limited to allochthonous units above the Rawhide fault. Underlying mylonitic rocks were similarly affected in in their upper levels are also either allochthonous or para-autochthonous.

POST DETACHMENT UNITS

Osborne Wash Formation

Unconformably overlying the deformed Gene Canyon and Copper Basin sedimentary rocks, as well as older crystalline rocks and metasedimentary rocks, is the Osborne Wash Formation. It is composed of olivine-bearing basalt flows interbedded with agglomerate, tuff, and alluvial fan deposits; intrusive dikes and plugs are also present. Alluvial fan deposits and air-fall tuffs at the base of the Osborne Wash Formation have filled old valleys with a relief of at least 300 m. North-trending feeder dikes are abundant in the northern Buckskin Mountains and cut all underlying rocks. Interbedded flows and alluvial deposits of the formation are more than 400 m thick and cover a nearly contiguous area of about 200 km^2, called "The Mesa," in the western Buckskin Mountains. Within the southern Whipple Mountains, a few thin basalt flows are present, but most of the preserved formation consists of well-indurated fanglomerates which crop out in areas of subdued topography and steep-walled arroyos.

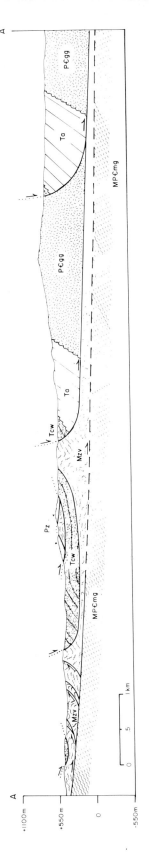

Figure 28. Cross-section through northern Rawhide Mountains. Symbols are the same as in Figure 27.

The widespread occurrence of the Osborne Wash Formation in the Whipple and Buckskin Mountains suggests that the present-day level of exposure is very similar to the level of exposure at the time the basal Osborne units were being deposited. Fanglomerates of the Osborne Wash Formation form extensive deposits in the southern and southeastern Whipples and crop out within a few hundred metres of the detachment surface. In the southern Buckskin Mountains (Mammon Mine area), Osborne Wash flows were deposited directly on the detachment surface.

Osborne Wash flows and alluvial units appear to be large undeformed, except by small faults probably related to compaction and intrusive activity within the volcanic sequence. A very uniform regional dip of 5° to 8° to the southwest is in marked contrast to the steep dips of the underlying faulted units. Major, northwest-trending faults in the upper plate are unconformably overlain by the Osborne Wash Formation and exhibit no post–Osborne Wash reactivation throughout most of the area. However, along the low hills south of the Bill Williams River, 5 km southeast of Lake Havasu, Osborne Wash flows are offset ~250 m across a northwest-trending normal fault. Osborne Wash feeder dikes are also cut by the Havasu Springs fault near the Central Arizona Project pumping plant at the southern end of Lake Havasu. Although much of the faulting ceased before Osborne Wash time, some movement occurred along the normal faults during the Pliocene.

The age of the Osborne Wash Formation is suggested by a 13.5 ± 1.0-m.y. (K-Ar, whole-rock) date on one of the basalt flows in the lower alluvial unit (Kuniyoshi and Freeman, 1974). Ages for similar "mesa-capping" basalts west of the Rawhide Mountains are 13.1 ± 0.2, 13.0 ± 0.5, and 9.2 ± 0.2 m.y. (Suneson and Lucchitta, 1979). The hypersthene-olivine Cobwebb basalt of Lasky and Webber (1949) with an age of 13 ± 2.1 m.y. (Eberly and Stanley, 1978) also seems correlative. Best and Brimhall (1974) have studied similar rocks to the north and suggest that they were derived from depths of 65 to 95 km. Leeman (1974) studied the strontium isotopic composition of these same rocks and found values for most to range from 0.7029 to 0.7041; this fact supports the mantle origin suggested by Best and Brimhall (1974). The Osborne Wash flows and alluvial gravels seem to fit into a regional picture that indicates a profound change in the type of volcanism and associated tectonic environment that occurred near the end of the Miocene.

Bouse Formation

The Pliocene Bouse Formation originally described by Metzger (1968) crops out discontinuously in many of the lower areas near the Colorado River (Smith, 1970; Winterer, 1975). These deposits formed in a marine embayment of the Gulf of California. The formation consists of marine (oldest) to brackish lacustrine (youngest) deposits of clay, silt, and sand. A near basal tufa bed crops out as a bright white band around many of the hills near Parker, around the southern and northern Whipples, along Lake Havasu, and according to Gassaway (1977), south of the Rawhide Mountains. The formation has a gentle regional southward dip of as yet undetermined origin. East of Lake Havasu about 12 km north of Parker Dam, the Bouse Formation dips at 15° to 20° to the southwest, apparently along the major northwest-trending Havasu Springs fault. Motion along this fault obviously did not cease until after Bouse time (middle or late Pliocene).

CONCLUDING REMARKS ON TECTONIC HISTORY

The "metamorphic core complexes" of the southwestern United States present a number of problems to those who wish to study them. However, most attention seems to have been focused on the nature and origin of three typical "core-complex" elements: (1) the mylonitic gneisses and related rocks of the complexes; (2) the domal uplifts in which mylonitic "core" rocks are exposed; and (3) the low-angle detachment faults that flank the uplifted terranes. Some clear differences in interpretation regarding the temporal and kinematic relations between these different elements are evident in the papers of this volume, particularly with respect to the "core complexes" of southeastern California and western and southern Arizona. Workers in what might be informally regarded as the "Arizona school," including G. H. Davis, W. A. Rehrig, S. J. Reynolds, and P. J. Coney, are developing tectonic models that regard the three aforementioned structural elements as forming in a space-time continuum. Although their views differ in detail (compare papers of G. H. Davis, and Rehrig and Reynolds, this volume), the contributors to this "school" believe that a genetic relationship exists between mylonitization of crystalline rocks and higher level detachment (decollement) faulting. Development of the mylonitic basement terranes is regarded as, at least in part, older than the detachment of higher allochthonous sheets, but kinematically associated with it. In essence, upper-crustal extension by high- to low-angle normal faulting is thought to occur in response to deeper, more ductile crustal extension by mylonitic flow. Doming for various reasons (megaboudins, Davis, 1977 and this volume; differential isostatic uplift, Rehrig and Reynolds, this volume) facilitates detachment faulting by the attenuation of rocks across the crest of the developing domes (or boudins), or by providing slopes down which allochthonous units move (Davis and Coney, 1979).

Our studies to date of the detached terrane along the California-Arizona boundary yield markedly different conclusions. We are unable at this stage of our investigations to discern a fundamental interrelationship between the formation of the mylonitic terranes, detachment faulting, and doming associated with the uplift of individual ranges. For reasons amplified below, we believe that the detachment faulting in the Colorado River trough postdated significantly the formation of mylonitic rocks, but predated the arching and doming that led, through erosion, to exposure of the regional detachment surface and underlying rocks (both mylonitic and nonmylonitic).

The strikingly similar directions of northeast-southwest extension in mylonitic rocks (by semiductile flow) and generally higher units (by normal and low-angle faulting) is perplexing, yet field relations in the Whipple-Buckskin-Rawhide Mountains part of the detached terrane indicate to us that regional mylonitization significantly predated the Oligocene(?)-Miocene detachment events. The mylonitic rocks in our study area may have been subject to major erosion prior to Oligocene(?) or Miocene sedimentation and faulting (Fig. 13). Mylonitic cobbles and boulders of lower-plate rocks occur abundantly in some of the Oligocene(?) and Miocene fanglomerate units of the southern and northern Whipple Mountains (both Gene Canyon and Copper Basin Formations). Sphenes from a mylonitic gneiss in one such boulder (Fig. 10) have yielded an 82.9 ± 3 (2σ)-m.y.-old fission-track age (Dokka and Lingrey, 1979), which may be relict from the parent biotite adamellite. A mylonitic gneiss from the lower plate of the Rawhide Mountains has yielded semiconcordant hornblende and biotite K-Ar ages of 57.4 and 52.3 m.y., respectively (Shackelford, 1977; Rehrig and Reynolds, this volume).

The regional detachment fault in the Whipple-Buckskin-Rawhide Mountains cuts with pronounced discordance across tilted and folded mylonitic gneisses with pronounced discordance at many localities. These gneisses were shattered and sheared for distances as great as several hundred metres below the Whipple detachment surface in the Whipple Mountains. High- to low-angle detachment faults of the types seen in upper-plate position within the detachment terrane are also well developed in the lower-plate mylonitic rocks of the terrane. The mylonitic rocks are not restricted to positions below the Whipple and Rawhide faults; metavolcanic rocks that experienced the mylonitic deformational event occur widely above the Rawhide fault, and mylonitized granitic rocks have now been recognized at one locality above the Whipple fault. Collectively, these field relations indicate that the mylonitic rocks, at least at levels near the Whipple detachment surface, were kinematically "dead" and cold, or cooling, at the onset of detachment faulting. The occurrence of mylonitic boulders in the earliest Tertiary deposits in the area indicates that the mylonitic front in the Whipple Mountains had been breached, at least locally, by erosion before the deposition of the Gene Canyon Formation. This explains why the

Oligocene(?)-Miocene volcanic and sedimentary units in the Colorado River trough did not experience either metamorphism or mylonitization; they are too young to have done so.

It is important to stress that low-angle faulting and northeastward extension of upper-plate rocks were not restricted in the Colorado trough to areas of mylonitic deformation immediately below the regional detachment fault. This is documented in the central Whipple Mountains where nonmylonitic, *lower-plate* gneisses and plutonic rocks are separated from structurally lower mylonitic equivalents by the abrupt mylonitic front (Figs. 15, 16, 17), and in the Chemehuevi Mountains (Fig. 2) where a nonmylonitized Mesozoic(?) quartz diorite comprises most of the lower plate. Might lower-plate extension synchronous with upper-plate normal faulting have occurred by means other than mylonitic flow? Anderson (1971) postulated that extreme "thin-skin" distension at shallow crustal levels in the Eldorado Mountains was compensated at depth by the intrusion of Miocene plutons. But, although present in the Eldorado Mountains, Miocene plutons have not been recognized in the lower plates of the detached terrane farther south, and if present, they cannot be extensively developed.

The concept of megaboudins developed by G. H. Davis (1977 and this volume) to explain interrelationships between crustal mylonitic deformation, doming, and high-level detachment faulting is provocative, and a concept that may prove applicable to the southeastern Arizona area for which it was developed. We have considerable difficulty, however, in applying the concept to the area of the Colorado River trough for a number of reasons. Perhaps most important is the consistent northeastward direction of transport of allochthonous units above the regional detachment surfaces in the Sacramento, Chemehuevi, Whipple, Buckskin, Rawhide, and Artillery Mountains (Fig. 2). One measure of this consistency is the characteristic northwest strike and southwest dip of Tertiary strata in the upper plate, strata that have been rotated along northeast-dipping listric normal faults (see generalized bedding attitude symbols, Fig. 2). The domed or arched configuration of the ranges listed above is the consequence of posttectonic uplift that has warped the regional detachment surfaces and associated structures. There is clearly not present in the area represented by Figure 2 a consistent regional orientation of the axes of these domes or arches, as one might expect if they are expressions of crustal megaboudins related to the northeast-southwest extensional directions of either the mylonitic rocks or the low-angle detachment terranes. Furthermore, the late-formed domes and arches of the Colorado trough ranges seem to be geographically independent of whether or not mylonitic rocks occur in their uplifted cores. Whatever the range-forming process in this area—and we admit to being puzzled by it—it seems to be the last major structural event of the region. That event seems to us to be temporally, spatially, and kinematically independent of either the earlier mylonitization of crustal rocks or the younger development of regional detachment faults across them.

The origin of the Colorado River trough structures and their relationship, if any, to conventional basin-range faulting in the western United States is also unclear. Mountain ranges in the area of Figure 2 are definitely not outlined by range-front faults of the type seen in the Great Basin. It is tempting to speculate that the low-angle fault structures of the area may represent the "bottoming-out" of typical basin-range faults into a transitional zone between an upper level of brittle deformation and deeper levels of crustal flow. However, field relations do not support this speculation (and seem to refute it). As just described, the Oligocene(?)-Miocene extension of upper-plate units seems to have been unaccompanied by regional and coeval extensional phenomena in the crystalline rocks of the lower plates.

Furthermore, the probable level of erosion in the Colorado River trough seems to be much too shallow (<5 km, <3 mi) to have exposed a brittle-ductile transitional zone of the Great Basin type postulated by Proffett (1977), among others. Several lines of evidence show that the detachment surface in the Whipple Mountains formed at very shallow crustal levels: (1) The stratigraphic thickness of rotated fault blocks—Tertiary strata plus Precambrian crystalline basement—does not apparently exceed 5 km. (2) Miocene erosion that occurred between two phases of movement on the Whipple fault locally exhumed and breached that fault prior to deposition of younger, now partly allochthonous strata (Figs. 11, 12, 13, 14). (3) Middle Miocene fanglomerate deposits and interbedded basalt flows (Osborne Wash Formation) partly buried the tilted fault blocks within 3 or 4 m.y. (maximum) after their last movements; these posttectonic surficial deposits locally lie on or only several hundred metres above the detachment surface, and relations (both field and temporal) indicate that the amount of erosion between faulting and fanglomerate deposition was not great.

It is likely that extensional faulting in the Colorado River trough predated, perhaps significantly, the major period of crustal extension by normal faulting in Great Basin areas to the north. A genetic relationship between the two faulted terranes cannot at present be demonstrated.

The origin of the low-angle faults of the Colorado River trough remains obscure. Our studies indicate the detachment of thin upper-crustal slabs beneath the area and their synchronous internal distension by normal, principally listric, faulting. Regional northeastward gravity sliding of the detached rocks is indicated (a conclusion first reached by Shackelford, 1975), as we have found no evidence that lower-plate rocks experienced a synchronous, regional extension. A hypothesis of regional gravity sliding (as opposed to one of in situ crustal extension) requires the existence of a slide toe (or toes) where the base of the sliding allochthon breaks through to the Earth's surface and rides out across it. We believe that candidate "toes" exist.

The Artillery Mountains lie east of the Rawhide Mountains (Fig. 2) and are the locale of the Artillery "thrust fault" of Lasky and Webber (1949). This southwest-dipping low-angle fault has placed Precambrian(?) augen gneiss atop a sedimentary and volcanic section now known to be as young as middle Miocene (on the basis of a 16.2-m.y.-old basalt flow in the lower plate, K-Ar, whole rock, Shackelford, 1976; Rehrig and Reynolds, this volume). The "thrust" plate may represent the leading edge of the Buckskin-Rawhide slide complex as it moved northeastward into a Miocene sedimentary basin (Shackelford, 1976).

In the Rawhide Mountains, geologic relations suggest that a still higher slab within the detached terrane may have overridden allochthonous units that overlie the Rawhide fault. Such overriding appears to represent major tectonic foreshortening, or telescoping, within the slide complex and could be another expression of a toe. The uppermost allochthonous slab in the Rawhide Mountains (Figs. 27, 28) consists of Tertiary strata deposited unconformably on Precambrian basement rocks. The basement rocks do not have a mylonitic overprint and hence were derived from a source terrane that lay above the mylonitic front of the region. From kinematic studies, we know that such a source terrane lay somewhere to the southwest. This upper slab, or plate, lies atop a chaotic assemblage of allochthonous units that includes Precambrian or Mesozoic metavolcanic and Paleozoic metasedimentary rocks. These units have been affected (like those of the lower plate) by metamorphism and deformation accompanying the mylonite-forming event. Their source terrane was located, therefore, below or astride the mylonitic front and included a stratigraphic section much more complete than that found in the higher plate. This source area also lay to the southwest, but not as far in that direction as the source terrane for the higher Tertiary-Precambrian nonmylonitic allochthon. We conclude, therefore, that the more coherent highest plate in the Rawhide Mountains has traveled farther northeastward than the allochthonous rocks it overrode for a distance of at least 10 km. These relations suggest that the higher plate broke through to the surface somewhere to the west and subsequently overrode allochthonous units emplaced by earlier transport along the Rawhide detachment fault. Such a process is indicative of major telescoping between elements of the detached terrane, not distension.

The not uncommon juxtaposition of older units above younger units along low-angle faults in the Rawhide and Artillery Mountains is in marked contrast to the consistent younger-over-older fault geometries in the Whipple Mountains to the west. These geometric differences are compatible with a regional gravity sliding model in which the Whipple Mountains (and the Mopah Mountains which lie immediately to the west) are a distending source area and the Rawhide-Artillery Mountains a more distal, partly foreshortened portion of the slide complex. The head, or breakaway zone, of the crustal slide complex is apparently defined by northeast-dipping normal faults in the Mopah Range (Fig. 2). The breadth of this detached terrane is at least 100 km, and northeastward displacements of upper-plate units within it must exceed several tens of kilometres in eastern areas. This estimate is supported by the mismatch of upper- and lower-plate plutonic rocks in the Whipple Mountains and by the problems of source area for allochthonous units in the Rawhide Mountains that rest on stratigraphically younger rocks.

In closing, we reiterate that the opinions and conclusions offered here may change as our field and laboratory studies progress. Our current impression is that different parts of the southern Nevada, southeastern California, and western and southern Arizona zone of "metamorphic core complexes" (Fig. 1) seem to have evolved differently and with differing chronologies. We are impressed (and puzzled) with what seem to be genuine differences between the metamorphic and tectonic evolution of "core complexes" and detachment terranes in the Colorado River trough area and in the southeastern Arizona region centered around Tucson (G. H. Davis, this volume).

ACKNOWLEDGMENTS

We thank the editors of this volume for encouraging us to submit, at an inconveniently late date for them, this manuscript. Our field-based investigations in the Whipple-Buckskin-Rawhide Mountains detached terrane have been generously supported by the geology program of the National Science Foundation Grants EAR77-09695 and GA-43309. In addition to studies by the authors, these grants have supported theses by Karl Evans, Steve Lingrey, John Lopez, and James Podruski, all completed or to be completed in 1979, and the ongoing investigations of Valerie Krass, Gary Osborne, Mark Rowley, Linda Thurn, and Robert Woodward. Paul Adams was helpful with some of the electron microprobe analyses. Greg Benson, Whitney Moore, and Mark Rowley all helped with the whole-rock analyses. Karl Frost provided able assistance in the field. We are appreciative of the individual and collective contributions of these students to our continually evolving understanding of the detachment terrane. Keith Howard and the U.S. Geological Survey kindly provided instrumental time for some of our petrological investigations. Finally, we are indebted to Keith Howard, Max Crittenden, and John Rodgers for their helpful reviews of this paper.

REFERENCES CITED

Anderson, J. L., and Cullers, R. L., 1978, Geochemistry and evolution of the Wolf River batholith, a late Precambrian rapakivi massif in north Wisconsin, U.S.A.: Precambrian Research, v. 7, p. 287–324.

Anderson, J. L., Podruski, J. A., and Rowley, M. C., 1979, Petrological studies in the "suprastructural" and "infrastructural" crystalline rocks of the Whipple Mountains of southeastern California: Geological Society of America Abstracts with Programs, v. 11, no. 3, p. 66.

Anderson, R. E., 1971, Thin-skinned distension in Tertiary rocks of southeastern Nevada: Geological Society of America Bulletin, v. 82, p. 43–538.

Armstrong, R. L., 1972, Low-angle (denudation) faults, hinterland of the Sevier orogenic belt, eastern and western Utah: Geological Society of America Bulletin, v. 83, p. 1729–1754.

Best, M. G., and Brimhall, W. H., 1974, Late Cenozoic alkalic basaltic magmas in the western Colorado Plateau and the Basin and Range transition zone, U.S.A., and their bearing on mantle dynamics: Geological Society of America Bulletin, v. 85, p. 1677–1690.

Best, M. G., and others, 1974, Mica granites of the Kern Mountains pluton eastern White Pine County, Nevada: Remobilized basement of the Cordilleran miogeosyncline?: Geological Society of America Bulletin, v. 85, p. 1277–1286.

Boetcher, A. L., and Wyllie, P. J., 1968, Melting of granite with excess water to 30 kilobars: Journal of

Geology, v. 76, p. 235–244.
Bull, W. B., and Ku, T. L., 1975, Age dating of the late Cenozoic deposits in the vicinity of the Vidal nuclear generating station site: Consulting report for Southern California Edison, Los Angeles, Woodword-Clyde Consultants, 38 p.
Carr, W. J., and Dickey, D. D., 1977, Cenozoic tectonics of eastern Mojave Desert: U.S. Geological Survey Professional Paper 1000, p. 75.
Chatterjee, N. D., and Johannes, W., 1974, Thermal stability and standard thermodynamic properties of synthetic $2M_1$-muscovite, $KAl_2(AlSi_3O_{10}(OH)_2)$: Contributions to Mineralogy and Petrology, v. 48, p. 89–114.
Coney, P. J., 1974, Structural analysis of the Snake Range decollement, east-central Nevada: Geological Society of America Bulletin, v. 85, p. 973–978.
Coonrad, W. L., 1960, Geology and mineral resources of Township 6 North, Ranges 23 and 24 East, San Bernardino base and meridian, San Bernardino County, California: Report submitted to Southern Pacific Land Company, 23 p.
Davis, G. A., and others, 1977, Enigmatic Miocene low-angle faulting, southeastern California and west-central Arizona—Suprastructural tectonics(?): Geological Society of America Abstracts with Programs, v. 9, no. 7, p. 943–944.
Davis, G. H., 1975, Gravity-induced folding off a gneiss dome complex, Rincon Mountains, Arizona: Geological Society of America Bulletin, v. 86, p. 979–990.
——1977, Characteristics of metamorphic core complexes, southern Arizona: Geological Society of America Abstracts with Programs, v. 9, no. 7, p. 944.
——1980, Structural characteristics of metamorphic core complexes, southern Arizona: Geological Society of America Memoir 153 (this volume).
Davis, G. H., and Coney, P. J., 1979, Geologic development of the Cordilleran metamorphic core complexes: Geology, v. 7, p. 120–124.
Dokka, R. K., and Lingrey, S. H., 1979, Fission track evidence for a Miocene cooling event, Whipple Mountains, southeastern California, in Armentrout, J. M., and others, eds., Cenozoic paleogeography of the western United States: Society of Economic Paleontologists and Mineralogists, Pacific Section, p. 141–145.
Eberly, L. D., and Stanley, T. B., Jr., 1978, Cenozoic stratigraphy and geologic history of southwestern Arizona: Geological Society of America Bulletin, v. 89, p. 921–940.
Ernst, W. G., 1963, Significance of phengitic micas from low-grade schists: American Mineralogist, v. 48, p. 1357–1373.
Evans, B. W., and Guidotti, C. V., 1966, The sillimanite-potash feldspar isograd in western Maine, U.S.A.: Contributions to Mineralogy and Petrology, v. 12, p. 25–62.
Gassaway, J. S., 1972, Geology of the Lincoln Ranch basin, Buckskin Mountains, Yuma County, Arizona [undergraduate thesis]: San Diego, California, San Diego State University, 64 p.
——1977, A reconnaissance study of Cenozoic geology in west-central Arizona [M.S. thesis]: San Diego, California, San Diego State University, 120 p.
Guidotti, C. V., 1978, Muscovite and K-feldspar from two-mica adamellite in northwestern Maine: Composition and petrogenetic implications: American Mineralogist, v. 63, p. 750–753.
Guidotti, C. V., and Sassi, F. P., 1976, Muscovite as a petrogenetic indicator mineral in pelitic schists: Neues Jahrbuch für Mineralogie Abhandlungen, v. 127, p. 97–142.
Guidotti, C. V., Cheney, J. T., and Guggenheim, S., 1976, Distribution of titanium between coexisting muscovite and biotite in pelitic schists from northwestern Maine: American Mineralogist, v. 62, p. 438–448.
Hamilton, W. B., 1964, Geologic map of the Big Maria Mountains NE quadrangle, Riverside County, California and Yuma County, Arizona: U.S. Geological Survey Geologic Quadrangle Map GQ-350.
——1971, Tectonic framework of southeastern California: Geological Society of America Abstracts with Programs, v. 3, no. 2, p. 130–131.
Hose, R. K., and Danes, Z. F., 1973, Development of the Late Mesozoic to Early Cenozoic structures in the eastern Great Basin, in De Jong, K. A., and Scholton, R., eds., Gravity and tectonics: New York, John Wiley & Sons, Inc., p. 429–442.
Kemnitzer, L. E., 1937, Structural studies in the Whipple Mountains, southeastern California [Ph.D. dissert.]: Pasadena, California Institute of Technology, 150 p.
Kuniyoshi, S., and Freeman, T., 1974, Potassium-argon ages of Tertiary volcanic rocks from the eastern Mojave Desert: Geological Society of America Abstracts with Programs, v. 6, no. 3, p. 204.
Lasky, S. G., and Webber, B. N., 1949, Manganese resources of the Artillery Mountains region, Mohave County, Arizona: U.S. Geological Survey Bulletin 961, 86 p.
Leeman, W. P., 1974, Late Cenozoic alkalic basalt from the western Grand Canyon area, Utah and Arizona: Isotopic composition of strontium: Geological Society of America Bulletin, v. 82, p. 1691–1696.
Lingrey, S. H., Evans, K. V., and Davis, G. A., 1977, Tertiary denudational faulting, Whipple Mountains, southeastern San Bernardino County, California: Geological Society of America Abstracts with Programs, v. 9, no. 4, p. 454–455.
Lopez, J. A., 1980, Geology of the northeast Whipple Mountains: Lake Havasu City South and Whipple

Wash Quadrangles [M.S. thesis]: Los Angeles, University of Southern California.

Lucchitta, I., and Suneson, N., 1977, Cenozoic volcanism and tectonism west-central Arizona: Geological Society of America Abstracts with Programs, v. 9, no. 4, p. 457-458.

McNair, A. H., 1951, Paleozoic stratigraphy of part of northwestern Arizona: American Association of Petroleum Geologists Bulletin, v. 35, p. 503-541.

Metzger, D. G., 1968, The Bouse Formation (Pliocene) of the Parker-Blythe-Cibola area, Arizona and California: U.S. Geological Survey Professional Paper 600-D, p. D126-D136.

Miller, F. K., 1970, Geologic map of the Quartzite quadrangle, Yuma County, Arizona: U.S. Geological Survey Geologic Quadrangle Map GQ-841.

Misch, P., 1960, Regional structural reconnaissance in central-northeast Nevada and some adjacent areas: Observations and interpretations: Intermountain Association of Petroleum Geologists 11th Annual Field Conference Guidebook, p. 17-42.

——— 1971, Geotectonic implications of Mesozoic decollement thrusting in parts of eastern Great Basin: Geological Society of America Abstracts with Programs, v. 3, no. 2, p. 164-166.

Miyashiro, A., 1974, Volcanic rock series in island arcs and active continental margins: American Journal of Science, v. 274, p. 321-355.

Piwinskii, A. J., 1968, Experimental studies of igneous rock series: Central Sierra Nevada batholith, California: Journal of Geology, v. 76, p. 549-570.

Podruski, J. A., 1979, Petrology of the upper plate crystalline complex in the eastern Whipple Mountains, San Bernardino County, California [M.S. thesis]: Los Angeles, University of Southern California, 193 p.

Proffett, J. M., Jr., 1977, Cenozoic geology of the Yerington district, Nevada, and implications for nature and origin of basin and range faulting: Geological Society of America Bulletin, v. 88, p. 247-266.

Ransome, F. L., 1931, Geological reconnaissance of the revised Parker route through the Whipple Mountains: Consulting report to the Metropolitan Water District of Southern California, 10 p.

Rehrig, W. A., and Reynolds, S. J., 1977, A northwest zone of metamorphic core complexes in Arizona: Geological Society of America Abstracts with Programs, v. 9, no. 7, p. 1139.

——— 1980, Geologic and geochronologic reconnaissance of a northwest-trending zone of metamorphic core complexes in southern and western Arizona: Geological Society of America Memoir 153 (this volume).

Reynolds, M. W., 1974, Geology of the Grapevine Mountains, Death Valley: Summary, in Guidebook, Death Valley Region, California and Nevada: Shoshone, California, Death Valley Publishing Co., p. 91-97.

Shackelford, T. J., 1975, Late Tertiary gravity sliding in the Rawhide Mountains, western Arizona: Geological Society of America Abstracts with Programs, v. 7, no. 3, p. 372-373.

——— 1976, Structural geology of the Rawhide Mountains, Mohave County, Arizona [Ph.D. dissert.]: Los Angeles, University of Southern California, 175 p.

——— 1977, Late Tertiary tectonic denudation of a Mesozoic(?) gneiss complex, Rawhide Mountains, Arizona: Geological Society of America Abstracts with Programs, v. 9, no. 7, p. 1169.

Sherborne, J. E., Jr., and others, 1979, Major uranium discovery in volcaniclastic sediments Basin and Range province, Yavapai County, Arizona: American Association of Petroleum Geologists Bulletin, v. 63, p. 621-646.

Silver, L. T., and others, 1977, The 1.4-1.5-b.y. transcontinental anorogenic plutonic perforation of North America: Geological Society of America Abstracts with Programs, v. 9, no. 7, p. 1176-1177.

Skippen, G. B., 1977, Dehydration and decarbonation equilibria, in Greenwood, H. J., ed., Application of thermodynamics to petrology and ore deposits, Mineralogical Association of Canada Short Course: Toronto, Canada, Evergreen Press, p. 66-83.

Smith, P. B., 1970, New evidence for Pliocene marine embayment along the lower Colorado River area, California and Arizona: Geological Society of America Bulletin, v. 81, p. 1411-1420.

Smith, R. B., 1978, Seismicity, crustal structure and intraplate tectonics of the interior of the western Cordillera, in Smith, R. B., and Eaton, G. P., eds., Cenozoic tectonics and regional geophysics of the western Cordillera: Geological Society of America Memoir 152, p. 111-144.

Stormer, J. C., Jr., 1975, A practical two feldspar geothermometer: American Mineralogist, v. 60, p. 667-674.

Suneson, N., and Lucchitta, I., 1979, K-Ar ages of Cenozoic volcanic rocks west-central Arizona: Isochron/West, no. 24, p. 25-29.

Terry, A. H., 1972, The geology of the Whipple Mountains thrust fault, southeastern California [M.S. thesis]: San Diego, California, San Diego State University, 90 p.

Thompson, A. B., 1974, Calculation of muscovite-paragonite-alkali feldspar phase relations: Contributions to Mineralogy and Petrology, v. 44, p. 173-194.

Thompson, A. B., and Algor, J. R., 1977, Model systems for anatexis in pelitic rocks, I. Theory of melting reactions in the system $KAlO_2$-$NaAlO_2$-Al_2O_3-SiO_2-H_2O: Contributions to Mineralogy and

Petrology, v. 63, p. 247–269.
Thompson, J. B., Jr., and Waldbaum, D. R., 1969, Mixing properties of sanidine crystalline solution, III. Calculations based on two phase data: American Mineralogist, v. 54, p. 811–838.
Varga, R. J., 1977, Geology of the Socorro Peak area, western Harquahala Mountains: Arizona Bureau of Mines Circular 20.
Velde, B., 1965, Phengitic micas: Synthesis, stability, and natural occurrence: American Journal of Science, v. 263, p. 886–913.
——1967, Si^{+4} content of natural phengites: Contributions to Mineralogy and Petrology, v. 14, p. 250–58.
——1972, Celadonite mica: Solid solution and stability: Contributions to Mineralogy and Petrology, v. 37, p. 235–247.
Whitney, J. A., and Stormer, J. C., Jr., 1977, The distribution of $NaAlSi_3O_8$ between coexisting microcline and plagioclase and its effect on geothermometric calculations: American Mineralogist, v. 62, p. 687–691.
Wilson, E. D., 1960, Geologic map of Yuma County, Arizona: Arizona Bureau of Mines.
——1962, A resume of the geology of Arizona: Arizona Bureau of Mines Bulletin 171, 140 p.
Wilson, E. D., and Moore, R. T., 1959, Geologic map of Mohave County, Arizona: Arizona Bureau of Mines.
Winkler, H.G.F., 1976, Petrogenesis of metamorphic rocks: New York, Springer-Verlag, Inc., 334 p.
Winterer, J. I., 1975, Biostratigraphy of Bouse Formation: A Pliocene Gulf of California deposit in California, Arizona, and Nevada [M.S. thesis]: Long Beach, California State University.

MANUSCRIPT RECEIVED BY THE SOCIETY JUNE 21, 1979
MANUSCRIPT ACCEPTED AUGUST 7, 1979

Printed in U.S.A.

Geologic and geochronologic reconnaissance of a northwest-trending zone of metamorphic core complexes in southern and western Arizona

WILLIAM A. REHRIG
Conoco, Inc.
555 17th Street
Denver, Colorado 80202

STEPHEN J. REYNOLDS
Department of Geosciences
University of Arizona
Tucson, Arizona 85721

ABSTRACT

Reconnaissance mapping indicates that parts of nine mountain ranges previously considered to be Precambrian basement are instead variations of Tertiary metamorphic core complexes. From southeast to northwest, these ranges include the Pinaleno, Picacho, South Mountains, parts of the Buckeye, White Tank, Harquahala, Harcuvar, Buckskin, and Rawhide Mountains. Together with the already recognized Santa Catalina–Rincon–Tortolita complex, these ranges define a broad northwest-trending belt through Arizona.

The northeast-trending Buckskin-Harcuvar-Harquahala Mountains are transverse foliation arches, the latest expression of a huge, northwest-elongated metamorphic area herein named the Harcuvar metamorphic core complex. This Tertiary phenomenon is superimposed on an ill-defined center of late Mesozoic metamorphism. Traverses into the complex from its unmetamorphosed southwestern margin reveal progressive Cretaceous conversion of Mesozoic sedimentary rocks into migmatites. Metamorphism just preceded intrusion of the Tank Pass batholith, an Upper Cretaceous pluton which itself became foliated and involved in early Tertiary migmatization and intrusion in the Harcuvar Mountains.

A marginal zone of penetrative mylonitization, capped by a more brittlely deformed dislocation surface, flanks the Harcuvar complex on its upper and broadly arcuate northeastern margins. Resting on this tectonic surface are highly tilted, unmetamorphosed, layered rocks (Paleozoic to Tertiary). Geochronologic and geologic data place the time of mylonitization as Tertiary, perhaps as recently as 25 to 20 m.y. B.P. This deformation (flattening and northeast-southeast extension) was closely followed by development of chlorite breccia, the dislocation surface, thick wedges of coarse clastic sediment, and listric faulting. Finally, the core complex was arched and uplifted.

A model for this sequence of events is predicated on mobile northeast-directed extension of a flat upper-crustal layer facilitated by intense mid-Tertiary plutonism in an actively tensile stress field. Tectonism in a brittle, surficial upper plate is governed by listric faulting and detachment as the plate fragments and extends "piggyback" style upon subjacent, ductilely stretched layer.

INTRODUCTION

Distributed throughout the interior of the North American Cordillera from Canada to Mexico is a zone of unique metamorphic centers that have been referred to as "metamorphic core complexes" (Crittenden and others, 1978). Lithologic, metamorphic, structural, and temporal characteristics between widely separated complexes are sufficiently similar to warrant their isolation for study as an intriguing and fundamental tectonic unit of the Cordillera. These complexes typically reveal prolonged plutonic, metamorphic, and deformational histories, which culminated in Tertiary time. The latest events in the complexes are particularly interesting and have a profound regional significance that is only now becoming appreciated.

The complexes we will discuss are contained within a northwest-trending zone through southern and western Arizona (Fig. 1). Additional complexes which exist southwest of this belt have been studied and described by Davis (1977 and this volume). The intent of this paper is to briefly characterize the metamorphic core complexes within the northwest-trending belt in southern Arizona in terms of distribution, physiography, rock types, structure, and geochronology. One particular example, the Harcuvar metamorphic core complex of western Arizona, will then be singled out for more comprehensive description. Finally, we will speculate on matters of mechanics and origin.

REGIONAL SETTING

The metamorphic core complexes in Arizona are located within the Basin and Range physiographic province (Wilson and Moore, 1959; Hayes, 1969). Exposures in the province consist of Precambrian granitic and metamorphic basement, upper Precambrian to Paleozoic stratified rocks, and an assortment of Mesozoic-Cenozoic sedimentary, volcanic, plutonic, and metamorphic rocks.

The region was tectonically active intermittently between 1.8 and 1.4 b.y. B.P (Anderson and Silver, 1976), but became more stable by later Precambrian time (Wilson, 1962). During the Paleozoic, southern Arizona was evidently *entirely* within the continental cratonic environment of thin Paleozoic sedimentation (McKee, 1951; Bryant, 1968; Peirce, 1976). This Paleozoic tectonic setting differs markedly from that of metamorphic core complexes north of Arizona which were emplaced in regions that were transitional between craton and ocean basin during Paleozoic time.

During Mesozoic time, southern Arizona was the site of magmatism, metamorphism, tectonic unrest, and the accumulation of locally thick sections of stratified rock. A Triassic(?)-Jurassic volcanic arc trended northwest through southern Arizona (Hayes and Drewes, 1968; Coney, this volume). In the Early Cretaceous, southeast Arizona was the site of clastic and carbonate deposition that formed the Bisbee Group (Hayes, 1969). In addition, extremely thick sections of impure arkose of Mesozoic age are widespread throughout western Arizona and attest to unstable tectonic conditions (Miller, 1970; Harding, 1978; Rehrig and Reynolds, this paper; S. Marshak, 1978, oral commun.).

Laramide time (80 to 50 m.y. B.P.) was characterized by compressional deformation (Drewes, 1978; Davis, 1979), plutonism, and volcanism (Damon and Mauger, 1966). During the Eocene, widespread

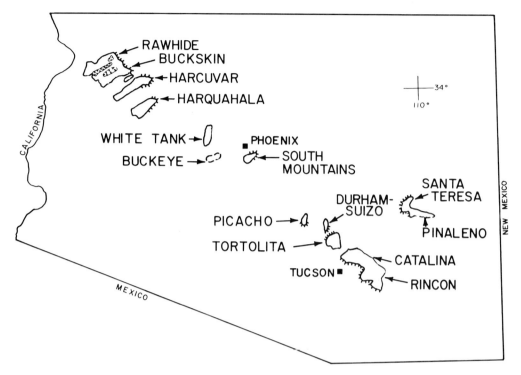

Figure 1. Generalized outlines of metamorphic core complexes in a northwest-trending zone in southern Arizona. Dislocation surfaces flanking parts of the core complexes are shown by hachures.

erosion and sparse continental sedimentation may have occurred over most of the region (Eberly and Stanley, 1978; H. W. Peirce, 1978, oral commun.). During Oligocene to Miocene time, tectonism, plutonism, metamorphism, silicic volcanism (mostly ignimbrites), and continental sedimentation dominated southern Arizona (Shafiqullah and others, 1976a; Eberly and Stanley, 1978; Coney and Reynolds, 1977; Keith, 1978). Basin-range block-faulting, which formed the present-day ranges (Eberly and Stanley, 1978), is superimposed on the mid-Tertiary events.

GENERAL DESCRIPTION

During the last several years, our reconnaissance mapping and regional tectonic studies have led us to a familiarity with most of the metamorphic core complexes of the western United States. Much of this work has been carried out in close association with researchers at the University of Arizona. Isotopic dating at the Laboratory of Isotope Geochemistry, University of Arizona, has been carried out in many of these metamorphic teranes, and this has led to a much better understanding of the problematical areas considered herein. The data collected during these joint studies allow the following general descriptions to be made regarding metamorphic core complexes in the southwestern United States.

Physiography

In most cases a distinctive, broad, bulbous or domical physiography characterizes the mountain ranges of a metamorphic core complex. Individual ranges tend to have long, flat-topped topographic profiles and elongate northeast-, north-, or northwest-trending plan outlines. Most basin-range high-angle faulting evidently postdates the development of these archlike features, but only in a few areas (for example, in the Santa Catalina–Rincon–Tortolita and Durham-Suizo terranes) does the high-angle faulting seriously disrupt the arch morphology.

Another common but by no means universal property of the metamorphic terranes is their high elevation. The Santa Catalina–Rincon–Tortolita and Pinaleno–Santa Teresa Mountains reach elevations of about 3,000 m and are among the highest in the Basin and Range province of the state. The Picacho, White Tank, Harcuvar, and Harquahala Mountains exceed 1,500 m in height and conspicuously rise above adjacent valley basins and nearby ranges. In contrast, the Tortolita and South Mountains and the Durham and Buckeye Hills have relatively low relief.

Lithologic and Structural Characteristics

Although the rocks and structures associated with metamorphic core complexes vary systematically both laterally and vertically, certain distinctive ones are commonly present (Fig. 2). These variations are not always exposed in a single complex, but may be reconstructed by examining different structural levels from several complexes.

Rocks exposed in the structurally deepest parts of the complexes are either amphibolite-grade gneisses or variably deformed plutonic rocks. The gneisses are commonly ductilely deformed and locally exhibit passive-flow to flexural-flow folds (Fig. 3B) and a weakly developed mineral lineation. The plutonic rocks range in composition from muscovite-garnet–bearing granite, alaskite, and pegmatite to a broad spectrum of quartz diorite to granite (igneous rock nomenclature in this paper is that adopted by the International Union of Geological Sciences; see Streckeisen, 1976). Rocks in the deeper structural levels are significantly less mylonitic than rocks at higher levels. This is best seen in the South Mountains (Reynolds and Rehrig, this volume) and has been documented in complexes outside Arizona (Waters and Krauskopf, 1941; Snook, 1965; Chase, 1973; Compton and others, 1977).

Upward, both igneous and metamorphic rocks of the core progressively exhibit increasingly mylonitic fabrics.[1] These fabrics are manifest in a characteristically moderate- to low-angle foliation (Fig. 3C) and consistently oriented mineral lineation and pervasive smearing out of mineral grains. The presence of this mylonitic fabric, in forms which vary from the subtlest foliation in granitic hosts to dark, finely laminated mylonite, is one of the most distinctive characteristics of the metamorphic core complexes that we describe.

Typically on one or several sides of an individual complex, uppermost parts of the mylonitic gneisses become increasingly chloritic and brecciated up structural section. Within the upward transition to a structureless breccia, the well-developed mylonitic foliation becomes progressively disrupted and destroyed (Fig. 3D). The chloritic breccia (or microbreccia) is overlain by a gently dipping surface of dislocation[2] (Fig. 3E). The breccia may attain a 50- to 100-m thickness, but in places is very thin; this nearly results in direct truncation of the most intensely mylonitic rocks by the dislocation surface.

Overlying the dislocation surface, in stark contrast to the brecciated and mylonitic rocks of the footwall, are tilted allochthonous rocks of highly variable age (Precambrian to Miocene) and type. Moderately thick sections of Tertiary conglomerate and sandstone are the most common upper-plate

---

[1]The term "mylonitic" is used strictly in a descriptive, nongenetic sense. In thin-section, rocks described by this term exhibit generally comminuted feldspars but recrystallized quartz. The rocks are very similar to the photographs of hand specimens and photomicrographs of protomylonite, mylonite, and mylonite gneiss presented in the classification of Higgins (1971).

[2]The term "dislocation surface" is intended to be purely descriptive. It signifies dislocation between rocks above and below the surface. No genetic connotation is attached to the term, and large amounts of movement are not necessarily implied by use of the term.

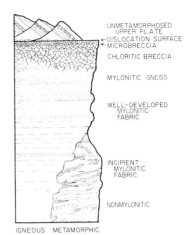

Figure 2. Schematic structural section on certain sides of metamorphic core complexes.

rocks. These rocks are typically not metamorphosed and are devoid of the distinctive mylonitic fabric. They are, however, cut by a myriad of faults that have various orientations, show generally normal separation (Fig. 3F), and merge downward into a basal dislocation surface.

The basal dislocation surface is an integral and important structural entity of metamorphic core complexes and locally has a large areal extent (Shackelford, 1976; Davis and others, 1977; Davis, this volume; Davis and others, this volume). In most Arizona complexes, this surface does not form an outer structural boundary on all sides of the complex. Instead, a diffuse mylonitic transition defines some of the margins; along these margins, the upper limit of the mylonitic zone (in contrast to the abrupt dislocation-surface contact) is represented simply by a gradational transition to nonmylonitic rocks at higher levels.

Geochronology

The metamorphic core areas of the Arizona complexes repeatedly yield K-Ar biotite dates of between 20 and 30 m.y. B.P. This was first discovered in the Catalina Mountains (Damon and others, 1963; Creasey and others, 1977), but is consistently supported by data from other complexes, including those described in this paper. The interpretation of these dates varies; some workers (Shakel, 1974) have maintained that the mid-Tertiary dates are only cooling ages having little to do with major pre-Tertiary development of the complexes. Others (Creasey and others, 1977) have argued that the ages date nearly synchronous plutonism and intense deformation (mylonitization), which are inferred to be the main events.

A primary intent of both this paper and our accompanying South Mountains paper (Reynolds and Rehrig, this volume) is to constrain the numerous temporal uncertainties regarding the timing of one or more metamorphic, plutonic, and structural events. Geochronologic data and interpretations are included below under discussion of individual complexes.

PINALENO-SANTA TERESA MOUNTAINS

The easternmost currently identified terrane of metamorphic core complexes in Arizona (Rehrig and Reynolds, 1977) is exposed in the Pinaleno and Santa Teresa Mountains (Fig. 1). The Pinaleno Mountains are a high range (several peaks over 3,000 m); the Santa Teresa Mountains are lower, but equally as rugged. Rocks and structures typical of metamorphic core complexes occur mostly in the

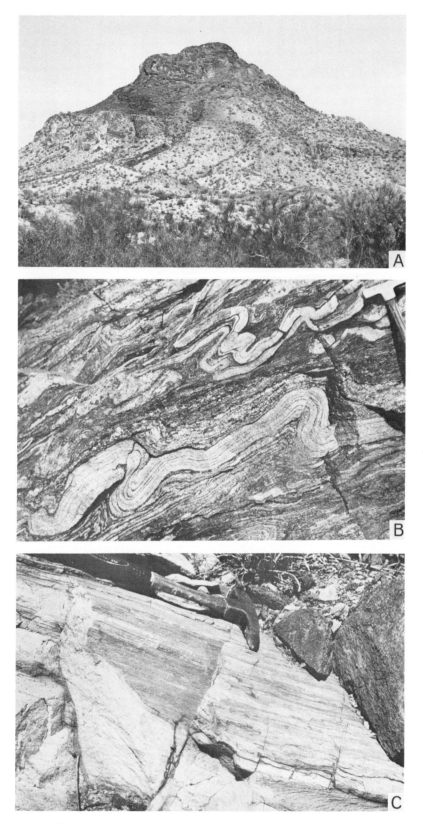

Figure 3. Photographs of typical rock types and structures as depicted in Figure 2. (A) Exposure in Rawhide Mountains of mylonitic gneissic basement (foreground), chloritic breccia ledge, dislocation surface, and faulted upper-plate rocks (the darker rocks in the upper part of hill). (B) Metamorphic basement of Harcuvar Mountains; note hammer in upper corner for scale. (C) Mylonitic foliation, Pinaleno Mountains; note character of pronounced lineation.

Figure 3 (continued). (D) Typical chlorite breccia below the dislocation surface. (E) Dislocation surface, Rawhide Mountains; surface is top of ledge; below ledge is chlorite breccia and above ledge are upper-plate Paleozoic carbonate rocks. (F) Normal faults in Miocene upper-plate sandstones, Rawhide Mountains.

northern part of the Pinaleno Mountains and the southern and eastern parts of the Santa Teresa Mountains. Much of the central Pinaleno Mountains is composed of the relatively undeformed, ~1.4-b.y.-old Pinaleno granite which intrudes older Precambrian gneisses and schists that generally possess a north- to east-striking steep foliation (Swan, 1976). The granite has been affected by later thermal events, as evidenced in part by an 874-m.y. K-Ar apparent age obtained from biotite (Table 1). Along the northeastern edge of the range (see Swan, 1976) are gneisses that possess a gently dipping (mostly to the north and northeast) cataclastic or mylonitic foliation and pronounced northeast-trending lineation. Protoliths of the mylonitic rocks include the Pinaleno granite and older gneisses (G. H. Davis, 1978, oral commun.; S. B. Keith, 1978, oral commun.), small mafic intrusions (now amphibolites), and leucocratic, equigranular to pegmatitic intrusions. A 28.3-m.y. whole-rock K-Ar age (Table 1) on an ultramylonitic derivative of probable Precambrian basement (M. Swan and S. Keith, 1978, oral commun.) suggests that some mid-Tertiary thermal and deformational activity affected these rocks.

Farther northwest in the region between the Pinaleno and Santa Teresa Mountains (Fig. 1), a granite possibly equivalent to the Pinaleno granite is overlain by a gently west-dipping dislocation surface, above which lie mid-Tertiary volcanic rocks and a several-kilometre-thick section of fanglomerate that strike northwest and dip steeply to moderately to the southwest (Blacet and Miller, 1978). A mylonitic zone is exposed in the footwall of the dislocation surface. A northeast-trending swarm of felsic dikes dissects the basement, but is truncated abruptly by the dislocation surface. One of these dikes yielded a K-Ar biotite date of 24.7 m.y. B.P. (Table 1), which together with Oligocene dates on volcanic rocks (Creasey and Krieger, 1978) correlative with those in the upper plate indicates Oligocene or younger activity along the dislocation surface. Slickensides associated with the surface suggest movement toward the southwest or northeast.

Farther north in the Santa Teresa Mountains near Jackson Mountain, quartzitic, metamorphic, and granitic rocks possess a moderate to gently dipping mylonitic foliation and northeast-trending lineation (G. H. Davis, 1977; 1978, oral commun.; this volume). These mylonitic rocks are overlain by a low-angle dislocation surface above which are exposed mid-Tertiary sedimentary and volcanic rocks (Blacet and Miller, 1978). Orientation of the dislocation surface (Blacet and Miller, 1978) suggests that it is synformal with a northeast-trending axis and probably extended over a larger area than current exposure would indicate. The upper-plate Tertiary rocks are cut by northwest- to west-striking normal faults that fail to cut the basal dislocation surface (Blacet and Miller, 1978). No isotopic dates are reported for the lower-plate foliated rocks, but the undeformed Goodwin Canyon Quartz Monzonite (Simons, 1964) a short distance to the northwest has yielded a K-Ar biotite date of 24.9 m.y. B.P. (Table 1). The adjacent Santa Teresa Granite (Simons, 1964) sends out fine-grained dikes that cut dikes presumably of Oligocene "turkey track" andesite; these relationships suggest that the Santa Teresa Granite is also of mid-Tertiary age (W. A. Rehrig, unpub. data). Therefore, there is evidence for considerable mid-Tertiary plutonism adjacent to exposures of the mylonitic rocks.

In summary, several points should be emphasized. Lower-plate mylonitic rocks have yielded a late Oligocene cooling age, and plutons of possibly similar age are present adjacent to or within the complex. In two areas, the mylonitic gneisses are overlain by a low-angle dislocation surface which displaces rocks as young as late Oligocene. In both exposures, the surface is synformal along a northeast-trending axis and undoubtedly once extended over a large area. Tertiary beds above the fault strike northwest, orthogonal to the northeast trend of lineation below the dislocation surface.

SANTA CATALINA–RINCON–TORTOLITA MOUNTAINS

The Santa Catalina–Rincon–Tortolita Mountains—referred to collectively herein as the Catalina complex—have received repeated study, principally by researchers at the University of Arizona (see references in Budden, 1976; Davis, 1975 and this volume; Keith and others, this volume). In addition, the U.S. Geological Survey has carried out mapping and radiometric dating throughout the Catalina complex (Creasey and others, 1977; Drewes, 1974, 1977; Banks, this volume). Our comments are necessarily brief and general because the complex is discussed elsewhere in this volume (see the papers by Banks, Davis, and Keith and others).

The Catalina complex consists of several broad arches of mylonitic rocks surrounded by a terrane that consists of less-metamorphosed rocks of variable type including Precambrian granitic rocks, upper Precambrian–Paleozoic strata, sparse Mesozoic strata, and thick sequences of Oligocene-Miocene clastic and volcanic rocks (Davis, this volume). The boundary between mylonitic rocks and less-metamorphosed rocks is commonly a low-angle dislocation surface. Our discussion will be concerned with mylonitically foliated basement of the Catalina Mountains.

Structurally lowest in the Catalina Mountains (Shakel, 1974; Creasey and others, 1977; Banks, this volume) is a biotitic, mylonitic gneiss, parts of which were originally Precambrian granite (Creasey and others, 1977; Shakel and others, 1977). The gneiss and abundant interlayered alaskite and pegmatite generally exhibit a low-angle, mylonitic foliation and conspicuous northeast-trending lineation. The mylonitic gneiss–pegmatite–alaskite sequence is overlain by a batholithic sill(?) of muscovite-bearing granite which is locally more than 5 km thick. This pluton is at least partly of early Tertiary age (Shakel and others, 1977; Keith and others, this volume) and is locally mylonitic, with foliation being best developed in *lower parts* of the batholith and progressively decreasing in intensity up structural section until the rock is undeformed. The granite sill is only locally foliated higher in the section, especially along its contacts with other rock types. At the top of the batholith, pegmatite dikes probably related to the granite intrude Precambrian-Paleozoic rocks and a quartz diorite (Banks, this volume) of possible Late Cretaceous–early Tertiary age. Overlying the quartz diorite pluton are Precambrian-Paleozoic strata that become less metamorphosed up section to the north.

Along the southern flank of the overall west-northwest–trending Catalina arch, mylonitic gneisses are locally chloritic, brecciated, and overlain by a low-angle dislocation surface, above which lie tilted

TABLE 1. RADIOMETRIC AGE DATES

Sample	Rock type	General location	Mineral dated	Apparent age (m.y.)
1	Rhyodacite dike	Pinaleno Mtns.	Biotite	24.7 ± 0.6
2	Goodwin Canyon Quartz Monzonite	Santa Teresa Mtns.	Biotite	24.9 ± 0.7
3	Mylonite	Northeast Pinaleno Mtns.	Whole rock	28.3 ± 2.0
4	Gneissic granite	Pinaleno Mtns.	Biotite	874 ± 20
5	Granodiorite	Picacho Mtns.	Biotite	24.4 ± 0.8
6	Granodiorite	White Tank Mtns.	Biotite	19.6 ± 0.6
7	Porphyritic granite	Southern Little Harquahala Mtns.	Biotite	140.4 ± 3.2
8	Granite	Granite Wash Mtns.	Biotite	55.1 ± 1.3
9	Granite	Tank Pass, Western Harcuvar Mtns.	Biotite	47.6 ± 1.1
10	Granite	Western Harcuvar Mtns.	Biotite	44.1 ± 1.3
11	Foliated granite	Harcuvar Mtns.	Biotite	51.1 ± 1.2
12	Andesite dike	Harquahala Mtns.	Hornblende	28.6 ± 1.7
			Biotite	22.1 ± 2.0
13	Muscovite schist	Granite Wash Mtns.	Muscovite	64.7 ± 1.7
14	Meta-tuff	Granite Wash Mtns.	Biotite	67.0 ± 1.6
15	Porphyritic granite	Little Harquahala Mtns.	Biotite	66.0 ± 2.0
16	Gneiss	Harcuvar Mtns.	Biotite	25.3 ± 0.6
17	Amphibolite	Harcuvar Mtns.	Hornblende	70.3 ± 2.6
18	Gneiss	Rawhide Mtns.	Hornblende	57.4 ± 1.3
			Biotite	52.3 ± 1.4
19	Basalt	Artillery Mtns.	Whole rock	16.2 ± 0.4

Note: Samples 1, 2, 5, 6, 7, 8, 9, 11, 14, 16, 18, and 19 determined at University of Arizona; others determined by Geochron, Inc. Samples are located in Figure 4. $\lambda_e = 0.581 \times 10^{-10}$ yr^{-1}; $\lambda_\beta = 4.963 \times 10^{-10}$ yr^{-1}; $\lambda = 5.544 \times 10^{-10}$ yr^{-1}; $^{40}K/K = 1.167 \times 10^{-4}$ atomic ratio.

mid-Tertiary sediments. The attitude of breccia and dislocation surface generally conforms to foliation in underlying mylonitic rocks. No such chloritic breccia and associated dislocation surface overlies the rocks on the north flank of the arch, but instead, the mylonitic fabric progressively dies out upward. Thus, this complex displays a characteristic asymmetry (in that the complex is not completely surrounded by a dislocation surface) which is exhibited to varying degrees by other metamorphic core complexes in Arizona and the western United States (see Hyndman, this volume).

PICACHO MOUNTAINS AND VICINITY

The Picacho Mountains lie northwest of the Catalina complex and exhibit many characteristics of metamorphic core complexes. Additionally, the nearby and more topographically subdued Durham and Suizo Hills and the isolated Desert Peak share similar characteristics (Fig. 1).

In the Durham-Suizo area, mylonitic foliation is superimposed on probable Tertiary and Precambrian plutonic rocks (Banks and others, 1977) and dips shallowly eastward. A low-angle dislocation surface separates chloritic and brecciated mylonites from upper-plate, mid-Tertiary sedimentary and volcanic rocks. Westward, down structural section below the dislocation surface, pervasiveness of mylonitic fabric decreases and becomes intense only along thin (1 to 10 cm thick) zones or on discrete, low-angle fracture surfaces.

North-northwest–trending dikes of quartz diorite similar to dikes dated at 23 to 24 m.y. B.P. by K-Ar in the nearby Picacho (Table 1) and Tortolita Mountains (Banks and others, 1978) cut weakly mylonitic lower-plate rocks of the Durham Hills. Nowhere to the west of this area does mylonitic foliation reverse its prevailing eastward dip to expose an archlike structure such as is exposed in other core complexes. In the Durham Hills, northward-striking high-angle normal faults displace the dislocation surface, so it appears likely that more recent block faulting has disrupted possible earlier archlike morphology in this area.

Mylonitic gneisses and foliated plutonic rocks in the Picacho Mountains (Fig. 1) have been compared to Tertiary gneisses in the Catalina Mountains by Yeend (1976). Although isotopic dates have not yet been obtained from the exposed rocks, the presence of low-angle mylonitic foliation with northeast-trending lineation supports this qualitative comparison (see Davis, this volume). In addition, a K-Ar biotite date of 25 m.y. B.P. from gneiss in a nearby drill hole (Eberly and Stanley, 1978) indicates mid-Tertiary cooling and possibly deformation. In parts of the range, foliated, coarsely porphyritic granitic rocks may be deformed Precambrian granites, a contention supported by Precambrian Rb-Sr model ages from gneiss in the previously mentioned drill hole (Eberly and Stanley, 1978).

The south end of the Picacho Mountains exposes the axis of a broad, northward-trending, elevated arch that plunges gently southward. The southeastern flank of the arch is marked by a zone of intense mylonitization, overlain by chloritic breccia, which is in turn overlain in an isolated exposure by a dislocation surface whose upper plate contains trachytic volcanic rock. This upper-plate remnant is correlated with the alkalic volcanic rocks of nearby Picacho Peak, which are dated at 21 to 22 m.y. B.P (Shafiqullah and others, 1976b). The Picacho Peak volcanic and sedimentary section is strongly tilted northeast toward the Picacho complex, and we consider it likely that this section is floored by the same dislocation surface as is exposed nearby.

To the north, the continuity of the Picacho arch is lost as foliation swings to a predominant east-northeast strike and shallow southerly dip. Still farther north where the range narrows and becomes lower in relief, a granodiorite dated by K-Ar (biotite) at 24.4 m.y. B.P. (Table 1) sharply intrudes the mylonitic gneisses along northerly trends. This intrusion is similar in rock type and apparent age to north-northwest–striking dikes in the Durham Hills and Tortolita Mountains.

SOUTH AND WHITE TANK MOUNTAINS

On the southern and western outskirts of the Phoenix basin, two ranges expose flat-lying mylonitic crystalline rocks that we interpret as characteristic of metamorphic core-complex terranes. In the South Mountains, south of Phoenix, structural and geochronologic relationships are of sufficient importance to be described in separate detail (see Reynolds and Rehrig, this volume). For the White Tank Mountains and areas to the south (Fig. 1), limited reconnaissance mapping permits but a few brief comments.

In the White Tank Mountains, our studies are incompete (S. J. Reynolds, work in progress) but suggest that mylonitic fabrics are most common in the eastern half of the range. Most of the northeast edge of the range is composed of granodiorite and granite that are variably foliated. Where best developed, the foliation dips to the east or southeast at shallow to moderate angles and contains a penetrative east- to northeast-trending lineation. Along the east-central edge of the range, amphibolite gneisses of probable Precambrian protoliths also exhibit the gently east-dipping mylonitic foliation and northeast-trending lineation. Near the southeast corner of the range, amphibolite gneisses are present that have been extensively intruded by small granitic-pegmatitic bodies. Foliation in these rocks is largely nonmylonitic, but nevertheless low-angle.

In the higher, central part of the range, schist and gneiss containing various proportions of biotite, muscovite, feldspar, and quartz are interlayered with biotite-bearing granitic pods and layers. The foliation in the metamorphic rocks strikes northeast and dips *steeply* to moderately to the southeast, much like unperturbed Precambrian terranes elsewhere in Arizona. Foliation orientation, rock type, and metamorphic grade, when considered together, suggest that the gneisses are Precambrian. The steep foliation is in places cut by a moderately southwest-dipping mylonitic foliation which contains a west- to southwest-trending lineation. This fabric has similar characteristics to that developed along the eastern side of the range. Farther west, massive plutonic rocks ranging from muscovite-garnet alaskites to biotite granodiorites are exposed. In the northwest corner of the range, mid-Tertiary(?) hypabyssal rocks intrude and overlie older, coarser-grained, holocrystalline granitic rocks. In addition, there is a north-northwest–oriented equigranular granodiorite which has yielded a 19.6-m.y. K-Ar apparent age on biotite (Table 1).

Brief reconnaissance in parts of the Buckeye Hills immediately south of the White Tank Mountains indicates the presence of abundant granitic and metamorphic rocks which only locally exhibit a gently dipping mylonitic foliation and east-northeast–trending lineation. The areal extent of rocks containing this mylonitic foliation is unknown and may be relatively small. Ages of various plutons, some of which are muscovite- and garnet-bearing is also unknown, but preliminary date indicate that *one* biotite granodiorite is Precambrian (Rehrig and Reynolds, unpub. data).

HARCUVAR COMPLEX

The Harcuvar metamorphic core complex is located in west-central Arizona (Fig. 1) and includes much of the Harquahala, Harcuvar, Buckskin, and Rawhide Mountains (Fig. 4). In addition, rocks important to discussion of the complex are exposed in the Granite Wash and Little Harquahala Mountains.

The Harquahala and Harcuvar Mountains are relatively high ranges for west-central Arizona; both reach elevations of well over 1,500 m. The Rawhide and Little Harquahala Mountains are much lower (generally less than 1,000 m), and the Buckskin and Granite Wash Mountains are intermediate in elevation. The Harquahala, Harcuvar, Buckskin, and to a lesser extent the Rawhide Mountains exhibit the smooth bulbous topographic profiles characteristic of metamorphic core complexes. These ranges are distinctive because of their northeasterly trend, which is anomalous amid the prevailing northwesterly trends of mountain chains in the Arizona Basin and Range province. The anomalous northeast trend is due to long, usually doubly plunging foliation arches in gently dipping quartzofeldspathic gneisses (Figs. 5, 6). The broach arches are arranged in en echelon fashion and collectively give the entire metamorphic core complex a northwest elongation.

Previously, the geology of the region had been interpreted to be largely Precambrian gneisses and schist intruded by Mesozoic and early Cenozoic plutons and fringed on the north and south by allochthonous Paleozoic, Mesozoic, and Cenozoic sedimentary rocks (Wilson, 1962; Wilson and others, 1969). Our recent and more detailed work indicates that most of the gneissic "basement" is not due to Precambrian deformation, but is the result of Mesozoic and Cenozoic plutonic, metamorphic, and tectonic events. The protoliths of the metamorphic rocks are believed to range in age from Precambrian to Tertiary.

We will first present some important geologic aspects of each range and then summarize general characteristics of the complex. We will then suggest possible inferences that can be made from our observations. Geologic study in the complex is continuing, and additional discussions will be presented elsewhere at later dates. Davis and others (this volume) describe the geology of the Rawhide Mountains and adjoining area to the west of the Harcuvar complex.

Harquahala Mountains

The main central and eastern mass of the Harquahala Mountains is composed of granitic and high-grade metamorphic rocks flanked on the south and southeast by Paleozoic and Mesozoic rocks locally of low to moderate metamorphic grades (Fig. 6). Plutonic rocks in the central crystalline core are locally foliated and consist mostly of biotitic granodiorite, porphyritic granite, equigranular muscovite-garnet granite, and abundant small tabular bodies of leucocratic alaskite and pegmatite. Other parts of the crystalline core are composed of migmatite, amphibolite, and biotite-bearing quartzofeldspathic schist and gneiss. Foliation in the metamorphic and plutonic rocks is generally gently dipping and defines a broad arch whose east-northeast–trending axis parallels the eastern crest of the range (Fig. 5). At least two directions of lineations are present in the rocks. A variable north- to northwest-trending metamorphic lineation occurs in central and western parts of the range and a younger northeast-trending mylonitic lineation is extensively developed in structurally high parts toward the northeast end of the range (Fig. 5). The older(?) metamorphic fabric is cut by a northwest-trending swarm of microdiorite-andesite dikes that yield somewhat discordant K-Ar hornblende and biotite ages of 28.6 and 22.1 m.y., respectively (Table 1).

South of the crystalline rocks, recognizable Paleozoic and Mesozoic sedimentary rocks are preserved (Fig. 6). Along the southern flank of the range, interbedded feldspathic sandstones and siltstones are exposed that are similar to Mesozoic strata in adjacent ranges (for example, Granite Wash). As these strata are traced northward toward the crystalline rocks, a progressive increase in metamorphism occurs until the two rock types ultimately merge. Farther to the east, metamorphosed calcareous sandstones and pure quartzites are overlain successively by siltstone-quartzite, marble, and interlayered pelitic schists and metasiltstones. This section, although metamorphosed and highly folded, may be in normal stratigraphic sequence and possibly represents parts of the upper Paleozoic Supai, Coconino, and Kaibab Formations and parts of the overlying impure Mesozoic arkose. Where adjacent to or overlain by granitic rocks which form the high crest of the range, the metamorphosed Mesozoic(?) rocks are locally gneissic and somewhat "granitic" in their appearance.

In the southwestern part of the range, large folds in a less-metamorphosed section of Paleozoic rocks (including Coconino and Kaibab Formations) have been mapped by Varga (1976). In and adjacent to

Varga's area, there are exposures of slightly metamorphosed impure clastic rocks of probable Mesozoic age. The contact between Paleozoic and Mesozoic rocks is locally complexly sheared, contorted, or covered by aluvium.

Recent mapping by S. B. Keith and S. J. Reynolds (1980, oral commun.) reveals that major rock units in this range are separated by multiple flat thrust faults. The impact of these structures on the pre-Tertiary history of this reigon is profound.

The youngest rocks of the range are probably those exposed at its extreme northeast end (Fig. 6). Here a sequence of dark, mid-Tertiary(?) volcanic breccia, conglomerate, sandstone, and siltstone dips 30°SW and is underlain by a dislocation surface that dips gently eastward. Below the dislocation surface are chloritic mylonitic gneisses that contain a pronounced N60°E-trending cataclastic lineation.

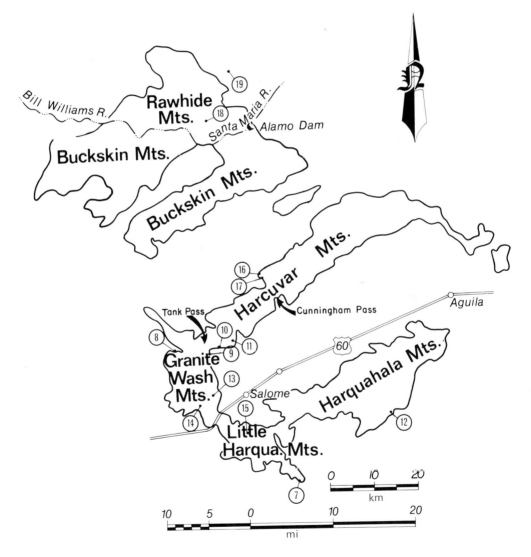

Figure 4. Geography and locations of samples used for dating, Harcuvar metamorphic core complex. Numbers refer to samples in Table 1.

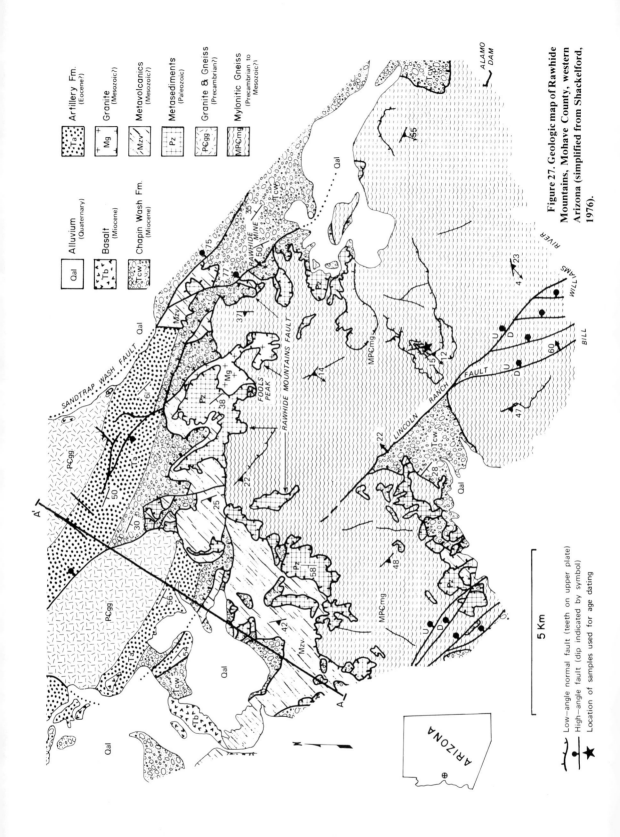

Figure 27. Geologic map of Rawhide Mountains, Mohave County, western Arizona (simplified from Shackelford, 1976).

Little Harquahala–Granite Wash Mountains

Southwest of the Harquahala Mountains lie the Little Harquahala Mountains, a range which although not a part of the metamorphic core complex per se, has some important geologic relationships that should be discussed. The folded Paleozoic section studied by Varga (1976) in the Harquahala Mountains continues southwest along strike into the Little Harquahala Mountains. As in Varga's area, the section is only slightly metamorphosed, but intensely faulted and folded (locally overturned); it partly consists of rocks tentatively correlated with the Cambrian Bolsa Quartzite, mid-Paleozoic carbonates (Martin and Redwall Formations), and the upper Paleozoic Supai Formation. A chloritic, feldspar-megacrystic, porphyritic granite underlies these sedimentary rocks in both the Little Harquahala and southwestern Harquahala Mountains. The contact between the Paleozoic rocks and the

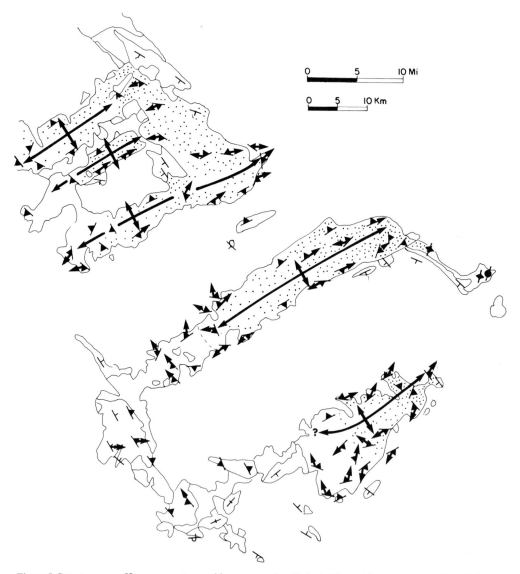

Figure 5. Structure map, Harcuvar metamorphic core complex. Refer to Figure 6 for explanation of symbology.

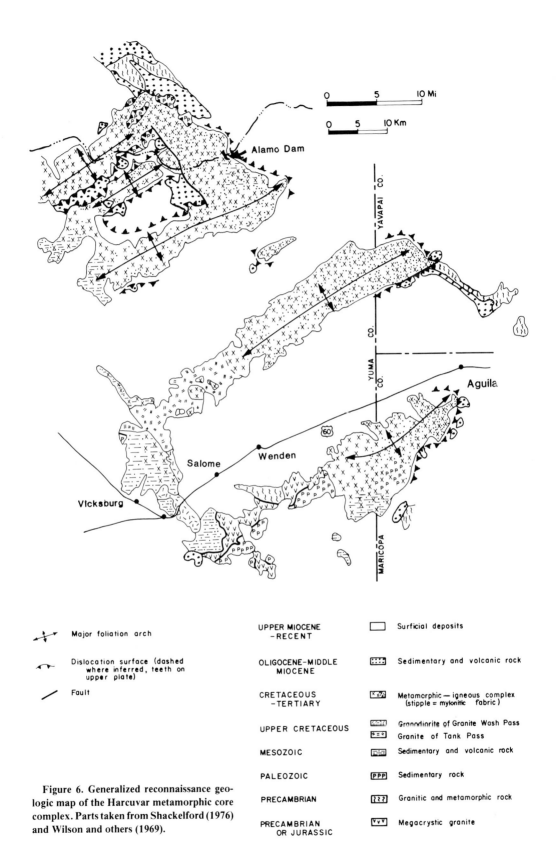

Figure 6. Generalized reconnaissance geologic map of the Harcuvar metamorphic core complex. Parts taken from Shackelford (1976) and Wilson and others (1969).

granite is generally a gently inclined fault. The granite below the fault is generally highly sheared and locally mylonitic. The age of the granite is uncertain, but is probably either Precambrian or Jurassic based on similarities to stratigraphically or radiometrically dated rocks elsewhere. A similar granite crops out in the Sore Fingers area at the southeast end of the Little Harquahala Mountains and yields a K-Ar biotite date of 140.4 m.y. B.P. (Table 1).

Throughout much if not all of the Little Harquahala Mountains, the western contact of the porphyritic granite is a low-angle fault surface. Along the southern part of the contact, the granite and fault overlie a faulted, heterogeneous sequence of Jurassic(?) silicic volcanic flows, ash-flow tuffs, and assorted clastic rocks ranging from quartzitic sandstones to volcanic conglomerates. Similarly, to the north, the granite *rests upon* an east- to northeast-dipping, low-angle fault. A K-Ar date of 66.0 m.y. B.P. (Table 1) was determined on recrystallized(?) biotite from the sheared granite directly above the fault surface. Below the fault and extending to the west and south are units of schistose sandstone, conglomeratic arkose, and fine-grained volcanic rocks in imbricate-fault juxtaposition. Farther west, probable Mesozoic quartzites, feldspathic sandstones, and calcarenites crop out. Although details of this complex area have not been worked out, the presence of imbricate faulting is well demonstrated.

Farther to the northwest, the Mesozoic(?) strata are intruded by the granodiorite of Granite Wash Pass (Fig. 6), which has yielded K-Ar biotite ages of 65 m.y. (Damon, 1968) and 69 m.y. (Eberly and Stanley, 1978). This undeformed Upper Cretaceous granodiorite continues northward into the Granite Wash Mountains where it also intrudes faulted and deformed Mesozoic strata. In their most westerly exposures in the Granite Wash Mountains, the Mesozoic strata are only slightly metamorphosed and mainly consist of interbedded quartzose and feldspathic sandstone, shaly siltstone, minor conglomerate, and rare limestone. As these gently east-dipping strata are traced eastward toward the granodiorite, they progressively become slaty, schistose, and finally reach amphibolite grade. K-Ar apparent ages of 67.0 m.y. on biotite and 64.7 m.y. on coarse muscovite (different locations, see Fig. 4, Table 1) have been obtained from the metamorphosed Mesozoic section. Metamorphic layering and different metamorphic grades in the metasedimentary units are discordantly intruded by the granodiorite and its locally dioritic border phase. Locally in the northwestern Granite Wash Mountains, the Mesozoic section paraconcordantly overlies upper Paleozoic(?) quartzites and carbonates, but just to the north lies below fault slices of Paleozoic(?) limestone and dolomite.

Another pluton, the granite of Tank Pass, crops out in much of the northeastern Granite Wash Mountains (Fig. 6). The equigranular biotite granite intrudes the Mesozoic section and is itself intruded by the dioritic border phase of the granodiorite of Granite Wash Pass. In addition, the Tank Pass granite is heavily intruded by a north-northwest–trending swarm of felsic and andesitic dikes of probable mid-Tertiary age. K-Ar biotite dates from the granite vary as follows from west to east: 55.1, 47.6, 44.1, and 51.1 m.y. B.P. (Fig. 4, Table 1). These dates are thought to represent minimum ages, the true emplacement age being somewhat older than the Late Cretaceous age of the granodiorite of Granite Wash Pass.

Harcuvar Mountains

The main northeast-trending part of the Harcuvar Mountains is composed predominantly of granitic rocks and amphibolite-grade metamorphic rocks (Fig. 6). The western third of the range is composed of the granite of Tank Pass, abundant roof pendants of gneiss, and numerous north-northwest–striking dikes of garnetiferous pegmatite-aplite, rhyolite, andesite, and diorite. The granite is undeformed to the southwest, but becomes progressively foliated toward the northeast. This foliation and its attendant northwest-trending lineation are different in style from the mylonitic fabrics that contain the distinctive northeast-trending lineation so characteristic of Tertiary activity in metamorphic core complexes (see Reynolds and Rehrig, this volume). Orientation of the foliation is variable, but generally dips about 30° to the southwest (Fig. 5). This orientation of foliation dictates that structurally lower rocks are exposed to the northeast. Indeed, toward the northeast, the granite is more intensely deformed (locally by flow-folding) and is involved in migmatization to a much larger extent. Near Cunningham Pass, the granite is extensively intruded by garnet-muscovite–bearing alaskite and pegmatite. The foliated granite and these later intrusions are all discordantly intruded by a swarm of northwest-trending Tertiary(?) dikes of andesite, microdiorite, and coarse, hornblende-bearing diorite.

Roof pendants in the granite and migmatites adjacent to the granite consist of amphibolite, quartzofeldspathic gneiss, micaceous schist, foliated granodiorite-granite, and a variety of alaskite and pegmatite. These metamorphic rocks generally contain a northwest-trending metamorphic (nonmylonitic) lineation that is at least locally parallel to axes of flexural-flow and passive-flow folds. This fabric is largely pre-Tertiary, as K-Ar apparent ages of 70.3 m.y. on hornblende and 51 m.y. on biotite have been obtained from rocks containing the northwest-trending lineation (Table 1).

Near Cunningham Pass (Fig. 4), the northwest-trending folds and lineation are distinctly cut by low-angle structural surfaces exhibiting a southwest-trending mylonitic lineation. In addition, the northwest-trending folds and lineation are intruded discordantly by leucocratic pegmatites that are themselves cut by the mylonitic foliation and southwest-trending lineation. There is, therefore, evidence for two deformational-metamorphic events separated by the intrusion of leucocratic pegmatites and alaskites. The earlier metamorphic event produced migmatites and ductile folds with northward- and northwestward-trending lineation, whereas the second event was accompanied by the development of mylonitic foliation and conspicuous southwest-trending lineation. A K-Ar biotite apparent age of 25.3 m.y. (Table 1) on the mylonitic gneisses indicates mid-Tertiary cooling and possibly an equally young age for the mylonitization. This assemblage of rocks and structures makes up most of the remainder of the range to the northeast of Cunningham Pass, where the northeast-trending lineation and mylonitic foliation are progressively better developed. This later mylonitic foliation is also evidently most intensely developed at structurally higher levels along the top and outer margins of the range; thus an arched carapace is formed.

At the northeast end of the range, the mylonitic rocks described above are chloritic, brecciated, and overlain by a dislocation surface whose gentle dip mimics that of the underlying gneisses. In isolated hills above the dislocation surface are allochthonous mid-Tertiary volcanic and sedimentary strata. An additional interesting assemblage of upper-plate rocks apparently overlies the dislocation surface where the large northwest-trending ridge departs from the main range (Fig. 6). In this area, gneiss and schist of almost certain Precambrian age display a near-vertical, northeast-striking, nonmylonitic (crystalloblastic) foliation that is cut by a siliceous and brecciated dislocation surface underlain by gently east-dipping mylonitic gneisses containing a northeast-trending lineation. Above the upper-plate metamorphic rocks in depositional contact is a section of basal arkosic conglomerates overlain by probable mid-Tertiary trachytic welded ash-flow tuffs. The Tertiary section dips moderately to the southwest and is in part the same section exposed in the aforementioned allochthonous blocks.

Buckskin and Rawhide Mountains

The following discussion is based both on the excellent study by Shackelford (1976) in the Rawhide Mountains and our own observations in both the Rawhide and Buckskin ranges. At least three large northeast-trending folation arches in metamorphic rocks govern the structure and outcrop distribution in the Buckskin and Rawhide Mountains (Figs. 5, 6). Gently inclined metamorphic and mylonitic rocks are exposed in the arches, and remnants of overlying, mostly nonmetamorphosed, Precambrian(?) to Miocene dislocated rocks crop out in northeast-trending belts along the axis of the synforms. The metamorphic rocks in the "basement" arches consist of quartzofeldspathic gneiss, mica schist, quartzite, marble, and granitic rocks of variable mineralogy and structural disposition.

In numerous areas in the Buckskin Mountains, the apparent metamorphic grade of the rocks increases toward the axis of the arch (down structural section). The continuous gradation from slightly metamorphosed impure arkose, siltstone, and thin (1 to 5 m) beds of carbonate, down section to their amphibolite-grade, migmatitic equivalents is best exemplified in traverses (W. A. Rehrig, unpub. data) across the northern flank and western nose of the southernmost Buckskin foliation arch. The increasing metamorphic gradation from west to east in the southwestern Buckskin Mountains is analogous to and perhaps correlative with similar transitions to amphibolite-grade metasedimentary units across the Granite Wash Mountains (Fig. 6). Metamorphism of the thick section of impure arkosic rocks has produced a rather uniform sequence of quartzofeldspathic gneisses interstratified with rare, thin beds of tectonite marble. These rocks are common in the basement of the Buckskin and Rawhide Mountains and suggest a sedimentary parentage, in part.

These basement rocks in most areas exhibit a variably developed, gently dipping, mylonitic foliation with a characteristic northeast-trending lineation. In at least several exposures in both ranges, this mylonitic fabric cuts across an earlier "ductile" foliation that is associated with northwest-trending fold axes. As in the Harcuvar and Harquahala Mountains, the mylonitic fabric appears to be best developed in northeastern parts of the Buckskin Mountains or in structurally high exposures along the flanks or top of the arches. The mylonitic fabric in these rocks become progressively brecciated, disrupted, and chloritic as the surface beneath overlying dislocated rocks is approached. The dislocation surface occurs nearly continuously around the edges of the arches except along the southwest margin where it is conspicuously absent. Much of the exposed crystalline basement of the Rawhides (Shackelford, 1976) is composed of this chloritic and variably brecciated mylonite. Above the dislocation surface lie Precambrian or Mesozoic metavolcanic rocks, Paleozoic platform strata, possible Mesozoic plutons, and abundant Oligocene(?)-Miocene coarse breccia and conglomerate, sandstone, siltstone, limestone, and volcanic rocks. The Tertiary sedimentary units [Chapin Wash and Artillery(?) Formations] are believed correlatable with the Copper Basin Formation in the nearby Whipple Mountains of California (Davis and others, this volume). Bedding in stratified rocks generally dips moderately to the southwest (Shackelford, 1976), with dips evidently being more gentle in northeastern exposures.

Shackelford (1976) has assessed in detail the kinematics of both the lower-plate metamorphic rocks and upper-plate dislocated rocks. He concluded that the fabric in the basement formed by flattening perpendicular to the gently dipping foliation and extension parallel to the northeast-trending lineation. The upper-plate rocks likewise extended in a southwest-northeast direction, but by brittle, listric normal-fault dislocation with tectonic transport to the northeast. The southwest dips of upper-plate rocks are consistent with tilting by antithetic rotation on downward-flattening faults during northeastward transport. Our observations *throughout* the Harcuvar complex are generally consistent with Shackelford's conclusion of northeast-directed extension in both upper- and lower-plate rocks.

The only isotopic age determinations come from "lower-plate" metamorphic rocks of the Rawhide Mountains, where slightly discordant K-Ar hornblende and biotite ages of 57.4 and 52.3 m.y., respectively (Table 1; Shackelford, 1977) were obtained. These apparent ages probably reflect metamorphism and/or deformation of older rocks, some of which may represent Mesozoic clastic sedimentary rocks. However, Precambrian or, locally, even Paleozoic parent rocks may have been present. Depending on the intensity and temperature of variably superimposed mylonitization, the slightly discordant K-Ar dates may even give the age of mylonitic deformation. This alternative is favored by G. A. Davis and others (this volume).

Upper-plate rocks cut by the low-angle dislocation surface are as young as Miocene (Shackelford, 1976). Flat-lying basalt flows dated at about 10 m.y. B.P. by K-Ar are unaffected by rotational normal faulting on the dislocation surface (Shackelford, 1976).

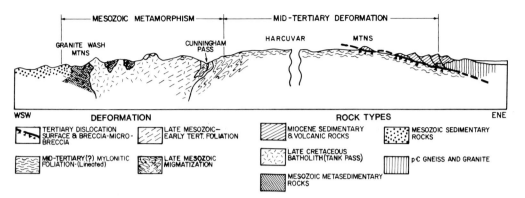

Figure 7. Schematic, diagrammatic cross-section across the Harcuvar and Granite Wash Mountains.

Summary of the Harcuvar Metamorphic Core Complex

Certain fundamental attributes and geologic relationships are common to most of the Harcuvar metamorphic core complex as schematically shown by a section through the Harcuvar Mountains (Fig. 7). The structurally lowest rocks are paragneiss and orthogneiss that exhibit textures suggesting intense metamorphism and plutonism, migmatization, and ductile deformation. These rocks commonly exhibit a gently dipping foliation which contains a metamorphic lineation of variable, but predominantly northwest trend. In the northeastern half of the complex, in structurally high exposures, this ductile fabric is cut and transposed by a low-angle mylonitic foliation that contains a conspicuous northeast-trending lineation.

The mylonitically foliated rocks are apparently exposed as an east-northeast–trending arched sheath or carapace overlying the older ductile metamorphic rocks. In most exposures of mylonitic rocks, a set of tension fractures is present that strikes north-northwest, perpendicular to the lineation. In addition, numerous Tertiary(?) dikes cut the mylonitic rocks and fill these fractures.

In their structurally highest exposures in the northeastern parts of the complex, the mylonitic rocks are brecciated, chloritic, and overlain by a major dislocation surface whose orientation mimics the gently dipping deformational foliation in the underlying metamorphic rocks. Above the surface lie moderately southwest-dipping (where bedded) rocks of Precambrian(?) to Miocene age. These upper-plate rocks strike northwest, perpendicular to the lineation in the underlying mylonitic rocks. They are generally unmetamorphosed and nonmylonitic.

Flanking the metamorphic complex to the west and south are unfoliated granitic plutons and slightly metamorphosed Paleozoic and Mesozoic sedimentary rocks. Both the plutons and sedimentary rocks can be traced eastward into the complex where they are highly metamorphosed and deformed.

Isotopic dates and geologic relationships place some constraints on the various metamorphic, plutonic, and tectonic events. A metamorphic and plutonic event of probable Precambrian age produced a steep, northeast-striking foliation in rocks exposed outside the complex (that is, the northwest-trending ridge in the easternmost Harcuvar Mountains). Paleozoic and Mesozoic sedimentation was followed by metamorphism that locally changed these rocks into amphibolite gneisses and lower-grade schists and slates. In the Granite Wash Mountains, foliation in these rocks and its N50°E-trending metamorphic lineation are cut discordantly by both the granite of Tank Pass and the granodiorite of Granite Wash Pass. K-Ar ages on the metasedimentary rocks and the crosscutting plutons suggest a minimum age of near 65 m.y. for the metamorphism. This metamorphism is possibly middle to Late Cretaceous, as it must be younger than the Paleozoic and Mesozoic strata and is *somewhat* older than the two plutons.

Farther to the east, the granite of Tank Pass is involved in a younger(?) metamorphic event in the central Harcuvar Mountains that produced migmatites, gneisses, and ductile deformation. The minimum age of this event is provided by three K-Ar dates between 51 and 57 m.y. B.P. The event must have occurred during or after emplacement of the Tank Pass pluton, probably in Late Cretaceous time, as suggested by the K-Ar date of 70 m.y. B.P. on synkinematic hornblende. This metamorphic event is locally postdated by intrusion of leucocratic pegmatite and alaskite that are possibly early Tertiary and responsible for the 44- to 55-m.y. K-Ar biotite cooling ages on the gneisses and the Tank Pass pluton.

All of the metamorphic and plutonic rocks described above are cut by the mylonitic foliation and northeast-trending lineation—a relationship that suggests a Tertiary age for this deformation. The mylonitization with its associated north- to northwest-striking tensional fractures and dikes is essentially the last deformational event (except brecciation) to affect the crystalline basement. *If* the mylonitic fabric is correlative with *identical* fabric in the South Mountains near Phoenix (Reynolds and Rehrig, this volume) and is represented by the 25.3 m.y. biotite cooling age at Cunningham Pass, then it is mid-Tertiary in age. Davis and others (this volume) favor an older age for the mylonitization. Data from the Harcuvar complex simply require that mylonitization postdated the latest Cretaceous–early Tertiary metamorphism and leucocratic plutonism and predated or was synchronous with mid-Tertiary cooling.

The mylonitic fabric is cut by and must therefore be postdated by the chloritic breccia. The dislocation surface associated with the chloritic breccia cuts rocks as young as Miocene (Shackelford, 1976), and consequently some movement must be Miocene. Dislocation must predate flat-lying 10- to 13-m.y.-old basalts (Shackelford, 1976; Davis and others, this volume). Slickensides and low-angle, listric normal faulting above and at the dislocation surface indicate intense northeast-southwest extension, as do mylonitic fabrics that underlie the dislocation surface. The mylonitic foliation, chloritic breccia, and low-angle faulting all exhibit broad northeast-trending arch morphologies. This may suggest that arching is perhaps the most recent phenomenon, probably Miocene in age. North-northwest–trending slickensides locally present on the dislocation surface may reflect structural adjustments of these late uplifts.

In a simplified and somewhat diagrammatic sense, we present a southwest-northeast geologic cross-section across the Harcuvar metamorphic core complex (Fig. 7). The section illustrates the main points of our summary. Note particularly that the dislocation surface and tilted upper-plate rocks on the northeast flank of the complex pass northeastward into less-tilted Tertiary basin deposits and volcanic rocks outside the confines of the metamorphic core complex.

THOUGHTS ON ORIGIN OF METAMORPHIC CORE COMPLEXES

Before presenting our ideas regarding core-complex evolution, we would like to summarize some salient characteristics of "typical" metamorphic core complexes as described in this paper. Any realistic model for origin of the complexes must explain these characteristics.

1. The complexes generally contain a core of granitic rocks, or metamorphic rocks whose fabric may be either Precambrian (as in the Santa Teresa and South Mountains) or unequivocally Mesozoic or Cenozoic (Harcuvar complex).

2. Intrusive rocks of variable age, composition, and structural setting are invariably present with the metamorphic rocks in the basement. Muscovite- and garnet-bearing granite, alaskite, and pegmatite are common, much more so than in adjacent "non–core-complex" terranes. Evidence of early to middle Tertiary plutonism, heating, or cooling is ubiquitous in the complexes. Cooling ages and some

plutonic emplacement ages in the complexes are nearly identical to ages of widespread middle Tertiary volcanism and plutonism outside of the complexes.

3. Plutons and adjacent metamorphic rocks within the complexes are overprinted in structurally high exposures by a characteristically low-angle mylonitic fabric that contains a conspicuous lineation of regionally consistent trend (N60°E ± 20°). The mylonitic fabric is commonly accompanied by tension fractures, ductile normal faults (Davis and others, 1975), and dikes whose north-northwest trend is perpendicular to the direction of the lineation. Mylonitization and dike intrusion are both *locally* well dated as Oligocene–early Miocene, with the dikes being intruded during and after the later stages of mylonitization (Reynolds and Rehrig, this volume). Additionally, certain lines of evidence might argue for as-yet-undetermined amounts of early Tertiery (Eocene) mylonitic fabrics (Davis and others, this volume; Keith and others, this volume). In either case, mylonitization appears nearly synchronous with widespread magmatic pulses in the crust.

The mylonitic fabric formed by flattening perpendicular to foliation and extension parallel to the lineation (Davis and others, 1975; Davis and others, this volume; Davis, this volume; Reynolds and Rehrig, this volume). Strain ratios recorded in the mylonitic rocks of approximately 9:2:1 indicate that the amount of extension is significant.

4. The complexes are asymmetrical with structurally high mylonitic rocks being chloritic, brecciated, and overlain by a low-angle dislocation surface on some sides and other sides being characterized by a progressive upward and downward decrease in the intensity of mylonitic fabric. Overlying the dislocation surface and subjacent chloritic breccia are generally unmetamorphosed and nonmylonitic rocks whose age is as young as Miocene. These detached rocks almost invariably strike perpendicular to lineation in underlying mylonitic rocks and commonly exhibit unidirectional dips, the result of systematic listric normal faulting. Core-complex asymmetry is further expressed by a concentration of plutonic phenomena along those parts of the complex opposite the dislocation-breccia zones.

Movement indicators, including slickensides, folds, and direction of listric fault rotation, suggest mostly unidirectional translation parallel to the aforementioned east-northeast–trending mylonitic lineation in the lower plate. In other cases, *latest* movement of upper-plate rocks was clearly down the present dip of the fault and not parallel to the east-northeast–trending lineation (Davis, 1975; Rehrig and Reynolds, unpub. data for the Harcuvar and Buckskin Mountains). In a general way, the dislocation surface and attendant chloritic breccia zone are concordant in attitude to the underlying zone of mylonitic rocks. It should be emphasized that the intensity of mylonitization increases toward the dislocation surface (Fig. 2). However, on an outcrop and locally larger scale, the dislocation surface discordantly truncates and clearly postdates the mylonitic foliation. This discordance appears particularly evident in the Whipple Mountains just west of the Buckskin Ranges (Davis and others, this volume). These opposing relationships imply an *overall* genetic association between mylonitic rocks and the dislocation fault, yet underscore their development as two somewhat sequential events.

5. Closely associated with the complexes are middle Tertiary fanglomerates, sedimentary breccia, sandstone, and somewhat finer grained clastic rocks. The sedimentary rocks and interbedded volcanic rocks, which are as young as ~16 m.y. (Shackelford, 1976; I. Lucchitta, 1978, oral commun.), have been displaced by the dislocation surface yet contain *rare* clasts of chloritic mylonitic rocks similar to those which underlie the fault. Interesting yet perplexing are the apparently more common occurrences of mylonite clasts in these sedimentary units in the neighboring Whipple Mountains (Davis and others, this volume). These substantially thick (locally thicker than 3,000 m) sedimentary rocks exhibit chaotic characteristics and rapid facies variations indicative of highly energetic accumulation in steep-sided, tectonically active basins. The fact that these basins existed at the very sites of the now-uplifted core complexes argues that the present physiography developed late, an inference supported by the arched morphology of the mylonitic foliation, breccia, and dislocation surface. Relatively flat-lying basalt flows dated between 10 and 13 m.y. B.P. unconformably overlie tilted upper-plate rocks. These basalts

are not found as remnants on top of the uplifted parts of the complexes. Thus, timing of the *latest* arching event in the metamorphic complexes may thus be confined between about 15 and 13 m.y. B.P. Most of the arches have their long axes oriented parallel to lineation in the mylonitic rocks that comprise them. This relationship remains unexplained.

From the above discussion, the key events contributing to evolution of the core complexes are (1) metamorphism and plutonism, which *may* significantly predate mylonitization; (2) mylonitization with locally synchronous pluton emplacement, dike intrusion, and cooling *within* the complexes and volcanism, plutonism, and sedimentation *outside* the complexes; (3) development of the chloritic breccia and overlying dislocation surface, probably accompanied by accumulation of thick sequences of coarse clastic sediment; and (4) arching and uplift. The sequence of events from mylonitization to arching may have been spaced over a period of 35 m.y. (50 to 15 m.y. B.P.) in some locations (Keith and others, this volume; Davis and others, this volume), but was closely sequential and compressed into a 10-m.y. interval from 25 to 15 m.y. B.P. in other areas (Reynolds and Rehrig, this volume).

If the above sequence is in proper order, an important point becomes apparent: the mylonitic zones developed as relatively thin, flat-lying crustal sheets of intense east-northeast–west-southwest extension and vertical flattening. Extension-flattening in the mylonitic zone was accommodated by a combination of mechanical comminution of certain minerals (feldspar), recrystallization of others (especially quartz), and locally significant amounts of intrusive dilation (north-northwest–trending dikes and plutons). Extension above the dislocation surface is manifest mainly as listric normal faults and accompanying antithetic rotation of the upper-plate rocks.

In areas *outside* the complexes, the mid-Tertiary east-northeast–west-southwest extension is evidenced by north-northwest–trending dike swarms and pluton alignment, north-northwest–striking tension fractures in mid-Tertiary plutons, and minor normal faulting of less rotational character (Rehrig and Heidrick, 1976). Therefore, areas both within and outside the complexes experienced widespread Tertiary east-northeast–west-southwest extension but responded to the strain in different manners with the areas outside the complexes behaving more brittlely.

In order to stimulate thought regarding evolution of the complexes, we propose the following speculative model as one possible explanation for their origin. The complexes can be envisioned in terms of a three-layer model (Fig. 8A) in which the mylonitic rocks represent a relatively thin, subhorizontal zone of extreme extension and flattening which is overlain and underlain by layers which responded to extension in different manners. Above the mylonitic zone, a relatively rigid plate of rocks extended brittlely by listric normal faulting and concurrent antithetic rotation (Davis and Coney, 1979). Below the mylonitic zone, rocks extended by various combinations of recrystallization, ductile deformation and metamorphism, pluton and dike intrusion, and brittle fracturing. This complete model is shown diagrammatically in Figure 8.

We regard the inception of mylonitic deformation to be a consequence primarily of crustal heating concurrent with an active tensional stress (Rehrig and Heidrick, 1976) of regional extent. The close temporal and spatial association between mylonitization and magmatism (Eberly and Stanley, 1978; Keith and others, this volume) suggests such a relationship. Heat introduced into a shallow level of the crust by extensive magma emplacement (thick sills?) probably helped initiate crustal extension and thinning (Fig. 8B). The combined application of heat and horizontal tensile stress to the crust would cause necking down of those parts of the crust which were especially hot (Davis and Coney, 1979); tensional failure of more brittle segments would also result. The mylonitic zone thus is a manifestation of the somewhat ductile necking (Davis, 1977 and this volume). Actual physical separation of more rigid blocks below the zone of mylonitization (Davis, 1977; Davis and Coney, 1979) would result in magma ingress and intrusion of both plutons and dikes. Wright (1976) has suggested such a mechanism for crustal extension beneath broad zones of listric faulting in the Death Valley region. In this perspective, it is interesting to note that plutons which are nearly the same age as mylonitization are distributed on the side of the complexes which is opposite from the inferred direction of tectonic transport of upper-plate rocks. In other words, the plutons may have preferentially been intruded on the side of the complex from which everything was moving away (see Fig. 8).

In most complexes, extension within the mylonitic zone continued after the rocks were too cool to permit *penetrative* deformation; this resulted in brittle fracture accompanied by dike intrusion (Reynolds and Rehrig, this volume). As mid-Tertiary thermal effects diminished (near 20 m.y. B.P.) and the mylonitic zone became cooler and more brittle, detachment between the mylonitic zone and the overlying, extending block was taken up by a discrete dislocation surface and attendant footwall breccia (Fig. 8E).

The specifics of this change from pervasive mylonitic deformation with tightly constrained directional fabrics to formation of a nonfoliated breccia and dislocation surface are the most enigmatic part of our model. The progressive increase in mylonitic fabric toward the dislocation zone (Fig. 2) strongly suggests a dependent relationship between the two phenomena. Yet, as we have stated, there is

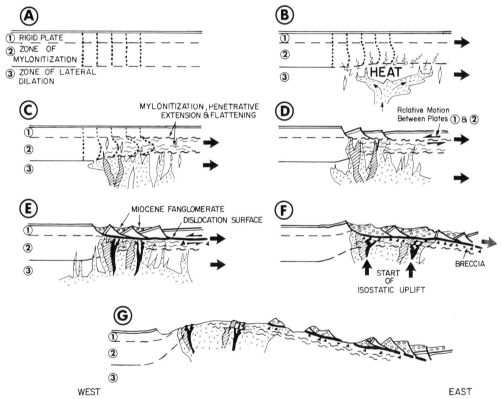

Figure 8. Model for Tertiary evolution of metamorphic core complexes. (A) Concept of three crustal layers, each responding distinctly to Tertiary heating and tensional stress. (B) Magma emplacement and east northeast-directed extension initiate necking (stretching) in zone 2. (C) Continued extension in zone 3 leads to tensional separation and intrusive dilation; mylonitization develops in more ductile zone 2. (D) Differences in response to crustal extension between zones 1 and 2 create differential strain and fragmentation of rigid plate. (E) Continued differential strain between zones 1 and 2 and dissipation of heat creates the dislocation of surface and further rotational faulting of upper-plate blocks; rapid fanglomerate-breccia accumulation within fault-bounded troughs; clasts mainly from nonmylonitic plate 1, but exposure and erosion of mylonite zone may occur locally. (F) As tectonic denudation proceeds, area of maximum thinning responds to removal of zone 1 by isostatic uplift and deformational thinning in zone 2. (G) Present configuration of complex.

abundant evidence that the dislocation surface distinctly postdates the underlying mylonitization. This apparent paradox we cannot totally explain and provides a challenging subject for future research.

Movement along the dislocation surface, deposition of coarsely clastic sedimentary-wedge deposits, and rotational faulting were nearly coincident processes. The interrelationship of these events is depicted in Figures 8E and 8F, which show coarse sediments accumulating in tectonic troughs that developed behind rotating fault blocks. These blocks are bounded by downward flattening faults (listric normal) that merge with the basal dislocation surface at relatively shallow depth. We are assuming that the east-northeast–directed extension (Fig. 8) continued during this stage. Although relative tectonic transport of upper-plate rocks is well documented, the total amount of movement need not have been large. The rigid blocks could have simply collapsed, extended, and rotated nearly in place in response to a more mobile, laterally extending basement. As rotational faulting continued, the rigid surface plate was tectonically removed laterally (Fig. 8D). This tectonic denudation in combination with the somewhat earlier thinning of rock within the mylonitic zone "instantly" disrupted isostatic equilibrium and promoted uplift (see Hyndman, this volume). Initial phases of uplift (related to mylonitic thinning) may have been northwest-oriented and localized where ductile thinning was greatest. As uplift (arching) proceeded, it would facilitate denudational removal of upper-plate rocks by creating a gravity gradient. Some formation of chloritic breccia may be due to this late gravitational adjustment stage.

We believe that the above model, although uncertain in some specific aspects, provides a conceptual basis for understanding the complexes. The model explains nicely the major characteristics observed from complex to complex. Following Davis (1977), we envision the mylonitic zone as an extensional-flattening feature akin to necking-down of a somewhat ductile layer. Acceptance of an extensional origin of the complexes enables comprehension of their local and regional characteristics. The above model differs somewhat from that presented by Davis (1977) and Davis and Coney (1979) in that we portray the complexes as areas between separating blocks while Davis considers these terranes to be *culminations* of "megaboudins." Nevertheless, the two concepts are variations of the same major theme.

The complexes are such fundamental features of the North American Cordillera that they must have origins related to plate tectonics. Their close temporal and spatial association with regions of Tertiary magmatism (Coney, this volume) suggests a genetic relationship. That relationship possibly is provided by the analysis of Coney and Reynolds (1977). They concluded that a Benioff zone which had dipped gently under the southwestern United States in early Tertiary time steepened and collapsed during mid-Tertiary time. The steepening and collapse of the slab would create tremendous instability in the mantle beneath the western United States, possibly providing the mechanisms of tensional deformation and magmatism apparently required for development of the metamorphic core complexes (see also Coney, this volume).

ACKNOWLEDGMENTS

Our curiosity and research into the complexes were kindled at the University of Arizona through an association with pioneers in this area such as Evans B. Mayo, Paul E. Damon, George H. Davis, and Peter J. Coney. In particular, we appreciate and have benefited greatly from field trips and discussions with Coney and Davis.

Many of our conclusions would not be possible without the unending support of Paul E. Damon, M. Shafiqullah, and Dan Lynch of the Laboratory of Isotope Geochemistry, University of Arizona (National Science Foundation Grant EAR 76-02590).

To Conoco, Inc., we owe thanks for much logistical support, cost-sharing of isotopic dating, and permission to report part of the geologic data contained in this paper. Reynolds is grateful for support from the University of Arizona and the Arizona Bureau of Geology and Mineral Technology. We wish to thank Greg A. Davis, Ed Dewitt, Warren Hamilton, Donald W. Hyndman, Stanley B. Keith, H. Wesley Peirce, Terry J. Shackelford, and Leon T. Silver for discussions and assistance during various stages of the work. Max D. Crittenden, Jr., G. A. Davis, Ed Dewitt, and Stanley B. Keith critically reviewed early drafts of the manuscript and suggested changes that promoted clarity of reason and expression.

REFERENCES CITED

Anderson, C. A., and Silver, L. T., 1976, Yavapai Series—A greenstone belt: Arizona Geological Society Digest, v. 10, p. 13–26.

Banks, N. G., 1980, Geology of a zone of metamorphic core compexes in southeastern Arizona: Geological Society of America Memoir 153 (this volume).

Banks, N. G., and others, 1977, Reconnaissance geologic map of the Tortolita Mountains quadrangle, Arizona: U.S. Geological Survey Miscellaneous Field Studies Map MF-864, scale 1:62,500.

——1978, Radiometric and chemical data for rocks of the Tortolita Mountains 15' quadrangle, Pinal County, Arizona: Isochron/West, no. 22, p. 17–22.

Blacet, P. M., and Miller, S. T., 1978, Reconnaissance geologic map of the Jackson Mountain quadrangle, Graham County, Arizona: U.S. Geological Survey Miscellaneous Field Studies Map MF-939, scale 1:62,500.

Bryant, D. L., 1968, Diagnostic characteristics of the Paleozoic formations of southeastern Arizona: Arizona Geological Society Guidebook III, p. 33–47.

Budden, R. T., 1976, The Tortolita and Santa Catalina Mountains—A spatially continuous gneissic complex [M.S. thesis]: Tucson, University of Arizona, 133 p.

Chase, R. B., 1973, Petrology of the northeastern border zone of the Idaho batholith, Bitterroot Range, Montana: Montana Bureau Mines and Geology Memoir 43, 28 p.

Compton, R. R., and others, 1977, Oligocene and Miocene metamorphism, folding, and low-angle faulting in northwestern Utah: Geological Society of America Bulletin, v. 88 , p. 1237–1250.

Coney, P. J., 1980, Cordilleran metamorphic core complexes: An overview: Geological Society of America Memoir 153 (this volume).

Coney, P. J., and Reynolds, S. J., 1977, Cordilleran Benioff zones: Nature, v. 270, p. 403–406.

Creasey, S. C., and Krieger, M. H., 1978, Galiuro volcanics, Pinal, Graham, and Cochise Counties, Arizona: U.S. Geological Survey Journal of Research, v. 6, p. 115–131.

Creasey, S. C., and others, 1977, Middle Tertiary plutonism in the Santa Catalina and Tortolita Mountains, Arizona: U.S. Geological Survey Journal of Research, v. 5, p. 705–717.

Crittenden, M. D., Jr., Coney, P. J., and Davis, G. H., 1978, Penrose Conference report: Tectonic significance of metamorphic core complexes in the North American Cordillera: Geology, v. 6, p. 79–80.

Damon, P. E., 1968, Correlation and chronology of ore deposits and volcanic rocks, in Annual progress report COO-689-100, Contract AT(11-1)-689 to U.S. Atomic Energy Commission: Tucson, University of Arizona, Geochronology Labs, 75 p.

Damon, P. E., and Mauger, R. L., 1966, Epeirogeny-orogeny viewed from the Basin and Range province: Society of Mining Engineers AIME Transactions, v. 235, p. 99–112.

Damon, P. E., Erickson, R. C., and Livingston, D. E., 1963, K-Ar dating of Basin and Range uplift, Catalina Mountains, Arizona: Nuclear Geophysics–Nuclear Science Journal, v. 38, p. 113–121.

Davis, G. A., and others, 1977, Enigmatic Miocene low-angle faulting, southeastern California and west-central Arizona—Suprastructural tectonics?: Geological Society of America Abstracts with Programs, v. 9, p. 943–944.

——1980, Mylonitization and detachment faulting in the Whipple-Buckskin-Rawhide Mountains terrane, southeastern California and western Arizona: Geological Society of America Memoir 153 (this volume).

Davis, G. H., 1975, Gravity-induced folding of a gneiss dome complex, Rincon Mountains, Arizona: Geological Society of America Bulletin, v. 86, p. 979–990.

——1977, Characteristics of metamorphic core complexes, southern Arizona: Geological Society of America Abstracts with Programs, v. 9, p. 944.

Davis, G. H., 1979, Laramide folding and faulting in southeastern Arizona: American Journal of Science, v. 279, p. 543–569.
——1980, Structural characteristics of metamorphic core complexes, southern Arizona: Geological Society of America Memoir 153 (this volume).
Davis, G. H., and Coney, P. J., 1979, Geologic development of the Cordilleran metamorphic core complexes: Geology, v. 7, p. 120–124.
Davis, G. H., and others, 1975, Origin of lineation in the Catalina-Rincon-Tortolita gneiss complexes, Arizona: Geological Society of America Abstracts with Programs, v. 7, p. 602.
Drewes, H., 1974, Geologic map and sections of the Happy Valley quadrangle, Cochise County, Arizona: U.S. Geological Survey Miscellaneous Inventory Map I-832, scale 1:48,000.
——1977, Geologic map and sections of the Rincon Valley quadrangle, Pima County, Arizona: U.S. Geological Survey Miscellaneous Inventory Map I-997, scale 1:48,000.
——1978, The Cordilleran orogenic belt between Nevada and Chihuahua: Geological Society of America Bulletin, v. 89, p. 641–657.
Eberly, L. D., and Stanley, T. D., Jr., 1978, Cenozoic stratigraphy and geologic history of southwestern Arizona: Geological Society of America Bulletin, v. 89, p. 921–940.
Harding, L. E., 1978, Petrology and tectonic setting of the Livingston Hills Formation, Yuma County, Arizona [M.S. thesis]: Tucson, University of Arizona, 89 p.
Hayes, P. T., 1969, Geology and topography, *in* Mineral and water resources of Arizona: Arizona Bureau of Mines Bulletin 180, p. 35–57.
Hayes, P. T., and Drewes, H., 1968, Mesozoic sedimentary and volcanic rocks of southeastern Arizona, *in* Titley, S. R., ed., Southern Arizona Guidebook III: Tucson, Arizona Geological Society, p. 49–58.
Higgins, M. E., 1971, Cataclastic rocks: U.S. Geological Survey Professional Paper 687, 97 p.
Hyndman, D. W., 1980, Bitterroot dome–Sapphire tectonic block, an example of a plutonic-core gneiss-dome complex with its detached suprastructure: Geological Society of America Memoir 153 (this volume).
Keith, S. B., 1978, Paleosubduction geometries inferred from Cretaceous and Tertiary magmatic patterns in southwestern North America: Geology, v. 6, p. 516–521.
Keith, S. B., and others, 1980, Evidence for multiple intrusion and deformation within the Santa Catalina–Rincon–Tortolita crystalline complex, southeastern Arizona: Geological Society of America Memoir 153 (this volume).

McKee, E. D., 1951, Sedimentary basins of Arizona and adjoining areas: Geological Society of America Bulletin, v. 62, p. 481–506.
Miller, F. K., 1970, Geologic map of the Quartzite quadrangle, Yuma County, Arizona: U.S. Geological Survey Map GQ-841, scale 1:62,500.
Peirce, H. W., 1976, Elements of Paleozoic tectonics in Arizona: Arizona Geological Society Digest, v. 10, p. 37–57.
Rehrig, W. A., and Heidrick, T. L., 1976, Regional tectonic stress during the Laramide and late Tertiary intrusive periods, Basin and Range province, Arizona: Arizona Geological Society Digest, v. 10, p. 205–228.
Rehrig, W. A., and Reynolds, S. J., 1977, A northwest zone of metamorphic core complexes in Arizona: Geological Society of America Abstracts with Programs, v. 9, p. 1139.
Reynolds, S. J., and Rehrig, W. A., 1980, Mid-Tertiary plutonism and mylonitization, South Mountains, central Arizona: Geological Society of America Memoir 153 (this volume).
Shackelford, T. J., 1976, Structural geology of the Rawhide Mountains, Mohave County, Arizona [Ph.D. dissert.]: Los Angeles, University of Southern California, 176 p.
——1977, Late Tertiary tectonic denudation of a Mesozoic(?) gneiss complex, Rawhide Mountains, Arizona: Geological Society of America Abstracts with Programs, v. 9, p. 1169.
Shafiqullah, M., Damon, P. E., and Peirce, H. W., 1976a, Late Cenozoic tectonic development of Arizona Basin and Range province [abs.]: International Geological Congress, 25th, Syndey, v. 1, p. 99.
Shafiqullah, M., Lynch, D. J., and Damon, P. E., 1976b, Geology, geochronology, and geochemistry of the Picacho Peak area, Pinal County, Arizona: Arizona Geological Society Digest, v. 10, p. 311.
Shakel, D. W., 1974, The geology of layered gneisses in part of the Santa Catalina forerange, Pima County, Arizona [M.S. thesis]: Tucson, University of Arizona, 233 p.
Shakel, D. W., Silver, L. T., and Damon, P. E., 1977, Observations on the history of the gneissic core complex, Santa Catalina Mountains, southern Arizona: Geological Society of America Abstracts with Programs, v. 9, p. 1169–1170.
Simons, F. S., 1964, Geology of the Klondyke quadrangle, Graham and Pinal Counties, Arizona: U.S. Geological Survey Professional Paper 461, 174 p.
Snook, J. R., 1965, Metamorphic and structural history of "Colville batholith" gneisses, north-central Washington: Geological Society of America Bulletin, v. 76, p. 759–776.
Streckeisen, A., 1976, To each plutonic rock its proper

name: Earth-Science Reviews, v. 12, p. 1-33.

Swan, M. M., 1976, The Stockton Pass fault—An element of the Texas lineament [Ph.D. thesis]: Tucson, University of Arizona, 119 p.

Varga, R. J., 1976, Stratigraphy and superposed deformation of a Paleozoic and Mesozoic sedimentary sequence in the Harquahala Mountains, Arizona [M.S. thesis]: Tucson, University of Arizona, 61 p.

Waters, A. C., and Krauskopf, K., 1941, Protoclastic border of the Colville batholith: Geological Society of America Bulletin, v. 52, p. 1355-1417.

Wilson, E. D., 1962, A resume of the geology of Arizona: Arizona Bureau of Mines Bulletin, v. 171, 140 p.

Wilson, E. D., and Moore, R. T., 1959, Structure of Basin and Range province in Arizona: Arizona Geological Society, Southern Arizona Guidebook II, p. 89-105.

Wilson, E. D., Moore, R. T., and Cooper, J. R., 1969, Geologic map of Arizona: Washington, D.C., Arizona Bureau of Mines and U.S. Geological Survey, scale 1:500,000.

Wright, Lauren, 1976, Late Cenozoic fault patterns and stress fields in the Great Basin and westward displacement of the Sierra Nevada block: Geology, v. 4, p. 489-494.

Yeend, W., 1976, Reconnaissance geologic map of the Picacho Mountains, Arizona: U.S. Geological Survey Miscellaneous Field Studies Map MF-778, scale 1:62,500.

MANUSCRIPT RECEIVED BY THE SOCIETY JUNE 21, 1979

MANUSCRIPT ACCEPTED AUGUST 7, 1979

Printed in U.S.A.

Mid-Tertiary plutonism and mylonitization, South Mountains, central Arizona

STEPHEN J. REYNOLDS
Department of Geosciences
University of Arizona
Tucson, Arizona 85721

WILLIAM A. REHRIG
Conoco, Inc.
555 17th Street
Denver, Colorado 80202

ABSTRACT

Rocks in the South Mountains of central Arizona are representative of rocks found in metamorphic core complexes elsewhere in Arizona. These core-complex terranes are in part characterized by low-angle mylonitic foliation that contains penetrative northeast-trending mineral lineations and pervasive smearing out of mineral grains. In the South Mountains, mylonitic rocks form a doubly plunging, northeast-trending foliation arch and have been derived from Precambrian amphibolite gneiss and a composite mid-Tertiary pluton. The pluton is undeformed in the core of the arch, but shows a progressive increase in pervasiveness of mylonitic fabric up structural section. Mylonitic plutonic rocks are exposed as a carapace overlying their less-deformed equivalents. Mylonitically foliated Precambrian amphibolite gneiss is restricted to a zone underlain and overlain by nonmylonitic (crystalloblastic) gneisses that are lithologically identical and largely retentive of their Precambrian foliation.

Fabrics in mylonitic rocks indicate extension parallel to the east-northeast–trending lineation and flattening perpendicular to the gently dipping foliation. The fabric is well dated as late Oligocene to early Miocene (25 to 20 m.y. B.P.). Mylonitic deformation was followed by more brittle deformation which produced a chloritic breccia that overlies mylonitic plutonic rocks in the northeast half of the arch. The chloritic breccia is probably related to normal faulting along a low-angle dislocation surface.

INTRODUCTION

Through the center of the western United States runs a discontinuous belt of mountain ranges that contain a distinctive assemblage of rocks and structures (Coney, this volume; Rehrig and Reynolds, this volume). These ranges, recently referred to as metamorphic core complexes (Crittenden and others, 1978; Coney, 1978), have striking similarities in lithologic, structural, and geochronologic aspects (Coney, this volume; Davis, this volume; Rehrig and Reynolds, this volume). The complexes generally contain metamorphic rocks whose gently dipping foliation defines asymmetrical arches or domes. Deeper structural levels of the complexes typically consist of metamorphic, migmatitic, and plutonic rocks (Fig. 1). These rocks commonly grade upward into gneissic rocks predominated by mylonitic textures. On some gently dipping sides of the domes, upper levels of the mylonitic rocks are locally brecciated, jointed, chloritic, and overlain by a gently dipping surface of dislocation. Above the dislocation surface lie highly faulted, generally unmetamorphosed sedimentary and igneous rocks.

Major unsolved problems that plague understanding of individual complexes involve ages of protoliths of metamorphosed and mylonitic rocks; ages of metamorphism, mylonitization, and dislocation; spatial distribution and kinematics of deformation; and regional tectonic setting and significance. Probable solutions to some of these problems have recently been suggested by results of reconnaissance mapping and detailed geochronology of a metamorphic core complex terrane in the South Mountains, an approximately 18-km-long, northeast-trending range located immediately south of Phoenix, Arizona (Figs. 2, 3). Rocks in the South Mountains are well exposed and easily accessible; they provide insight into the importance of *mid-Tertiary* tectonism in development of the mylonitic rocks in the Arizona complexes (Reynolds and others, 1978).

Igneous rock nomenclature used in this paper is that adopted by IUGS (Streckeisen, 1976). The term

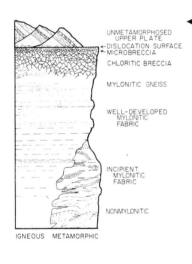

Figure 1. Schematic structural section across one flank of a typical metamorphic core complex.

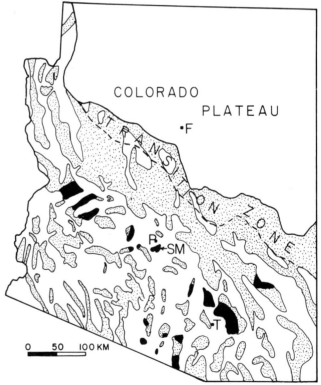

Figure 2. Regional geologic setting and location of South Mountains. Depicted in the Basin and Range province are ranges of bedrock older than late Tertiary (stippled) surrounded by upper Tertiary-Quaternary sedimentary and volcanic rocks (unpatterned). Ranges possessing characteristics of metamorphic core complexes are shown as black. Positions of the South Mountains (SM), Flagstaff (F), Phoenix (P), and Tucson (T) are shown.

Figure 3. Aerial photograph of South Mountains looking southwest. Note archlike topography of closest half of the range.

"mylonitic" is used strictly in a descriptive, nongenetic sense. In thin section, rocks described as mylonitic exhibit generally comminuted, brittlely deformed feldspars but recrystallized quartz. The rocks are very similar to photographs of hand specimens and photomicrographs of protomylonite, mylonite, and mylonite gneiss as presented in the classification of Higgins (1971). The term "dislocation surface" is intended to be purely descriptive and to signify dislocation between rocks above and below the surface. No genetic connotation is attached to the term, and its use does *not* necessarily imply large amounts of movement.

REGIONAL GEOLOGIC SETTING

The South Mountains (also referred to as South Mountain or the Salt River Mountains) are one of a series of metamorphic core complexes that are exposed in a northwest-trending belt across southern and western Arizona (Rehrig and Reynolds, 1977 and this volume). These complexes are situated within the Basin and Range province of Arizona (Fig. 2), a region characterized by north- to

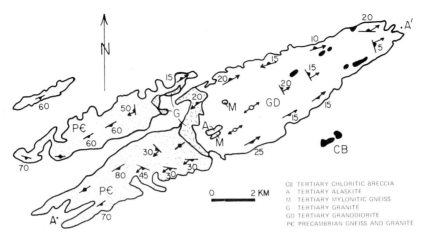

Figure 4. Generalized geologic map of the South Mountains. Numerous dikes and small outcrops are omitted and upper Tertiary–Quaternary surficial deposits are unpatterned. Location of cross section (Fig. 9) is labeled A and A'. Attitude of foliation and trend of lineation are shown, with subhorizontal foliation being depicted by open circles.

Figure 5 (facing pages). Photographs of typical rock types. (A) Precambrian amphibolite gneiss whose older, steep foliation (as defined by compositional layering) is cut by more gently dipping mylonitic foliation. (B) Mylonitic granodiorite with foliated aplite and quartz veins. (C) Mylonitic gneiss composed of mylonitic granodiorite and interlayered alaskite (light colored). (D) Chloritic breccia. (E) Microbreccia ledge that is light colored on weathered surfaces but is dark gray on fresh surfaces; above ledge is a planar surface of dislocation.

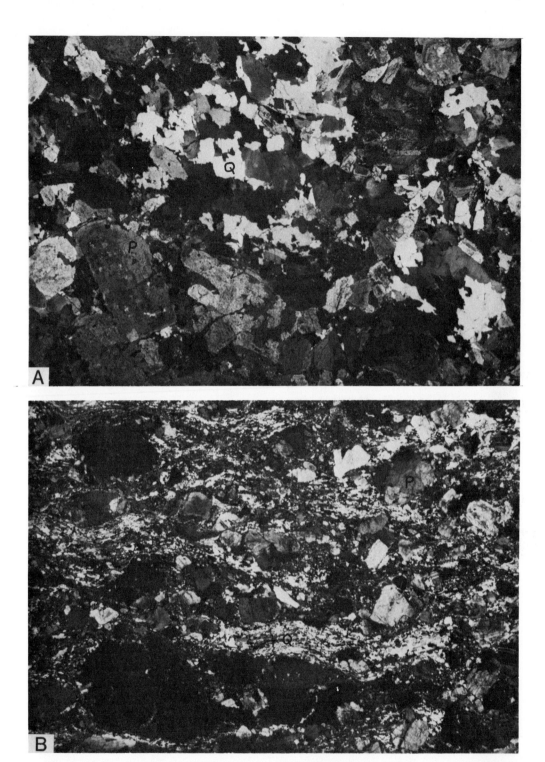

Figure 6. Photomicrographs of granodiorite in various structural conditions. Width of photomicrographs represents about 1.7 cm. (A) Relatively undeformed granodiorite with slight cataclasis of plagioclase (P) and weak undulatory extinction in quartz (Q). (B) Moderately mylonitic granodiorite with finely crystalline aggregates of quartz (Q) and brittlely deformed porphyroclasts of K-feldspar (dark) and plagioclase (P). (C) Intensely mylonitic granodiorite (mylonitic gneiss) with lenses of finely crystalline quartz (light) interlayered with plagioclase, K-feldspar, and biotite (all dark) (note normal fault cutting mylonitic foliation). (D) Chloritic breccia with angular fragments of quartz, K-feldspar, and plagioclase from original mylonitic granodiorite. (E) Microbreccia showing reduced size of fragments.

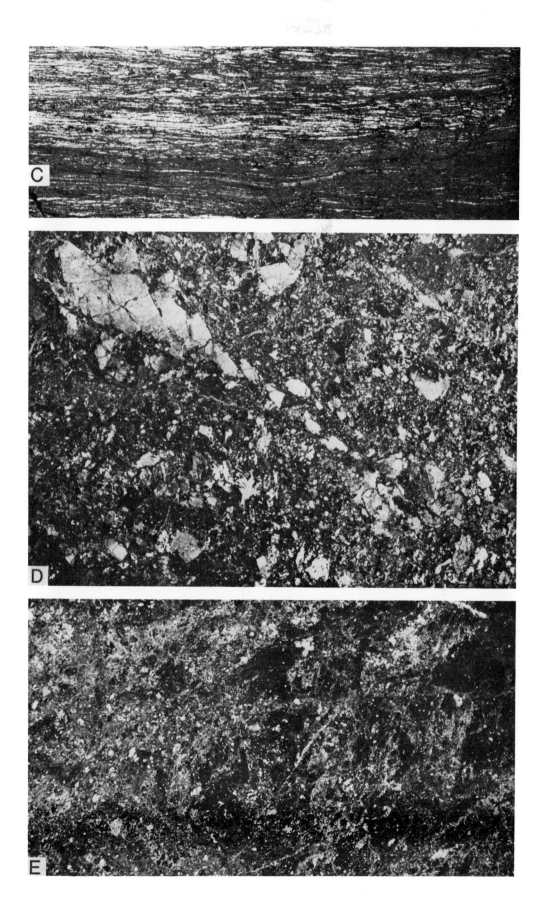

northwest-trending fault-block mountains and basins whose configuration has largely developed since 15 m.y. B.P. (Eberly and Stanley, 1978; Peirce, 1976; Shafiqullah and others, 1976). Exposures in ranges adjacent to the South Mountains are mostly Precambrian (1.4 to 1.8 b.y. B.P.) metamorphic and granitic rocks, Late Cretaceous and Tertiary intermediate to felsic plutons, mid-Tertiary clastic and volcanic rocks, and other terranes typical of metamorphic core complexes. Many intermountain basins are filled by thousands of metres of semiconsolidated, variably sized clastic material (Eberly and Stanley, 1978). In addition, large thicknesses of evaporites are present in some basins such as the Phoenix Basin, immediately north and northwest of the South Mountains (Eaton and others, 1972; Peirce, 1976).

GENERAL GEOLOGIC RELATIONSHIPS

Seven rock units were mapped in the range (Fig. 4). Precambrian rocks exposed in the western half of the range (Avedisian, 1966) are amphibolite-grade gneiss and schist with local intrusive masses. A dominant Precambrian rock type is amphibolite (Fig. 5A) composed of plagioclase and hornblende with variable amounts of quartz and biotite. Another common rock type consists of gneiss whose relative proportions of feldspar, quartz, amphibole, and biotite range from one extreme to another. Schist that is rich in biotite and muscovite constitutes an additional minor rock type. Near the western end of the range, variably foliated intermediate to granitic plutons intrude the metamorphic rocks.

Almost the entire eastern half of the range is occupied by Tertiary granodiorite (Figs. 5B, 6A, 6B) that generally displays a weakly to strongly developed mylonitic foliation. In structurally low exposures, the granodiorite lacks any mylonitic fabric and is essentially undeformed. The granodiorite is equigranular to slightly porphyritic and has a fairly uniform mineral content consisting of 40% to 50% plagioclase, 30% to 35% quartz, 15% to 20% K-feldspar, and 5% biotite.

In the center of the range, a Tertiary granite has intruded between the Precambrian amphibolite gneiss and the granodiorite. The granite's western contact with the amphibolite gneiss dips steeply or moderately to the west except in the northernmost outcrops where the contact is subhorizontal and the roof of the pluton is exposed. Along its eastern boundary, the granite both sharply intrudes and grades into the granodiorite. The contact dips moderately to the west. Overall, the granite has a thick, tabular form that dips moderately to the west and almost everywhere occupies a position between the Precambrian amphibolite gneiss and the granodiorite.

Where undeformed, the granite contains 30% to 40% quartz, 30% to 40% plagioclase, 20% to 30% K-feldspar, and less than 5% biotite (Avedisian, 1966). The granite is locally highly altered with abundant sericite and limonite, both as disseminations and concentrations along steep, north-northwest–striking fractures. In its higher structural levels, the granite has a well-developed, low-angle mylonitic foliation.

In the center of the range, north-northwest–striking dikes of variable grain size and composition (andesitic to rhyolitic) intrude all three rock units described above (Fig. 7). The dikes are as thick as 10 m and vary from undeformed to well foliated and lineated (mylonitic).

Another unit, mylonitic gneiss, is exposed in several topographically high areas in the center of the range (Fig. 7). The mylonitic gneiss is composed of strongly mylonitic layers and lenses of granodiorite, alaskite, aplite, and amphibolite (Figs. 5C, 6C). Discrete layers in the mylonitic gneiss are as thin as 1 cm and locally thicker than 2 m. The mylonitic gneiss *grades downward* into less mylonitic granodiorite and is overlain in one exposure by a mappable outcrop of alaskite. The alaskite is generally mylonitic and is composed of only feldspar and quartz.

Chloritic breccia, the last bedrock unit to be discussed, gradationally overlies the mylonitic granodiorite in numerous small knobs in the eastern third of the range. The breccia (Figs. 5D, 6D) is green-gray and is composed of chloritic, locally hematitic, highly jointed and brecciated masses of the mylonitic granodiorite. The breccia is characterized by closely spaced, curviplanar joints that exhibit anastomosing patterns. Angular fragments of granodiorite are seen on both outcrop (Fig. 5D) and microscopic (Fig. 6D) scales. Biotite in the breccia is generally totally altered to chlorite.

In several places, the chloritic breccia is overlain by a distinctive ledge of dense microbreccia (Fig. 5E). The microbreccia is brown or tan where weathered, but characteristically dark gray on fresh surfaces. The microbreccia is remarkably fine grained and contains small, angular fragments of quartz and feldspar (Fig. 6E). The top of the microbreccia ledge is a smooth, locally slickensided, planar surface that strongly resembles dislocation surfaces exposed in other metamorphic complexes.

Bedrock exposures of the range are surrounded by upper Tertiary–Quaternary semiconsolidated sand and gravel. In addition, talus from bedrock exposures forms aprons along the base of many slopes.

STRUCTURAL RELATIONSHIPS

Most rocks in the range exhibit a gently dipping mylonitic foliation that contains a pervasive N60°E-trending lineation. The foliation is defined by planar mineral aggregates and thin bands of intensely granulated and recrystallized rock. Plagioclase and K-feldspar are generally brittlely deformed and commonly form porphyroclasts that are enveloped in a foliated matrix of finer-grained quartz,

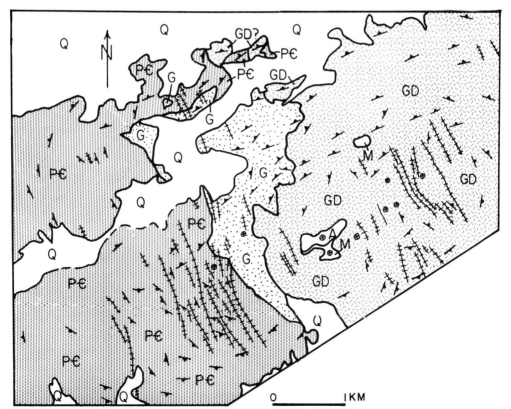

Figure 7. Reconnaissance geologic map of the central part of the South Mountains. Map units are same as those in Figure 4 and include upper Tertiary–Quaternary surficial deposits (Q) and Tertiary dikes (hatched lines).

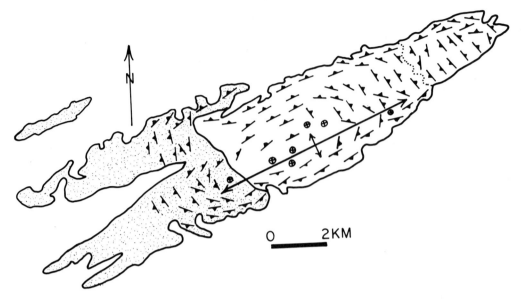

Figure 8. Attitude of mylonitic foliation where it is present. Precambrian rocks are stippled, and all other rocks are unpatterned. Axis of major arch is indicated by large arrow. See Figure 4 for average dips of foliation in various areas.

feldspar, and biotite. Some isolated, single crystals of quartz are brittlely deformed, but the majority of quartz in the rock forms foliated, largely recrystallized and sutured aggregates that interfinger with adjacent, less-foliated rock. Mylonitic foliation is dominantly formed by these flattened quartz aggregates. Lineation in mylonitic rocks is defined by elongate mineral aggregates and discontinuous streaks of granulated feldspar. Lineation with a slickensiding aspect is common on aplites, quartz veins, and fractures contained within the mylonitic rocks.

Mylonitically foliated rocks contain joints, quartz-filled tension fractures, and "ductile normal faults" that mostly strike north-northwest, perpendicular to lineation. Inclusions in deformed plutonic rocks are elongate parallel to lineation and flattened perpendicular to foliation with average axial ratios for 15 inclusions of 10:2.5:1. Folds are rare in mylonitic plutonic rocks, but are more abundant in mylonitically deformed Precambrian amphibolite gneiss.

Gently dipping mylonitic foliation defines an asymmetrical northeast-trending, doubly plunging arch or dome (Fig. 8). The foliation generally dips less than 20° where it is contained within plutonic rocks, but is more steeply dipping where it affects Precambrian amphibolite gneiss (Fig. 9). The simple pattern of the arch is interrupted on its northeast end, where southwest-dipping foliation is present. This attitude of foliation is restricted to structurally high rocks that are chloritic, jointed, and brecciated (northeast of dotted line in Fig. 8). Foliation in nonchloritic, nonjointed, mylonitic granodiorite below the chloritic rocks is northeast-dipping; this orientation is consistent with the form of the major arch.

Excellent exposures in the range display the three-dimensional distribution of mylonitic fabric and variations in intensity of mylonitization. Both the granite and granodiorite are undeformed in the core of the arch except for jointing and minor faulting. However, both plutons exhibit a gradual increase in intensity of mylonitization toward the top and margins of the arch (up structural section). The structurally lowest manifestation of this deformation *may* be the development of small joints and normal faults that cut nonfoliated plutonic rock and have east-northeast–trending slickensides. These faults locally have thin selvages of crushed, highly slickensided rock. Farther up section, the intervening rock gradually becomes more strongly foliated until the entire rock is best described as mylonitic.

Where the rock is moderately to well foliated, thin and discontinuous quartz veins parallel to foliation are abundant and exhibit east-northeast–trending lineation on their boundaries. Quartz veins are much more prevalent in well-foliated rock than in underlying, less-deformed equivalent rock units.

A similar distribution of mylonitic fabric is revealed where mylonitization affects Precambrian amphibolite gneiss. In the core of the range near the granite contact, the amphibolite gneiss possesses a foliation that is generally nonmylonitic, northeast-striking, and steeply dipping. This crystalloblastic foliation is typically defined by compositional layering and generally contains no lineation even though folds are locally abundant. The foliation is interpreted to be of Precambrian age because it is similar in style and orientation to that exhibited by Precambrian metamorphic terranes elsewhere in central Arizona. Even in the lowest exposures in the range, the steep foliation is locally cut and transposed (or rotated) by thin zones of moderately dipping mylonitic foliation. However, the intensity of mylonitic deformation increases upward from the core to several 15-m-thick zones of northeast-lineated mylonitic gneiss that cut equally thick zones of much less mylonitic amphibolite gneiss. It is important to note that the mylonitic fabric also decreases in intensity *upward* from the main zones of mylonitic rock. At high structural levels in the western parts of the range, foliation in the amphibolite gneiss is again *nonmylonitic*, generally east to northeast striking, and steeply dipping.

North-northwest–striking dikes are likewise undeformed in the core of the arch. They are also generally undeformed where they intrude rocks with moderately well-developed mylonitic fabric. However, in structurally high parts of the range where adjacent rocks are intensely deformed, the dikes locally exhibit a gently dipping mylonitic foliation and an east-northeast–trending lineation. Undeformed dikes are commonly near well-foliated dikes of similar rock type and strike.

Another important lithologic and structural transition is exposed along the northeast end of the range where mylonitic granodiorite grades upward into chloritic breccia. Structurally lowest exposures of the granodiorite in this area are nonchloritic and well foliated. Up section, chlorite and curviplanar joints are present in the granodiorite. The rocks are progressively more jointed and brecciated higher in the section where they ultimately grade into chloritic breccia. Remnants of mylonitic foliation in the granodiorite are preserved in the breccia. Relict mylonitic foliation in the breccia generally dips to the southwest: this indicates that total disorientation and random rotation of the foliation did not occur, except locally. Joints, breccia zones, and normal faults (northeast side down) have variable northwest strikes. Slickensides in the breccia and microbreccia have scattered but dominantly northeast trends.

GEOCHRONOLOGY

A relative chronology can be inferred from field relationships in the area. The oldest rocks are the amphibolite gneisses that were metamorphosed and deformed; they now bear a steep, northeast-

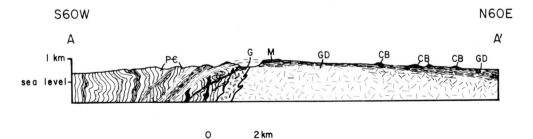

Figure 9. Northeast-trending cross section along axis of range. See Figure 4 for location of section and explanation of rock units.

striking foliation. These rocks were intruded successively by the granodiorite and granite plutons. The granite locally intrudes and is therefore somewhat younger than the granodiorite. In other exposures, however, the contact between the plutons is gradational and thus indicates their partial contemporaneity.

Nearly synchronous mylonitization and dike intrusion postdated emplacement of the plutons. The granite, in some places, is apparently slightly less deformed than the granodiorite where the two are in contact; therefore, mylonitization may have begun during emplacement of the granite. Formation of most mylonitic foliation predates the initial emplacement of the dikes from the fact that the dikes cut across moderately well-developed fabric. However, significant mylonitization also occurred later than emplacement of the dikes, as shown by their locally high degree of mylonitic deformation adjacent to mylonitic rocks. Clearly, the dikes and some mylonitization are nearly the same age, an inference supported by the presence of undeformed dikes near highly deformed dikes of similar, if not identical, strike and rock type.

Chloritic breccia deforms and postdates mylonitic fabric in the granodiorite. Mylonitic foliation in the granodiorite is generally jointed, brecciated, faulted, and rotated where it is partially converted to the chloritic breccia. In addition, discrete fragments of the *mylonitic* granodiorite are present in outcrops and thin sections of the breccia. The chloritic breccia is younger than most or all of the dikes from the facts that no dikes have been observed cutting the breccia and some of the youngest dikes are clearly brecciated.

Relative ages of the chloritic breccia and arching of foliation in the mylonitic rocks are less certain. The chloritic breccia generally conforms to the arch, so either it has been arched or it formed on a previously arched surface. Slickensides in the breccia commonly trend northeast; this indicates that movement was mostly unidirectional and *not everywhere* down the present structural dip of the breccia zone and underlying mylonitic rocks. Some northwest- and southeast-trending slickensides are present along with northeast-trending slickensides on some exposures of the breccia on the flanks of the arch; therefore, some movement may have occurred down the flanks of the arch as it formed.

Radiometric age determinations of rocks in the South Mountains permit assignment of absolute ages to the foregoing relative chronology. Geochronologic research in the complex is continuing and will be discussed in greater detail elsewhere (Reynolds and others, in prep.). Isotopic studies of the amphibolite gneisses are incomplete, but are consistent with a Precambrian age as suggested by similarities in rock type and structure to isotopically dated Precambrian rocks (1.7 b.y. B.P.) elsewhere in Arizona (Anderson and Silver, 1976). The granodiorite and granite plutons involve rock types, mineralized fracture patterns (see Rehrig and Heidrick, 1976), and aplite orientations similar to other mid-Tertiary plutons of southern and central Arizona. K-Ar biotite ages determined by M. Shafiqullah and P. E. Damon at the University of Arizona are 20.3 and 20.1 m.y. on relatively undeformed granodiorite and 19.2 m.y. on undeformed granite. Rb-Sr whole-rock isochrons indicate ages of about 25 m.y. for both the granodiorite and granite (Figs. 10A, 10B). Included on the granodiorite isochron plot are four granodiorite points and three aplite points. Included on the granite isochron are two granite points, one granitic dike analysis, and one aplite point. Mineralogic and trace-element analyses strongly support a cogenetic relationship between the granodiorite, granite, granitic dikes, and aplites (Reynolds, in prep.). All data, therefore, require a mid-Tertiary (late Oligocene) age for both the granodiorite and granite.

The abundant dikes are also mid-Tertiary; they intrude the plutons and are strikingly similar in trend and rock type to mid-Tertiary dikes elsewhere (Rehrig and Heidrick, 1976; Rehrig and Reynolds, this volume). The dikes are probably no younger than the 19-m.y. K-Ar age on the granite; they are abundant where the granite sample was collected and might have reset the age of the sample. In addition, preliminary Rb-Sr analyses of the dikes suggest that many are cogenetic with the pluton. Mylonitization must also be mid-Tertiary as it only slightly postdates the pluton and is nearly the same

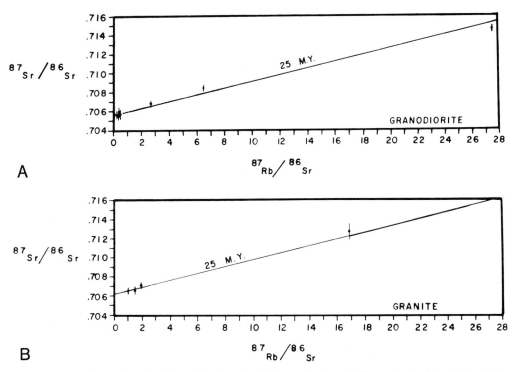

Figure 10. Rb-Sr whole-rock reference isochrons for (A) granodiorite and (B) granite. Analytical data will be published elsewhere (Reynolds and others, in prep.).

age as the dikes. Mylonitization took place before final cooling in the complex, which is defined by the K-Ar ages. Penetrative mylonitization in this range was therefore mid-Tertiary (between 25 and 20 m.y. B.P.).

The chloritic breccia must be younger than the mylonitic fabric and is similar to chloritic breccias and related dislocation surfaces in western Arizona, which are about *as young as* 15 m.y. old (Shackelford, 1976; Davis and others, 1977; Otten, 1978). Basin-range faulting, which mainly formed the *present-day* mountains and basins, postdated brecciation and occurred mainly between 14 and 8 m.y. B.P. (Peirce, 1976; Shafiqullah and others, 1976; Eberly and Stanley, 1978).

SUMMARY

Some of the more important observations should be summarized before inferences and speculations can be discussed.

1. The complex is asymmetrical: gently dipping mylonitic granodiorite is overlain by chlorite breccia in the northeast half of the range, whereas no chlorite breccia is associated with moderately dipping mylonitic amphibolite gneiss on its southwest side.

2. Mylonitization in the amphibolite gneiss, plutons, and dikes increases in intensity upward from the nonmylonitic core of the arch.

3. Mylonitic fabric cuts a broad zone through the Precambrian amphibolite gneiss. Rocks above and below the zone are lithologically identical, and most retain their Precambrian structure.

4. Fabrics in all rock types indicate that mylonitization resulted from extension parallel to lineation and flattening perpendicular to foliation. Lineation contained within the foliation trends N60°E,

orthogonal to dikes, joints, quartz veins, "ductile normal faults," and other extensional features in the mylonitic rocks. Dikes in particular are perpendicular to lineation, are essentially the same age as mylonitic deformation, and are almost totally confined to positions either within or below the zone of mylonitization. They presumably represent fillings of tensional features that formed as rocks extended parallel to lineation.

5. The chlorite breccia deforms and postdates the mylonitic fabric. Mylonitization took place under conditions of biotite stability. Mylonitization resulted in recrystallization and flow of quartz and formed a well-defined foliation and lineation. The deformation that formed the breccia occurred under chlorite-stable conditions and was accompanied by very minor recrystallization of quartz. It destroyed the earlier mylonitic fabric and generally formed no foliation. The chlorite breccia is not present everywhere that mylonitic rocks are. The two clearly are drastically different in structural style and are at least partly different in relative age.

6. Mylonitization is of mid-Tertiary (late Oligocene to early Miocene) age and is exposed in a range characterized by abundant evidence of mid-Tertiary plutonism. Mylonitization is only slightly later in age than the plutonism and is not a protoclastic phenomenon from the fact that the mylonitic fabric is superimposed across cooling joints, dikes, aplites, and quartz veins which postdate solidification of the plutons.

7. Foliation in the Precambrian amphibolite gneiss was apparently little affected by Mesozoic and Laramide tectonics. It maintained its steep, northeast-striking orientation from its formation 1.7 b.y. ago until mid-Tertiary time, when the plutons were intruded discordantly across it.

SPECULATIONS

Some speculations regarding origin of the complex can be made after considering the foregoing observations and inferences. The spatial and temporal association of mylonitization and the mid-Tertiary plutons implies a genetic relationship. However, because mylonitic fabric cuts through the amphibolite gneiss, the event is not strictly a case of "granite tectonics." Perhaps mylonitization was facilitated by heat that the pluton brought into upper levels of the crust. As the rocks were heated or while they were still hot, they extended parallel to the east-northeast–trending lineation and were flattened perpendicular to the subhorizontal foliation. This was possibly in response to the regional mid-Tertiary stress field of σ_1 (maximum compression) vertical, σ_2 north-northwest, and σ_3 (minimum compression) east-northeast that was proposed by Rehrig and Heidrick (1976). Davis and others (1975) have documented similar orientations for strain and possible stress axes for mylonitic rocks in the Tortolita Mountains near Tucson, Arizona.

All evidence indicates that east-northeast–directed extension slightly predated, was synchronous with, and slightly postdated development of the mylonitic fabric. Hydrothermally altered and mineralized fractures in the mid-Tertiary plutons, when interpreted in the manner suggested by Rehrig and Heidrick (1972, 1976) indicate that σ_3 was east-northeast during late magmatic to postmagmatic cooling and fracturing of the plutons. North-northwest–trending dikes the *same age* as mylonitization indicate east-northeast–west-southwest extension. Mid-Tertiary dikes outside the complex also trend north-northwest and thus indicate that this is a *regional* extension (Rehrig and Heidrick, 1976) and not one related to *local* strains accompanying mylonitization! Dikes that postdate mylonitization in the complex also strike north-northwest and imply that east-northeast–west-southwest extension existed after mylonitization. In addition, fabrics in the chloritic breccia imply that some form of northeast-southwest extension continued after mylonitization ceased.

It is important to emphasize that the mylonitization is not believed to be a part of classic Basin and Range province deformation. Mylonitic rocks evidently occur both in the present-day ranges and the bottoms of the basins. Mylonitization clearly predates essentially all faulting within the past 14 m.y. which formed the *present* basins and ranges (Eberly and Stanley, 1978; Peirce, 1976; Shafiqullah and others, 1976). Therefore, mylonitization is believed to be a manifestation of a poorly understood Oligocene to early Miocene flattening and extensional event that took place along low-angle surfaces and zones. Mylonitic fabric of this type and orientation in southern Arizona is unrelated to Laramide thrusting, as has been hypothesized by some workers (Drewes, 1977; Thorman, 1977).

Exact temperatures of mylonitization are not known, but several observations provide some constraints. Temperature and fluid content must have been sufficient so that, for the strain rates involved, quartz and some biotite could recrystallize while plagioclase and most K-feldspar deformed brittlely. Silica must have been mobile because quartz veins are more abundant where the rock is mylonitic than in nonmylonitic rocks. A similar observation was noted by Chase (1973) in mylonitic gneiss along the Bitterroot front, Montana.

Confining pressure during deformation cannot have been excessive. Plutons slightly predating mylonitization are not deep-level stocks but are instead hydrothermally altered and fractured in a fashion suggestive of fairly low confining pressure. Dikes that are the same age as mylonitization are generally aphanitic; again this situation indicates fast cooling at shallow depths. These dikes and plutons appear to be the only satisfactory way to convey sufficient heat to these high crustal levels.

If the complex is analogous to similar complexes in southern and western Arizona, the chloritic breccia was formed in association with an overlying dislocation surface, above which would lie highly extended rocks as young as early to middle Miocene. In the South Mountains, slickensides, observed displacements of low-angle normal faults in the breccia, and antithetic rotation of mylonitic foliation in the granodiorite all indicate relative transport related to the chloritic breccia as being dominantly to the northeast. The chloritic breccia and overlying dislocation surface are well exposed in the southern foothills of the South Mountains (south of the area shown in Fig. 4). In this area, upper-plate rocks above the dislocation surface are faulted and brecciated amphibolite gneisses. Relict mylonitic foliation in the gneisses dips southwest and is cut by northeast-dipping normal faults. Additional possible upper-plate rocks are brecciated mid-Tertiary volcanic units, tilted and faulted mid-Tertiary sedimentary and volcanic units, and deformed Precambrian(?) granites locally exposed near Tempe (3 to 10 km northeast of the northeasternmost outcrops of mylonitic rocks in the South Mountains). The sedimentary and volcanic units dip moderately to the southwest; this is consistent with antithetic rotation that would have accompanied southwest-to-northeast transport along low-angle, listric normal faults.

Finally, we would like to emphasize that rocks with mylonitic fabrics nearly identical to those in the South Mountains compose large parts of the metamorphic core complexes of Arizona (Davis, this volume; Rehrig and Reynolds, this volume). The *majority* of these ranges are depicted on the geologic map of Arizona (Wilson and others, 1969) as Precambrian gneiss, but several are shown as Tertiary-Cretaceous gneiss. However, studies on the South Mountains indicate that the mylonitic fabric in many of these ranges may largely be *mid-Tertiary* in age.

ACKNOWLEDGMENTS

Our sincere appreciation goes to Continental Oil Company for logistical support and to Paul E. Damon, M. Shafiqullah, and Dan Lynch of the Laboratory of Isotope Geochemistry, Department of Geosciences, University of Arizona, for unending cooperation on radiometric age dates (National Science Foundation Grant EAR76-02590), partly as a cost-sharing, joint research program with Conoco. We are especially grateful to George H. Davis and Peter J. Coney, Department of Geosciences, University of Arizona and Stanley B. Keith and Wesley Peirce of the Arizona Bureau of Geology and Mineral Technology for participating in numerous enlightening discussions and offering insight and sound advice. Max D. Crittenden, Jr., George H. Davis, Ed Dewitt, and Stanley B. Keith critically reviewed the manuscript and improved it significantly. We also thank Gregory A. Davis, Ed Dewitt, Warren Hamilton, Donald Hyndman, and Leon T. Silver for sharing many of their ideas. Arend Meijer, Steve Rooke, Kim Wilson, and Gary Kolbasuk of the University of Arizona contributed significantly to early phases of Rb-Sr analysis. Sally Adams and Janee Stempel assisted in preparation of the manuscript. Finally, Reynolds gratefully acknowledges financial support from the Department of Geosciences of the University of Arizona and the Arizona Bureau of Geology and Mineral Technology.

REFERENCES CITED

Anderson, C. A., and Silver, L. T., 1976, Yavapai Series—A greenstone belt: Arizona Geological Society Digest, v. 10, p. 13–26.

Avedisian, G. E., 1966, Geology of the western half of Phoenix South Mountain Park, Arizona [M.S. thesis]: Tempe, Arizona State University, 52 p.

Chase, R. B., 1973, Petrology of the northeastern border zone of the Idaho batholith, Bitterroot Range, Montana: Montana Bureau Mines and Geology Memoir 43, 28 p.

Coney, P. J., 1978, Mesozoic-Cenozoic Cordilleran plate tectonics, in Smith, R. B., and Eaton, G. P., eds., Cenozoic tectonics and regional geophysics of the western Cordillera: Geological Society of America Memoir 152, p. 33–50.

Coney, P. J., 1980, Cordilleran metamorphic core complexes: An overview: Geological Society of America Memoir 153 (this volume).

Crittenden, M. D., Jr., Coney, P. J., and Davis, G. H., 1978, Penrose Conference report: Tectonic significance of metamorphic core complexes in the North American Cordillera: Geology, v. 6, p. 79–80.

Davis, G. A., and others, 1977, Enigmatic Miocene low-angle faulting, southeastern California and west-central Arizona—Suprastructural tectonics?: Geological Society of America Abstracts with Programs, v. 9, p. 943–944.

Davis, G. H., 1980, Structural characteristics of metamorphic core complexes, southern Arizona: Geological Society of America Memoir 153 (this volume).

Davis, G. H., and others, 1975, Origin of lineation in the Catalina-Rincon-Tortolita gneiss complexes, Arizona: Geological Society of America Abstracts with Programs, v. 7, p. 602.

Drewes, H., 1977, Geologic map and sections of the Rincon Valley quadrangle, Pima County, Arizona: U.S. Geological Survey Miscellaneous Geological Investigations Map I-997.

Eaton, G. P., Peterson, D. L., and Schumann, H. H., 1972, Geophysical, geohydrological, and geochemical reconnaissance of the Luke salt body, central Arizona: U.S. Geological Survey Professional Paper 753, 28 p.

Eberly, L. D., and Stanley, T. B., Jr., 1978, Cenozoic stratigraphy and geologic history of southwestern Arizona: Geological Society of America Bulletin, v. 89, p. 921–940.

Higgins, M. E., 1971, Cataclastic rocks: U.S. Geological Survey Professional Paper 687, 97 p.

Otten, J. K., 1978, Tertiary geologic history of the Date Creek basin, west-central Arizona: Geological Society of America Abstracts with Programs, v. 10, p. 140–141.

Peirce, H. W., 1976, Tectonic significance of Basin and Range thick evaporite deposits: Arizona Geological Society Digest, v. 10, p. 325–339.

Rehrig, W. A., and Heidrick, T. L., 1972, Regional fracturing in Laramide stocks of Arizona and its relationship to porphyry copper mineralization: Economic Geology, v. 67, p. 198–213.

Rehrig, W. A., and Heidrick, T. L., 1976, Regional tectonic stress during the Laramide and late Tertiary intrusive periods, Basin and Range province, Arizona: Arizona Geological Society Digest, v. 10, p. 205–228.

Rehrig, W. A., and Reynolds, S. J., 1977, A northwest zone of metamorphic core complexes in Arizona: Geological Society of America Abstracts with Programs, v. 9, p. 1139.

—— 1980, Geologic and geochronologic reconnaissance of a northwest-trending zone of metamorphic core complexes in southern and western Arizona: Geological Society of America Memoir 153 (this volume).

Reynolds, S. J., Rehrig, W. A., and Damon, P. E., 1978, Metamorphic core complex terrain at South Mountain, near Phoenix, Arizona: Geological Society of America Abstracts with Programs, v. 10, p. 143–144.

Shackelford, T. J., 1976, Juxtaposition of contrasting structural and lithologic terrains along a major Miocene gravity detachment surface, Rawhide Mtns., Arizona: Geological Society of America

Abstracts with Programs, v. 8, p. 1099.
Shafiqullah, M., Damon, P. E., and Peirce, H. W., 1976, Late Cenozoic tectonic development of Arizona Basin and Range province [abs.]: International Geological Congress, 25th, Sydney, v. 1, p. 99.
Streckeisen, A., 1976, To each plutonic rock its proper name: Earth-Science Reviews, v. 12, p. 1–33.
Thorman, C. H., 1977, Gravity-induced folding off a gneiss dome complex, Rincon Mountains, Arizona: Discussion: Geological Society of America Bulletin, v. 88, p. 1211–1212.
Wilson, E. D., Moore, R. T., and Cooper, J. R., 1969, Geologic map of Arizona: Arizona Bureau of Mines and U.S. Geological Survey, scale 1:500,000.

MANUSCRIPT RECEIVED BY THE SOCIETY JUNE 21, 1979
MANUSCRIPT ACCEPTED AUGUST 7, 1979

Printed in U.S.A.

Geology of a zone of metamorphic core complexes in southeastern Arizona

NORMAN G. BANKS
U.S. Geological Survey
345 Middlefield Road
Menlo Park, California 94025

ABSTRACT

An elongate northwest-trending zone of batholith-size metamorphic core complexes extends some 130 km from the Rincon Mountains to the Picacho Mountains in southeastern Arizona. The complexes are characterized by undeformed to gneissic granitic intrusions, gneissic to phyllonitic xenoliths and wall rocks derived mainly from Precambrian granitic rock, shallow-dipping foliation, and remarkably uniform directions of lineation. Parts of this zone have been recognized and studied extensively for more than 30 yr, but there remains a divergence of opinion about the age, depth of emplacement, and origin of the complexes. Field relations indicate that host rocks as young as or younger than Mesozoic were involved in cataclasis. K-Ar and fission-track ages indicate that the complexes were at temperatures uniformly in excess of 400 °C in the middle Tertiary (20 to 30 m.y. ago) and that the bedrock between and very near the complexes was not thermally affected. Tertiary plutons characteristically associated with the high-grade metamorphic rocks are also cataclastically deformed, and stratigraphic depths to the top of metamorphic terranes were no more than 6 km and possibly less than 3 km. These and other data suggest to some that the complexes developed during intrusion of composite batholiths at shallow depth in an anisotropic stress field during the middle Tertiary. On the other hand, Rb-Sr and U-Th-Pb techniques yielded older ages (≥44 m.y.) for some samples. These and additional data suggest to others that major development of cataclasis preceded the middle Tertiary and included regional thrusting.

INTRODUCTION

An elongate northwest-trending zone of metamorphic core complexes extends from the Rincon Mountains to the Picacho Mountains, Arizona (Fig. 1), and other core complexes occur along strike of the zone as well as elsewhere in Arizona (Shackelford, 1976; Rehrig and Heidrick, 1976; Rehrig and Reynolds, 1977 and this volume; Shackelford, 1977; G. A. Davis and others, 1977). The zone of complexes discussed here lies within the boundaries of an elongate northwest-trending tectonic block that is part of a series of such blocks making up a structural belt that coincides with the Texas lineament or Texas zone (Titley, 1976). The tectonic block was active in late Paleozoic and Mesozoic time (Titley,

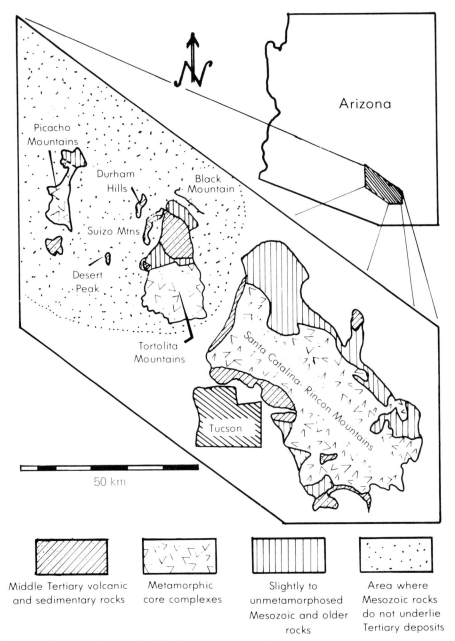

Figure 1. Index map of a zone of middle Tertiary metamorphic core complexes, southeastern Arizona.

1976), coincides with an arch defined by the attitude of adjacent middle Tertiary volcanic rocks (Rehrig and Heidrick, 1976), and, along with similar arches in the belt of deformed rocks, has been recently interpreted as heat- or magmatic-induced doming related to plate movements (Rehrig and Heidrick, 1976).

Some of the complexes have been studied extensively for more than 30 yr, but their age, depth of emplacement, and the origin of their characteristic gneissic and cataclastic textures continue to be subjects of great interest and some controversy. Much of the controversy about complexes discussed here concerns divergence of opinions about the identification of some of the rock units in the complexes and about the interpretation of radiometric data.

The metamorphic grade and fabric of the complexes in Arizona are similar to those of core complexes to the south in Mexico (Anderson and others, 1977), as well as farther north in the North American Cordillera. Decollement surfaces (Davis, 1975; Davis and others, 1975; W. A. Rehrig, 1977, oral commun.) and discrepancies between K-Ar and Rb-Sr ages (Creasey and others, 1977; Shakel and others, 1977) are additional features shared by the complexes discussed here and those to the north (for example, Armstrong and Hansen, 1966; Compton and others, 1977; Hazzard and Turner, 1959; Lee and others, 1970; Kistler and Willden, 1967; Simony and others, 1977; Thorman, 1965). However, the geologic relations and radiometric ages for the complexes described here appear for the most part to indicate younger ages and intrusion at shallower depths than is the case for the complexes to the north (Armstrong, 1976; Campbell and Reesor, 1977; Fox and others, 1976, 1977; Hazzard and Turner, 1959; Kistler and Willden, 1967; Lee and others, 1970; Nielsen and Duncan, 1977; Reesor, 1970; Reesor and Moore, 1971; Simony and others, 1977; and others in this volume).

Geologic relations in the complexes described here indicate that metamorphism and cataclasis in at least two complexes could not have been older than Permian. In the Santa Catalina–Rincon Mountains complex (Fig. 1), the Leatherwood Quartz Diorite of Bromfield (1952), which intrudes Mesozoic sedimentary rock (Creasey and Theodore, 1975) was involved in the metamorphism (Moore and others, 1949; Bromfield, 1952; Peirce, 1958; Dubois, 1959a; Creasey and others, 1977; Banks, 1976). In the Tortolita Mountains (Fig. 1), rocks intruding Permian sedimentary rocks were involved (Banks and others, 1977). Granitic intrusive bodies and crosscutting pegmatite dikes and sills together make up the bulk of all of the complexes and are the youngest rocks involved in the metamorphism and cataclasis.

The metamorphosed intrusive bodies yield middle Tertiary (20 to 30 m.y.) K-Ar and fission-track (FT) ages from many widely spaced samples throughout the zone, and they grade, locally abruptly, into mildly deformed or undeformed rock that yields essentially the same K-Ar and FT ages as the deformed parts of the plutons. Xenoliths and septa of Precambrian igneous rock throughout the complexes yield middle Tertiary K-Ar ages that are essentially identical to those of the intrusive phases, which indicates that the thermal event(s) responsible for the essentially concordant ages was strong and uniform over large areas, and either occurred in, lasted until, or reoccurred in the middle Tertiary. Some (but not all) of the few Rb-Sr ages for the complexes (Damon and Giletti, 1961; Livingston and others, 1967; Shakel and others, 1972; Shakel, 1972, 1974) and U-Th-Pb ages for the cataclastically deformed intrusive phase of the Santa Catalina–Rincon Mountains complex (Catanzaro and Kulp, 1964; Shakel and others, 1977) seem to favor the last two of the preceding three alternatives. However, the Rb-Sr ages show "poor systematics reflecting original differences or lack of isotopic equilibrium" (Shakel and others, 1977), and the U-Th-Pb ages, while reaffirming that a strong and widespread thermal event occurred in or lasted until the middle Tertiary, are not necessarily definitive of the age of the cataclasis or the age of emplacement of the major intrusive bodies involved. Additional petrographic and K-Ar data limit the nature of the thermal events, as host rocks near and even less than 1 km from the individual complexes are thermally unaffected by the presence of the complexes (Drewes, 1974, 1977; Marvin and others, 1973; Banks, 1977; Banks and others, 1977, 1978; Creasey and others, 1977). K-Ar ages of unfoliated dikes and plutons cutting the cataclastic rocks indicate that the metamorphism ended no later than 20 m.y. ago.

In the complexes having sufficient stratigraphic data for reconstruction (Santa Catalina–Rincon and Tortolita Mountains), cataclasis occurred at depths of < 3 to 6 km, barring tectonic thickening of the cover. This range in reconstructed stratigraphic depth is due to uncertainties whether Mesozoic or Tertiary strata covered the complexes during cataclasis, which depends on the unresolved exact time of cataclasis, and to the extremely variable thickness of these strata. Deeper burial of the complexes by stacked, but now absent, thrust plates cannot be ruled out (Drewes, 1973, 1974, 1976, 1977), but as pointed out by Rehrig and Heidrick (1976) and Davis (1977b), large-scale regional overthrusting is not required to explain the geologic relations in nearby mountain ranges, and, according to Davis (1977b), regional overthrusting may not be supported by facies and structural studies in southern Arizona. Additionally, neither stacking of a reasonable number of thrust sheets over the complexes nor resorting to geologically creditable vertical tectonics substantially reduces potential problems of the very high pressures required to explain the experimentally determined stabilities of the mineral assemblages observed in some of the rocks. Thus, these mineral assemblages cannot be used as indicators of depth of emplacement of the complexes, but must be explained by metasomatism or crystallization prior to emplacement of the magmas. Estimation of the depths of intrusion and deformation in and east of the Suizo Mountains and the Durham Hills, at Desert Peak, and in the Picacho Mountains complex (Fig. 1) is not possible by stratigraphic reconstruction. However, the character and thickness of the adjacent Mesozoic and middle Tertiary sedimentary deposits do not suggest exhumation of these complexes from depths any deeper than those for the complexes farther southeast in the zone.

DESCRIPTION OF THE COMPLEXES

Major metamorphic core complexes crop out in three mountain ranges in the area of Figure 1, the Santa Catalina-Rincon Mountains (including the Tanque Verde Mountains), the Tortolita Mountains, and the Picacho Mountains. Petrographic and structural data for smaller isolated outcrops of similar rocks in the Durham Hills, Suizo Mountains, Desert Peak, and possibly part of Black Mountain suggest the presence of at least one other complex. In addition to their geographic proximity, decollement features, 20- to 30-m.y. K-Ar and FT ages, similar fabric, telescoped thermal regimes, and shallow stratigraphic emplacement, the rocks in the complexes have in common (1) petrographically very similar suites of intrusive rocks, (2) petrographically similar granitic gneiss, mylonite, and schistose mylonite (including blastomylonite and phyllonite) derived from granitic rocks, (3) generally gently dipping foliation and preferred orientation of lineation, and (4) abundant pegmatite dikes and sills that for the most part lack fluorine-rich minerals characteristic of pegmatites derived from differentiation of igneous rocks. Sedimentary rocks occur in two of the complexes and often display spectacular deformational structures.

The most common undeformed plutons associated with all of the complexes are petrographically very similar appearing, coarsely crystalline, porphyritic- to granitic-textured, sphene-bearing quartz monzonite that locally grades or ranges from granodiorite to granite in composition (Fig. 2, examples A, B, C; Tables 1, 2, samples 19 through 22 and 28). All complexes also have similar appearing, northwest-trending postfoliation dikes of quartz diorite porphyry and of porphyritic to equigranular quartz monzonite (Fig. 3A; Tables 1, 2, samples 34 through 39). In addition, all but the Picacho Mountains are spatially associated with dioritic and biotite granodioritic plutons (Fig. 2, examples D, E; Tables 1, 2, samples 11 through 18), and the Tortolita Mountains complex and Desert Peak include late intrusions of only slightly foliated fine-grained equigranular quartz monzonite (Fig. 3B; Tables 1, 2, samples 31, 32, 33).

The gneissic rock and most of the schistose mylonite are derived, in approximate order of abundance, from medium to coarsely crystalline, porphyritic- and granitic-textured, quartz monzonitic to granitic intrusive bodies (Fig. 4B, 4C; Tables 1, 2, samples 23 through 27), xenoliths and wall rocks of Precambrian Y granitic rock (Fig. 4A; Table 1, samples 2, 3, 4, 6, 10; Table 2, sample 40), pegmatite dikes and sills, and the above-mentioned dioritic and biotite granodioritic plutons. Microscopically, the gneiss and mylonitic gneiss are composed of anastomosing and intertwining laminae of ribbon and

Figure 2. Examples of major intrusive rock types in the metamorphic core complexes. (A) Porphyritic quartz monzonite from the pluton of Samaniego Ridge, Santa Catalina Mountains. (B) Porphyritic quartz monzonite of Wild Burro Canyon, Tortolita Mountains. (C) Porphyritic quartz monzonite of Picacho Mountains. (D) Quartz diorite, early intrusive phase of Durham Hills. (E) Quartz diorite, early intrusive phase of Desert Peak.

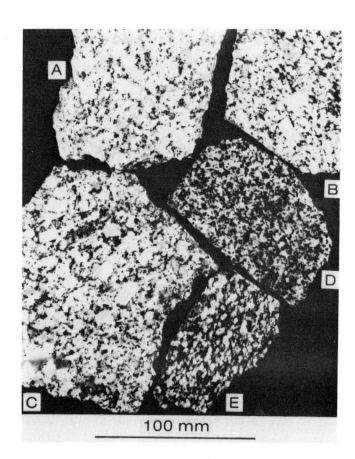

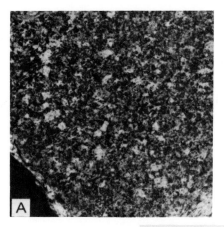 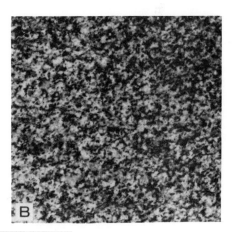

Figure 3. Stained rock slabs from Tortolita Mountains. (A) Postfoliation dike of seriate porphyritic quartz monzonite; K-feldspar is light, plagioclase and quartz are gray, mafic minerals are black. (B) Fine-grained hypidiomorphic equigranular quartz monzonite of Tortolita Mountains.

TABLE 1. MODAL DATA FOR INTRUSIVE PHASES, HOST ROCKS, AND LATE DIKES AND PLUTONS IN AND NEAR METAMORPHIC CORE COMPLEXES IN SOUTHEASTERN ARIZONA

	1	1a	2	3	4	5	6	7	8	9	10	11	12	13	14	15	16	17	18	19
Quartz	28	32	32.4	31	27.5	29.0	33.1	39.3	31.2	31	29		22	20-24	3.6	2-20	30	12	14.0	16.3
K-feldspar and perthite	28	29	10.6	6	20.5	30.0	24.7	39.2	11.3	12	36		5	4-8	7.4	tr-22	29	19	2.0	30.5
Plagioclase	33	28	33.8	31	23.5	30.3	19.7	18.5	31.4	40	32		42	45-50	38.2	35-65	32	33	64.0	41.0
Biotite	11	9	21.4	29		8.5	4.2	2.0	14.6	14	3		19	14-24	26.0	10-26	2	18	11.6	9.1
Hornblende			0.2							1				0-5	17.8	6-18		10	6.0	1.4
Pyroxene														tr	tr			<1		
Opaque minerals			1.0	..	1.0	1.3	1.2	0.5	0.5	2	1		tr	1-2	1.0	1-4	tr	4	tr	1.0
Sphene			tr	..			tr	tr	3.5	tr	..		tr	<1	1.6	1-4	7	tr	1.2	0.3
Apatite			0.6	.1	tr	..	0.2	tr	tr	tr	4		<1	<1	tr	tr	tr	tr	tr	0.4
Muscovite			tr	<1	20.5	1.0	16.7	0.5	7.5	tr	..			..	4.4			tr	0.4	
Garnet		2	tr		tr		0.2			..	..			..	..					
Other	..		tr	tr	tr	tr	tr	tr	tr	tr	tr		10	..	..	..	tr	tr	tr	tr
An content	30±5	20	22-34	20-30	15-25	35-40	35±5	20-35	30±5	20-40	7-12		20-30	35-45	35-45	30-50	15-25	30-45	48	20
Quartz	30.4	33.0	15-30	23	24	25	26	27	28	29	30		32	33	34	35	36	37	38	39
Quartz	30.4	33.0	15-30	33.2	31	31	28	24	17	20-40	25		26-35	38.5	15-20	35	36	0.5-5	28	15-30
K-feldspar and perthite	24.1	38.3	25-40	22.8	22	26	27	32	25	25-55	37		26-33	31.5	30-60	4	1.7	0.5-5	38	30-55
Plagioclase	37.7	24.4	25-40	35.8	39	35	29	36	45	25-40	35		26-40	27.5	30-65	28	39.7	38-55	30	20-50
Biotite	6.1	2.9	3-10	3.8	5	5	4	3	5				2-4	1.5	2-10	18	12.7	9-20	3	2-10
Hornblende	0.3		0-2												0-2	46	35.6	20-50		
Pyroxene				0-2																
Opaque minerals	0.5	0.9	0.5-1	0.2	<1	<1	<1	1	2		3		0.5-1	0.5	tr	1	1.0	0.5-1	1	0.5-2
Sphene	0.6	0.3	tr-3	tr	tr	tr	tr	1	<1				tr	tr	tr	tr	3.0	tr-1	tr	tr
Apatite	0.3	0.2	<1	tr	tr	tr	tr	<1	<1				tr	0.5	tr	tr	3.6	tr-4	tr	tr
Muscovite	..	..	..	4.2	2	2	2	<1					..	..	..	..	..	..	..	..
Garnet	..	..	..	tr	<1	<1	<1	1												
Other	tr	tr	tr	tr	tr	tr	tr	tr	tr		tr		tr	tr	tr	tr	tr	tr	tr	tr
An content	20		10-35	20	15-25	20-25	20-25	15-25	25-40	20-25			20-20	22	15-25	35-45	50-70	40-70	40	20-40

Note: Trace = tr; leaders (..) = not present.
Sample notes for each column as follows:

Oracle Granite of Peterson (1938)
1. Average of 4 analyses (Creasey, 1965), coarsely crystalline porphyritic quartz monzonite, north of Santa Catalina Mountains at San Manuel.
1a. Average of 38 thin-section analyses (Banerjee, 1957), coarsely crystalline porphyritic quartz monzonite, north of Santa Catalina Mountains at San Manuel.
2. Black to dark-gray mylonitic augen gneiss (field sample ML62D), Sabino Canyon (lat. 32°20.2'N, long. 110°47.7'W), forerange of Santa Catalina Mountains.
3. Black to dark-gray mylonitic augen gneiss, average of 24 analyses (Sherwonit, 1974), various localities in forerange of Santa Catalina Mountains.
4. Ultramylonite (field sample TN9) near surface of Catalina fault (lat. 32°21.4'N, long. 110°57.4'W), forerange of Santa Catalina Mountains.
5. Very fresh coarsely crystalline hypidiomorphic-granular quartz monzonite (field sample TMN-51B), at Antelope Peak (lat. 32°40.3'N, long. 111°01.4'W), north of Tortolita Mountains.
6. Mylonitic gneiss (field sample TMN-94), Guild Wash mylonite zone (lat. 32°34.7'N, long. 111°06.4'W), Tortolita Mountains.
7. Slightly altered hypidiomorphic-granular granite (field sample TMN30; lat. 32°44.2'N, long. 111°03.3'W), northeast of Suizo Mountains.
8. Slightly sheared, coarsely crystalline porphyritic granodiorite (field sample TMN46; lat. 32°43.0'N, long. 111°02.7'W), northeast of Suizo Mountains.
9. Coarsely crystalline seriate porphyritic granodiorite (field sample YA1), average of 4 samples, northern Picacho Mountains.
10. Granitic gneiss, average of 2 samples (field sample YA2), southern Picacho Mountains.

Leatherwood Quartz Diorite of Bromfield (1952)
11. Medium- to fine-grained, seriate porphyritic to intersertal diorite (field sample BR2), Lombar Hill (lat. 32°37.8'N, long. 110°43.7'W), Santa Catalina Mountains.
12. Average of 26 samples, Santa Catalina Mountains (Hanson, 1966); other includes +10% epidote.
13. Range of composition, Santa Catalina Mountains (Banks, 1976).

Other early quartz dioritic intrusive phases
14. Pluton of Chirreon Wash (field sample TMJ28), early intrusive phase (lat. 32°30.7'N, long. 111°00.7'W), Tortolita Mountains; medium-grained hypidiomorphic-granular syenodiorite border phase.
15. Range in composition of pluton of Chirreon Wash, Tortolita Mountains (Banks and others, 1977).
16. Average of 2 samples of mylonitic gneiss of pluton of Cottonwood Canyon, early intrusive phase, Tortolita Mountains.
17. Average of 2 samples of early intrusive phase from northern and southern Durham Hills.
18. Slightly foliated syenodiorite (field sample TMN59A), lat. 32°35.8'N, long. 111°14.7'W), Desert Peak.

Porphyritic and gneissic granodiorite-granite intrusive phases
19. Coarsely crystalline porphyritic quartz monzonitic facies of pluton of Samaniego Ridge (field sample BR21; lat. 31°31.5'N, long. 110°50'W), Santa Catalina Mountains (Creasey and others, 1977).
20. Coarsely crystalline seriate porphyritic quartz monzonite facies of pluton of Samaniego Ridge (field

21. sample ML61; lat. 32°37.5'N, long. 110°52'W), Santa Catalina Mountains (Creasey and others, 1977). Medium-grained granitic facies of pluton of Samaniego Ridge (field sample BR16; lat. 32°31.5'N, long. 110°48'W), Santa Catalina Mountains (Creasey and others, 1977).
22. Range of composition of pluton of Samaniego Ridge, Santa Catalina Mountains (Banks, 1976).
23. Quartz monzonite and augen granitic gneiss at Windy Point (field sample GGNS1; lat. 32°22.1'N, long. 110°43'W), Santa Catalina Mountains (Creasey and others, 1977).
24. Average of 2 samples, quartz monzonite to granitic augen or mylonitic gneiss of Windy Point, forerange of Santa Catalina Mountains.
25. Average of 76 samples, quartz monzonite to granitic augen or mylonitic gneiss of Windy Point, forerange of Santa Catalina Mountains (Sherwonit, 1974).
26. Average of 10 samples, quartz monzonite augen gneiss of Windy Point, main range Santa Catalina Mountains (Pilkington, 1962).
27. Average of 2 samples, quartz monzonite to granitic augen or mylonitic gneiss, northern and southern Suizo Mountains.
28. Composite of 2 samples (field sample YA3), coarsely crystalline seriate porphyritic granodiorite to quartz monzonite, central Picacho Mountains.

Pegmatite dikes
29. Range of composition of major minerals in pegmatite dikes, main range, Santa Catalina Mountains (Pilkington, 1962).
30. Pegmatite dikes at Marshall Gulch, Santa Catalina Mountains (Matter, 1969).

Late fine-grained quartz monzonite plutons
31. Slightly foliated, fine-grained hypidiomorphic-granular quartz monzonite of the Tortolita Mountains (field sample ML105; lat. 32°27.5'N, long. 110°58'W; Creasey and others, 1977).
32. Range of composition of fine-grained hypidiomorphic-granular quartz monzonite of the Tortolita Mountains (Banks, 1976; Banks and others, 1977).
33. Slightly foliated, fine-grained hypidiomorphic-granular quartz monzonite (field sample TMN59B; lat. 32°36.3'N, long. 111°14.8'W), Desert Peak.

Postfoliation dikes
34. Range of composition of unfoliated dikes of quartz monzonite porphyry with large K-spar phenocrysts, cut pluton of Chirreon Wash, Tortolita Mountains (Banks and others, 1977).
35. Postfoliation dike of quartz diorite porphyry, near Lombar Hill (field sample BR1; lat. 32°37.8'N, long. 111°03.5'W), Santa Catalina Mountains.
36. Postfoliation dike of quartz diorite porphyry (field sample TMN58; lat. 32°43.7'N, long. 111°09.3'W), Tortolita Mountains (Banks and others, 1977).
37. Range in composition of postfoliation dikes and sills of quartz diorite porphyry, Tortolita Mountains (Banks and others, 1977).
38. Postfoliation dike of quartz latite to equigranular quartz monzonite (field sample TMN83E) that cuts dikes of quartz diorite porphyry (lat. 32°31.8'N, long. 111°03.5'W), Tortolita Mountains.
39. Range in composition of postfoliation dikes of quartz latite and equigranular quartz monzonite that cut dikes of quartz diorite porphyry, Tortolita Mountains (Banks and others, 1977).

TABLE 2. CHEMICAL AND NORMATIVE DATA FOR INTRUSIVE PHASES, HOST ROCKS, AND LATE DIKES AND PLUTONS IN AND NEAR METAMORPHIC CORE COMPLEXES IN SOUTHEASTERN ARIZONA

	1	9	10	40	11	14	19	20	21	23	28	31	35	36	38
						Rapid rock analyses (in percent)									
SiO_2	63.1	70.0	67.0	73.6	62.7	49.2	68.6	67.0	73.2	74.1	66.6	73.2	50.3	49.5	75.3
Al_2O_3	14.4	15.4	14.6	14.5	16.1	17.4	15.3	15.5	13.6	15.1	15.6	14.4	15.7	16.0	13.4
Fe_2O_3	6.6	1.6	3.7	1.0	2.1	1.9	1.3	2.0	0.89	0.33	2.0	0.72	3.5	2.4	0.79
FeO		0.92	2.2	0.08	2.7	6.1	1.3	1.9	0.64	0.52	1.3	0.52	5.6	5.1	0.24
MgO	1.4	0.52	1.5	0.17	2.9	5.5	0.9	1.7	0.50	0.08	1.8	0.24	7.8	8.1	0.22
CaO	1.5	1.9	2.6	0.96	4.7	7.7	2.2	3.0	1.1	1.3	3.1	1.1	6.6	7.9	0.85
Na_2O	2.1	3.4	3.5	4.6	3.7	3.3	3.8	4.2	3.6	4.3	4.2	3.5	3.6	3.5	3.4
K_2O	3.6	4.8	2.9	3.7	2.8	2.8	4.6	3.8	4.6	3.9	3.8	5.0	2.0	2.2	4.8
H_2O^+	1.8	0.74	0.94	0.45	0.94	1.7	0.79	0.72	0.32	0.49	0.62	0.42	2.0	1.9	0.25
H_2O^-	0.30	0.26	0.17	0.14	0.26	0.14	0.19	0.08	0.22	0.05	0.21	0.17	0.37	0.16	0.14
TiO_2	0.96	0.34	0.94	0.16	0.60	1.6	0.38	0.62	0.20	0.01	0.46	0.10	1.6	1.2	0.09
P_2O_5	0.32	0.20	0.33	0.13	0.22	0.91	0.19	0.28	0.11	0.04	0.27	0.07	0.41	0.71	0.03
MnO	0.04	0.04	0.06	0.01	0.08	0.12	0.05	0.09	0.03	0.02	0.03	0.02	0.14	0.13	0.03
CO_2	n.d.	0.01	0.04	0.02	0.08	0.16	0.01	0.02	0.02	0.02	0.02	0.02	0.05	0.20	0.01
Cl		0.01	0.01	0.01	0.01		0.01	0.01	0.01	0.01	0.01	<.01	0.01		
F		<.01	0.05	0.02	0.02		0.01	0.06	0.03	<.01	0.03	<.01	0.02		
Total	+100-	100+	100+	100-	100-	99-	100+	101-	99+	100+	100+	99+	100-	99	100-
					Quantitative emission spectrographic analyses (in parts per million)										
Ba		2,400	720	1,400			1,100	640	330	1,100	850	1,100			
Be		<1					5	6	7	2		2			
Co		8	10	8			8	11	<2	<2		<2			
Cr		<2	19	31			8	12	<2	<2	<2	<2			
Cu		65	140	4			9	9	13	5	4	5			
Ni		7	120	26			8	14	<2	<2	<2	2			
Sc		7	16	9			7	8	<2	<2	<2	<2			
Sr		550	270	770			350	380	170	180	190	180			
V		39	69	64			44	72	22	16	8	16			
Y		26	69	17			19	26	10	10	25	10			
Zr		50	160	76			36	110	48	82	56	82			
Ga		17	20	22			18	21	18	16	19	16			
Yb		1	11	2			2	2	1	1	2	1			
						Norms (in percent)									
Quartz	41.6	28.8	31.1	27.6	16.4		23.7	19.5	32.2	31.0	19.4	31.4	11.9	13.2	35.3
Corundum	7.4	1.9	1.6	1.7			0.6		1.0	1.6		1.5			1.2
Orthoclase	11.2	17.1	22.0	28.4	16.6	16.8	27.3	22.3	27.5	23.0	22.5	29.8	30.7	30.0	28.5
Albite	18.7	25.4	39.1	28.8	31.4	28.4	32.3	35.2	30.8	36.3	35.5	29.8	20.9	21.7	28.9
Anorthite	5.6	10.3	3.7	8.1	19.1	24.8	9.7	12.1	4.7	6.0	12.6	4.9	3.8	5.0	4.0
Wollastonite					1.0	2.9	0.3				0.4		10.8	4.4	
Enstatite	3.7	3.7	0.4	1.3	7.3	3.7	2.2	4.2	1.3	0.2	4.5	0.6	2.8	1.2	0.6
Ferrosilite					2.4	2.0	0.8	1.0		0.7		0.2	5.1	3.5	
Magnetite	0.9	4.5	0.2	2.1	3.1	2.8	1.9	2.9	1.3		2.9	1.1	3.1	2.3	0.6
Ilmenite	0.9	1.8	1.0	0.6	1.1	3.1	0.7	1.2	0.4	0.5	0.9	0.2			0.2
Hematite	6.3	0.6		0.2						tr					0.4
Apatite	0.8	0.8	0.8	0.5	0.5	2.2	0.5	0.7	0.3	0.1	0.6	0.2	1.0	1.7	0.1
Other		tr	tr	tr	0.2	11.7	tr	0.1	0.1	0.1	0.1	0.1	7.9	15.2	tr

Notes: Numbered columns (except 1 and 40) as in Table 1; Not detected = n.d.; leaders (...) = not determined; trace = tr.
+Includes 5.5% CuO + SO₃ + Fe${S_2}$.
Sample notes for each column as follows (for additional information, see Table 1).
1. Mineralized Oracle Granite, churnhole 67 of San Manuel (Schwartz, 1953).
9. and 10. Analysts: F. Brown, R. Moore, and C. Heropoulos. Analysis courtesy of Warren Yeend.
40. Mylonitic sample (laboratory sample ML62R) of Oracle Granite of Peterson (1938) in Sabino Canyon (lat 32°20.2'N, long 110°47.7'W), Santa Catalina Mountains (Creasey and others, 1977).
11. Analysts: R. Moore, B. McCall, and C. Heropoulos.
14. Analysts: Z. A. Hamlin and F. Brown.
19-21, 23. From Creasey and others, 1977.
28. Analysts: F. Brown, R. Moore, and C. Heropoulos. Analysis courtesy of Warren Yeend.
31. From Creasey and others, 1977.
35. Analysts: R. Moore, B. McCall, and C. Heropoulos.
36 and 38. Analysts: Z. A. Hamlin and F. Brown.

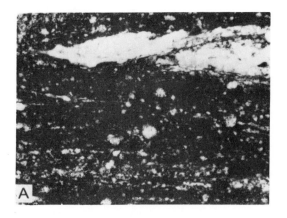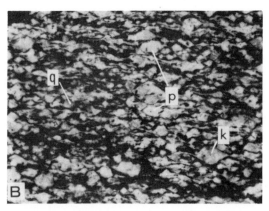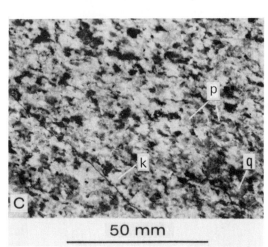

Figure 4. Rock slabs of cataclastically deformed igneous rocks from the metamorphic core complexes. (A) Blastomylonite derived from Precambrian Y granitic rock; xenolith in gneiss of Windy Point, Santa Catalina Mountains; cataclastically deformed pegmatite occurs in plane of foliation. (B) Mylonitic gneiss of Windy Point, Santa Catalina Mountains, derived from porphyritic quartz monzonite, intrusive into Mesozoic or younger rocks; k = K-feldspar porphyroclast, q = quartz porphyroclast, p = plagioclase porphyroclast; dark grains are biotite, muscovite, and garnet. (C) Gneissic intrusive phase of Suizo Mountains; specimen is stained, light grains are K-feldspar (k), gray grains are quartz (q) and plagioclase (p), dark grains are biotite, muscovite, and garnet.

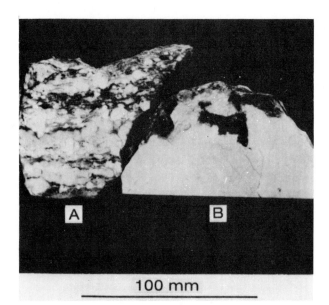

Figure 5. Pegmatite rock samples from the Santa Catalina–Rincon Mountains metamorphic core complex. (A) Cataclastically deformed. (B) Undeformed; large white area is part of a single oligoclase crystal.

granulated quartz, granulated feldspar, and shredded and bent biotite and muscovite with augen of feldspar and (depending on degree of cataclasis) quartz. Garnet, magnetite, apatite, and locally hornblende are obvious accessory minerals. Zircon, sphene, and monazite (Shakel and others, 1977) are local accessories. The schistose rock displays much more neomineralization than the gneissic rock, and in general, cataclasis, ductility, and neomineralization are more pronounced in the Precambrian granitic host rocks than in the intruding granitic rock (Fig. 4). The Precambrian rock also tends to have more biotite or muscovite than the gneissic phases of the younger granitic rocks. Deformation of both Precambrian host and younger granitic rock near decollement surfaces is more brittle and extreme than elsewhere in the complexes.

The pegmatite dikes and sills are usually less than 2 m thick and are most commonly several to several tens of millimetres thick. They postdated (Fig. 5B) and predated or accompanied cataclasis (Figs. 4A, 5A) and are mostly quartz monzonitic in composition (Table 1, samples 29, 30). Their most common occurrence is in the plane of foliation (Fig. 4A) in host rocks and in the immediately adjacent intrusive phases. They are rare in undeformed plutons and parts of plutons.

Santa Catalina-Rincon Mountains

The Santa Catalina-Rincon Mountains complex includes approximately 1,600 km^2 of igneous and metamorphic rocks exposed in an irregularly elongate mass bounded on the south by a gently dipping decollement surface, the Catalina fault (Fig. 6). Several gentle arches occur in the complex. The highest elevation is Mount Lemmon (~1,800 m above the surrounding alluvial pediment) on the flank of one of these arches (the forerange of the Santa Catalina Mountains, Fig. 6). The complex has been the subject of a considerable amount of recent study (Drewes, 1974, 1976, 1977; Creasey and Theodore, 1975; Davis, 1975; Davis and Frost, 1976; Banks, 1976, 1977; Creasey and others, 1977; Shakel and others, 1977), many theses (principally at the University of Arizona), and a number of less recent studies (Moore and others, 1949; Bromfield, 1952; Dubois, 1959a, 1959b; Damon and Giletti, 1961; Damon and others, 1963; Mayo, 1964; Catanzaro and Kulp, 1964; Livingston and others, 1967; Mauger and others, 1968; Shakel, 1972; Shakel and others, 1972; and several more local studies). Despite this effort, the number and ages of intrusions present and the number and ages of deformations involved are still undecided, probably in part due to divergence of opinion about the identity of some of the rock types, inconclusive results of the radiometric studies, and lack of complex-wide, detailed mapping and petrology.

Four rock units are dominant in the complex (Fig. 6): (1) generally light colored gneissic to mylonitic quartz monzonite and granite that intrude rocks as young as late or post-Mesozoic (hereafter this granitic gneiss is referred to as gneiss of Windy Point); (2) generally darker colored gneissic rocks derived mainly from Precambrian granitic rocks (granodiorite and quartz monzonite); (3) black to tan mylonitic, blastomylonitic, and phyllonitic rocks also derived mainly from Precambrian Y granite rocks; and (4) undeformed granodioritic to granitic (mainly quartz monzonitic) rocks that likewise intrude rock as young as late or post-Mesozoic along the northern margin of the complex (hereafter, when a specific pluton is not under discussion, the undeformed intrusive rock is referred to as the Tertiary intrusive rock). The first three of these four main rock units are foliated, lineated, and intruded by locally abundant pegmatite dikes and sills, and together they comprise the bulk of the complex (Fig. 6).

Also present in the complex are northwest-striking dikes that crosscut the foliation; metamorphosed sedimentary rocks of Precambrian, Paleozoic, and Mesozoic age; and several quartz diorite and biotite granodiorite intrusive bodies. The latter, the Leatherwood Quartz Diorite of Bromfield (1952) (hereafter referred to simply as Leatherwood Quartz Diorite; Fig. 7), occurs mostly as sill-like bodies along the northern margin of the complex (Figs. 6, 8), cuts Mesozoic deposits (Bromfield, 1952; Creasey and

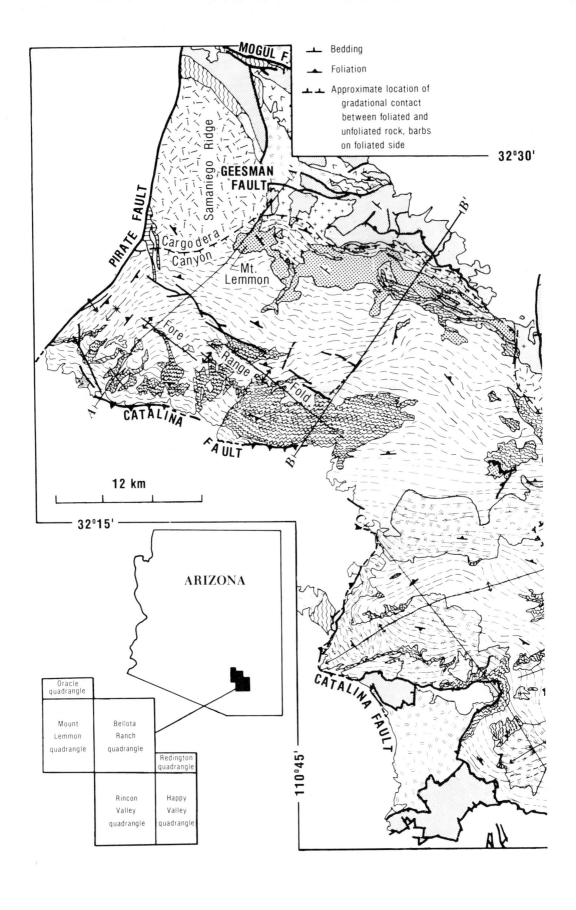

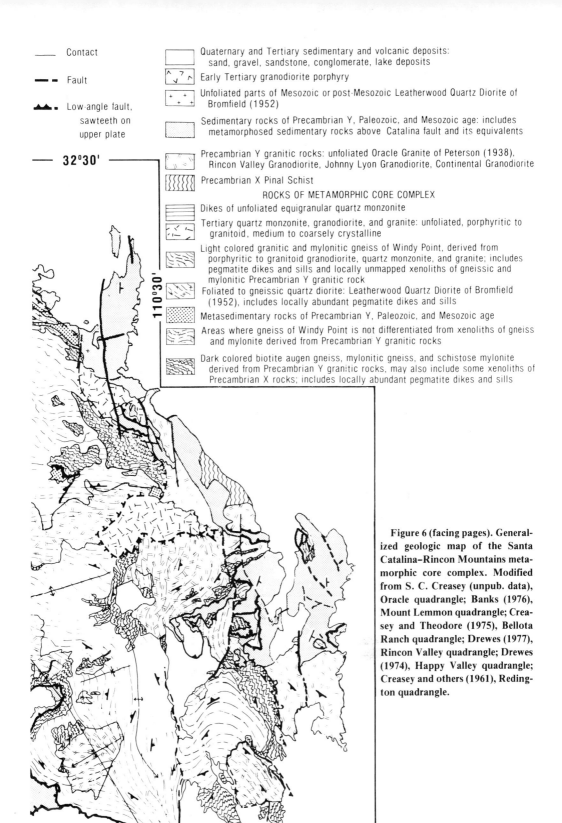

Figure 6 (facing pages). Generalized geologic map of the Santa Catalina–Rincon Mountains metamorphic core complex. Modified from S. C. Creasey (unpub. data), Oracle quadrangle; Banks (1976), Mount Lemmon quadrangle; Creasey and Theodore (1975), Bellota Ranch quadrangle; Drewes (1977), Rincon Valley quadrangle; Drewes (1974), Happy Valley quadrangle; Creasey and others (1961), Redington quadrangle.

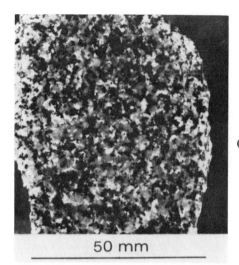

Figure 7. Leatherwood Quartz Diorite of Bromfield (1952) (Tables 1, 2, sample 11).

Theodore, 1975), is locally foliated, and is intruded by both the gneiss of Windy Point and undeformed Tertiary intrusive rock (Moore and others, 1949; Peirce, 1958; Dubois, 1959a, 1959b; Creasey and Theodore, 1975; Banks, 1976, 1977; Creasey and others, 1977). The actual age of the quartz diorite has not been established; the published ages are Tertiary and essentially concordant with those of the intruding rocks (Table 3). Modal and chemical data for the Leatherwood occur in Tables 1 and 2 (samples 11, 12, 13).

The complex is flanked to the north, east, and south (Figs. 6, 8, cross sections B–B', C–C') by several sheets of folded Precambrian through Tertiary sedimentary rocks (see Creasey and Theodore, 1975; Drewes, 1974, 1977; Davis, 1975, 1977b; Thorman, 1977). Detailed structural studies suggest that the folds in the sheets around the Rincon Mountains developed by gravity gliding off the gneiss in Miocene time (McColly, 1961; Arnold, 1971; Davis, 1975, 1977b; Davis and Frost, 1976). This is also an attractive mechanism to explain the similar complex structure and low-angle tectonic features north of the complex (Figs. 5, 8, cross section B–B'). Drewes (1977) also documented that there was rapid shedding of cover from the complex into adjacent basins 20 to 30 m.y. ago.

The bulk of the gneiss of Windy Point occurs as 2-km-thick to greater than 5-km-thick sheetlike bodies intruded at or below the contact between the Precambrian Y granitic rocks and the overlying sedimentary rocks (Fig. 8). Deformation of the gneiss is variable and most intense near inclusions and contacts with older rocks, particularly near the base and top of the sill-like masses. Where least deformed, coarse- to fine-grained granitic as well as coarsely crystalline porphyritic igneous textures are easily recognized. The fairly monotonous mineralogy is quartz, plagioclase, K-feldspar, biotite, and muscovite with accessory apatite, opaque minerals, garnet, and local sphene, zircon and monazite. Modal and chemical data for the gneiss are presented in Tables 1, 2 (samples 23 through 27); K-Ar and FT ages are in Table 3 (samples 23 and 48 through 52).

It is unlikely that the occurrence of muscovite with biotite in the gneiss of Windy Point indicates the depth of emplacement of the intrusion. In thin section, much of the muscovite appears to have replaced biotite rather than to have coexisted with it in a melt (Fig. 9; also Dubois, 1959b). Furthermore, the indicated stratigraphic depth of emplacement of 3 to 6 km for the gneiss is so much less than the >12-km depth required experimentally (Winkler, 1974) that, given its late to post-Mesozoic age, much more than a creditable amount of vertical or horizontal tectonics would be needed to explain their equilibrium crystallization from the magma after its emplacement. Thus, the muscovite seems better explained

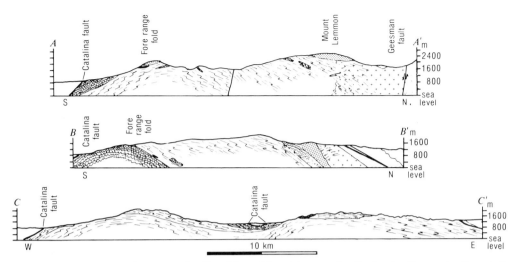

Figure 8. Cross sections of the Santa Catalina–Rincon Mountains complex. Section locations and unit-symbol identifications are shown in Figure 6.

as metasomatic or hydrothermal in origin, or else it crystallized before the magma was emplaced at its present stratigraphic level. Alternatively, iron, magnesium, and trace metals in the muscovite allowed its crystallization with biotite at depths consistent with geologic relations.

Both on the complex scale and outcrop scale, the wall rock, xenoliths, and pendants of Precambrian granite are much darker and more cataclastically and ductilely deformed (Fig. 4A) than the intruding and enclosing gneiss of Windy Point (Fig. 4B). The latter is generally recognizable in hand specimen as having been derived from igneous rock (Fig. 4B), but the dark mylonitic gneiss and gradational blastomylonitic schist frequently is not (Fig. 4A). Megascopically, the dark gneiss and derivative

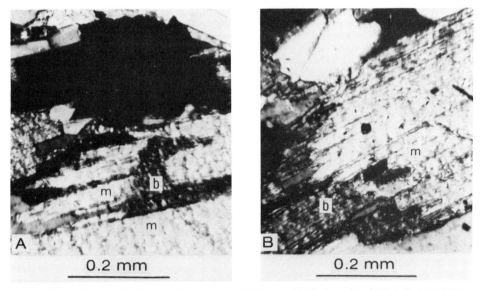

Figure 9. Photomicrographs of muscovite replacement of igneous biotite in gneiss of Windy Point at Wilderness of Rocks (A) and Windy Point (B), Santa Catalina Mountains.

TABLE 3. K-Ar AND FISSION-TRACK (F-T) AGES

Sample no.	Laboratory no.	Calculated age (m.y.)	Method	Location Area	Lat (N), long (W)	Reference
				Leatherwood Quartz Diorite of Bromfield (1952)		
41	PED-1-68	30.3 ± 0.6	K-Ar biotite	Santa Catalina Mts.	32°26.2', 110°45.5'	Damon and others (1969)
				Pegmatite that cuts Leatherwood Quartz Diorite of Bromfield (1952)		
42	PED-15-59	48.6 ± 2.1	K-Ar muscovite	Santa Catalina Mts.	32°26.1', 110°45.5'	Damon and others (1963)
				Granodiorite of Happy Valley (Drewes, 1974)		
43	81	27.0 ± 0.9	K-Ar biotite	Rincon Mountains	32°07 ', 110°25 '	Marvin and others (1973)
44	82	28.0 ± 1.1	K-Ar biotite	Rincon Mountains	32°12 ', 110°27 '	Marvin and others (1973)
				Pegmatite that cuts granodiorite of Happy Valley (Drewes, 1974)		
45	83	37.7 ± 1.6	K-Ar muscovite	Rincon Mountains	32°12 ', 110°27 '	Marvin and others (1973)
				Pluton of Chirreon Wash, early intrusive phase, Tortolita Mountains		
46	UAKA-75-86	25.1 ± 0.5	K-Ar biotite	Tortolita Mountains	32°32.9', 111°01.9'	Banks and others (1978)
				Pluton of Samaniego Ridge, Santa Catalina Mountains		
47	PED-16-59	25.6 ± 0.8	K-Ar biotite	Santa Catalina Mts.	32°26.6', 110°52.1'	Damon and others (1963)
19	BR21	23.9 ± 1.2	K-Ar biotite	Santa Catalina Mts.	31°31.5', 110°50 '	
		23.7 ± 0.7	K-Ar hornblende	Santa Catalina Mts.	31°31.5', 110°50 '	
		30.0 ± 3.0	F-T sphene	Santa Catalina Mts.	31°31.5', 110°50 '	
		28.9 ± 3.3	F-T zircon	Santa Catalina Mts.	31°31.5', 110°50 '	
		23.5 ± 2.8	F-T apatite	Santa Catalina Mts.	31°31.5', 110°50 '	
20	ML61	27.9 ± 0.7	K-Ar hornblende	Santa Catalina Mts.	32°37.5', 110°52 '	
		24.7 ± 0.7	K-Ar biotite	Santa Catalina Mts.	32°37.5', 110°52 '	Creasey and others (1977)
		28.3 ± 3.1	F-T sphene	Santa Catalina Mts.	32°37.5', 110°52 '	
		27.1 ± 3.4	F-T zircon	Santa Catalina Mts.	32°37.5', 110°52 '	
		20.8 ± 2.1	F-T apatite	Santa Catalina Mts.	32°37.5', 110°52 '	
21	BR16	23.8 ± 0.7	K-Ar biotite	Santa Catalina Mts.	32°31.5', 110°48 '	
		28.0 ± 3.0	F-T sphene	Santa Catalina Mts.	32°31.5', 110°48 '	
		25.9 ± 2.5	F-T zircon	Santa Catalina Mts.	32°31.5', 110°48 '	
		21.7 ± 2.1	F-T apatite	Santa Catalina Mts.	32°31.5', 110°48 '	
				Granodiorite-granite gneiss of Windy Point, Santa Catalina-Rincon Mountains		
48	80(71DA)	24.1 ± 0.9	K-Ar biotite	Rincon Mts.	32°07 ', 110°28 '	
		25.3 ± 0.9	K-Ar muscovite	Rincon Mts.	32°07 ', 110°28 '	Marvin and others (1973)
		231 ± 38	F-T zircon	Rincon Mts.	32°07 ', 110°28 '	C. W. Naeser and Harald Drewes*
		21.6 ± 2.5	F-T apatite	Rincon Mts.	32°07 ', 110°28 '	
49	71D194	27.7 ± 0.9	K-Ar biotite	Rincon Mts.	32°15.6', 110°28.8'	Harald Drewes*
		28.6 ± 0.6	K-Ar muscovite	Rincon Mts.	32°15.6', 110°28.8'	
50	73D52	25.0 ± 0.8	K-Ar biotite	Rincon Mts.	32° 8.9', 110°31.8'	
		20-30 Precambrian?	F-T zircon	Rincon Mts.	32° 8.9', 110°31.8'	
			F-T zircon	Rincon Mts.	32° 8.9', 110°31.8'	C. W. Naeser and Harald Drewes*
		24.4 ± 5.0	F-T apatite	Rincon Mts.	32° 8.9', 110°31.8'	
		23.9 ± 5.0	F-T apatite	Rincon Mts.	32° 8.9', 110°31.8'	
23	PED-4a-58	25.4 ± 1.0	K-Ar biotite	Santa Catalina Mts.	32°22.1', 110°43 '	Damon and others (1963)
		30.2 ± 0.9	K-Ar muscovite	Santa Catalina Mts.	32°22.1', 110°43 '	
23		32 ± 3	K-Ar muscovite	Santa Catalina Mts.	32°22.1', 110°43 '	Catanzaro and Kulp (1964)
23	GGN-S1	23.3 ± 0.7	K-Ar biotite	Santa Catalina Mts.	32°22.1', 110°43 '	
		24.8 ± 0.7	K-Ar muscovite	Santa Catalina Mts.	32°22.1', 110°43 '	Creasey and others (1977)
		19.3 ± 2.7	F-T apatite	Santa Catalina Mts.	32°22.1', 110°43 '	
51	PED-18-62L	25.6 ± 1.0	K-Ar biotite	Santa Catalina Mts.	32°20.3', 110°41.4'	
		26.1 ± 1.0	K-Ar muscovite	Santa Catalina Mts.	32°20.3', 110°41.4'	Livingston and others (1967)
		27.5 ± 0.8	K-Ar orthoclase	Santa Catalina Mts.	32°20.3', 110°41.4'	
		30.0 ± 1.0	K-Ar plagioclase	Santa Catalina Mts.	32°20.3', 110°41.4'	
52	UAKA-71-11	23.1 ± 0.5	K-Ar biotite	Santa Catalina Mts.	32°20.6', 110°55.4'	P. E. Damon and M. Shafiqullah*
				Gneissic and mylonitic Precambrian Y granitic rocks		
53	PED-29-60	27.1 ± 1.0	K-Ar biotite	Rincon Mts.	32°13.8', 110°31.0'	Damon and others (1963)
54	84	29.1 ± 1.1	K-Ar biotite	Rincon Mts.	32°12 ', 110°27 '	
55	85	34.7 ± 1.2	K-Ar biotite	Rincon Mts.	32°13.5', 110°25.5'	Marvin and others (1973)
		29.8 ± 0.9	K-Ar muscovite	Rincon Mts.	32°13.5', 110°25.5'	
		29.7 ± 1.0	K-Ar muscovite	Rincon Mts.	32°13.5', 110°25.5'	
56	PED-4-58	26.6 ± 1.0	K-Ar muscovite	Santa Catalina Mts.	32°18.4', 110°48.6'	Damon and others (1963)
57	PED-18-62D	28.3 ± 0.9	K-Ar biotite	Santa Catalina Mts.	32°20.3', 110°31.0'	Livingston and others (1967)
58	PED-27-57	39.4 ± 1.2	K-Ar biotite	Santa Catalina Mts.	uncertain	Damon and others (1963); P. E. Damon (1978, written commun.)
59	Locality 5	26	K-Ar biotite	Several miles west of Picacho Mts.	32°45 ', 111°30 '	Eberly and Stanley (1978)
				Pegmatite dikes cutting gneiss of Windy Point and older rocks, Santa Catalina-Rincon Mountains		
60	PED-30-60	33.5 ± 1.1	K-Ar muscovite	Rincon Mts.	32°12.4', 110°33.2'	Damon and others (1963)
61	UAKA-74-80	25.4 ± 0.5	K-Ar biotite	Rincon Mts.	32°11.5', 110°28.2'	P. E. Damon and M. Shafiqullah*
62	RM-1-66	32.0 ± 2	K-Ar muscovite	Santa Catalina Mts.	32°20.3', 110°41.4'	
63	PED-56-66	31.9 ± 0.9	K-Ar muscovite	Santa Catalina Mts.	32°20.6', 110°55.4'	Mauger and others (1968)
				Porphyritic Quartz Monzonite of Wild Burro Canyon, Tortolita Mountains		
64	RC3	21.6 ± 0.6	K-Ar hornblende	Tortolita Mts.	32°29.5', 111°04.0'	Creasey and others (1977)
		21.1 ± 0.6	K-Ar biotite	Tortolita Mts.	32°29.5', 111°04.0'	
				Late equigranular Quartz Monzonite of Tortolita Mountains		
65	PED-20-62	28.0 ± 0.9	K-Ar biotite	Tortolita Mts.	32°28 ', 111°05 '	Mauger and others (1968)
66	RC25	22.7 ± 0.7	K-Ar biotite	Tortolita Mts.	32°28 ', 111°02 '	
		18.5 ± 2.4	F-T apatite	Tortolita Mts.	32°28 ', 111°02 '	Creasey and others (1977)
31	ML105	17.0 ± 2.1	F-T apatite	Tortolita Mts.	32°27.5', 110°52 '	
				Undeformed porphyritic Quartz Monzonite of Picacho Mountains		
67	UAKA-76-18	24.6 ± 0.5	K-Ar biotite	Picacho Mts.	32°50 ', 111°22 '	P. E. Damon and M. Shafiqullah*
				Dikes that cut foliated rocks		
68	UAKA-74-83	24.3 ± 0.5	K-Ar whole rock	Rincon Mts.	32°11.5', 110°28.2'	P. E. Damon and M. Shafiqullah*
69	UAKA-72-21	21.0 ± 0.3	K-Ar whole rock	Santa Catalina Mts.	32°21.6', 110°52.6'	Shakel (1974)
70	PED-17-59	24.5 ± 0.5	K-Ar biotite	Santa Catalina Mts.	32°26.6', 110°52.1'	Damon and others (1963)
38	UAKA-75-87	24.0 ± 0.5	K-Ar biotite	Tortolita Mts.	32°31.8', 111°03.3'	Banks and others (1978)

Note: Previously published K-Ar ages recalculated with the following constants: $\lambda_\beta = 4.963 \times 10^{-10}$ yr^{-1}; $\lambda_\epsilon = 0.581 \times 10^{-10}$ yr^{-1}; $\lambda = 5.544 \times 10^{-10}$ yr^{-1}; ^{40}K/K $= 1.167 \times 10^{-4}$ atom/atom. Published fission-track ages recalculated 3% older to conform with new K-Ar constants.

*1977, written communication.

blastomylonitic and phyllonitic schist resembles and has been mistaken for some facies of the Precambrian X Pinal Schist, offering one reason for the long history of controversy concerning the complex; however, the gradation from essentially undeformed Precambrian Y granitic rock through augen mylonitic gneiss into blastomylonite and phyllonite is well exposed at many localities, including the rinds of car- and house-sized blocks intruded and enclosed by pegmatite and gneiss of Windy Point just below the Precambrian Y sedimentary rocks on Mount Lemmon (Fig. 8, cross section A–A'). Similar mylonitic rinds on the much larger xenoliths of gneiss derived from Precambrian granitic rock in the Rincon Mountains are apparent in Figure 6.

A recently published (Shakel and others, 1977) U-Th-Pb age of 1,440 ± 10 m.y. for zircon from such xenoliths in the Santa Catalina Mountains confirms the above field observations (Creasey and Theodore, 1975; Banks, 1976; Creasey and others, 1977; Banks, 1977) that the dark schistose rock and biotite augen gneiss intruded by and included in the gneiss of Windy Point are in large part the Precambrian Y Oracle Granite of Peterson (1938) (hereafter simply referred to as Oracle Granite). Gneiss of still older Precambrian Y granitic rocks, the Continental, Johnny Lyon, and Rincon Valley Granodiorites, occur in and around the Rincon Mountains (Drewes, 1974, 1977) and are part of the protoliths of the schistose and dark gneissic rocks in that part of the complex. The presence of large Pinal Schist xenoliths in the deformed Precambrian Y granitic rocks cannot be ruled out but has yet to be confirmed by structural analyses or U-Th-Pb studies.

The chemistry of a very large sample collected over several layers of gneissic and more mylonitic Precambrian Y granite (Table 2, sample 40) is similar to the chemistry of the gneiss of Windy Point. Likewise, the overall mineralogy of the deformed Precambrian granite is essentially the same as that of the gneiss of Windy Point. However, individual layers of the Precambrian rock contain much more muscovite and biotite and proportionately less K-feldspar than the intruding gneiss (Table 1, samples 2, 3, 4). Middle Tertiary K-Ar and FT ages of widely spaced samples from the Precambrian granitic rock within the complex (Table 3, samples 53 through 58) attest to the strength and uniformity of the thermal anomaly in the complex 20 to 30 m.y. ago.

The pegmatite dikes and sills (intrusive into both the gneiss of Windy Point and Precambrian host rocks) are granitic to quartz monzonitic in composition (Table 1, samples 29, 30) and are composed of albite-oligoclase, orthoclase-microcline, quartz, muscovite, biotite, magnetite, and locally abundant garnet, particularly at contacts with host rocks. Plagioclase crystals in pegmatite at some localities are as much as 500 mm long (for example, see Fig. 5B). The pegmatite dikes predated (Figs. 4A, 5A) and postdated or accompanied deformation (Fig. 5B). They appear to be most abundant at contacts of the gneiss of Windy Point with wall rocks and the larger xenoliths.

As noted above, the most intense deformation of the host Precambrian granitic rocks and the intruding gneiss of Windy Point and pegmatite occurs at the base and top of the intrusive mass. Foliation is generally shallow, defined by parallelism of micas and aligned cataclastically deformed feldspar and quartz, and is oriented roughly parallel to the top and the bottom of the intrusive sheets (Figs. 8, 10). Lineation consists of mineral alignment, mineral rods, fold crenulations, and intersections of compositional layers with poorly developed axial-plane foliation. The lineation occurs in the plane of foliation and is strongly concentrated along a N70°E trend (Fig. 10). Lineation and foliation in the gneiss of Windy Point and the Precambrian Y wall rocks and xenoliths are strikingly similar (Fig. 10). Thus, although the Precambrian rocks are more deformed (ductilely and cataclastically) than the gneiss of Windy Point, foliation and lineation in both were apparently formed in the same stress field.

The largest of the Tertiary intrusive bodies along the northern edge of the complex is the pluton of Samaniego Ridge, which occurs along the projected strike of the Geesman fault (Fig. 6). Radial and concentric jointing in the pluton (McCullough, 1963) is suggestive of a subvertical cooling center and is centered on the projected intersection of the Geesman and Pirate faults (Fig. 6). The center of jointing is bisected by the Pirate fault, which has faulted Tertiary gravel and conglomerate against the pluton.

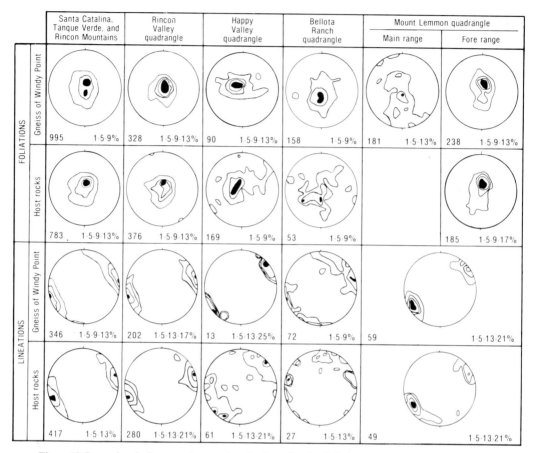

Figure 10. Lower-hemisphere equal-area net projections showing foliation and lineation in the Santa Catalina–Rincon Mountains complex. Contours are percent per one percent area. Data sources: Banks (1976); Creasey and Theodore (1975); and Drewes (1974, 1977).

The Pirate and Geesman faults are normal and have large displacement (Figs. 6, 8).

The texture and composition of the pluton of Samaniego Ridge (Banks, 1976; Creasey and others, 1977) are variable, and some facies are so similar in composition (Tables 1, 2, compare samples 1, 5, 7, 9 with samples 19 through 22) and appearance with undeformed Precambrian Y granitic rocks (Fig. 11) that the pluton was once considered to be Precambrian in age—a fact that further complicated the interpretation of the complex. Rocks in the pluton grade from medium-grained granitoid quartz monzonite-granite to coarsely porphyritic granodiorite-quartz monzonite containing 20-mm, euhedral to irregular K-feldspar phenocrysts in a groundmass of 5- to 10-mm-diameter quartz, plagioclase, biotite, local hornblende, and locally abundant sphene (Figs. 11B, 12; Tables 1, 2, samples 19, 20, 21, 22). Apatite, allanite, and zircon are the accessory minerals. These rock types grade upward toward the crest of Samaniego Ridge (Fig. 6) into a thick porphyro-aphanitic phase that is suggestive of chilling. Multiple-mineral K-Ar and FT ages for four widely spaced samples (Table 3, samples 19, 20, 21, 47) from the pluton indicate it was emplaced about 25 to 29 m.y. ago and cooled below the closure temperature of apatite about 20 to 23 m.y. ago (Creasey and others, 1977).

On its northern and northeastern sides, the pluton of Samaniego Ridge intruded the Precambrian Oracle Granite; to the east it passively intruded the Leatherwood Quartz Diorite; and to the west, as noted above, it is faulted against Tertiary gravel along the Pirate fault (Fig. 6). However, parallelism

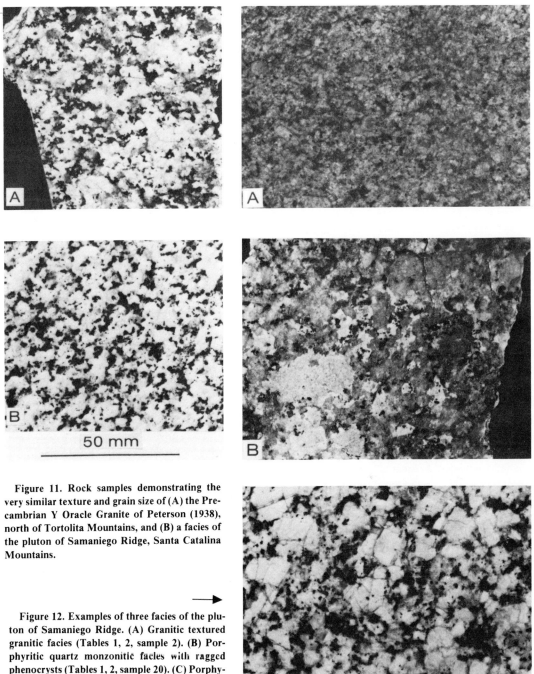

Figure 11. Rock samples demonstrating the very similar texture and grain size of (A) the Precambrian Y Oracle Granite of Peterson (1938), north of Tortolita Mountains, and (B) a facies of the pluton of Samaniego Ridge, Santa Catalina Mountains.

Figure 12. Examples of three facies of the pluton of Samaniego Ridge. (A) Granitic textured granitic facies (Tables 1, 2, sample 2). (B) Porphyritic quartz monzonitic facies with ragged phenocrysts (Tables 1, 2, sample 20). (C) Porphyritic quartz monzonitic facies with well-formed phenocrysts (Tables 1, 2, sample 19). Specimens are stained rock slabs; light grains are K-feldspar, gray are quartz and plagioclase, black are hornblende, biotite, sphene, and opaque minerals.

with the Pirate fault of pegmatites and igneous foliation and bisection of the radial and concentric jointing by the fault suggest that the structure might have been a zone of weakness before or during intrusion of the pluton. In the Cargodera Canyon area (Fig. 6), the pluton includes abundant xenoliths or pendants of metasedimentary rock, which are aligned parallel to foliation and xenoliths of the gneissic rock to the south. The pluton is crosscut by north-northwest–striking dikes of fine-grained granodiorite-quartz monzonite that is petrographically similar to the youngest pluton in the Tortolita Mountains (see Fig. 3B) to the west. North of Cargodera Canyon, pegmatite dikes are rare, and aplite dikes are scarce. South of the canyon, igneous alignment gives way to progressively more shear zones and cataclasis of the quartz monzonite. Concurrent with the southward increase in deformation, muscovite and garnet appear, in part at the expense of the biotite, and pegmatite dikes and sills become increasingly abundant, mostly in the plane of foliation.

Published descriptions are meager for the two occurrences of undeformed Tertiary quartz monzonite east of the pluton of Samaniego Ridge (Fig. 6). Brief visits to the areas suggest they are petrographically identical to some facies of the pluton of Samaniego Ridge, and the quartz monzonite pluton in the Bellota Ranch quadrangle grades westward into cataclastically deformed, two-mica garnet-bearing gneiss (T. G. Theodore, 1977, oral commun.; B. N. Moore, unpub. manuscript).

Given the great volume of the gneiss of Windy Point (Figs. 6, 8), it would seem likely that more than one intrusive pulse was involved in its emplacement. However, mapping to date, which is largely reconnaissance in nature, has not defined intrusive contacts within the gneiss of Windy Point other than those with its host rocks, xenoliths, and roof pendants (Moore and others, 1949; Drewes, 1974, 1977; Creasey and Theodore, 1975; Banks, 1976). Likewise, as described above, at two of the three areas along the northern margin of the complex, current published mapping suggests a dominantly gradational contact between the Tertiary intrusive rock and the gneiss of Windy Point (Moore and others, 1949; Bromfield, 1952; McCullough, 1963; Creasey and Theodore, 1975; Banks, 1976, 1977; Creasey and others, 1977). Furthermore, as pointed out by Damon and others (1963), Damon (1968), Livingston and others (1967), Mauger and others (1968), Creasey and others (1977), and Banks (1977), there is a tight grouping of 20 to 30 m.y. in multiple-mineral K-Ar and FT ages from widely spaced samples throughout the gneiss of Windy Point, from thermally reset gneissic and mylonitic xenoliths of Precambrian Y granitic rock in the gneiss of Windy Point, and from the largest mass of Tertiary intrusive rock, the pluton of Samaniego Ridge (Table 3).

The above listed relations (lack of obvious or persistent intrusive contacts within the gneiss of Windy Point, apparently gradational contacts between the sill-like masses of this gneiss and the undeformed Tertiary intrusive rock to the north, and the tight grouping and general concordance of the mineral ages throughout the gneiss, its xenoliths, and the Tertiary intrusive rock) when combined with the earlier-mentioned late to post-Mesozoic field age of the gneiss and the field and K-Ar evidence for telescoped thermal boundaries of the complex, led Banks (1977) and Creasey and others (1977) to suggest that the areas of Tertiary intrusive rock were conduits, perhaps with some resurgence, for the compositionally similar but cataclastically deformed sill-like masses of gneiss of Windy Point. The same data also led them to suggest that the cataclasis of the gneiss of Windy Point and its contiguous wall rock and xenoliths occurred during and was related to the relatively short-lived emplacement of a partially floored, composite batholith.

There were some recognized problems and discrepancies at the time of these suggestions. The main problems were the shallow depth of deformation implied by the middle Tertiary age, the mechanism by which the consistently oriented lineation was obtained, and how the pegmatite formed. However, the same geology and geochronology provide equally perplexing problems to postulates of an older age of the complex. Discussions of these problems are deferred to a later section. The major discrepancies were that some of the few Rb-Sr data for the pluton of Samaniego Ridge, gneiss of Windy Point, and its xenoliths suggested ages much older (Damon and Giletti, 1961; Shakel, 1972, 1974; Shakel and others,

1972) than the K-Ar and FT ages, as did also the then existing U-Th-Pb data for zircon from the gneiss of Windy Point (Catanzaro and Kulp, 1964).

However, these Rb-Sr data were not internally consistent for the same rock unit or even the same sample locality (Damon and Giletti, 1961; Livingston and others, 1967; Shakel, 1972, 1974; Shakel and others, 1972), included pegmatite rock that was not necessarily derived by fractional crystallization of the other samples on the isochrons, and did not define tight isochrons. Additionally, U-Th-Pb ages of 27 ± 2 m.y. from the pluton of Samaniego Ridge (Shakel and others, 1977) confirmed the already well-documented K-Ar and FT age of the pluton as middle Tertiary and questioned its Rb-Sr age as well as the other published isochron ages. Moreover, the Catanzaro and Kulp (1964) data could not be considered to be definitive of the original age of emplacement of the gneiss of Windy Point, given the large discordance of the U-Th-Pb ages, two populations of zircon in the sample, and the late to post-Mesozoic field age of the gneiss.

Four additional discrepancies occur in the 12 ages produced since the suggestions of Banks (1977) and Creasey and others (1977). Two are in previously unpublished data presented here through the courtesy of Harald Drewes, C. W. Naeser, P. E. Damon, and M. Shafiquallah (Table 3, samples 48, 49, 50, 52). These are the 231-m.y. and Precambrian(?) FT ages of zircon in samples 48 and 50 (Table 3). However, these ages can be attributed to the likely presence of zircon derived from assimilated Precambrian Y rocks, as suggested by the data of Catanzaro and Kulp (1964), and partial annealing of such zircon. The two other much more problematical ages are the concordant 44- to 47-m.y. U-Th-Pb ages for monazite from two widely separated samples of gneiss of Windy Point (Shakel and others, 1977). These older ages are supported by some of the ages for pegmatite dikes cutting the complex (Table 3, samples 42, 45, 60, 61, 62, 63) and suggest that (1) the gneiss of Windy Point is appreciably older than the pluton of Samaniego Ridge and (2) as favored by Damon and others (1963), Shakel (1972), Drewes (1977), Shakel and others (1977), Mauger and others (1968), and Damon (1968), the K-Ar and FT ages are cooling ages set at least 15 to 20 m.y. after the emplacement of the gneiss of Windy Point.

Additional Rb-Sr data that have become available since submittal of this paper (Keith and others, this volume) also suggest that the gneiss of Windy Point could be about 20 m.y. older than the pluton of Samaniego Ridge. However, the pegmatite, U-Th-Pb, and new Rb-Sr ages are not exempt from interpretation and thus are not unequivocally indicative of the minimum age of the gneiss of Windy Point. For example, only some of the pegmatite ages are older than the host gneiss, and those that are might be explained by leaching of potassium after closure of the minerals to argon loss or to presence of extraneous radiogenic argon. Location of the potentially water-saturated pegmatites in zones of structural weakness would facilitate the former mechanism, and extraneous radiogenic argon derived from degassing of minerals in the Precambrian granite was potentially available to all of the minerals formed during metamorphism. There likewise is a possibility that the monazite ages might be anomalously old, because there is evidence that the mineral is reset at middle to upper amphibolite-facies metamorphism (L. T. Silver, 1978, oral commun.). Thus, given the intimate intrusive and assimilative relations between the Oracle Granite and gneiss of Windy Point, the monazite data, like the zircon data of Catanzaro and Kulp (1964), could be reflecting partial resetting of Precambrian Y xenocrysts in the gneiss. Furthermore, the Rb-Sr isochron age (Keith and others, this volume) also is subject to interpretation because the isochron is constructed between a loose cluster of points near the intercept and a few points with high Sr^{87}/Sr^{86} ratios from pegmatite dikes that are assumed but not proven to have developed in an isotopically closed system by fractional crystallization of the gneiss. On the other hand, L. T. Silver (1978, oral commun.) noted that monazite is not a common accessory mineral of Precambrian Y granite, and lack of proof of cogenesis of gneiss and pegmatite does not remove the possibility. Thus, the U-Th-Pb ages of the monazites, the older ages for the pegmatites, and the Rb-Sr isochron age cannot be ignored. If they indicate the minimum or actual age of the gneiss

(Shakel and others, 1977; Keith and others, this volume), a continuous intrusive contact was overlooked and needs to be mapped between the Tertiary intrusive rock and gneiss of Windy Point along the northern margin of the complex (Fig. 6).

Field checking in Cargodera Canyon by S. B. Keith (1978, oral commun.) suggests that a contact may be present. However, a continuous contact has yet to be documented, the area is characterized by a >2-km-wide screen of xenoliths of variably granitized and deformed sedimentary and possibly Precambrian granitic rock that complicates interpretation of the contact, and establishment of a continuous contact in itself would not demonstrate lack of comagmatism between the Tertiary intrusive rock and gneiss of Windy Point. Clearly, future progress on interpretation of the complex requires detailed mapping supported by U-Th-Pb studies in the zone between the undeformed granitic rock and the granitic gneiss to the south.

If a ≥20-m.y. age gap between the two rock types is documented in the future, geologically viable mechanisms are required to explain (1) how reheating (amphibolite grade or higher) of the >5,000 km^3 of cataclastic terrane was achieved at least 20 m.y. after emplacement of the gneiss or, alternatively, how high heat flow was maintained in the cataclastic terrane for at least 20 m.y., and (2) as required by the telescoped thermal and metamorphic boundaries of the complex, how the reheating or maintenance of heat was kept specific to the cataclastic terrane or, alternatively, how vertical or horizontal tectonics moved the complex or wall rocks to the present juxtaposition sometime after 30 m.y. ago. A logistical problem with reheating is whether, given the 3- to 5-km thickness and shallow stratigraphic emplacement of the gneiss, a buried source could achieve the grade and uniformity of heat to reset the K-Ar and FT ages, while keeping the heat restricted to the cataclastic terrane for the length of time required. The pegmatite dikes, the only exposed potential heat sources, are unlikely to have done such reheating because of their granite-minimum composition, their cataclasis requires a superimposed second event of cataclasis as well as reheating, and if 30 m.y. old, they cannot be plotted on isochrons that indicate the age of the gneiss of Windy Point. Long-term maintenance of telescoped thermal boundaries on the complex is not as problematical if, in agreement with the geology, the complex developed at shallow levels where cooling of adjacent host rocks was achieved by ingress of ground water. Concerning the tectonic emplacement of the complexes, comparison of the thickness, stratigraphy, and lithology of the adjacent and more distant Mesozoic and Tertiary basin does not reveal evidence suggesting that the complexes experienced anomalous burial to, or exhumation from, the great depths that would be required to insulate them for 20 m.y. or more, nor do geologic maps of the region postulate regional thrusts that might have emplaced the complexes ≤30 m.y. ago. Thus, the age of the complex is not unequivocally resolved.

Tortolita Mountains

Intrusive and metamorphic rocks of the Tortolita Mountains complex crop out in a roughly equant area of about 225 km^2 about 30 km northwest of Tucson (Figs. 1, 13). Field relations (Banks and others, 1977) indicate that similarly oriented foliation and lineation developed in the complex during two events, the last in post-Permian time, and that the two events were likely closely spaced in time. The complex is much smaller, and the relief (800 m above the surrounding alluvial plain) is not nearly as spectacular as in the Santa Catalina–Rincon Mountains. However, rock types in the Tortolita Mountains are similar to those in the larger complex to the east, and the relations between rock types and the processes that occurred in the Tortolita Mountains complex are more easily ascertained, although it has received much less study than the larger complex (Budden, 1975; Davis and others, 1975; Creasey and others, 1977; Banks and others, 1977).

At least five major intrusive rock types occur in the complex. Two are associated with an early pulse of magma and were in part penetratively foliated and lineated during or before intrusion of two other

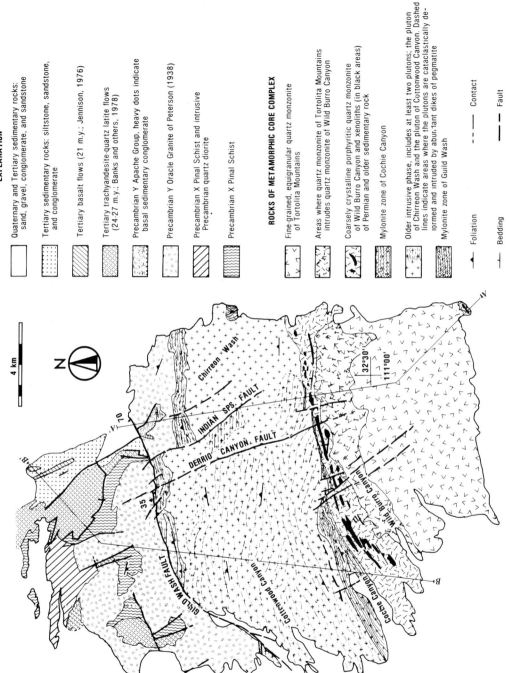

Figure 13. Generalized geologic map of the Tortolita Mountains metamorphic core complex and vicinity (modified from Budden, 1975; Banks, 1976; Banks and others 1977; R. P. Ashley, unpub. data).

major rock types associated with a late pulse of magma. Foliation and lineation in the two younger rock types are weak and local but are oriented parallel to foliation and lineation in the two early rock types. The fifth major type is represented by pegmatite dikes of at least two ages that mainly intrude the first two plutons. Two major zones of mylonite occur in the complex, and dikes (not shown in Fig. 13) representing at least three minor intrusive rock types were emplaced after cataclasis. Dikes of one of these (Fig. 3A) cut the last major intrusive body (Fig. 3B; Banks and others, 1977; Budden, 1975). About 6 km north of the complex (4 km east of the Suizo Mountains) a 1,438 ± 10-m.y. biotite K-Ar age was obtained from Precambrian quartz diorite (Banks and others, 1978), which indicates that, like the Santa Catalina–Rincon Mountains, the Tortolita Mountains complex was tectonically juxtaposed with its host rocks on as yet unrecognized structures, or the heating was a local rather than a regional event. The latter is favored here.

The two early plutons occur along the northern half of the complex (Fig. 13). The contact between them is irregular and characterized by a wide zone where abundant xenoliths of the older pluton (hereafter the pluton of Chirreon Wash) occur in a matrix of the younger rock (hereafter referred to as the gneiss of Cottonwood Canyon). The contact zone occurs in the area of cataclasis and flooding by pegmatite dikes, making delineation of the two plutons impractical during reconnaissance mapping of the complex (Banks and others, 1977). The pluton of Chirreon Wash is east and south of the gneiss of Cottonwood Canyon, is bounded on the north and south by zones of mylonite, intrudes the northernmost mylonite zone (hereafter referred to as the Guild Wash mylonite zone), and grades into the southern mylonite zone (hereafter the Cochie Canyon mylonite zone; Fig. 13). Cataclastic deformation of the pluton occurs in most of the exposures west of Derrio Canyon (Fig. 13). In the essentially undeformed exposures east of Derrio Canyon, the pluton grades from fine- or medium-grained diorite and quartz diorite (Fig. 14; Tables 1, 2, samples 14, 15) to sphene-bearing, biotite-hornblende, porphyritic granodiorite and quartz monzonite. The porphyritic quartz monzonite is similar to some facies of the pluton of Samaniego Ridge, and parts of the quartz dioritic facies are texturally and mineralogically similar to the Leatherwood Quartz Diorite in the Santa Catalina–Rincon Mountains complex (compare Fig. 14 with Fig. 7, and Table 1, samples 11, 12, 13 with samples 14, 15, 16), although the chemistry of the two analyzed samples is significantly dissimilar (Table 2, samples 11, 14).

The gneiss of Cottonwood Canyon occurs entirely within the area of cataclasis and, like the pluton of

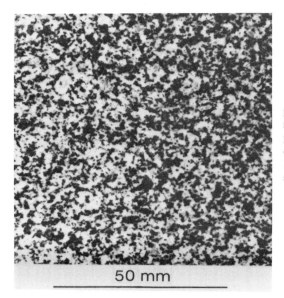

Figure 14. Stained slab of quartz dioritic border phase of the pluton of Chirreon Wash, the earliest intrusive phase of the Tortolita Mountains complex; light grains are K-feldspar, gray grains are dominantly plagioclase and quartz (the latter is 10% of rock), black grains are biotite, hornblende, and opaque minerals.

Chirreon Wash, intrudes the Guild Wash mylonite and grades southward to form part of the mylonite in Cochie Canyon. The least-deformed parts of the gneiss indicate it was derived from porphyritic quartz monzonite. Its derivative two-mica, garnet-bearing, penetratively foliated and lineated gneiss is not distinguishable in hand specimen from samples of the gneiss of Windy Point in the Santa Catalina–Rincon Mountains complex.

The Guild Wash mylonite is derived from the Oracle Granite of Precambrian Y age that occurs along the northern boundary of the complex (Fig. 13). The conversion of the granite to mylonite is gradational (Fig. 15), and small to very large remnants of augen mylonitic gneiss clearly derived from the Oracle Granite occur in the mylonite (Fig. 13). East of the Indian Springs fault (Fig. 13), the intruding pluton of Chirreon Wash is unfoliated or only slightly foliated (Fig. 15F), and the schistose host rock grades northward, in places gradually and in places abruptly, into unsheared but thermally metamorphosed Precambrian Oracle Granite that is overlain by sedimentary rocks of the Precambrian Apache Group, which dip moderately to the north away from the complex (Figs. 13, 16, sec. A–A′). West of the Indian Springs fault, postmetamorphic movement along the Guild Wash fault (in part parallel to the foliation) resulted in brittle cataclasis and juxtaposition of schistose Precambrian granite against essentially unmetamorphosed Precambrian granite (Fig. 16, sec. B–B′).

The Cochie Canyon or southern mylonite zone crosses the approximate middle of the complex (Fig. 13) and is oriented parallel to and about 6 to 8 km south of the Guild Wash mylonite zone. As noted above, the Cochie Canyon mylonite is derived from the gneiss of Cottonwood Canyon and the pluton of Chirreon Wash, both of which grade southward from clearly recognizable igneous rock into folded schistose rock (Fig. 17). The Cochie Canyon mylonite thins eastward approximately in proportion to the decrease in cataclasis of the plutons from which it was derived (Fig. 13); it is intruded on the south by the first rock type of the second pulse of magma, essentially undeformed porphyritic sphene-bearing quartz monzonite (Figs. 13, 16) that is very similar to the most common facies of the pluton of Samaniego Ridge in the Santa Catalina Mountains (compare examples A and B in Fig. 2).

This porphyritic quartz monzonite (hereafter the quartz monzonite of Wild Burro Canyon) crops out in a band approximately 3 km wide. Included in the northern half of the band (Figs. 13, 16) are abundant lineated and folded xenoliths of metasedimentary rocks ranging in age from Precambrian Y (Apache Group; Budden, 1975) to Permian (Concha Limestone; Banks and others, 1977). Petrographically identical porphyritic quartz monzonite, also with xenoliths of Paleozoic rocks, crops out at the southeasternmost part of the complex (Figs. 13, 16, sec. A–A′). The geographic and structural locations of these two outcrops of porphyritic quartz monzonite suggest that they represent an arched porphyritic roof or border phase of the large pluton of fine- to medium-grained quartz monzonite-granite (hereafter the quartz monzonite of the Tortolita Mountains; Fig. 3B; Tables 1, 2, samples 31, 32) that occupies the exposed southern one-third of the complex between the outcrops of porphyritic quartz monzonite (Fig. 13). At the present level of exposure, the contact between the porphyritic and equigranular quartz monzonites is intrusive over a zone as much as 2 km wide (Fig. 13), and both rock types are only slightly foliated locally and are cut by only a few pegmatite dikes.

In contrast, pegmatite dikes are abundant in the mylonite in Guild Wash, occur most frequently in the plane of foliation, and are increasingly more frequent southward toward the intrusive gneiss of Cottonwood Canyon and the pluton of Chirreon Wash. As noted earlier, pegmatite dikes also locally flood these cataclastically deformed plutons. The pegmatites postdated and predated or accompanied cataclasis (Fig. 15E).

For three reasons, the pegmatites seem more likely to be the result of metamorphism than of differentiation of either the early-phase plutons or the younger unfoliated plutons: (1) They are abundant and ubiquitous in deformed rock but are rare in undeformed parts of the pluton of Chirreon Wash and in the younger undeformed plutons. (2) Their volume, 10% to 90% of half of the exposed outcrop of the complex, is much greater than can be expected to be produced by differentiation of the

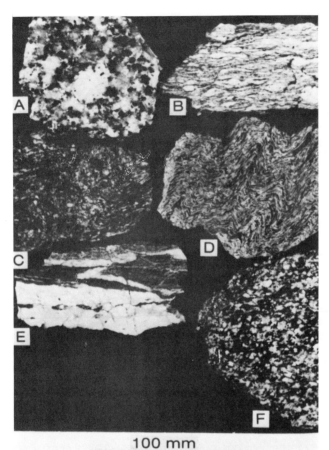

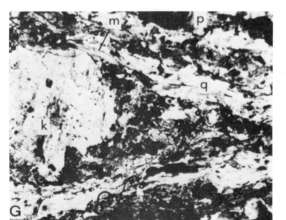 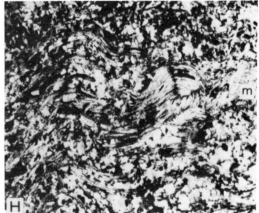

Figure 15. Series of samples collected on a north-to-south traverse through progressively more cataclastically deformed Precambrian Oracle Granite of Peterson (1938), northern border of the Tortolita Mountains metamorphic core complex. (A) Undeformed Oracle Granite of Peterson (1938). (B) Augen gneiss–protomylonite derived from the granite of A; augen are K-feldspar porphyroclasts. (C) Mylonite derived from the granite of A; K-feldspar crystal fragments are still megascopically recognizable. (D) Contorted blastomylonite schist derived from the granite of A. (E) Ultramylonite schist derived from the granite of A; two pegmatite "sills" occur in the plane of foliation, one is only slightly deformed and the other is cataclastically deformed and folded. (F) Slightly foliated quartz diorite border facies of the pluton of Chirreon Wash; sample was collected several metres from sample E. (G) Photomicrograph of specimen B; k = K-feldspar, m = muscovite, p = plagioclase, q = quartz. (H) Photomicrograph of specimen D; m = muscovite.

existing plutons without postulation of a much larger source pluton and fortuity of their restricted emplacement in deformed rock. (3) They very seldom (tourmaline in one sample of float) contain the more exotic fluorine-rich minerals common to pegmatites derived by differentiation.

In addition to the pegmatite dikes, the plutons of Chirreon Wash and Cottonwood Canyon are cut by dikes of unfoliated quartz monzonite porphyry (Table 1, sample 34) that are petrographically very similar to the porphyro-aphanitic rock along the crest of Samaniego Ridge in the Santa Catalina Mountains. Except for an aplitic rather than a medium-grained groundmass, these dikes likewise closely resemble the quartz monzonite of Wild Burro Canyon (Fig. 13). Also present north of the Cochie Canyon mylonite are dikes and sills of unfoliated biotite-hornblende diorite porphyry (Table 1, sample 37) that are very similar in appearance to dikes of unfoliated diorite in the Santa Catalina Mountains (Table 1, sample 35). These dikes, both mylonite zones, and all of the plutons in the complex are cut by dikes of quartz latite–rhyolite porphyry that grade, toward their centers and southward along strike, into fine-grained quartz monzonite-granite (Tables 1, 2, samples 38, 39) that is very similar in texture and mineralogy to the quartz monzonite of the Tortolita Mountains (Fig. 3).

Folding, foliation, and lineation are penetrative in both zones of schistose mylonite; are well developed in the pegmatite, metasedimentary rocks, and the early intrusive types; and although weakly developed, also occur in the two younger major intrusive types. The folds and lineation have been studied in some detail by Davis and others (1975). The most conspicuous folds are overturned and recumbent, shallow-dipping flexural slip and flexural flow folds whose fold axes trend east-northeast. Most of the folds are moderately tight, but some are isoclinal and rootless with axial surfaces subparallel to the foliation. Amplitudes and wavelengths of these folds are generally less than 1 m. The overturned and recumbent folds have been systematically refolded by upright open folds characterized by wavelengths and amplitudes of about 6 and 3 m respectively, and axial surfaces that strike north-northwest. Foliation (Fig. 16) is generally much steeper than in the Santa Catalina Mountains and is defined by schistosity in the mylonitized rocks; by planar alignment of feldspar, mica, and quartz in the intrusive phases; and by bedding-plane cleavage, schistosity, and stretched pebbles in the metasedimentary rocks.

The lineation occurs in the plane of foliation and is displayed by mineral alignment, slickensides, stretched pebbles, fold crenulations, and intersections of compositional layers with schistosity and poorly developed axial-plane foliation. Like that in the Santa Catalina–Rincon Mountains, the trend of the lineation is restricted in orientation to N50° to 80°E, averaging N65° to 70°E. The foliation and lineation is transposed by shallow-dipping, relatively wide spaced, mostly north-northwest–striking surfaces and faults, some of which are occupied by the unfoliated dikes. The Guild Wash mylonite zone is offset along some of these faults, the largest indicating 1.5 km of apparent right-lateral offset of the eastern one-third of the complex on the Indian Springs fault (Fig. 13). After development of these faults, postcataclastic movement along the Guild Wash mylonite zone west of the Indian Springs fault has resulted in a shallow-dipping fault similar in mode of deformation and appearance to the decollement Catalina fault flanking the south of the Santa Catalina–Rincon Mountains complex. Where the Guild Wash fault departs from the mylonite zone east of Indian Springs fault (Fig. 13), its dip steepens from 30° to 70° and the cataclasite is not present below the steepened shear zone.

The geologic relations indicate that the oldest plutons postdate Precambrian Y sedimentary rock and that the younger, mildly deformed rock types intrude Permian rocks. Radiometric data indicate a 25.1 ± 0.5-m.y. K-Ar age for biotite from the pluton of Chirreon Wash, the oldest exposed pluton in the complex (Table 3, sample 46), and a 24.0 ± 0.5-m.y. K-Ar age for biotite from one of the youngest unfoliated dikes cutting the complex (Table 3, sample 38). However, there is not agreement in detail between radiometric ages and geologic relations. Rocks geologically older than the above-mentioned dike give younger radiometric ages. Creasey and others (1977) obtained a concordant biotite-hornblende K-Ar age for the quartz monzonite of Wild Burro Canyon south of the Cochie Canyon

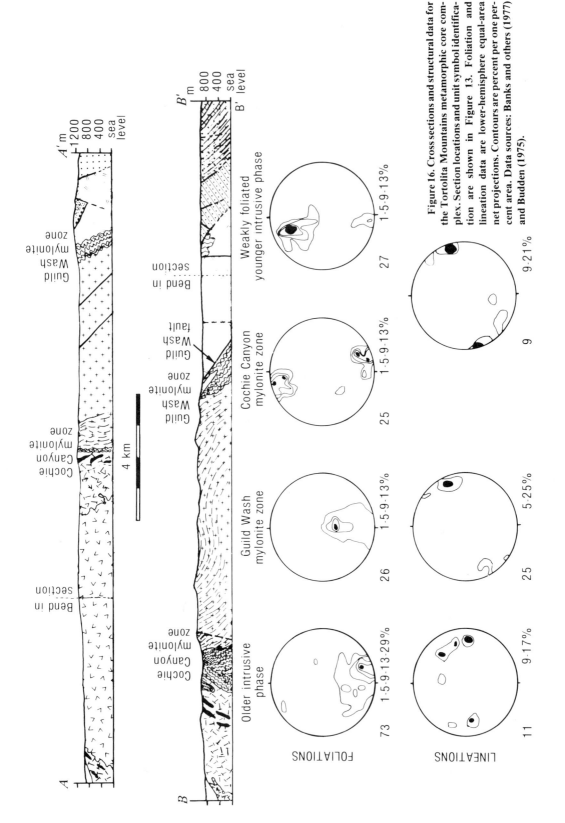

Figure 16. Cross sections and structural data for the Tortolita Mountains metamorphic core complex. Section locations and unit symbol identification are shown in Figure 13. Foliation and lineation data are lower-hemisphere equal-area net projections. Contours are percent per one percent area. Data sources: Banks and others (1977) and Budden (1975).

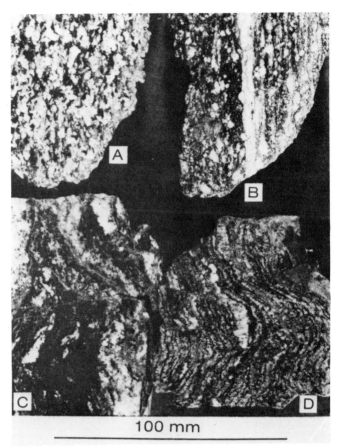

Figure 17. Series of samples collected on a north-to-south traverse through progressively more deformed rock of the southern mylonite zone of the Tortolita Mountains metamorphic core complex. (A) Mildly foliated quartz dioritic facies of the pluton of Chirreon Wash. (B) Gneissic quartz diorite derived from quartz diorite of sample A, pegmatite "sill" occurs in the plane of foliation. (C) Blastomylonite derived from pegmatite and the quartz diorite of sample A. (D) Phyllonitic schist near the southern contact of the southern mylonite zone; wider white layers are quartz at anticline hinge zones.

mylonite zone of 21.1 ± 0.6 and 21.6 ± 0.6 m.y., respectively (Table 3, sample 64), a biotite K-Ar age for the quartz monzonite of the Tortolita Mountains of 22.7 ± 0.7 m.y., and an apatite FT age of 18.5 ± 2.4 m.y. for the same sample (Table 3, sample 66). They also reported a 17.0 ± 2.1 m.y. FT age for apatite from a slightly foliated sample of the same pluton (Table 3, sample 31). Additionally, the oldest radiometric age obtained for the complex is for a gneissic variety of the quartz monzonite of the Tortolita Mountains (28.0 ± 0.9 m.y., Table 3, sample 65) rather than for the pluton of Chirreon Wash. Thus the radiometric ages indicate an intense 30- to 20-m.y.-old thermal event with a cooling history similar to the Santa Catalina–Rincon Mountains complex; however, definitive ages for each pluton and the two events of deformation have not been obtained.

On the other hand, several observations suggest that both magmatic events and both metamorphic events were related. First: the early and late magma pulses have very similar compositional and textural facies that are likewise very similar to the gneissic and undeformed plutons in neighboring complexes, and all intrusive phases in and throughout each complex yield essentially the same K-Ar and FT ages. Second: the northern contacts of the older plutons parallel the northern mylonite, cataclasis of the northern mylonite increases progressively southward toward the older plutons, and cataclasis of these plutons is unrelated to cataclasis of the northern mylonite (Figs. 13, 15), which suggests that there were two events of cataclasis and that cataclasis of the northern mylonite predated or accompanied intrusion of the first two plutons. Third: the northern contacts of the younger plutons parallel foliation in the southern mylonite and in the older plutons, and cataclasis of the older plutons decreases northward away from the younger plutons and eastward in proportion to the decrease in thickness of the southern mylonite zone, which suggests contemporaneity of cataclasis of the older plutons and southern mylonite and that the cataclasis predated or accompanied intrusion of the younger plutons. Fourth: parallelism of pluton elongation, northern contacts of plutons, and both mylonite zones suggest that intrusion and deformation occurred in the same stress field. Fifth: although of differing degrees, foliation and lineation in the early and late plutons and the two mylonites are very similarly oriented, which suggests that cataclasis in all occurred within the same stress field.

The stratigraphic level of emplacement of the Tortolita Mountains complex appears to be about the same as that of the Santa Catalina–Rincon Mountains complex, because the pluton of Chirreon Wash underlies the Apache Group (Fig. 16, cross section A–A′), and the quartz monzonite of Wild Burro Canyon includes xenoliths of sedimentary rocks as young as Permian (Banks and others, 1977). However, the Tortolita Mountains lie within the area where Mesozoic sedimentary deposits do not underlie Tertiary deposits (Titley, 1976; see Fig. 1); therefore, Mesozoic cover over the complex may have been much less than over the Santa Catalina–Rincon Mountains complex and definitely was much less during the 30- to 20-m.y. thermal event required by the radiometric ages. Thus, as for the larger complex to the southeast, the coexistence of muscovite and biotite in rocks is unlikely to indicate the depth of emplacement of the complex because an unreasonable and unsupported amount of vertical or horizontal tectonic thickening would be required to postulate coexistence of the minerals with melt.

Again, by comparison with the larger complex to the southwest, there was possibly a cover of middle Tertiary rocks several kilometres thick over the complex (Figs. 13, 16) during the thermal event in the complex (Banks and others, 1977, 1978). Similarly, the complex appeared to be a topographic high and possibly a volcanic center 20 to 30 m.y. ago. Volcanic rocks in the Tertiary section have chemistry similar to the intrusive rocks in the complex (Jennison, 1976; Banks and others, 1978), and there is evidence in the coarsening of Tertiary conglomerate toward the complex and in the presence of large glide blocks of Paleozoic rocks in the adjacent Tertiary section that the complex was a highland or site of inflation during the 20- to 30-m.y. thermal event (Banks, 1977; Banks and others, 1977, 1978). Tectonic movement at and parallel to the base of the Tertiary volcanic rocks north of the complex (Fig. 16, sec. B–B′; see also Banks and others, 1977) might also be explained by gravitational movement of the Tertiary deposits on subjacent exfoliation planes in the Precambrian basement away from a domal area over the complex.

Suizo Mountains

Cataclastically deformed granitic rock crops out over about 12 km^2 in and northeast of the Suizo Mountains (Figs 1, 18). The total exposed strike length of the metamorphosed rock is about 9 km. Two main rock types occur in these exposures: (1) a buff- to cream-colored, medium-grained, two-mica, garnet-bearing, quartz monzonitic mylonite gneiss and protomylonite (Table 1, sample 27) that is petrographically very similar to gneisses of Windy Point in the Santa Catalina–Rincon Mountains (Table 1, samples 23, 24, 25, 26) and Cottonwood Canyon in the Tortolita Mountains; and (2) black to gray schistose blastomylonite and phyllonite that is petrographically indistinguishable from the xenoliths of schistose Precambrian granite in the forerange of the Santa Catalina Mountains. Both rock types are penetratively foliated and lineated, and both are host to coarse-grained pegmatite dikes intruded predominantly on planes of foliation. Although folded, foliation dips predominantly east-southeast, and the mineral alignment and streaking is in the plane of foliation and oriented east-northeast (Fig. 18). In cross section, the cataclastically deformed younger granitic rock of undetermined thickness (but at least 500 m thick) underlies the mylonitic Precambrian granite. The intrusive, less-deformed granite also occurs as sills in the plane of foliation in the overlying schistose mylonite; the thickest occurs at the highest elevations of the Suizo Mountains, 200 m above the surrounding alluvial and pediment plain.

There are many well-exposed outcrops at the north end of the Suizo Mountains where the schistose mylonite is in gradational contact with undeformed Precambrian granitic rock (Banks and others, 1977). On the other hand, except for monotony of composition and texture, the schistose mylonite is megascopically similar to, and has been previously considered to be, Precambrian X Pinal Schist. Thus, presence of Pinal Schist within the schistose mylonite cannot be ruled out, particularly because large xenoliths of the Pinal occur in undeformed Precambrian granite within 2 km of the south end of the Suizo Mountains (Fig. 18).

The cataclastic rocks are cut by undeformed dikes of diorite porphyry and fine-grained quartz monzonite–granite that are petrographically correlatable with the two types of postfoliation dikes in the Santa Catalina Mountains and with two of the three young dikes in the Tortolita Mountains. Geochronology is presently unavailable for the rocks of the Suizo Mountains, and determination of depth of its emplacement is not possible by stratigraphic reconstruction, because the only host rock is Precambrian granite and the thickness of granite and thickness or even presence of sedimentary and volcanic cover during metamorphism is at present unknown. Except for some brief descriptions by

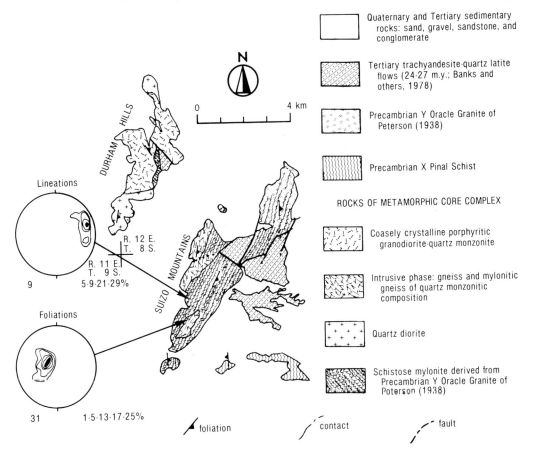

Figure 18. Generalized geologic map of the Suizo Mountains and Durham Hills (after Banks and others, 1977). Foliation and lineation are lower-hemisphere equal-area net projections (data source: Banks and others, 1977). Contours are percent per one percent area.

Barter (1962) and Iles (1967), no studies other than the reconnaissance mapping of Banks and others (1977) have been published for the metamorphic rocks in the Suizo Mountains.

Black Mountain

Penetratively foliated and east-northeast–lineated schist was mapped in the northeast corner of the Tortolita Mountains quadrangle in the foothills of Black Mountain (Banks and others, 1977). Foliation in the schists dips at steep to moderate angles under a quartz diorite border phase of a porphyritic, sphene-bearing, quartz monzonite intrusion. Both rock types are similar to those in the Tortolita Mountains and the Santa Catalina Mountains. As in the Tortolita Mountains complex, the foliation plane is the site of abundant pegmatite dikes that decrease abruptly in abundance in the intrusive rock northeastward away from the contact with Precambrian host rocks. The similarity of lineation, petrography of the intrusive rock, and metamorphism in these outcrops to those in the Tortolita Mountains complex suggests that part of the granitic rock in Black Mountain may be related to the zone of complexes under discussion.

Durham Hills

Metamorphic and intrusive rocks in the Durham Hills (Figs. 1, 18) are tentatively correlated with those in the zone of complexes on the basis of similarity of field relations and petrography, although no radiometric ages are presently available. The bulk of the outcrop (~8 km^2) is porphyritic, sphene-bearing, biotite-hornblende quartz monzonite that is similar petrographically to the quartz monzonite of Wild Burro Canyon in the Tortolita Mountains and the pluton of Samaniego Ridge in the Santa Catalina Mountains. The quartz monzonite intrudes quartz diorite (Fig. 2D) that is similar to some facies of the border phase of the pluton of Chirreon Wash in the Tortolita Mountains and also intrudes strongly foliated and lineated schistose blastomylonite and phyllonite that is similar to the schist derived from Precambrian granite in the Suizo, Tortolita, and Santa Catalina Mountains. The quartz monzonite is cut by generally north-northwest–oriented dikes of diorite porphyry (Tables 1, 2, sample 36) and quartz monzonite porphyry that are similar to two of the three types of late dikes in the Tortolita Mountains. Middle Tertiary volcanic and clastic rocks are in fault contact with the quartz monzonite, and relations that might indicate the depth of intrusion are not exposed.

Desert Peak

Desert Peak is a small (2.5-km^2) alluvium-isolated, north-trending, 200-m-high hill located halfway between the Tortolita Mountains and the Picacho Mountains (Fig. 1). It consists of mildly foliated quartz diorite (Fig. 2E; Table 1, sample 18) that is similar to that in the Durham Hills and the border facies of the pluton of Chirreon Wash in the Tortolita Mountains. Foliation at Desert Peak dips moderately to the southeast, and the foliation planes were loci for intrusion of abundant dikes less than a metre thick of pegmatite and fine-grained quartz monzonite, particularly on the northern and western slopes of the peak. The quartz monzonite (Table 1, sample 33) is petrographically very similar to the quartz monzonite of the Tortolita Mountains. No age or chemical data are presently available for the rocks at Desert Peak.

Picacho Mountains

About 70 km^2 of lineated metamorphic rocks and middle Tertiary intrusive rock crop out in the Picacho Mountains (Figs. 1, 19). The complex stands 800 m above the surrounding alluvial plain, is

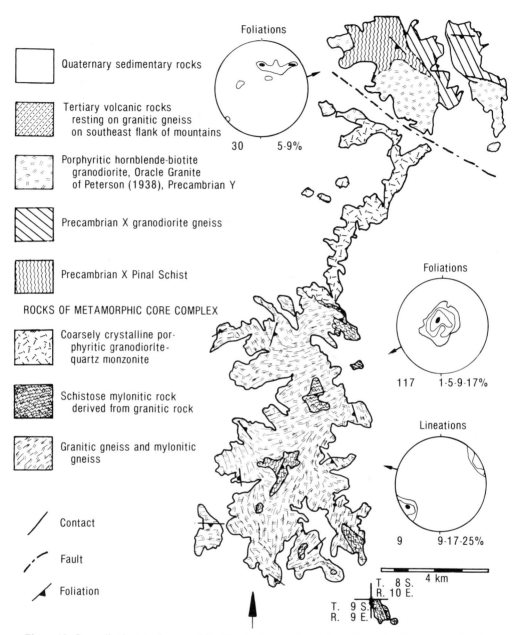

Figure 19. Generalized geologic map of the Picacho Mountains, Arizona (modified from Yeend, 1976), with lower-hemisphere equal-area net projections showing orientation of foliation in the northern Picacho Mountains and foliation and lineation in the southern Picacho Mountains (foliation and lineation data from Yeend, 1976). Contours are percent per one percent area.

overlain in at least one locality by 21-m.y.-old volcanic rocks (Yeend, 1976; Shafiqullah and others, 1976), and is separated from middle Tertiary clastic and volcanic rocks to the south by alluvium and from Precambrian X and Y basement rock by an inferred fault to the north (Yeend, 1976).

Little work has been published on the complex other than the reconnaissance geologic map of Yeend (1976), which shows a body of coarsely crystalline, porphyritic, sphene-bearing, biotite-hornblende granodiorite–quartz monzonite (Fig. 2C) in the northern exposed part of the complex intruding lineated gneiss and schist characteristic of the higher parts of the range to the south (Fig. 19). The intrusive rock is very similar in age (Table 3, sample 67), chemistry (Table 2, sample 28), and petrography (Table 1, sample 28) to the coarsely porphyritic quartz monzonitic intrusive rock in the Santa Catalina Mountains, Tortolita Mountains, and Durham Hills. In thin sections, it shows mild but progressively more cataclasis southward toward the gneiss.

The gneiss (Tables 1, 2, sample 10) is dominantly quartz monzonitic in composition and in a drill hole to the west has yielded a middle Tertiary age (Table 3, sample 59) for a sample that Rb-Sr data indicate is of Precambrian Y age (Eberly and Stanley, 1978). However, too little information is currently available to determine whether the gneiss is wholly derived from Precambrian granitic rock. The schistose rocks are similar both megascopically and in thin secton to the schistose rocks derived from granitic rocks in the Santa Catalina–Rincon Mountains, Suizo Mountains, and Durham Hills. They overlie the granitic gneiss at the crest of the range and at lower elevations on the southeast flank of the complex (Fig. 19), and Yeend (1976) shows the schist plunging beneath the Tertiary intrusive rock along part of the northern boundary of the gneissic rock. Given the scarcity of detailed study, it is presently unknown whether the schist layers at the crest and southeast flank of the range indicate the former presence of an overlying sill-like or laccolithlike pluton.

The schist and gneissic rock are cut by dikes of unfoliated, equigranular quartz monzonite and quartz diorite porphyry that are petrographically identical to those described for the Santa Catalina–Rincon and Tortolita Mountains complexes, and the intrusive body is cut by porphyro-aphanitic quartz monzonite dikes that are likewise similar to those in the Santa Catalina–Rincon and Tortolita Mountains.

When viewed from the distance, the complex suggests a broad domal structure with gently dipping foliation (Fig. 19), but the foliation is more complex internally (Yeend, 1976). Although based on scanty data, the foliation in the domal structure appears distinguishable in trend from the foliation in the Precambrian basement rock mapped by Yeend (1976) north of the complex (Fig. 19). Lineation south of the Tertiary intrusive body is in the plane of the foliation and shows approximately the same N55° to 60°E preferred orientation observed in the rocks to the southeast along the zone of complexes.

The depth of emplacement of the complex is currently not definable by direct stratigraphic means because it is not in contact with rocks older than the middle Tertiary volcanic rocks that, according to W. A. Rehrig (1978, oral commun.), rest on a decollement surface. Furthermore, no Precambrian, Paleozoic, or Mesozoic sedimentary rocks underlie the Tertiary deposits in the areas immediately to the west (Shafiqullah and others, 1976), southwest (P. M. Blacet, unpub. data), and south (Banks and Dockter, 1976; R. P. Ashley, 1977, oral commun.) of the complex, and only 300 m or less of Tertiary clastic debris was deposited above the Precambrian granitic basement in these areas before exposure of the gneiss. Thus, depth of metamorphism of the complex is not easily defined no matter what its age. On the other hand, if the complex developed in middle Tertiary time, judging from the thickness of the adjacent middle Tertiary clastic deposits, it was not buried more deeply than the complexes to the southeast.

SUMMARY AND INTERPRETATION

Sixty and forty years ago, respectively, C. F. Toulmin, Jr., and B. N. Moore (Moore and others, 1949) concluded from field criteria that the Santa Catalina–Rincon Mountains consist largely of Cretaceous or younger quartz monzonite intrusive rocks that grade into cataclastic gneiss. These conclusions were substantiated during later mapping (Bromfield, 1952; McCullough, 1963; Creasey and Theodore, 1975; Banks, 1976, 1977; Creasey and others, 1977).

Several field observations suggest further that cataclasis was related to intrusion of the large quartz monzonite bodies and that development of the pegmatite dikes was related to the cataclasis: (1) Foliation in the complex is generally shallow and parallel to contacts between host rocks and the intruding bodies, which are commonly sheetlike and more than a kilometre in thickness. (2) Cataclasis, which is significantly less intense in the intrusive rocks than in the intruded or included rocks, increases progressively and markedly in both toward intrusive contacts. (3) Despite differences in degree of development, foliation and lineation are very similarly oriented in both host and intrusive rocks. (4) Undeformed and cataclastically deformed pegmatite bodies are abundant in the cataclastic terrane, particularly near intrusive contacts as thin sills in the plane of foliation, and are rare in undeformed plutons and parts of plutons.

Stratigraphic levels of emplacement of the two largest complexes in the zone were 3 to 6 km, and stratigraphy and lithology of the adjacent Mesozoic and Tertiary sedimentary-volcanic basins do not indicate significantly greater exhumation of the smaller complexes, which lack direct stratigraphic control.

Radiometric studies indicate that (1) intruded and intrusive rocks yield essentially the same 20- to 30-m.y. K-Ar and FT ages throughout the Santa Catalina–Rincon complex, as well as throughout very similar complexes in a zone 30 km wide extending at least 60 km northwest, and (2) in agreement with visual appearances of adjacent terrane, Tertiary and older rocks between and as near as 1 km from the complexes were thermally unaffected by the complexes. These data require that (1) 20 to 30 m.y. ago all parts of each complex were at temperatures almost uniformly in excess of 400 °C (Damon, 1968), and (2) either the thermal anomalies were restricted to the cataclastic terrane or the complexes and hosts were juxtaposed tectonically at or after 20 to 30 m.y. ago.

Banks (1977) and Creasey and others (1977) concluded from the above data and observations that the complexes developed during intrusion of magma in shallow environments during middle Tertiary time, and Banks (1977) attributed the development of the pegmatite dikes (and their preferred locations adjacent to intrusive contacts) to heat supplied by the magma on granite-minimum minerals milled in the presence of water along planes of foliation (particularly at sites of high strain adjacent to intrusive contacts). The water is considered to have been introduced meteoric water; water from dehydration of kaolinite, montmorillonite, and chlorite in the host rocks; and an unknown but probably small proportion of magmatic water. Banks (1977) also noted (1) that such water would promote metasomatic or hydrothermal growth of garnet and muscovite in the cataclastic rock, (2) that introduction of ground water or its circulation adjacent to the complexes would account for their telescoped thermal boundaries, and (3) that two conditions were present in the complexes, in addition to weakening caused by the presence of water (Griggs and Blacic, 1965; Griggs, 1967; Jones, 1975; Murrell and Ismail, 1976), to explain how granitic rocks of approximately the same composition as their hosts caused intense deformation at such shallow depths. These conditions are (1) a large proportion of intrusive rock to host rock (satisfied by field distributions; Figs. 6, 8, 13, 16, 18, 19) and (2) intrusion temperatures of the magma above the granite minimum (satisfied by the overall composition of the rocks and suggested by the paucity of aplite and pegmatite dikes in the undeformed plutons and parts of plutons).

Mayo (1964) and Davis and others (1975) interpreted the preferred orientation of the lineation (Fig. 20) to indicate extension in the plane of the foliation parallel to σ_3 (N65°E) with σ_1 directed vertically, and Banks (1977) pointed out that the middle Tertiary dikes adjacent to and crosscutting the complexes displayed preferred orientation (Fig. 20) that demonstrated the presence of an anisotropic stress field with σ_3 directed about N65°E at the level of the complexes during the 20- to 30-m.y. thermal event in the complexes. Furthermore, the data of Rehrig and Heidrick (1976) demonstrated that the anisotropic middle Tertiary stress field was regional to southern Arizona. These data suggested to Banks (1977) that cataclasis occurred during intrusive doming or gravitational flattening of the

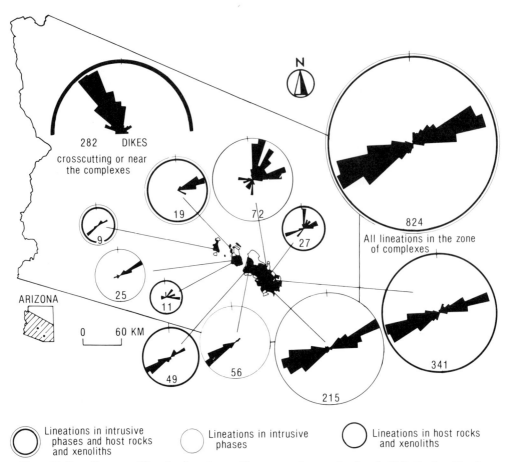

Figure 20. Rose diagrams of lineation in metamorphic core complexes and strike of middle Tertiary dikes in and near the complexes (after Banks, 1977). Data sources: Banks (1976); Banks and Dockter (1976); Banks and others (1977); Budden (1975); Creasey and Theodore (1975); Drewes (1974, 1977); Yeend (1976); R. P. Ashley (unpub. data).

partially solidified intrusive bodies and their thermally weakened host rocks, and that the lineation developed during the doming or flattening by extension parallel to σ_3 of the anisotropic middle Tertiary stress field.

Some of these suggestions were not new, and alternative explanations for the complexes are present in the literature. As noted earlier, Mayo (1964) and Davis and others (1975) suggested that the lineation was extensional in origin, and Davis (1977a, 1977b), without explaining the preferred orientation, likewise suggested the lineation might have resulted from gravitational flattening of the complexes. Davis (1975, 1977a), Davis and others, (1975), and Mayo (1964) pointed out that the lineation alternatively could have developed by extension during tectonic, diapiric, or intrusive uplift of the complexes. The lineation has also been attributed to regional thrusting of Late Cretaceous to middle Tertiary age (Darton, 1925; Moore and others, 1949; Drewes, 1976, 1977; Thorman, 1977), and the cataclasis has been interpreted by some to have occurred during one or more events, some at great depth, that occurred from 44 m.y. ago to Precambrian time (Moore and others, 1949; Dubois, 1959b; Damon and Giletti, 1961; Catanzaro and Kulp, 1964; Damon and others, 1963; Livingston and others, 1967; Mauger and others, 1968; Davis, 1975; Drewes, 1974, 1976, 1977; Shakel, 1972; Shakel and others, 1972, 1977; Thorman, 1977; Keith and others, this volume).

Most of the suggestions for a pre-Cretaceous age for the complexes were made in apparent disregard for the field observations of Moore and others (1949) and before recognition that (1) the schistose rocks in the complexes are largely cataclastically deformed granitic rocks rather than Precambrian X Pinal Schist, (2) that granitic rocks in many of the critical outcrops are post-Mesozoic rather than Precambrian Y in age, and (3) that available Rb-Sr isochrons included samples which were not in isotopic equilibrium or which developed in open or unrelated isotopic systems in violation of the requirements of the dating technique.

Recent but sparse U-Th-Pb data and new Rb-Sr data suggest to Shakel and others (1977) and Keith and others (this volume) that the complexes are ≥44 m.y. old and have more extended histories than suggested by data available to Banks (1977) and Creasey and others (1977). As noted earlier, these new ages are not themselves exempt from interpretation, and ages of 44 m.y. and older present substantial geologic and logistic problems, as discussed below.

If the complexes are of Late Cretaceous to early Tertiary age, the lineation must be reinterpreted to be tectonic in origin because regional σ_3 at that time was at right angles to the lineation in the complexes (Rehrig and Heidrick, 1976). Furthermore, development of the cataclastic terrane by Cretaceous–early Tertiary thrusting does not provide the mechanisms and reasons for (1) the subsequent superimposed reheating of the cataclastic terrane at 20 to 30 m.y., (2) the telescoped thermal boundaries of the complexes, (3) the structurally indistinguishable orientation of deformation in the gneissic intrusive rocks and the acknowledged 20- to 30-m.y.-old Tertiary plutons, and (4) preference of middle Tertiary plutons for the complexes. Additionally, as noted earlier, it has not been universally accepted that interpretation of geologic relations in southern Arizona requires regional Cretaceous thrust sheets.

If the complexes developed about 44 to 50 m.y. ago, either the complexes could have been brought to their current positions from great depths about 20 to 30 m.y. ago on speculative faults, or they were sites of very local and extreme burial and exhumation without leaving evidence of such anomalies in the adjacent Tertiary sedimentary basins, or the complexes were subjected at depths of ≤6 km for periods exceeding 15 m.y. to remarkably restricted long-term or recurrent impossibly high heating from unknown sources. In addition, fortuity must be postulated for the association of the acknowledged 20- to 30-m.y.-old plutons with the complexes and for the structurally indistinguishable deformation in the 20- to 30-m.y.-old plutons and those considered to be at least 15 m.y. older.

Clearly, studies of the complexes discussed here have progressed to a point that requires complex-wide detailed geologic and structural mapping, comprehensive petrofabric and petrochemical work, and multidiscipline isotopic studies that ask questions of the methods of dating as well as the ages of the rocks. When completed, such studies should decide if 50 or 30 m.y. records the time of initial deformation and whether a simple intrusion-related model, extended or multiple event models, or tectonic models explain the complexes. Until such work is concluded, there remain certain attractions for the interpretation of Banks (1977) and Creasey and others (1977) that cataclasis occurred during intrusion of partially floored middle Tertiary composite batholiths and plutons. (1) No fortuity is required to explain coincidence of middle Tertiary plutons and the complexes. (2) Anisotropic regional stress was present and appropriately oriented to explain the preferred orientation of the lineation during middle Tertiary time. (3) No fortuity of coincidence of space and orientation is required to explain the deformation in the acknowledged 20- to 30-m.y.-old plutons and the earlier intruded plutons. (4) Simultaneous deformation of host and intruding rock is completely consistent with the evidence of greater strain (deformation) in the thermally weakened solids (hosts) than in the gneissic (partially molten) intrusive rocks and, particularly, with the cataclastic envelopes on xenoliths with relatively undeformed interiors; multiple, superimposed, fortuitously identical or long-lived deformational events do not explain the xenolith rinds. (5) The middle Tertiary was a time of peak magmatism in southern Arizona (Damon and Bikerman, 1964; Damon and Mauger, 1966; Rehrig and Heidrick, 1976). (6) the 20- to 30-m.y.-old intrusive rocks, gneissic intrusive rocks, and the adjacent 20- to

30-m.y.-old volcanic rocks have similar chemistry and ranges in chemistry. (7) Basins and structures adjacent to the complexes indicate that the complexes were rapidly shedding highlands during the middle Tertiary, as expected for large volumes of magma injected at the indicated shallow depths of intrusion. (8) Concordance of multimineral K-Ar and FT ages throughout large volumes of host and intrusive rocks is creditably explained by intrusion of the observed gneissic rocks and contact metamorphism of their hosts but is logistically improbable if thought to be the result of long-term or fortuitously superimposed heating by hidden sources, particularly in view of the thermally telescoped boundaries of the complexes. (9) The intrusive model does not require speculative middle or late Tertiary tectonic juxtaposition of the complexes and adjacent terrane.

ACKNOWLEDGMENTS

Mapping of the complexes involved interaction between several field parties, and development of the concepts presented here had to have been influenced to a large extent by open discussions with these individuals and, later, with participants of the 1977 Penrose Conference on metamorphic core complexes. Among these associates are T. G. Theodore, G. H. Davis, and S. B. Keith. Citation of these individuals, however, does not imply a responsibility for the interpretations presented here. I also am greatly indebted to Warren Yeend for use of his thin sections and unpublished chemical and modal data for the Picacho Mountains, and to Harald Drewes, C. W. Naeser, P. E. Damon, and M. Shafiqullah for allowing incorporation of as yet unpublished radiometric data in Table 3. J. M. Schmidt aided compilation and drafting of the figures. Lowell Kohnitz took the photographs of the rock slabs. Various drafts of the manuscript benefited greatly by constructive reviews of T. G. Theodore, M. D. Crittenden, Jr., P. E. Damon, and J. H. Dover, although this acknowledgment does not imply that all were in agreement or satisfied with the presentation.

REFERENCES CITED

Anderson, T. H., Silver, L. T., and Salas, G. A., 1977, Metamorphic core complexes of the southern part of the North American Cordillera—Northwestern Mexico: Geological Society of America Abstracts with Programs, v. 9, p. 881.

Armstrong, R. L., 1976, Metamorphic rocks: Isochron/West, no. 15, p. 1-33.

Armstrong, R. L., and Hansen, Edward, 1966, Cordilleran infrastructure in the eastern Great Basin: American Journal of Science, v. 264, p. 112-127.

Arnold, L. C., 1971, Structural geology along the southeastern margin of the Tucson Basin [Ph.D. thesis]: Tucson, University of Arizona, 99 p.

Banerjee, A. K., 1957, Structure and petrology of the Oracle Granite, Pinal County, Arizona [Ph.D. thesis]: Tucson, University of Arizona, 112 p.

Banks, N. G., 1976, Reconnaissance geologic map of Mount Lemmon quadrangle, Arizona: U.S. Geological Survey Miscellaneous Field Studies Map MF-747, scale 1:62,500;

—— 1977, Geologic setting and interpretation of a zone of middle Tertiary igneous-metamorphic complexes in south-central Arizona: U.S. Geological Survey Open-file Report 77-376, 29 p.

Banks, N. G., and Dockter, R. D., 1976, Reconnaissance geologic map of the Vaca Hills quadrangle, Arizona: U.S. Geological Survey Miscellaneous Field Studies Map MF-793, scale 1:62,500.

Banks, N. G., and others, 1977, Reconnaissance geologic map of the Tortolita Mountains quadrangle, Arizona: U.S. Geological Survey Miscellaneous Field Studies Map MF-864, scale 1:62,500.

—— 1978, Radiometric and chemical data for rocks of the Tortolita Mountains quadrangle, Pinal County, Arizona: Isochron/West, no. 22, p. 17-22.

Barter, C. F., 1962, Geology of the Owl Head mining district, Pinal County, Arizona [M.S. thesis]: Tucson, University of Arizona, 73 p.

Bromfield, C. S., 1952, Some geologic features of the Santa Catalina Mountains, in Guidebook for field trip excursions in Southern Arizona: Arizona Geological Society Guidebook I, p. 51-55.

Budden, R. T., 1975, The Tortolita–Santa Catalina Mountains complex [M.S. thesis]: Tucson, University of Arizona, 133 p.

Campbell, R. B., and Reesor, J. E., 1977, The Shuswap metamorphic complex, British Columbia: Geological Society of America Abstracts with Programs, v. 9, p. 920.

Catanzaro, E. J., and Kulp, J. L., 1964, Discordant zircons from the Little Belt (Montana), Beartooth (Montana), and Santa Catalina (Arizona) Mountains: Geochimica et Cosmochimica Acta, v. 28, p. 87–124.

Compton, R. R., and others, 1977, Oligocene and Miocene metamorphism, folding, and low-angle faulting in northwestern Utah: Geological Society of America Bulletin, v. 88, p. 1237–1250.

Creasey, S. C., 1965, Geology of the San Manuel area, Pinal County, Arizona: U.S. Geological Survey Professional Paper 471, 64 p.

Creasey, S. C., and Theodore, T. G., 1975, Preliminary reconnaissance geologic map of the Bellota Ranch quadrangle, Arizona: U.S. Geological Survey Open-file Map 75-295, scale 1:31,250.

Creasey, S. C., Jackson, E. D., and Gulbrandsen, R. A., 1961, Reconnaissance geologic map of parts of the San Pedro and Aravaipa Valleys, south-central Arizona: U.S. Geological Survey Mineral Investigations Field Studies Map MF-238, scale 1:125,000.

Creasey, S. C., and others, 1977, Middle Tertiary plutonism in the Santa Catalina and Tortolita Mountains, Arizona: U.S. Geological Survey Journal of Research, v. 5, p. 705–717.

Damon, P. E., 1968, Application of the potassium-argon method to the dating of igneous and metamorphic rock within the Basin Ranges of the Southwest: Arizona Geological Society Guidebook III, p. 7–20.

Damon, P. E., and Bikerman, Michael, 1964, Potassium-argon dating of post-Laramide volcanic rocks within the Basin and Range province of southeastern Arizona and adjacent areas: Arizona Geological Society Digest, v. 7, p. 63–78.

Damon, P. E., and Giletti, B. J., 1961, The age of basement rocks of the Colorado Plateau and adjacent areas: New York Academy of Sciences, Annals, v. 91, p. 443–453.

Damon, P. E., and Mauger, R. L., 1966, Epeirogeny-orogeny viewed from the Basin and Range province: Society of Mining Engineers, Transactions, v. 235, no. 1, p. 99–112.

Damon, P. E., Ericson, R. C., and Livingston, D. E., 1963, K-Ar dating of Basin and Range uplift, Catalina Mountains, Arizona: National Academy of Sciences—National Research Council Publication 1075, p. 113–121.

Damon, P. E., and others, 1969, Correlation and chronology of ore deposits and volcanic rocks: U.S. Atomic Energy Commission Research Division, Annual Progress Report No. C00-689-120, Contract AT(11-1)689, p. 48.

Darton, N. H., 1925, A resume of Arizona geology: Arizona Bureau of Mines Bulletin 119, 298 p.

Davis, G. A., and others, 1977, Enigmatic Miocene low-angle faulting, southeastern California and western and west-central Arizona—Suprastructural tectonics?: Geological Society of America Abstracts with Programs, v. 9, p. 943–944.

Davis, G. H., 1975, Gravity-induced folding off a gneiss dome complex, Rincon Mountains, Arizona: Geological Society of America Bulletin, v. 86, p. 979–990.

——1977a, Characteristics of metamorphic core complexes, southern Arizona: Geological Society of America Abstracts with Programs, v. 9, p. 944.

——1977b, Gravity-induced folding off a gneiss dome complex, Rincon Mountains, Arizona—Reply: Geological Society of America Bulletin, v. 88, p. 1212–1216.

Davis, G. H., and Frost, E. G., 1976, Internal structure and mechanism of emplacement of a small gravity-glide sheet, Saguaro National Monument (East), Tucson, Arizona: Arizona Geological Society Digest, v. 10, p. 287–304.

Davis, G. H., and others, 1975, Origin of lineation in the Catalina-Rincon-Tortolita gneiss complex, Arizona: Geological Society of America Abstracts with Programs, v. 7, p. 602.

Drewes, Harald, 1973, Large-scale thrust faulting in southeastern Arizona: Geological Society of America Abstracts with Programs, v. 5, p. 35.

——1974, Geologic map and sections of the Happy Valley quadrangle, Cochise County, Arizona: U.S. Geological Survey Miscellaneous Investigations Map I-832, scale 1:48,000.

——1976, Laramide tectonics from Paradise to Hells Gate, southeastern Arizona: Arizona Geological Society Digest, v. 10, p. 151–168.

——1977, Geologic map of the Rincon Valley quadrangle, Pima County, Arizona: U.S. Geological Survey Miscellaneous Investigations Map I-997, scale: 1:48,000.

Dubois, R. L., 1959a, Geology of the Santa Catalina Mountains: Arizona Geological Society Guidebook II, p. 107–116.

——1959b, Petrography and structure of a part of the gneissic complex of the Santa Catalina Mountains, Arizona: Arizona Geological Society Guidebook II, p. 117–126.

Eberly, L. D., and Stanley, T. B., Jr., 1978, Cenozoic stratigraphy and geologic history of southwestern Arizona: Geological Society of America Bulletin, v. 89, p. 921–940.

Fox, K. F., Jr., Rinehart, C. D., and Engels, J. C., 1977, Plutonism and orogeny in north-central Washington—Timing and regional context: U.S. Geological Survey Professional Paper 989, 27 p.

Fox, K. F., Jr., and others, 1976, Age of emplacement of the Okanogan gneiss dome, north-central Washington: Geological Society of America Bulletin, v. 87, p. 1217-1224.

Griggs, D. T., 1967, Hydrolytic weakening of quartz and other silicates: Royal Astronomical Society Geophysical Journal, v. 14, p. 19-31.

Griggs, D. T., and Blacic, J . D., 1965, Quartz—Anomalous weakening of synthetic crystals: Science, v. 147, p. 292-295.

Hanson, H. S., 1966, Petrography and structure of the Leatherwood Quartz Diorite, Santa Catalina Mountains, Pima County, Arizona [Ph.D. thesis]: Tucson, University of Arizona, 104 p.

Hazzard, J. C., and Turner, F. E., 1959, Decollement-type overthrusting in south-central Idaho, northwestern Utah, and northeastern Nevada [abs.]: Geological Society of America Bulletin, v. 68, p. 1829.

Iles, C. D., 1967, Mineralization of a portion of the Owl Head mining district, Pinal County, Arizona [M.S. thesis]: Tucson, University of Arizona, 114 p.

Jennison, Margo, 1976, Miocene basalt in the "Pantano Formation," Three Buttes area, Owl Head mining district, Pinal County, Arizona [abs.]: Tucson, University of Arizona, 4th Annual Geoscience Daze, p. 16.

Jones, M. E., 1975, Water weakening of quartz and its application to natural rock deformation: Geological Society of London Journal, v. 131, p. 429-432.

Keith, S. B., and others, 1980, Evidence for multiple intrusion and deformation within the Santa Catalina-Rincon-Tortolita crystalline complex, southeastern Arizona: Geological Society of America Memoir 153 (this volume).

Kistler, R. W., and Willden, Ronald, 1967, Tectonic and igneous chronology of the southern Ruby Mountains, Nevada, in Abstracts for 1967: Geological Society of America Special Paper 115, p. 336.

Lee, D. E., and others, 1970, Modification of potassium-argon ages by Tertiary thrusting in the Snake Range, White Pine County, Nevada, in Geological Survey research 1970: U.S. Geological Survey Professional Paper 700-D, p. D92-D102.

Livingston, D. E., and others, 1967, Argon 40 in cogenetic feldspar-mica mineral assemblages: Journal of Geophysical Research, v. 72, p. 1361-1375.

Marvin, R. F., and others, 1973, Radiometric ages of igneous rocks from Pima, Santa Cruz, and Cochise Counties, southern Arizona: U.S. Geological Survey Bulletin 1379, 27 p.

Matter, Phillip, 1969, Petrochemical variations across some Arizona pegmatites and their enclosing rocks [M.S. thesis]: Tucson, University of Arizona, 173 p.

Mauger, R. L., Damon, P. E., and Livingston, D. E., 1968, Cenozoic argon ages on metamorphic rocks from the Basin and Range province: American Journal of Science, v. 266, p. 579-589.

Mayo, E. B., 1964, Folds in gneiss beyond North Campbell Avenue, Tucson, Arizona: Arizona Geological Society Digest, v. 7, p. 123-145.

McColly, R. A., 1961, Geology of the Saguaro National Monument, Pima County, Arizona [M.S. thesis]: Tucson, University of Arizona, 80 p.

McCullough, E. J., Jr., 1963, A structural study of the Pusch Ridge-Romero Canyon area, Santa Catalina Mountains [Ph.D. thesis]: Tucson, University of Arizona, 67 p.

Moore, B. N., and others, 1949, Geology of the Tucson quadrangle, Arizona: U.S. Geological Survey Open-File Report, 20 p.

Murrell, S.A.F., and Ismail, I.A.H., 1976, The effect of temperature on the strength at high confining pressure of granodiorite containing free and chemically-bound water: Contributions to Mineralogy and Petrology, v. 55, p. 317-330.

Nielsen, K. C., and Duncan, I. J., 1977, Structural and metamorphic constraints for tectonic models of the Shuswap complex, southern British Columbia: Geological Society of America Abstracts with Programs, v. 9, p. 1114-1115.

Peirce, F. L., 1958, Structure and petrography of part of the Santa Catalina Mountains [Ph.D. thesis]: Tucson, University of Arizona, 86 p.

Peterson, N. P. 1938, Geology and ore deposits of the Mammoth mining camp area, Pinal County, Arizona: Arizona Bureau of Mines Bulletin 144, Geology Series 11, 63 p.

Pilkington, H. D., 1962, Structure and petrology of a part of the east flank of the Santa Catalina Mountains, Pima County, Arizona [Ph.D. thesis]: Tucson, University of Arizona, 120 p.

Reesor, J. E., 1970, Some aspects of structural evolution and regional setting of the Shuswap metamorphic complex: Geological Society of Canada Special Paper 6, p. 73-86.

Reesor, J. E., and Moore, J. M., 1971, Petrology and structure of the Thor-Odin gneiss dome, Shuswap metamorphic complex, British Columbia: Canada Geological Survey Bulletin 195, 149 p.

Rehrig, W. A., and Heidrick, T. L., 1976, Regional tectonic stress during the Laramide and late Tertiary intrusive periods, Basin and Range province, Arizona: Arizona Geological Society Digest, v. 10, p. 205-228.

Rehrig, W. A., and Reynolds, S. J., 1977, A northwest zone of metamorphic core complexes in Arizona: Geological Society of America Abstracts with Programs, v. 9, p. 1139.

——1980, Geologic and geochronologic reconnaissance of a northwest-trending zone of metamorphic core complexes in southern and western Arizona: Geological Society of America Memoir 153 (this volume).

Schwartz, G. M., 1953, Geology of the San Manuel copper deposit, Arizona: U.S. Geological Survey Professional Paper 256, 63 p.

Shackelford, T. J., 1976, Juxtaposition of contrasting structural and lithologic terranes along a major Miocene gravity detachment surface, Rawhide Mountains, Arizona: Geological Society of America Abstracts with Programs, v. 8, p. 1099.

—— 1977, Late Tertiary tectonic denudation of a Mesozoic(?) gneiss complex, Rawhide Mountains, Arizona: Geological Society of America Abstracts with Programs, v. 9, p. 1169.

Shafiqullah, M., and others, 1976, Geology, geochronology, and geochemistry of the Picacho Peak area, Pinal County, Arizona: Arizona Geological Society Digest, v. 10, p. 303–324.

Shakel, D. W., 1972, "Older" Precambrian gneisses in southern Arizona: International Geological Congress, 24th, Montreal, Proceedings, Section 1, Precambrian Geology, p. 278–287.

—— 1974, The geology of layered gneiss in part of the Catalina forerange, Pima County, Arizona [M.S. thesis]: Tucson, University of Arizona, 233 p.

Shakel, D. W., Livingston, D. E., and Pushkar, P. D., 1972, Geochronology of crystalline rocks in the Santa Catalina Mountains, near Tucson, Arizona: Geological Society of America Abstracts with Programs, v. 4, p. 408.

Shakel, D. W., Silver, L. T., and Damon, P. E., 1977, Observations on the history of the gneissic core complex, Santa Catalina Mountains, southern Arizona: Geological Society of America Abstracts with Programs, v. 9, p. 1169–1170.

Sherwonit, W. E., 1974, A petrographic study of the Catalina gneiss in the forerange of the Santa Catalina Mountains, Arizona [M.S. thesis]: Tucson, University of Arizona, 165 p.

Simony, P., and others, 1977, Structural and metamorphic evolution of the northeast flank of Shuswap complex, southern Canoe River, British Columbia: Geological Society of America Abstracts with Programs, v. 9, p. 1177–1178.

Thorman, C. H., 1965, Mid-Tertiary K-Ar dates from late Mesozoic metamorphosed rocks, Wood Hills and Ruby–East Humboldt Range, Elko County, Nevada, in Abstracts for 1965: Geological Society of America Special Paper 87, p. 234–235.

—— 1977, Gravity-induced folding off a gneiss dome complex, Rincon Mountains, Arizona: Discussion: Geological Society of America Bulletin, v. 88, p. 1211–1212.

Titley, S. R., 1976, Evidence for a Mesozoic linear tectonic pattern in southeastern Arizona: Arizona Geological Society Digest, v. 10, p. 71–101.

Winkler, H.G.F., 1974, Petrogenesis of metamorphic rocks (third edition): New York, Springer-Verlag, 320 p.

Yeend, Warren, 1976, Reconnaissance geologic map of the Picacho Mountains, Arizona: U.S. Geological Survey Miscellaneous Field Studies Map MF-778, scale 1:62,500.

MANUSCRIPT RECEIVED BY THE SOCIETY JUNE 21, 1979

MANUSCRIPT ACCEPTED AUGUST 7, 1979

Printed in U.S.A.

Geological Society of America
Memoir 153
1980

Evidence for multiple intrusion and deformation within the Santa Catalina–Rincon–Tortolita crystalline complex, southeastern Arizona

STANLEY B. KEITH Bureau of Geology and Mineral Technology
University of Arizona, Tucson, Arizona 85721

STEPHEN J. REYNOLDS Department of Geosciences
University of Arizona, Tucson, Arizona 85721

PAUL E. DAMON AND MUHAMMAD SHAFIQULLAH
Laboratory of Isotope Geochemistry, Department of Geosciences
University of Arizona, Tucson, Arizona 85721

DONALD E. LIVINGSTON Bendix Field Engineering, Geochemical Division
P.O. Box 1569, Grand Junction, Colorado 81501

PAUL D. PUSHKAR Department of Geology
Wright State University, Dayton, Ohio 45431

The central feature of the range, as worked out by Mr. Tolman, is a great post-Carboniferous intrusive mass of siliceous muscovite granite modified to a gneissic rock near its margins, surrounded by an intense contact metamorphism in which rocks of widely different kinds have been conspicuously affected. The oldest rock cut by this granite is a coarse biotite granite which apparently as a result of the later granitic intrusion, grades into augen gneiss, and locally this rock in turn has been transformed into a thinly fissile schist.

Interpretation by C. F. Tolman of the geology
of the Santa Catalina Mountains as told by
Ransome, 1916, p. 144

ABSTRACT

Recent field work and accumulated Rb-Sr studies, when combined with previous U-Th-Pb and K-Ar investigations, allow a new synthesis of the crystalline terrane within the Santa Catalina–Rincon–Tortolita crystalline complex. When all the available data are integrated, it is apparent that the crystalline core is mainly a composite batholith that has been deformed by variable amounts of cataclasis. The batholith was formed by three episodes of geologically, mineralogically, geochemically, and geochronologically distinct plutons. The first episode (75 to 60 m.y. B.P.) consisted of at least two (and probably three) calc-alkalic, epidote-bearing biotite granodiorite plutons (Leatherwood suite). The Leatherwood suite is intruded by distinctive leucocratic muscovite-bearing peraluminous granitic plutons (Wilderness suite), which are 44 to 50 m.y. old. At least three Wilderness suite plutons are known, and their origin has been much debated. Leatherwood and Wilderness plutons are intruded by a third suite of four biotite quartz monzonite to granite plutons (Catalina suite) that mark the final consolidation of the batholith 29 to 25 m.y. ago.

Much of the mylonitic (cataclastic) deformation of the plutonic rocks and recrystallization of the enclosing host rocks may be related to intrusion of the various plutons. At least three episodes of mylonitization (cataclasis) may be delineated by observing relations between mylonitic and nonmylonitic crosscutting plutons. The southern part of the Leatherwood pluton bears a moderate to strong mylonitic foliation that is cut by undeformed leucogranites and pegmatite phases of the Wilderness pluton.

Elsewhere in the Santa Catalina–Rincon–Tortolita crystalline core, Wilderness suite plutons contain penetrative mylonitic foliation. Foliated Wilderness suite plutons are intruded by an undeformed portion of a Catalina suite pluton. In the Tortolita Mountains, however, intrusions of the Catalina suite themselves contain evidence for at least two events of mylonitic deformation. The most significant of these events is clearly constrained to the Catalina intrusive episode because it formed during or after the emplacement of Tortolita quartz monzonite (about 27 m.y. B.P.) but before the intrusion of postfoliation dikes (about 24 m.y. B.P.). All three episodes of mylonitization contain the distinctive and much discussed east-northeast–trending lineations. All events of mylonitization are constrained to a 50-m.y. interval of time from 70 to 20 m.y. ago. Although continuous mylonitization from 70 to 20 m.y. ago cannot be unequivocally disproved, the strong association of mylonitization with the three plutonic episodes suggests that deformation in the Santa Catalina–Rincon–Tortolita crystalline core, like intrusion, was episodic.

INTRODUCTION

The Santa Catalina–Rincon–Tortolita crystalline complex is located (Fig. 1) at the southeast end of a zone of crystalline complexes that trends northwest through southern Arizona (Rehrig and Reynolds, this volume; Banks, this volume; Davis and others, this volume; Davis, this volume). Crystalline complexes in this zone are in part characterized by chiefly mylonitic varieties of cataclastic rocks whose gently dipping foliation defines broad arches or domes (Davis, this volume; Coney, 1979 and this volume; Rehrig and Reynolds, this volume; Reynolds and Rehrig, this volume). Evidence of post-Paleozoic plutonism, metamorphism, mylonitization, and cataclasis abound in all of the complexes. Latest cooling ages in the Arizona crystalline complexes are generally middle Tertiary (Damon and others, 1963; Mauger and others, 1968; Creasey and others, 1977; Banks and others, 1978; Banks, this volume; Rehrig and Reynolds, this volume). On some gently dipping sides of the domes, the upper levels of the mylonitic gneisses are jointed, brecciated, chloritic, and hematitic. The mylonitic gneisses are overlain by a low-angle dislocation surface (for definition of this term, see Rehrig and Reynolds, this volume). Above the dislocation surface lie highly faulted and locally folded (Davis, 1975) rocks that are generally unmetamorphosed and lack the mylonitic textures that characterize the basement below the dislocation surface.

The Santa Catalina–Rincon–Tortolita crystalline complex is within the Basin and Range province, an area characterized by late Tertiary fault-block mountain ranges and valleys. The oldest rocks exposed in the ranges (Fig. 2) consist of middle Proterozoic metasedimentary and metaigneous rocks that underwent a major metamorphic-deformational-plutonic event about 1.65 b.y. ago and a later episode of granitic intrusion 1.45 b.y. ago (Silver, 1978). These rocks were beveled by erosion and

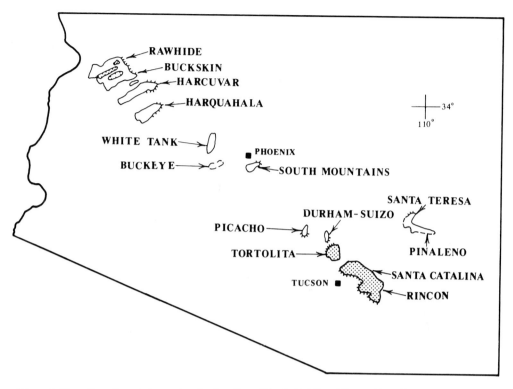

Figure 1. Map of southern Arizona showing location of Santa Catalina–Rincon–Tortolita crystalline complex (dot pattern). Similar complexes are outlined, and dislocation surfaces are depicted by hachures (from Rehrig and Reynolds, this volume).

overlain by 1.4- to 1.1-b.y.-old sedimentary rocks of the Apache Group and associated diabase (Shride, 1967). Apache Group rocks are overlain by a cratonic sequence of Paleozoic carbonate and fine-grained clastic rocks (Bryant, 1968; Peirce, 1976). Mesozoic rocks of the region represent a period of volcanism, plutonism, and tectonic instability (Hayes, 1970; Hayes and Drewes, 1968, 1978; Titley, 1976). The Late Cretaceous–early Tertiary Laramide orogeny was also a time of deformation (Drewes, 1976a; Davis, 1979) and alkali-calcic to calc-alkalic magmatism (Keith, 1978). After a period of *relative* quiescence, high potassium–calc-alkalic magmatism and deformation were renewed in middle Tertiary time (Damon and Mauger, 1966; Shafiqullah and others, 1978; Keith, 1978). These events were followed after 15 m.y. B.P. by block faulting and basaltic volcanism (Shafiqullah and others, 1976; Eberly and Stanley, 1978; Scarborough and Peirce, 1978).

Within this regional geologic framework, the Santa Catalina–Rincon–Tortolita crystalline complex has been an enigma since research started on it in the early 1900s. The purpose of this paper is to summarize published and unpublished geologic and geochronologic studies and to integrate these into a discussion concerning ages and correlations of major rock units within the complex. Particular attention will be given to post-Paleozoic intrusions that together constitute a composite batholith which dominates the geology of the complex.

Igneous rock nomenclature in this paper follows traditional usage of rock names within the Santa Catalina–Rincon–Tortolita crystalline complex except where noted in the text. Rocks described by the term "mylonitic" possess a foliation (fluxion structure of Higgins, 1971) and in thin section show comminuted and brittlely deformed feldspars (from 60% to 80% of rock) and recrystallized, sutured aggregates of quartz (as much as 40% of rock). The mylonitic rocks are similar to photographs of hand

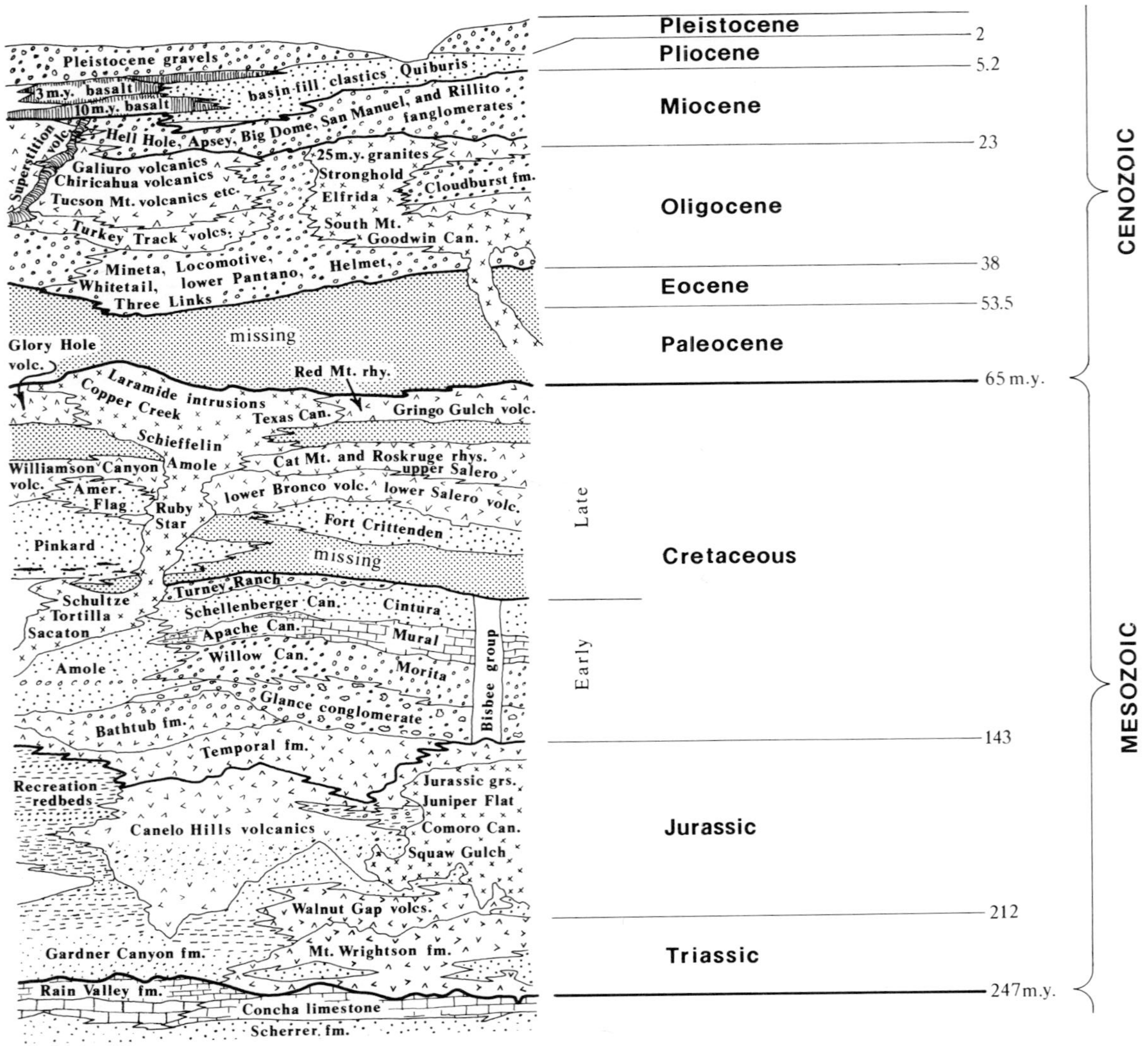
Pleistocene gravels
3 m.y. basalt
10 m.y. basalt
basin-fill clastics Quiburis
San Manuel, and Rillito fanglomerates
Hell Hole, Apsey, Big Dome,
Superstition volc.
25 m.y. granites
Galiuro volcanics
Chiricahua volcanics
Tucson Mt. volcanics etc.
Stronghold
Cloudburst fm.
Elfrida
South Mt.
Turkey Track volcs.
Goodwin Can.
Mineta, Locomotive,
Helmet,
Whitetail, lower Pantano,
Three Links
missing
Glory Hole volc.
Red Mt. rhy.
Laramide intrusions
Copper Creek
Texas Can.
Gringo Gulch volc.
Schieffelin
Williamson Canyon volc.
Amole
Cat Mt. and Roskruge rhys.
upper Salero
Amer.
Flag
lower Bronco volc.
lower Salero volc.
Ruby Star
Fort Crittenden
Pinkard
missing
Turney Ranch
Schultze
Tortilla
Sacaton
Schellenberger Can.
Cintura
Apache Can.
Mural
Willow Can.
Morita
Amole
Bisbee group
Glance conglomerate
Bathtub fm.
Temporal fm.
Recreation redbeds
Jurassic grs.
Juniper Flat
Canelo Hills volcanics
Comoro Can.
Squaw Gulch
Walnut Gap volcs.
Gardner Canyon fm.
Mt. Wrightson fm.
Rain Valley fm.
Concha limestone
Scherrer fm.
Pleistocene
Pliocene
Miocene
Oligocene
Eocene
Paleocene
Late
Cretaceous
Early
Jurassic
Triassic
2
5.2
23
38
53.5
65 m.y.
143
212
247 m.y.
CENOZOIC
MESOZOIC

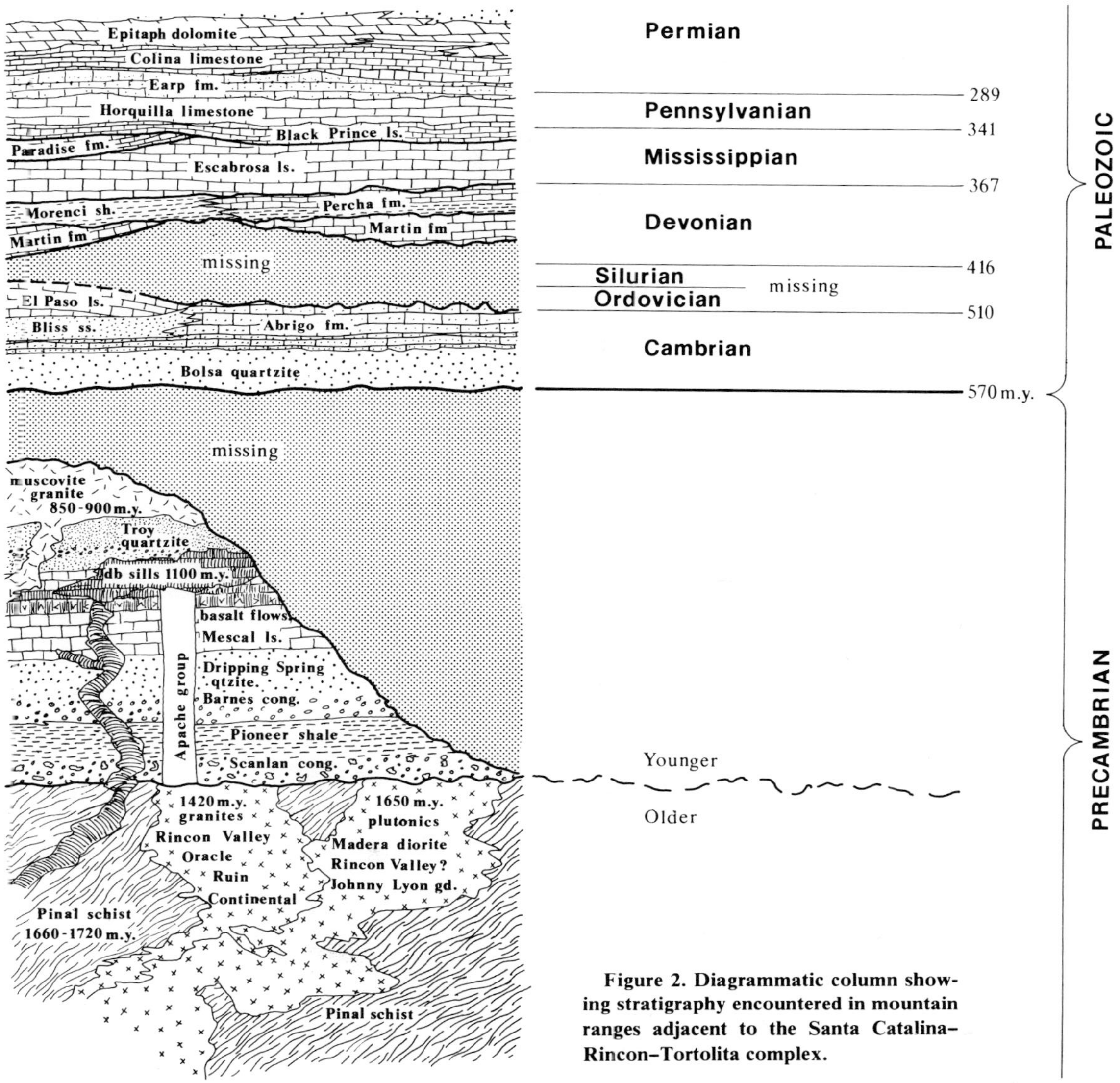

Figure 2. Diagrammatic column showing stratigraphy encountered in mountain ranges adjacent to the Santa Catalina–Rincon–Tortolita complex.

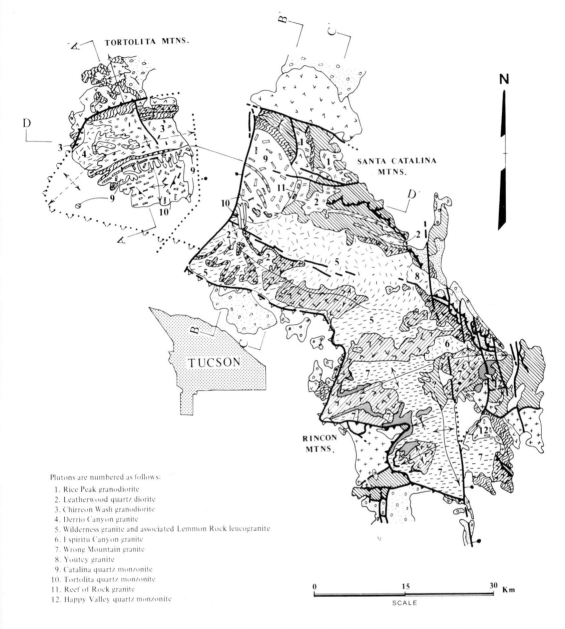

Figure 3. Generalized geologic map of Santa Catalina–Rincon–Tortolita crystalline complex showing locations of cross sections and plutons discussed in text. Sources of map data are as follows: for Tortolita Mountains—Budden (1975), Banks and others (1977), and Keith (unpub. mapping); for Santa Catalina Mountains—Tolman (1914, unpub. mapping as presented in Wilson and others, 1969), Creasey (1967), Shakel (1974), Creasey and Theodore (1975), Banks (1976), Hoelle (1976), Suemnicht (1977), Wilson (1977), and Keith (unpub. mapping); for Rincon Mountains—Drewes (1974, 1977) and Thorman and Drewes (1978). Aligned patterns in Late Cretaceous through Oligocene intrusions show deformed areas of mylonitic gneiss, and random patterns show undeformed areas. East-northeast-trending ruled lines show areas of mylonitically deformed porphyritic mesocratic gneiss believed to have been previously undeformed 1,400- to 1,450-m.y.-old biotite granitic rocks (shown by random pattern). Barbed heavy lines are low-angle faults with barbs in upper plate. Heavy lines are high-angle normal faults with bar and ball on downthrown side.

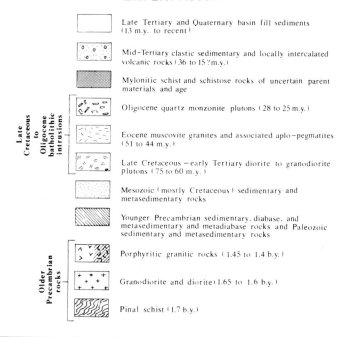

specimens and photomicrographs of protomylonite, mylonite, and mylonite gneiss as presented in the classification of cataclastic rocks proposed by Higgins (1971). Some areas of mylonitic gneiss contain very coarse-grained pegmatites concordantly interlayered with the mylonitic augen gneisses. The interiors of these pegmatites are commonly highly broken or brecciated, possess no foliation, and are more properly referred to as cataclasites in the classification scheme of Higgins (1971). Areas we call "mylonitic" may contain variable fractions (usually minor) of pegmatite cataclasites.

GENERAL GEOLOGY OF THE CRYSTALLINE COMPLEX

General geology of the Santa Catalina–Rincon–Tortolita crystalline complex is presented in simplified map form in Figure 3 and depicted diagrammatically in four cross sections (Figs. 4, 5, 6, and 7). Frequently mentioned localities are shown geographically in Figure 8. Readers are directed to discussions and references cited in Creasey and others (1977), Budden (1975), Shakel (1974, 1978), Davis (1975, 1977a, 1977b, 1978, and this volume), Drewes (1977), Drewes and Thorman (1978), Thorman (1977), Davis and Coney (1979), and Banks (this volume) for other recent perspectives.

In a general way, the Santa Catalina–Rincon–Tortolita complex is composed of a crystalline core that is dominated by Phanerozoic plutonic rocks (~75% of outcrop). The remainder of the crystalline core consists of middle Proterozoic plutonic rocks (~20% of outcrop) and subordinate amounts of middle Proterozoic and Phanerozoic metasedimentary rocks. The crystalline core is mostly fault-bounded, except for segments of its north and northeast margins which are intrusive in nature.

The Phanerozoic plutonic rocks form a large composite batholith within which at least 10 and possibly 12 or more individual plutons (Fig. 3) have been delineated (see App. 1 for discussion of nomenclature of these bodies). Individual plutons are generally compositionally zoned and commonly

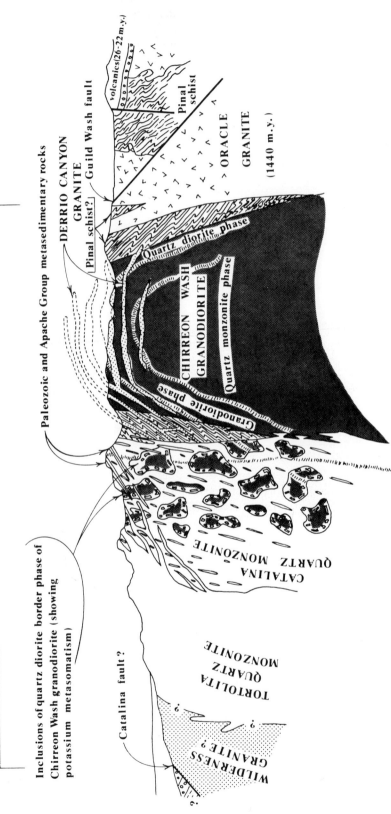

Figure 4. Diagrammatic cross section A–A' through Tortolita Mountains. Location of section shown in Figure 3.

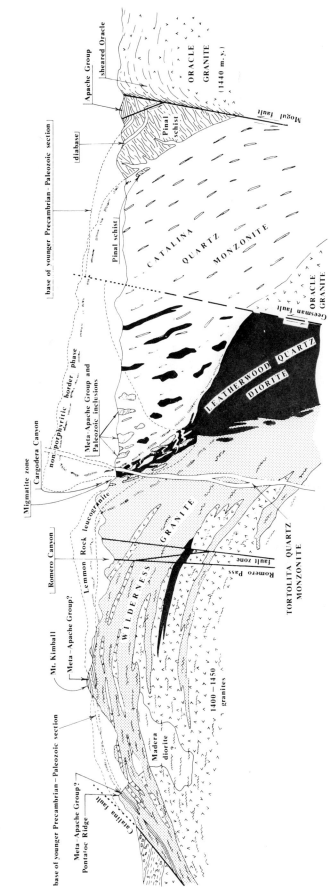

Figure 5. Diagrammatic cross section B–B' through western Santa Catalina Mountains. Location of section shown in Figure 3. Mylonitic rocks shown by wavy lines.

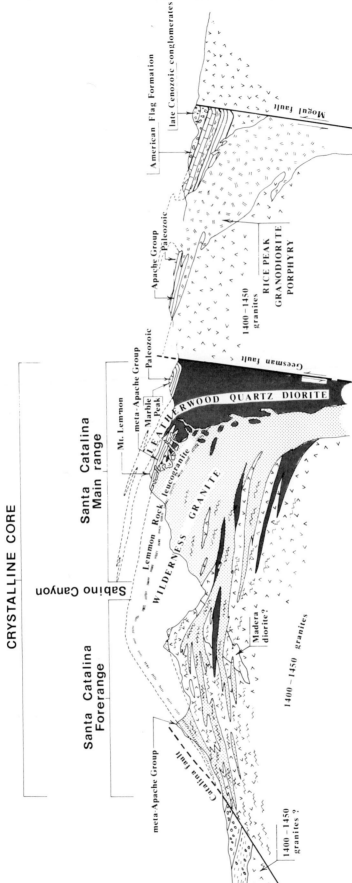

Figure 6. Diagrammatic cross section C–C' through central Santa Catalina Mountains. Location of section shown in Figure 3. Mylonitic rocks shown by wavy lines.

have asymmetric laccolithic shapes. The tops of many of the laccolithic bodies and sills occur just below or at the Precambrian Apache Group unconformity with older Precambrian rocks or within the Apache Group (Figs. 2, 5, 6). The asymmetric laccolithic geometry is spectacularly displayed in the western end of the Santa Catalina Mountains (Fig. 9) where a batholithic sill of Wilderness granite extends southward into the Catalina forerange from its root zone in Cargodera Canyon. The asymmetric, flat-lying parts of the various intrusions occupy large areas throughout the complex and, together with their host rocks, have been affected by a conspicuous mylonitization which has imposed on the rocks a penetrative, gently inclined mylonitic foliation. Although all three ranges of the complex have these general attributes in common, enough differences exist to merit brief discussions of the geology of each range.

Santa Catalina Mountains

The Santa Catalina part of the complex (Figs. 3, 5, 6, 7) has received the most intense study (see discussions and references in Shakel, 1974, 1978; Budden, 1975; Davis, 1975, 1978, and this volume; Creasey and others, 1977; Drewes, 1977; Banks, 1977 and this volume). The southern part of the range (Santa Catalina forerange in Fig. 8) is predominantly composed of interlayered dark and light gneissic bands of variable thickness. In order of decreasing abundance, the dark layers are composed of mylonitic biotite quartz monzonite gneiss, epidote-bearing and non–epidote-bearing mylonitic diorite gneiss, mylonite schist, and amphibolite. Layers and lenses of probable metasedimentary rocks are locally present. The above rocks are interlayered with, contained within, or injected by abundant light-colored layers and dikes of garnet- and biotite-bearing muscovite granite and pegmatites. Most of the rocks in the forerange are deformed by a pervasive low-angle mylonitic foliation which contains conspicuous lineation trending west-southwest. Gently dipping patterns of mylonitic foliation in gneisses of the forerange define a broad arch whose axis trends west-northwest. Mylonitic gneisses deep in the forerange arch are structurally the lowest of the rocks exposed in the complex. Up structural section to the south, the gneisses become jointed, brecciated, and chloritic. Ultimately, the highly broken mylonitic gneisses are overlain by the Catalina fault named by Pashley (1966), a dislocation surface that dips gently to the south. Up structural section to the north of the forerange, the proportion of dark-colored mylonitic quartz monzonite and diorite gneiss gradually decreases until the rock is entirely a light-colored mylonitic two-mica granite. Farther up section, mylonitization of the granite in a general way becomes less intense, and in many places the rock is essentially an undeformed biotite-muscovite-garnet–bearing granite (Wilderness granite of Shakel, 1978). To the north, the upper parts of the granite commonly grade into alaskite and pegmatite (Lemmon Rock leucogranite of Shakel, 1978). The entire sill sequence from its lowest exposed levels in the forerange on the south to the crest of the main range on the north is thicker than 4.5 km. In the Mount Lemmon area, pegmatite and alaskite associated with the leucogranite pervasively intrude the Leatherwood quartz diorite (Peirce, 1958; Hanson, 1966) and adjacent metamorphosed and locally highly deformed Apache Group rocks (Pilkington, 1962; Waag, 1968). Farther east, the granitic main phase of the Wilderness pluton directly intrudes the Leatherwood quartz diorite (Pilkington, 1962; Wilson, 1977). Many parts of the Leatherwood pluton have been deformed into mylonitic schist and gneiss, especially where the Leatherwood has concordantly intruded metamorphosed rocks of the Apache Group. Deformed Leatherwood quartz diorite is commonly intruded by undeformed Lemmon Rock leucogranite or Wilderness granite (Pilkington, 1962; Hanson, 1966; Wilson, 1977). The Leatherwood and its intruded cover of Apache Group and Paleozoic strata are terminated to the north by the Geesman fault and a low-angle normal fault mapped by Creasey and Theodore (1975). North of these faults, exposed rocks are more typical of southern Arizona geology—that is, ~1.7-b.y.-old Pinal Schist, 1.44-b.y.-old Oracle Granite (Shakel and others, 1977), and less-metamorphosed Apache Group and Paleozoic and

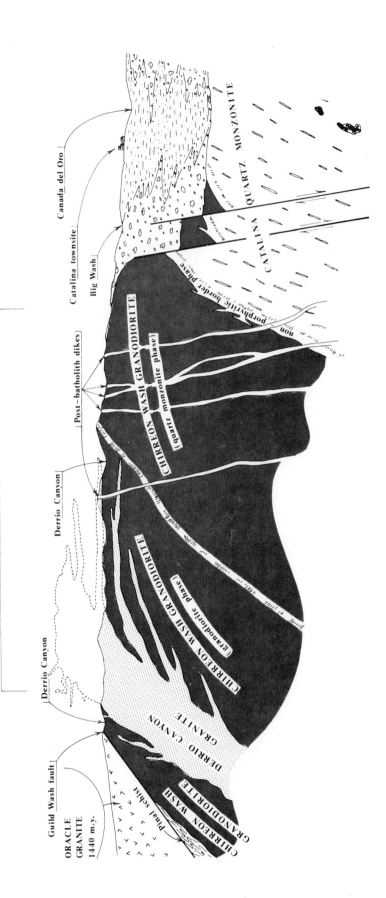

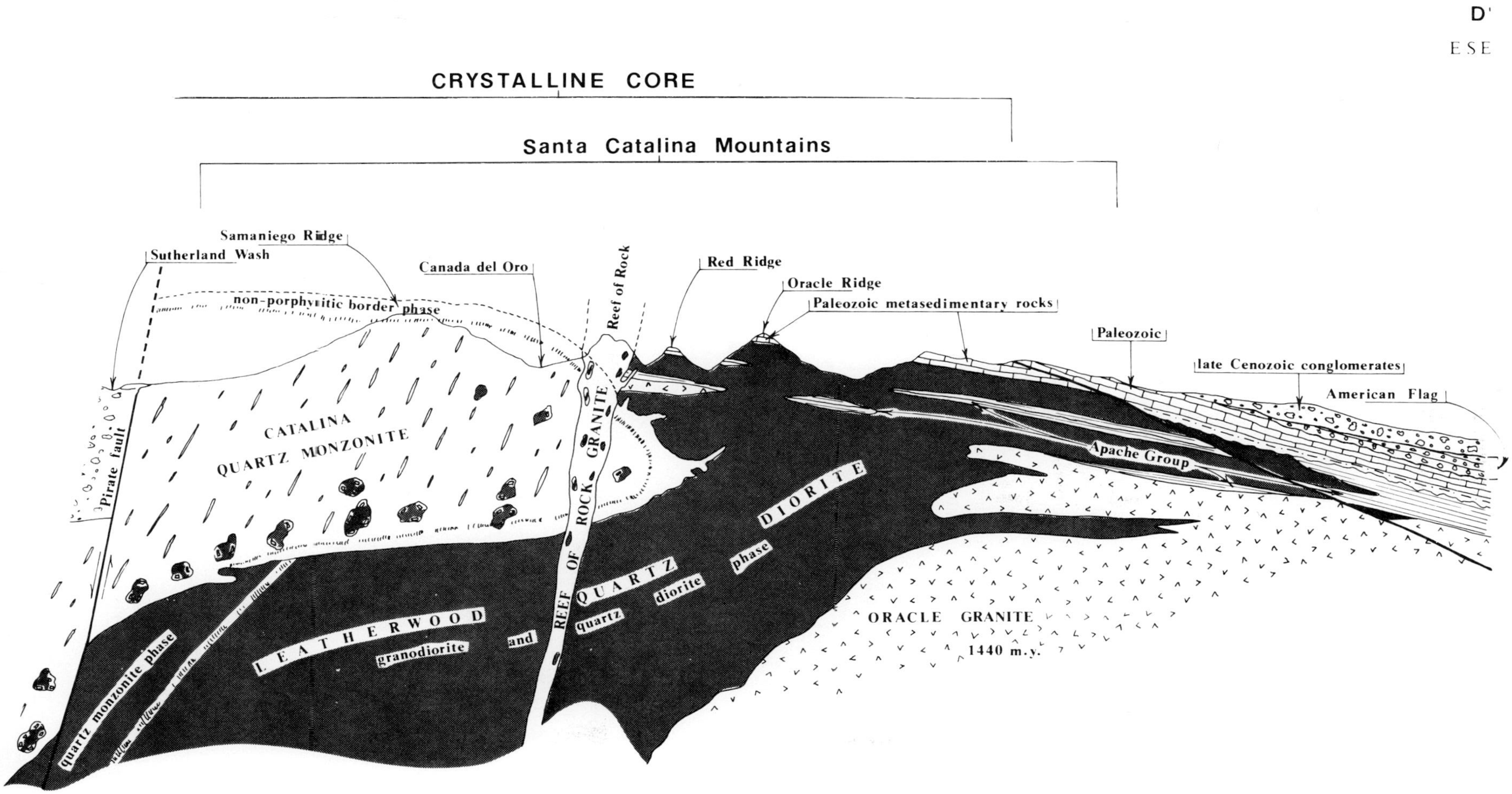

Figure 7 (facing pages). Diagrammatic cross section D-D' through central Tortolita Mountains and northern Santa Catalina Mountains.

Mesozoic strata—all intruded by the Rice Peak granodiorite porphyry of presumed Laramide age and middle Tertiary porphyritic sphene-bearing hornblende-biotite Catalina quartz monzonite (Hoelle, 1976; Suemnicht, 1977; Shakel, 1978). At the far north end of the range, the Mogul fault juxtaposes these rocks against Oracle Granite in the upthrown northern block. In the western part of the Santa Catalina Mountains, undeformed Catalina quartz monzonite intrudes foliated Wilderness granite in Cargodera Canyon. Catalina quartz monzonite also intrudes and contains inclusions of Leatherwood quartz diorite (Suemnicht, 1977).

Tortolita Mountains

The western termination of the Santa Catalina Mountains is along the Pirate fault, a north-northeast–trending fault of late Tertiary age. Except for the alluvium-covered interval west of the Pirate fault, much of the Santa Catalina Mountains geology continues (Fig. 7) westward into the Tortolita Mountains (Budden, 1975; Banks and others, 1977; Davis, this volume). In this range, arches in mylonitic foliation are difficult to place precisely, but probably exist. The northern part of the crystalline core in the Tortolita Mountains consists of the Chirreon Wash granodiorite, an east-northeast-to due-east–trending composite pluton with quartz diorite, granodiorite, and quartz monzonite phases (Fig. 7). Intruding the granodiorite in its western exposures are abundant tabular bodies of granite, pegmatites, and alaskite herein referred to as Derrio Canyon granite. These plutons are locally mylonitic and bordered by east-trending schistose bands on both the north and south. To the

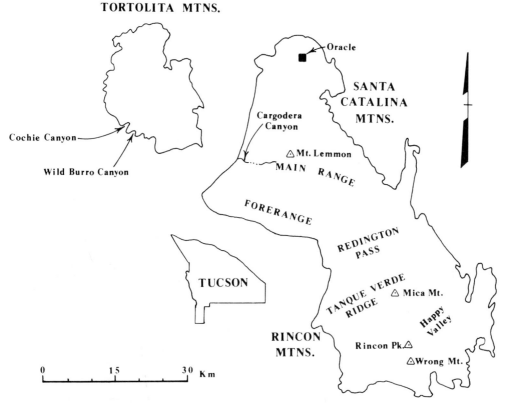

Figure 8. Outline map of the Tortolita, Santa Catalina, and Rincon Mountains showing localities frequently mentioned in text and Appendix 1.

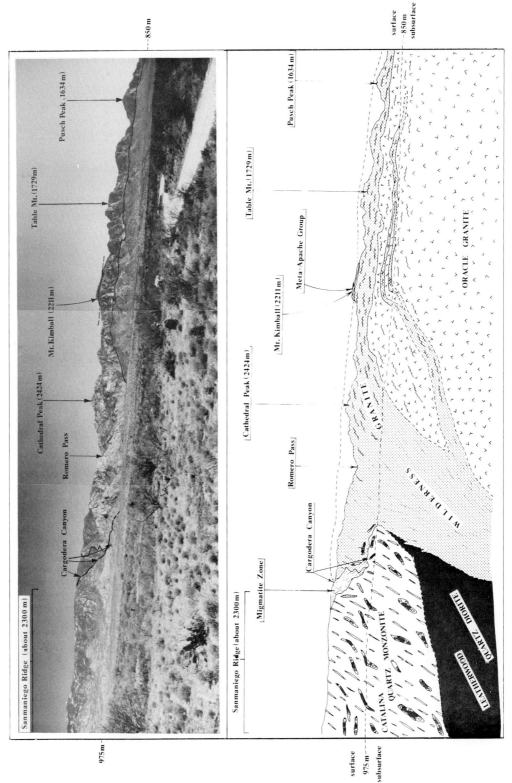

Figure 9. Panorama of the western Santa Catalina Mountains looking southeast. Photo and interpretative diagram show the asymmetric laccolithic geometry of the Wilderness granite batholith and, with less certainty, the inferred laccolithic geometry of the Catalina quartz monzonite intrusion. Mylonitic rocks are shown by wavy lines. Field of view is 25 km wide.

north, the schistose rocks are in contact with mylonitized Oracle Granite. Farther to the northwest, both Oracle Granite and the schistose rocks are chloritized, brecciated, and overlain by a dislocation surface (Guild Wash fault). The schistose band that borders the Chirreon Wash pluton on the south is intruded on its south side by an east-northeast–trending mass of quartz monzonite which is lithologically similar to and correlated by us with the Catalina quartz monzonite of the northwest Santa Catalina Mountains. The Catalina quartz monzonite locally contains large quartz diorite inclusions of presumed Chirreon Wash granodiorite and truncates pegmatite apophyses of Derrio Canyon granite in the east-central Tortolita Mountains. In turn, the Catalina quartz monzonite is intruded by numerous apophyses of the Tortolita quartz monzonite pluton that crops out to the south-southeast. Both the Catalina and Tortolita intrusions locally contain a low-angle mylonitic foliation. The mylonitic foliation in all plutons is crosscut by northwest-striking, high-angle normal faults and shears that in many places are intruded by northwest-striking, undeformed to locally foliated granodiorite, quartz monzonite, and quartz latite dikes.

Rincon Mountains

The Rincon Mountains are geologically more similar to the Santa Catalina Mountains than they are to the Tortolita Mountains. Much of the Rincon Mountains is composed of muscovite-garnet–bearing granite (Wrong Mountain Quartz Monzonite of Drewes, 1977). The granite commonly envelops a dark biotitic augen gneiss (Continental Granodiorite of Drewes, 1977). Some parts of the granite have abundant pegmatite and alaskite. Both the granite and the dark augen gneiss exhibit the distinctive low-angle mylonitic foliation. This mylonitic gneiss complex is overlain to the northeast by metamorphosed and locally highly deformed younger Precambrian and Paleozoic rocks which become lower grade and less deformed up section (Drewes, 1974; Frost, 1977; Davis, this volume). The western and southern boundaries of the mylonitic complex are—like those of the southern Santa Catalina forerange—highly jointed, brecciated, chloritized, and overlain by the Catalina fault, a dislocation surface which dips gently off the flanks of the range (Pashley, 1966; Davis, 1975; Davis and Frost, 1976; Drewes, 1977). The low-angle mylonitic fabric has been deformed into several broad west-southwest-plunging arches and one north-northwest–trending arch and is intruded by several north-northwest-striking undeformed dikes (Thorman and Drewes, 1978).

CORRELATION AND GEOCHRONOLOGY OF ROCK UNITS

We have correlated rock units throughout the Santa Catalina–Rincon–Tortolita crystalline complex by considering rock types and field relationships in conjunction with trace-element and isotope geochemistry. Where possible, we have suggested correlation of rocks exposed within the complex to those exposed outside it. Correlation of rock units within the complex has been hampered by problems in nomenclature, presence of a pervasive mylonitic overprint, intense metamorphism of sedimentary protoliths adjacent to plutons, and, for all rock types, ubiquitous 20- to 30-m.y. K-Ar and fission-track cooling ages that present difficulties in determining original emplacement ages.

Perhaps the most serious problem is nomenclature because genetic bias is inherent in most of the terminology (see App. 1 for detailed discussion). Terms that contain the term "gneiss" in the Santa Catalina–Rincon–Tortolita complex commonly carry a metamorphic connotation of in situ anatexis, partial melting, hydrothermal metamorphism, or alkali-silica metasomatism (see, for example, Hernon, 1932; DuBois, 1959a, 1959b; Mayo, 1964; Peterson, 1968; Shakel, 1974; Sherwonit, 1974; Drewes, 1977; Banks, this volume). Terms that carry a plutonic name (such as quartz monzonite) in the complex traditionally have the genetic connotation of being fundamentally intrusive in origin (see for example, Tolman, as reported in Ransome, 1916; Creasey and others, 1977).

Another serious obstacle for correlation has resulted from one or more pervasive events of mylonitization. Large areas of the complex have been affected by an event (or by events) of mylonitization that largely obliterated original rock textures. Recognition of the importance of mylonitization (Creasey and Theodore, 1975; Davis and others, 1975; Banks, 1976) represents a fundamental change from earlier workers who regarded the mylonitic fabric as a "minor" event that was secondary to processes of metamorphism, anatexis, and alkali-silica metasomatism (Hernon, 1932; DuBois, 1959a, 1959b; Pilkington, 1962; Mayo, 1964; Peterson, 1968; Sherwonit, 1974; Shakel, 1974). These earlier workers believed that original protolith compositions were greatly obscured by drastic chemical changes which accompanied a "metamorphic transformation." In contrast, implicit in a "mylonitic model" is that a rock's present-day fabric is only a *textural* modification of an earlier protolith and that the bulk-rock chemistry need not have been significantly affected. In other words, "metamorphism" (mylonitization) could have been largely isochemical. The "mylonitic model" is strongly supported by our data because we can match Rb-Sr geochemistry of mylonitically deformed rocks within the Santa Catalina–Rincon–Tortolita crystalline complex to nondeformed counterparts in or adjacent to it (see succeeding sections).

The mylonitic model provides an additional caveat to the interpretation of various schistose tracts within the Santa Catalina–Rincon–Tortolita crystalline complex. It was or is widely believed that many of the gneissic rocks were derived from possible Precambrian Pinal Schist (Blake, 1908a, 1908b; DuBois, 1959a, 1959b; Pilkington, 1962; Mayo, 1964; Peterson, 1968; Sherwonit, 1974; Drewes, 1974, 1977). As a result, many of the schistose exposures in the gneisses have been mapped as Pinal Schist (for example, Drewes, 1974, 1977). However, the distinct probability now exists that many of these schistose bodies are mylonite schists derived during mylonitization at contacts between igneous protoliths of different compositions (for example, Davis, 1978, this volume). Conversely, it is equally possible, as we shall attempt to show in the Tortolita Mountains, to mistake real Pinal Schist for mylonitic schist. It is also possible to mistake recrystallized clastic rocks of the Apache Group for Pinal Schist or vice versa.

Most K-Ar and fission-track dates fall within the range of 20 to 30 m.y. B.P. (Damon and others, 1963; Mauger and others, 1968; Marvin and others, 1973, 1978; Creasey and others, 1977; Banks and others, 1978; this paper). These dates, which represent the termination of a thermal event, cannot alone be used to distinguish and correlate intrusive events within the complex. However, they do provide a cooling history for the most recent thermal events which suggests that the structurally lowest part of the complex cooled from about 400 °C 28 m.y. ago to 100 °C about 21 m.y. ago (Damon, 1968; Creasey and others, 1977) under a steep geothermal gradient (Mauger and others, 1968).

In order to document events that occurred before the late cooling history, Rb-Sr whole-rock studies were instituted in 1962. Little has been reported (Shakel, 1972, 1974; Shakel and others, 1972; Hoelle, 1976) until now because the results obtained were difficult to interpret. Reported U-Th-Pb studies (Shakel and others, 1977), the rediscovery of Tolman's deformed laccolith model by Creasey and others (1977), and recent field observations by us have provided a new perspective from which to evaluate the Rb-Sr data. Consequently, this is the first discussion that has as its data base the recently reported U-Pb studies (Shakel and others, 1977; Shakel, 1978), the recently summarized K-Ar and fission-track data (Creasey and others, 1977; Marvin and others, 1978; Damon and others, 1980), and numerous, previously unpublished Rb-Sr analyses determined by workers of the Laboratory of Isotope Geochemistry, University of Arizona, during the past 18 yr. When data from all these methods are considered together, an internally consistent evolutionary picture of the complex emerges. We will discuss rocks of the complex from oldest to youngest, beginning with Pinal Schist. The reader is referred to Damon and others (1980) for location, rock type, Rb-Sr analytical data, and geologic significance of each data point and to Table 1 for a summary of all available isotopic dates. Selected

TABLE 1. SUMMARY OF ISOTOPIC DATA FOR PLUTONS IN AND ADJACENT TO THE SANTA CATALINA–RINCON–TORTOLITA CRYSTALLINE COMPLEX

Rock name and comments	Reference* and sample no.	Age (m.y.)	Method	Area
MIDDLE PROTEROZOIC GRANITIC ROCKS				
Madera Diorite Emplacement age is probably around 1,670 m.y. Younger ages are variously reduced by younger metamorphisms or intrusive events	11: DEL-9 – 13-65	1,695 ± 30	Rb-Sr whole-rock isochron	Pinal Mts.
	6: PED-2-59	1,665 ± 40	K-Ar biotite	Pinal Mts.
	1: 24H	1,645 ± 60	K-Ar hornblende	Ray mine area
	1: 24B	100 ± 2	K-Ar biotite	Ray mine area
	11: DEL-12 & 16-65	1,540 ± 50	Rb-Sr mineral isochron	Pinal Mts.
	2: TMN-93	1,440 ± 10	K-Ar biotite	Tortolita Mts.
Johnny Lyon Granodiorite 1,625 m.y. is probably emplacement age. 1,545-m.y. age is reduced age on Rincon Valley Granodiorite mapped by Drewes (1974). Correlated with Johnny Lyon Granodiorite by Silver (1978)	21: L-312, L-609	1,625 ± 10	U-Pb zircon	Johnny Lyon Hills
	14: 69D64	1,545 ± 60	K-Ar biotite	E. Rincon Mts.
Rincon Valley Granodiorite Rock petrographically resembles Johnny Lyon Granodiorite. K-Ar ages probably reduced (Silver, 1978)	14: 69D61	1,455 ± 50	K-Ar biotite	Rincon Valley
	14: 69D61	1,560 ± 100	K-Ar hornblende	Rincon Valley
Oracle Granite Emplacement age is about 1,440 m.y. Cenozoic ages reflect resetting by Late Cretaceous-Tertiary thermal events	6: PED-3-58	1,425 ± 40	K-Ar muscovite	near Oracle
	12: PED-2-58	1,410 ± 40	Rb-Sr biotite	near Oracle
	6: PED-2-58	1,430 ± 40	K-Ar biotite	near Oracle
	12: DEL-13-62	1,380 ± 39	K-Ar biotite	near Oracle
	12: DEL-13-62	1,425 ± 40	K-Ar plagioclase	near Oracle
	19: Unknown	1,440	U-Pb zircon	near Oracle
	11: Several	1,380	Rb-Sr whole-rock isochron	Oracle-Sierra Ancha regions
	7: PED-27-57	39.4 ± 1.2	K-Ar biotite	Oracle
Ruin Granite Emplacement age is about 1,425 m.y.	12: PED-32-61	1,470 ± 40	K-Ar biotite	Sierra Ancha
	12: DEL-3-62	1,415 ± 40	K-Ar biotite	Hess Canyon, Gila Mts.
	12: DEL-3-62	1,435 ± 40	K-Ar plagioclase	Hess Canyon, Gila Mts.
	12: DEL-3-62	1,455 ± 40	K-Ar biotite	Hess Canyon, Gila Mts.
	12: DEL-3-62	1,385	Rb-Sr biotite	Hess Canyon, Gila Mts.
Continental Granodiorite Most exposures are coarsely porphyritic and petrographically resemble other members of the 1,400- to 1,450-m.y. granitic suite. Silver (1978) showed that outcrops in type area in northern Santa Rita Mountains contain rocks of Johnny Lyon and and Oracle ages. Most of the Precambrian ages of various exposures suggest most outcrops of Continental Granodiorite are 1,400 m.y. old. We have restricted the term "Continental Granodiorite" to exposures of coarsely porphyritic 1,400-m.y.-old rocks and their mylonitic equivalents within the Rincon Mountain part of the crystalline core of the complex. Younger ages reflect subsequent thermal events	14: 66D46	1,360 ± 200	Pb-α zircon	N. Santa Rita Mts.
	14: 66D46	56.8 ± 2.5	K-Ar biotite	N. Santa Rita Mts.
	14: 65D914	1,450 ± 160	Pb-α zircon	N. Santa Rita Mts.
	14: 65D914	825 ± 80	Rb-Sr whole rock	N. Santa Rita Mts.
	13: 74D62	1,340 ± 60	K-Ar biotite	Tanque Verde Mts.
	13: 74D62	1,415 ± 50	K-Ar muscovite	Tanque Verde Mts.
	13: 74D62	1,410 ± 50	Rb-Sr whole rock	Tanque Verde Mts.
	13: 74D57	26.1 ± 3.5	Fission-track apatite	Tanque Verde Mts.
	13: 74D57	1,395 ± 50	K-Ar biotite	Tanque Verde Mts.
	13: 74D57	1,390 ± 50	K-Ar muscovite	Tanque Verde Mts.
	13: 74D57	~800	Fission-track zircon	Tanque Verde Mts.
	13: 74D57	55.0 ± 4.2	Fission-track apatite	Tanque Verde Mts.
	20: Unknown	1,430	U-Pb zircon	N. Santa Rita Mts.
	20: Unknown	1,600+	U-Pb zircon	N. Santa Rita Mts.
	20: Unknown	1,420	U-Pb zircon	Little Rincon Mts.
	9: PED-14-59	1,375 ± 40	K-Ar muscovite	N. Empire Mts.
POST-PALEOZOIC BATHOLITHIC ROCKS				
Quartz diorite-granodiorite (Leatherwood) suite				
Chirreon Wash Granodiorite	2: UAKA-75-86	25.1 ± 0.5	K-Ar biotite	N.-central Tortolita Mts.
Leatherwood quartz diorite Emplacement age is probably 75 to 65 m.y. Younger ages are interpreted as reduced ages (see text). 73-m.y. Rb-Sr whole-rock isochron is a composite of Leatherwood plus Chirreon Wash plutons. 70-m.y. Rb-Sr whole-rock isochron is Leatherwood only. 38-m.y. K-Ar sericite age is from quartz vein that cuts Leatherwood quartz diorite	5: BR 2	27.1 ± 0.5	K-Ar biotite	N. Santa Catalina Mts.
	5: BR 2	64.4 ± 1.0	K-Ar hornblende	N. Santa Catalina Mts.
	5: Unknown	27.1 ± 0.5	K-Ar biotite	N. Santa Catalina Mts.
	5: Unknown	38.4 ± 0.6	K-Ar hornblende	N. Santa Catalina Mts.
	4: ML-60†	23.5 ± 0.7	K-Ar biotite	N. Santa Catalina Mts.
	4: ML-60†	36.8 ± 1.0	K-Ar hornblende	N. Santa Catalina Mts.
	4: ML-60†	21.2 ± 0.7	Fission-track apatite	N. Santa Catalina Mts.
	4: ML-60†	28.6 ± 3.7	Fission-track sphene	N. Santa Catalina Mts.
	22: 24	24.9 ± 1.0	K-Ar whole rock	N. Santa Catalina Mts.
	22: 51	29.6 ± 1.1	K-Ar biotite	N. Santa Catalina Mts.
	22: 30	36.7 ± 2.0	K-Ar hornblende	N. Santa Catalina Mts.
	22: 45	32.2 ± 1.2	K-Ar biotite	N. Santa Catalina Mts.
	22: 52	28.9 ± 1.1	K-Ar biotite	N. Santa Catalina Mts.
	22: 69	28.5 ± 1.1	K-Ar biotite	N. Santa Catalina Mts.
	8: PED-1-68	30.2 ± 0.8	K-Ar biotite	N. Santa Catalina Mts.
	9: PED-1-68	25.7 ± 7.1	Rb-Sr mineral isochron	N. Santa Catalina Mts.
	9: UAKA-76-116	31.7 ± 0.7	K-Ar biotite	N. Santa Catalina Mts.
	9: Several	73	Rb-Sr w.-r. ref. isochron	N. Santa Catalina Mts.
	9: Several	70	Rb-Sr w.-r. ref. isochron	N. Santa Catalina Mts.
	9: UAKA-78-66	37.7 ± 0.8	K-Ar sericite	N. Santa Catalina Mts.
	12: PED-18-62D§	28.3 ± 0.9	K-Ar biotite	Santa Catalina forerange
Muscovite granite (Wilderness) suite				
Wilderness granite Emplacement age is probably 44 to 50 m.y. Younger ages are interpreted as variably reduced (see text)	7: PED-4-58	26.5 ± 1.0	K-Ar muscovite	Santa Catalina forerange
	7: PED-4a-58	25.4 ± 1.0	K-Ar biotite	
	7: PED-4a-58	30.2 ± 0.9	K-Ar muscovite	
	7: PED-4a-58	23.3 ± 1.4	Rb-Sr biotite	
	12: PED-4a-58	36.9 ± 1.0	Rb-Sr muscovite	
	3: 100	33 ± 3	K-Ar biotite	Windy Point, Santa Catalina main range
	3: 100	1,340 ± 30	$^{206}Pb/^{207}Pb$-zircon	
	3: 100	50	U-Pb zircon	
	4: GGN-S1	23.3 ± 0.7	K-Ar biotite	
	4: GGN-S1	24.8 ± 0.7	K-Ar muscovite	
	4: GGN-S1	19.3 ± 2.7	Fission-track apatite	
	12: PED-18-62L§	25.6 ± 1.0	K-Ar biotite	Santa Catalina forerange
	12: PED-18-62L§	26.1 ± 1.0	K-Ar muscovite	Santa Catalina forerange

TABLE 1. (Continued)

Rock name and comments	Reference* and sample no.	Age (m.y.)	Method	Area
Muscovite granite (Wilderness suite (continued)				
Wilderness granite (continued)	12: PED-18-62L§	27.5 ± 0.8	K-Ar orthoclase	Santa Catalina forerange
	12: PED-18-62L§	30.0 ± 1.0	K-Ar plagioclase	Santa Catalina forerange
	16: RM-1-66§	32.0 ± 0.3	K-Ar muscovite	Santa Catalina forerange
	16: PED-56-66§	31.9 ± 0.9	K-Ar muscovite	Santa Catalina forerange
	9: UAKA-71-11§	23.1 ± 0.5	K-Ar muscovite	Santa Catalina forerange
	9: UAKA-72-76§	24.0 ± 0.6	K-Ar K-feldspar	Santa Catalina forerange
	7: PED-15-59	47.9 ± 2.1	K-Ar muscovite	Santa Catalina main range
	15: 77-D-68	21.9 ± 0.8	K-Ar biotite	Redington Pass
	15: 77-D-68	35.1 ± 1.2	K-Ar muscovite	Redington Pass
	19: Unknown	44-47	U-Th-Pb monazite	Santa Catalina main range
	19: Unknown	44-47	U-Th-Pb monazite	W. Santa Catalina forerange
	9: Numerous	47	Rb-Sr w.-r. ref. isochron	Santa Catalina main range
	9: UAKA-71-21	46.5 ± 10.4	K-Ar garnet	Santa Catalina forerange
Youtcy granite Possibly an extension of Wilderness pluton; date may reflect age of emplacement	15: 77D81	45.8 ± 1.6	K-Ar muscovite	Redington Pass
Espiritu Canyon granite Ages are probably reduced	13: 71D194	27.6 ± 0.9	K-Ar biotite	Mica Mtn.
	13: 71D194	28.5 ± 0.6	K-Ar muscovite	Mica Mtn.
Wrong Mountain granite K-Ar ages are probably reduced. Rb-Sr sample plots on 47-m.y. Wilderness Granite isochron. Former older age reported by Drewes (1977) was a model age which assumed a 0.703 initial ratio for $^{87}Sr/^{86}Sr$	14: 71D17	24.1 ± 0.9	K-Ar biotite	E. Rincon Mts.
	14: 71D17	25.4 ± 0.9	K-Ar muscovite	E. Rincon Mts.
	13: 71D17	23.3 ± 5.8	Fission-track apatite	E. Rincon Mts.
	13: 71D17	24.6 ± 4.0	Fission-track zircon	E. Rincon Mts.
	13: 73D52	25.0 ± 0.8	K-Ar biotite	Central Rincon Mts.
	13: 73D52	25.3 ± 2.9	Fission-track apatite	Central Rincon Mts.
	13: 73D52	20-30	Fission-track zircon	Central Rincon Mts.
	9: 73D52	47	Rb-Sr w.-r. ref. isochron	Central Rincon Mts.
	7: PED-30-60	33.5 ± 1.1	K-Ar muscovite	Mica Mtn.
	9: UAKA-74-80	25.4 ± 0.5	K-Ar biotite	E. Rincon Mts.
Quartz monzonite (Catalina) suite				
Catalina quartz monzonite Large degree of concordance implies emplacement age of about 28 to 25 m.y. 90-m.y. isochron included samples of xenoliths (probably Leatherwood), contaminated border phase, and a radiogenically enriched aplite. 90-m.y. age is too old	7: PED-16-59	25.6 ± 0.8	K-Ar biotite	NW. Santa Catalina Mts.
	4: BR 21	23.9 ± 1.2	K-Ar biotite	NW. Santa Catalina Mts.
	4: BR 21	23.7 ± 0.7	K-Ar hornblende	NW. Santa Catalina Mts.
	4: BR 21	30.0 ± 3.0	Fission-track sphene	NW. Santa Catalina Mts.
	4: BR 21	28.9 ± 3.3	Fission-track zircon	NW. Santa Catalina Mts.
	4: BR 21	23.5 ± 2.8	Fission-track apatite	NW. Santa Catalina Mts.
	4: ML 61	22.9 ± 0.7	K-Ar hornblende	NW. Santa Catalina Mts.
	4: ML 61	24.7 ± 0.7	K-Ar biotite	NW. Santa Catalina Mts.
	4: ML 61	28.3 ± 3.1	Fission-track sphene	NW. Santa Catalina Mts.
	4: ML 61	27.1 ± 3.4	Fission-track zircon	NW. Santa Catalina Mts.
	4: ML 61	20.8 ± 2.1	Fission-track apatite	NW. Santa Catalina Mts.
	4: BR 16	23.8 ± 0.7	K-Ar biotite	NW. Santa Catalina Mts.
	4: BR 16	28.0 ± 3.0	Fission-track sphene	NW. Santa Catalina Mts.
	4: BR 16	25.9 ± 2.5	Fission-track zircon	NW. Santa Catalina Mts.
	4: BR 16	21.7 ± 2.1	Fission-track apatite	NW. Santa Catalina Mts.
	16: PED-20-62	28.0 ± 0.9	K-Ar biotite	SW. Tortolita Mts.
	4: RC 3	21.6 ± 0.6	K-Ar hornblende	SW. Tortolita Mts.
	4: RC 3	21.1 ± 0.6	K-Ar biotite	SW. Tortolita Mts.
	19: Unknown	27	U-Pb zircon	NW. Santa Catalina Mts.
	9: Numerous	26	Rb-Sr w.-r. ref. isochron	NW. Santa Catalina Mts.
	18, 10: Numerous	90	Rb-Sr whole-rock isochron	NW. Santa Catalina Mts.
Tortolita quartz monzonite Age of emplacement is about 25 m.y.	7: PED-17-59	24.5 ± 0.5	K-Ar biotite	Cargodera Canyon, Santa Catalina Mts.
	4: RC 25	22.7 ± 0.7	K-Ar biotite	Tortolita Mts.
	4: RC 25	18.5 ± 2.4	Fission-track apatite	Tortolita Mts.
	4: ML 105	17.0 ± 2.1	Fission-track apatite	Tortolita Mts.
	9: Numerous	26	Rb-Sr w.-r. ref. isochron	Tortolita Mts.
Happy Valley quartz monzonite 28 and 38 m.y. K-Ar dates on northern mass may be reduced if pluton is a member of the Wilderness suite	14: 69D93	28.0 ± 1.1	K-Ar biotite	Happy Valley
	14: 69D95	37.7 ± 1.6	K-Ar muscovite	Happy Valley
	14: 71D122	26.9 ± 0.9	K-Ar biotite	Happy Valley
MISCELLANEOUS ROCKS				
Postbatholith dikes Probable emplacement ages	9: UAKA-74-83	24.3 ± 0.5	K-Ar whole rocks	Rincon Mts.
	9, 17: UAKA-72-21	21.0 ± 0.3	K-Ar whole rock	Santa Catalina forerange
	2: UAKA-75-87	24.0 ± 0.5	K-Ar biotite	North-Central Tortolita Mts.
Schistose rocks Ages are probably reduced	7: PED-29-60	27.7 ± 1.0	K-Ar muscovite	Mica Mtn.
	14: 69D92	29.1 ± 1.1	K-Ar muscovite	Happy Valley
	14: 70D143	34.6 ± 1.2	K-Ar biotite	Happy Valley
	14: 70D143	29.6 ± 0.9	K-Ar muscovite	Happy Valley

Note: Constants used are $\lambda_\beta = 4.963 \times 10^{-10} yr^{-1}$; $\lambda_e = 0.581 \times 10^{-10} yr^{-1}$, $\lambda = 5.544 \times 10^{-10} yr^{-1}$; $^{40}K/K = 1.167 \times 10^{-4}$ atom/atom; $\lambda_{Rb} = 1.42 \times 10^{-11} yr^{-1}$. Fission-track dates reported by Creasey and others (1977) are recalculated 2.6% older to conform with new K-Ar constants. Rb-Sr w.-r. ref. isochron = Rb-Sr whole-rock reference isochron.

*References as follows: 1, Banks and others (1972); 2, Banks and others (1978); 3, Catanzaro and Kulp (1964); 4, Creasey and others (1977); 5, Creasey (1979, written commun.); 6, Damon and others (1962); 7, Damon and others (1963); 8, Damon and others (1969); 9, Damon and others (1980); 10, Hoelle (1976); 11, Livingston (1969); 12, Livingston and others (1967); 13, Marvin and Cole (1978); 14, Marvin and others (1973); 15, Marvin and others (1978); 16, Mauger and others (1968); 17, Shakel (1974); 18, Shakel and others (1972); 19, Shakel and others (1977); 20, Silver (1978); 21, Silver and Deutsch (1963); 22, Soderman (1979, written commun.).

†Mafic inclusion in Catalina quartz monzonite of unassigned protolith by Creasey and others (1977). Creasey (1979, written commun.) assigned this inclusion to the Leatherwood quartz diorite.

§These samples are found within the forerange mylonitic gneiss complex and are correlated with presumed nonmylonitic counterparts pending further work.

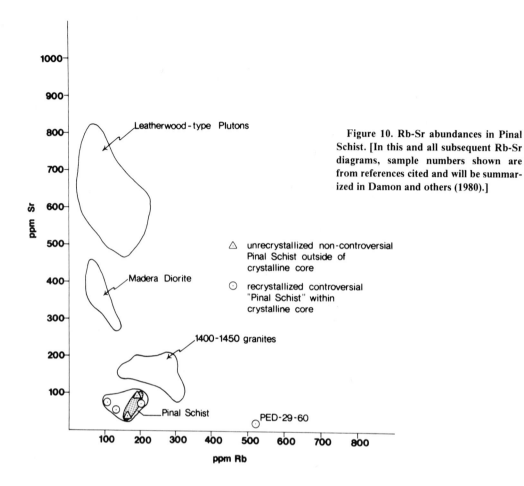

Figure 10. Rb-Sr abundances in Pinal Schist. [In this and all subsequent Rb-Sr diagrams, sample numbers shown are from references cited and will be summarized in Damon and others (1980).]

data points of particular interest are identified in the figures and are discussed in Damon and others (1980).

Pinal Schist

The oldest rocks within the complex are exposures of probable 1.7-b.y.-old Pinal Schist (Fig. 3). Different interpretations currently exist in published maps [see, for example, discussion in map text of Banks and others (1977) in the Tortolita Mountains and Davis (this volume) versus Drewes (1977) in the Tanque Verde Ridge area] regarding which exposures in the complex are indeed Pinal Schist and which are mylonitic schists derived from probable igneous protoliths. Exposures of unequivocal Pinal Schist in the complex are few and probably largely restricted to the northwestern Santa Catalina Mountains and the central Tortolita Mountains. As further support for these contentions, the Rb-Sr abundances and Sr-isotope ratios (Damon and others, 1980) from the schistose bands north and south of Chirreon Wash granodiorite in the Tortolita Mountains are similar to those for unequivocal Pinal Schist outside the complex (Fig. 10). The data do not support the proposal of Banks (this volume) that the schistose bands are mylonite schist derived entirely from the Chirreon Wash pluton and Oracle Granite (in the case of the northern band). Rb-Sr abundances in all of the other samples we have collected are unlike those of Pinal Schist.

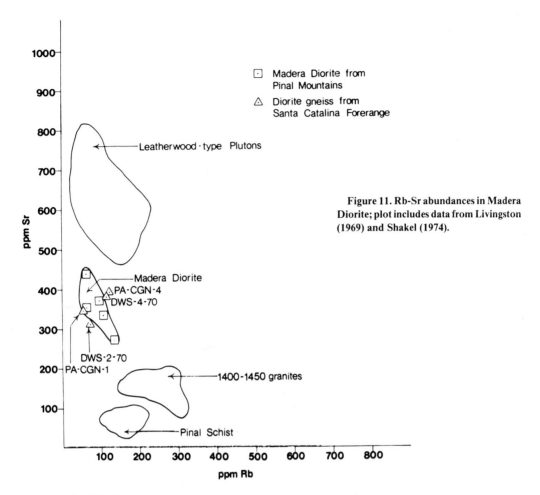

Figure 11. Rb-Sr abundances in Madera Diorite; plot includes data from Livingston (1969) and Shakel (1974).

Precambrian Diorite

Some exposures of foliated diorite (Fig. 3) *may* represent the next oldest rocks in the complex. These diorites lack the abundant epidote and Rb-Sr abundances (Fig. 11) characteristic of the Leatherwood quartz diorite of Late Cretaceous–early Tertiary age. Instead, the Rb-Sr data suggest similarities to 1.65-b.y.-old Madera Diorite of the Pinal Mountains (Ransome, 1903; Livingston and Damon, 1968; Livingston, 1969). This similarity is supported by isotopic data (Fig. 12) for the diorite that plots near a 1.65-b.y. reference isochron through Madera Diorite (Livingston, 1969) at the type locality in the Pinal Mountains. Although the data are somewhat equivocal, the possibility that a Maderalike protolith may constitute part of the dark, mylonitic augen gneisses should not be ignored.

Oracle Granite and Equivalent Granitic Rocks

Creasey and others (1977) suggested that dark augen gneisses of the Catalina forerange are mylonitic equivalents of 1.44-b.y.-old Oracle Granite, a contention which was confirmed by U-Pb systematics in zircons reported by Shakel and others (1977). Rb-Sr abundances (Fig. 13) strongly support this and further indicate that porphyritic varieties of Rincon Valley and Continental Granodiorites (Drewes, 1974, 1977) in the Rincon Mountains and rocks mapped as Oracle Granite in the Tortolita Mountains

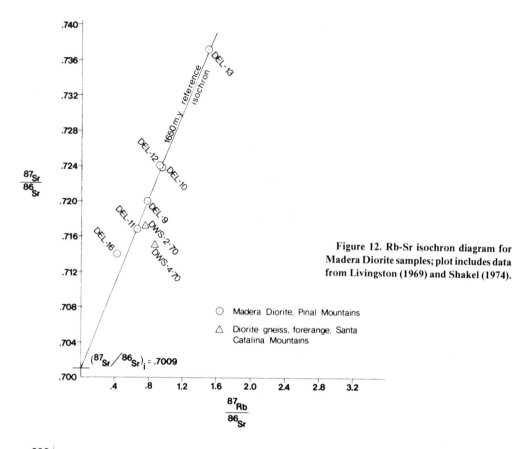

Figure 12. Rb-Sr isochron diagram for Madera Diorite samples; plot includes data from Livingston (1969) and Shakel (1974).

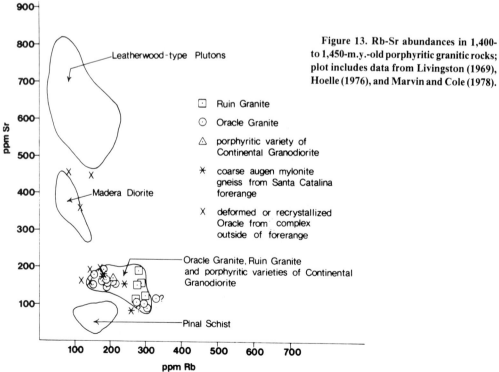

Figure 13. Rb-Sr abundances in 1,400- to 1,450-m.y.-old porphyritic granitic rocks; plot includes data from Livingston (1969), Hoelle (1976), and Marvin and Cole (1978).

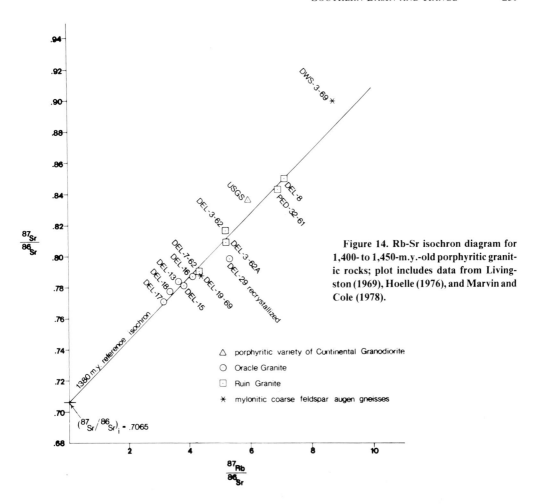

Figure 14. Rb-Sr isochron diagram for 1,400- to 1,450-m.y.-old porphyritic granitic rocks; plot includes data from Livingston (1969), Hoelle (1976), and Marvin and Cole (1978).

(Banks and others, 1977) are members of the 1.4- to 1.45-b.y.-old generation of porphyritic granitic plutons (Damon and Giletti, 1961; Giletti and Damon, 1961; Livingston and Damon, 1968; Silver, 1968, 1978). Isotopic data (Fig. 14) for these rocks, when compared to data for undeformed 1.4-b.y.-old Oracle and Ruin Granites (Livingston, 1969), strengthen the correlation. All available data, therefore, indicate that a *majority* of dark augen gneisses in the complex are deformed 1.4- to 1.45-b.y.old plutons (see App. 1 for further discussion).

Younger Precambrian Apache Group, Diabase, and Paleozoic Rocks

Younger Precambrian Apache Group, diabase, and Paleozoic strata (Fig. 3) occur within the complex and are variably metamorphosed and deformed (Waag, 1968; Budden, 1975; Frost, 1977). Although we have no Rb-Sr data on these rocks, we concur with most of the recent mapping in the Tortolita (Banks and others, 1977) and Santa Catalina Mountains (Creasey and Theodore, 1975; Banks, 1976) about the map position of Apache Group, diabase, and Paleozoic strata. Also, on lithologic grounds, we would add that parts of the southern Santa Catalina forerange rocks (Shakel, 1974, 1978) between Pontatoc and Ventana Canyons (Fig. 5) may equate to metadiabase (amphibolite lenses) or metamorphosed Apache Group strata (quartzite and schist lenses).

Post-Paleozoic Batholithic Rocks

The post-Paleozoic intrusive bodies in the Santa Catalina–Rincon–Tortolita complex can be divided into three suites with plutons of each suite possessing distinctive rock types, field relationships, trace-element and isotopic compositions, and age. The suites are as follows: (1) Late Cretaceous–early Tertiary quartz diorite–granodiorite (Leatherwood suite); (2) Eocene muscovite granite–pegmatite–alaskite with high $^{87}Sr/^{86}Sr$ initial ratios (Wilderness suite); and (3) middle Tertiary quartz monzonite (Catalina suite). The history of nomenclature for each pluton is discussed in Appendix 1, and a summary of the various published radiometric determinations on each pluton is presented in Table 1. Representative modes for plutons of each suite are given in Tables 2 through 5. Analytical data and sample descriptions for new Rb-Sr and K-Ar data will appear in Damon and others (1980).

Late Cretaceous–Early Tertiary (Laramide) Quartz Diorite–Granodiorite (Leatherwood) Suite. Three plutons of quartz dioritic to granodioritic compositions are currently assigned to the Leatherwood suite: the Chirreon Wash granodiorite, an east-elongated, ~40-km² pluton in the east-central Tortolita Mountains; the Rice Peak granodiorite, two probably interconnected stock and sill-like masses, about 12 km², in the northern Santa Catalina Mountains; and the Leatherwood quartz diorite, a 40-km² body in the northeast part of the Santa Catalina Mountains.

Leatherwood suite plutons are characterized by abundant biotite (15% to 25%) in more mafic phases and are mainly quartz dioritic to granodioritic in composition. Epidote is a characteristic accessory mineral of all Leatherwood suite plutons and occurs as several textural varieties (Hanson, 1966; Creasey, 1967; Banks and others, 1978). Representative modes of Leatherwood suite plutons are given in Table 2.

Crosscutting relationships with adjacent plutons indicate that plutons of the Leatherwood suite are the oldest post-Paleozoic intrusions in the complex. Leatherwood quartz diorite and Chirreon Wash granodiorite are both clearly cut by muscovite granite and related aplite, alaskite, and pegmatite apophyses of the Eocene Wilderness suite. In the Tortolita Mountains, Chirreon Wash granodiorite is cut by hundreds of pegmatite dikes and sheets that extend eastward from and grade into or cut the Derrio Canyon muscovite-garnet granite. The Derrio Canyon granite cuts across the Chirreon Wash granodiorite as a north-northeast–trending dikelike mass. The Chirreon Wash granodiorite is older than the middle Tertiary Catalina quartz monzonite, as shown by indirect relationships. The Catalina quartz monzonite occurs in a 2- to 3-km-wide east-northeast– to due-east–trending arcuate belt parallel to and south of the southern contact of the Chirreon Wash granodiorite (Fig. 3). The Chirreon Wash granodiorite is generally separated from the Catalina quartz monzonite by a narrow, arcuate, east-northeast–trending screen of schist tentatively correlated here with the Pinal Schist (see previous

TABLE 2. MODAL MINERALOGY OF LEATHERWOOD SUITE PLUTONS

	1	2	3	4	5	6
Quartz	22	21.0	28	16	19	2-20
K-feldspar	5	7.5	8	6	12	tr.-20
Plagioclase	42	47.0	31	41	51	36-65
	(An_{30-45})	(An_{38})				(An_{35-45})
Biotite	19	20.0	18	9	9	10-26
Hornblende	1	3.5	5	..	..	6-18
Epidote	1	..	..	4	9	1-2
Sphene	<1	tr.	..	1	1	1-4
Other	10	..	6	18	3	..
Opaque grains	..	2.0	..	tr.	1	1-4
Apatite	..	tr.	..	tr.	tr.	tr.

Note: Columns as follows: (1) average of 26 thin sections reported by Hanson (1966); (2) sample BR2 from Banks (this volume); (3) sample 5-2-1 from Suemnicht (1977); (4) sample AZ-LWQ-3 from Tom Heidrick (1979, written commun.); (5) sample AZ-LWD-1 from Tom Heidrick (1979, written commun.); (6) range in composition of Chirreon Wash pluton reported by Banks and others (1977). Trace = tr.

TABLE 3. CHEMICAL DATA FOR LEATHERWOOD QUARTZ DIORITE

	BR-2	AZ-LWD-1	AZ-LWD-3
SiO_2	62.7	64.8	62.1
Al_2O_3	16.1	16.3	16.6
Fe_2O_3	2.1	4.3*	4.9*
FeO	2.1	..	..
MgO	2.9	n.d.	n.d.
CaO	4.7	2.38	3.08
Na_2O	3.7	3.94	3.48
K_2O	2.8	2.40	2.28
P_2O_5	0.22	0.13	0.28
TiO_2	0.60	0.61	0.72
L.O.I.	..	0.52	1.0

Note: Samples AZ-LWD-1 and AZ-LWD-3 from Tom Heidrick (1979, written commun.). Sample BR 2 from Banks (this volume). Analysis of AZ samples by Rocky Mountain Geochemical Corp., Midvale, Utah. Loss on ignition (L.O.I.) determined gravimetrically. Other data determined by atomic absorption, except TiO_2 and P_2O_5, which were determined colorimetrically. Not determined = n.d.
*Total Fe expressed as Fe_2O_3.

discussion). Numerous pegmatites related to the Derrio Canyon granite of the Wilderness suite extend eastward across the Chirreon Wash granodiorite, intrude the schistose screen, and are truncated by the Catalina quartz monzonite. Also, large, partially metasomatized inclusions of epidote-bearing biotite quartz diorite are common within the Catalina quartz monzonite. Many of these inclusions bear a strong resemblance to an epidote-bearing biotite quartz diorite phase locally present along the northwest border of the Chirreon Wash granodiorite. If the quartz diorite inclusions correlate with the border phase, the Chirreon Wash granodiorite is older than the Catalina quartz monzonite.

Within the north-central Santa Catalina Mountains, the Leatherwood quartz diorite intrudes upper Paleozoic rocks (Creasey and Theodore, 1975; Banks, 1976) and is, thus, at least a young as Mesozoic. The Leatherwood quartz diorite sill sequence in the Apache Group is abruptly truncated for some 20

TABLE 4. MODAL MINERALOGY OF WILDERNESS GRANITE

	Quartz	Plagioclase	K-Feldspar*	Biotite	Muscovite	Opaque grains	Garnet	Others	Reference
Marshall Gulch pegmatite (part of Lemmon Rock leucogranite)	25	35 (An$_{7-12}$)	37 (m)	Minor	Minor	Minor	Minor	Minor	Matter (1969)
Control Road pegmatite (part of Lemmon Rock leucogranite)	35	31 (An$_{7-12}$)	29 (m)	Minor	More than above	Minor	Minor	Minor	Matter (1969)
Caseco pegmatite—about 30 m below upper contact between Wilderness Granite and metamorphosed Apache Groups	20	42 (An$_{7-13}$)	27	Trace	6	..	4	..	Matter (1969)
Aplite—same location as above	19	32 (An$_{7-13}$)	43	Minor	4	..	1	..	Matter (1969)
Wilderness Granite—various locations within upper half of main sill; average of 10 analyses	28	29 (An$_{20-25}$)	27 (m)	4	7	1 (magnetite)	1-2[†]	..	Pilkington (1962)
East Fork gneiss[#]—layer just below base of main Wilderness Granite sill; average of 7 analyses	31.3	26.6 (An$_{20-25}$)[§]	29.7 (o)	7.4	4.5	0.2	..	0.4	Sherwonit (1974)
Thimble Peak gneiss[#]— average of 16 analyses	32.9	33.6 (An$_{20-25}$)[§]	26.6	4.1	2.4	0.2	..	0.2	Sherwonit (1974)
Sabino Narrows gneiss[#]— light bands; average of 8 analyses	31.6	36.8 (An$_{20-25}$)[§]	28.2	1.7	1.6	0.1	..	..	Sherwonit (1974)
Gibbons Mountain gneiss[#]— average of 7 analyses	29.9	35.5 (An$_{20-25}$)[§]	26.0	7.7	0.3	0.4	..	0.2	Sherwonit (1974)
Soldier Canyon gneiss[#]— light bands; average of 15 analyses	30.4	37.8 (An$_{25-30}$)[§]	25.7	4.2	0.6	0.3	..	0.2	Sherwonit (1974)
Seven Falls gneiss[#]— average of 7 analyses	29.1	41.3 (An$_{25-30}$)[§]	21.8	7.2	0.1	0.3	..	0.3	Sherwonit (1974)

*Microcline = m; orthoclase = o.
[†]Included under "others" by Pilkington (1962).
[§]These An values are from Peterson (1968).
[#]Terminology for gneiss units in Santa Catalina forerange is from Peterson (1968). These units are interpreted by us as injection sheets on lower levels of the Wilderness sill complex.

TABLE 5. AVERAGE MODAL ANALYSES OF CATALINA SUITE PLUTONS AND ORACLE GRANITE

	1	2	3	4	5	6
Quartz	33	26.8	35.0	39.2	42.2	29.5
K-feldspar	28	34.3	40.2	31.2	41.0	29.8
Plagioclase	28	26.7 (An$_{26-35}$)	16.2 (An$_{28}$)	26.4 (An$_{20}$)	10.4 (An$_{25}$)	35.6 (An$_{30-35}$)
Biotite	9	7.5	7.4	3.2	3.2	0.6
Hornblende	..	0.8	0.2	..	..	..
Opaque grains	..	1.3	2.0	Trace	..	1.0
Sphene	..	0.0	0.6	..	..	..
Apatite	..	0.4	0.1	..	..	Trace
Muscovite	..	..	..	..	..	3.2

Note: Columns as follows:
1. Oracle granite (Banerjee, 1957), average of 38 samples reported under sample 50 by Hoelle (1976, Table 1).
2. Catalina quartz monzonite, main porphyritic phase; average of 11 samples as follows: one sample (48) from Erickson (1962), two samples (BR-21 and ML-61) from Creasey and others (1977), five samples (nos. 6, 9, 10, 11, and 12) from Hoelle (1976), 3 samples (5-3-2, 5-4-2, 5-4-3) from Suemnicht (1977).
3. Catalina quartz monzonite, border phase; average of one sample (BR-16) from Creasey and others (1977) and one sample from Suemnicht (1977).
4. Tortolita quartz monzonite; one sample (ML-105) from Creasey and others (1977).
5. Reef of Rock granite; average of four samples as follows: one sample (53) from Peirce (1958); three samples (5-6-3, 5-6-4, 5-6-5) from Suemnicht (1977).
6. Northern body of Happy Valley quartz monzonite; average of eight samples (A through G) from Miles (1965); the body is questionably assigned to Catalina suite.

km along its west-northwest–trending southern margin by Eocene Wilderness granite and related Lemmon Rock leucogranite. Aplites and pegmatites of the Lemmon Rock leucogranite abundantly intrude Leatherwood quartz diorite in the Mount Lemmon area as originally pointed out by Peirce (1958) and Hanson (1966). Similarly, pegmatite apophyses of presumed Wilderness granite intrude Leatherwood in the Korn Kob mine area 20 km east-southeast of Mount Lemmon (Wilson, 1977; Ted Theodore, 1979, oral commun.).

From the work of Banks (1976) and Suemnicht (1977), Leatherwood quartz diorite is intruded by middle Tertiary Catalina quartz monzonite 5 km north of Mount Lemmon. Here, numerous finer-grained apophyses and the coarse-grained main phase of Catalina quartz monzonite intrude Leatherwood, and large inclusions of Leatherwood are contained in Catalina quartz monzonite (Suemnicht, 1977). Also, the Leatherwood quartz diorite is intruded along its western contact by a large north-northeast–trending dikelike mass of the Catalina suite named Reef of Rock granite (Suemnicht, 1977). Numerous xenoliths of Leatherwood occur in Reef of Rock granite, according to Suemnicht (1977).

In summary, geologic relationships suggest that plutons of the Leatherwood suite are post-Paleozoic but predate granites of the Wilderness suite, which will be shown in the next section to be of Eocene age (44 to 50 m.y. old). The Leatherwood suite is therefore restricted to Mesozoic or earliest Tertiary.

We have Rb-Sr geochemistry for Leatherwood quartz diorite and Chirreon Wash granodiorite. High Sr abundances for both plutons are unique in the Santa Catalina–Rincon–Tortolita crystalline complex (Fig. 15). Poor spread in $^{87}Sr/^{86}Sr$ ratios prohibits a conclusive isochron for either pluton (Fig. 16). However, if the plutons are *assumed* to be comagmatic—a permissible assumption considering their impressively similar rock types, position in the intrusive sequence, and Rb-Sr geochemistry—the data for both plutons approximately conform to a Late Cretaceous (73 m.y. B.P.) reference

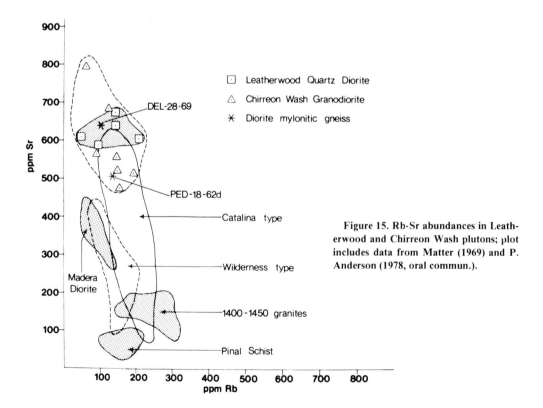

Figure 15. Rb-Sr abundances in Leatherwood and Chirreon Wash plutons; plot includes data from Matter (1969) and P. Anderson (1978, oral commun.).

isochron. The reference isochron, although not rigorously defensible, does not violate any independent geologic age constraints for either or both plutons.

Additional isotopic support for the Laramide age of the Leatherwood quartz diorite was generously provided to us by S. C. Creasey (1979, written commun.) of the U.S. Geological Survey, who has determined K-Ar isotopic ages of two hornblende-biotite mineral pairs. One sample of hornblende-biotite quartz monzonite (Leatherwood) came from Red Ridge about 5 km north of Mount Lemmon and about 150 m (500 ft) east of the contact with the Catalina quartz monzonite. The K-Ar isotopic ages are 38.4 ± 0.7 m.y. on hornblende and 26.1 ± 0.6 m.y. on biotite. These ages agree reasonably well with the K-Ar ages of hornblende (36.8 ± 1.0 m.y.) and biotite (23.5 ± 0.7 m.y.) from an inclusion of Leatherwood in the Catalina quartz monzonite near Cargodera Canyon (location IV in Fig. 2 of Creasey and others, 1977).

The second sample of hornblende-biotite quartz diorite came from near Lombar Hill about 10 km northeast of Mount Lemmon and about 3,960 m (13,000 ft) from the nearest known outcrop of Catalina quartz monzonite. K-Ar isotopic ages are 64.4 ± 1.0 m.y. on hornblende and 27.1 ± 0.5 m.y. on biotite. Creasey (1979, written commun.) indicated that the discordant ages of the mineral pairs are due to the differential Ar loss of the hornblende and biotite from reheating by the Catalina quartz monzonite and that the 64.4-m.y. age of the hornblende from the sample near Lombar Hill approximates the intrusion age of the Leatherwood. He concurred with our interpretation that the age of the Leatherwood is Laramide.

Creasey's K-Ar biotite data are similar to other K-Ar biotite ages of 30.2 (Damon and others, 1969) and 31.7 m.y. (this paper). These ages are complemented by 29.6, 32.2, 28.9, and 28.5 m.y. K-Ar biotite ages for probable Leatherwood sills in the Control mine area about 10 km northeast of Mount Lemmon (S. M. Soderman, 1979, written commun.). A whole-rock determination on one of these sills is 24.9 m.y., a hornblende concentrate yielded a discordant 36.7-m.y. K-Ar apparent age (S. M. Soderman, 1979, written commun.). Additionally, a Rb-Sr whole-rock–mineral isochron on

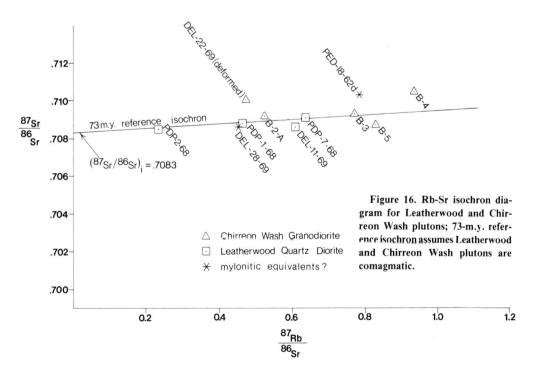

Figure 16. Rb-Sr isochron diagram for Leatherwood and Chirreon Wash plutons; 73-m.y. reference isochron assumes Leatherwood and Chirreon Wash plutons are comagmatic.

biotite, K-feldspar, plagioclase, epidote, and whole-rock from a single sample is 25.7 ± 7.1 m.y. (Damon and others, 1980). We interpret the 32- to 26-m.y. K-Ar biotite ages and Rb-Sr whole-rock–mineral isochron as reduced ages that reflect thermal resetting by Catalina suite plutons or uplift cooling of the complex between 30 and 20 m.y. B.P. The older hornblende ages are less reset because of their higher retentivity. With the exception of the 64-m.y. hornblende age, all K-Ar ages and the Rb-Sr whole-rock–mineral isochron age are reduced because they are all younger than 44 to 51 m.y., the age of Wilderness plutons that clearly intrude the Leatherwood (see next section). The 25-m.y. K-Ar biotite age reported by Banks and others (1978) for the Chirreon Wash pluton is similarly a reduced age.

Attempts to date the Leatherwood quartz diorite by U-Th-Pb techniques on zircons were unsuccessful and gave very discordant U-Th-Pb ages (Shakel, 1978). According to Shakel (1978), the discordant zircons indicated a "high degree of zircon inheritance," possibly from 1,440-m.y.-old Precambrian terrane.

Several additional indirect geologic relationships support the Laramide (75 to 50 m.y. B.P.) age for the Leatherwood suite. The Rice Peak granodiorite has been correlated on petrographic grounds with the Leatherwood quartz diorite by Waag (1968) and Creasey and Theodore (1975). If this correlation is valid, several intriguing implications arise for the age of the Leatherwood suite. The Rice Peak granodiorite clearly intrudes the lower part of the American Flag Formation from which Bromfield (1950) reported the freshwater pelecypod *Unio* (Triassic to Holocene) and the gastropod *Viviparus* (probably Cretaceous to Holocene). The rock type of the lower American Flag Formation is very suggestive of a correlation with the Cretaceous Fort Crittenden Formation (Hayes, 1970). Fort Crittenden equivalents are older than 75 to 72 m.y. on the basis of numerous radiometric ages for widespread overlying volcanic rocks, but are no older than 85 m.y. on the basis of the fossil assemblages contained within those equivalents (see Hayes and Drewes, 1978). *Assuming that all the correlations are valid,* Rice Peak granodiorite (and Leatherwood suite in general) is younger than ~72 m.y., the presumed minimum age of the lower American Flag Formation. Also of interest is the fact that Rice Peak granodiorite has been correlated with a petrographic analogue at San Manuel by Creasey (1967). Two samples of hydrothermal biotite from the granodiorite porphyry at San Manuel gave ages of 65 and 69 m.y. (Creasey, 1965; Rose and Cook, 1965).

Leatherwood quartz diorite is associated with significant porphyry copper–type skarn mineralization at Marble Peak (Braun, 1969) and possibly the Korn Kob mine farther southeast (Wilson, 1977). The time of most porphyry copper–type mineralization in southeast Arizona is essentially coeval with dates from various igneous minerals in spatially and temporally associated intrusions (Creasey and Kistler, 1962; Creasey, 1965; Rose and Cook, 1965; Livingston and others, 1967; Johnson, 1972). With the exception of Bisbee, Arizona, all porphyry copper mineralization is 70 to 50 m.y. old. A K-Ar age of 37.7 ± 0.8 m.y. (Damon and others, 1980) on sericite from a quartz vein that cuts Leatherwood indicates that Leatherwood is at least 38 m.y. old.

Another correlative device is chemical in nature. Three sets of major-element chemical analyses are now available for the Leatherwood quartz diorite (Table 3); K_2O/SiO_2 ratios indicate that the Leatherwood is calc-alkalic in the sense of Keith (1978). Data in the form of hundreds of chemical analyses suggest that true calc-alkalic plutons were emplaced in southeast Arizona only between 70 and 50 m.y. B.P. All other Arizona Phanerozoic plutonic rocks are more alkalic, according to the classification in Keith (1978).

A minimum age of about 50 m.y. for Leatherwood suite plutons is provided by the radiometric age of the Wilderness suite (next section), which clearly intrudes the Leatherwood suite in the Tortolita and Santa Catalina Mountains. Although no technique alone conclusively dates the Leatherwood suite, consideration of all available data converges on a 75- to 64-m.y. age for it.

Eocene Muscovite Granite (Wilderness) Suite. About 65% of the area of the Santa Catalina–Rincon–Tortolita complex is underlain by large sheetlike and laccolith-shaped muscovite granite plutons. Five asymmetric, laccolithic granitic intrusions are currently assigned to the Wilderness suite. They include the Derrio Canyon granite, a group of irregularly stacked sills in the northwestern Tortolita Mountains (Fig. 4); the Wilderness granite (and associated Lemmon Rock leucogranite), a west-northwest–trending batholith-sized laccolith in the main range of the central Santa Catalina Mountains, and several sills or injection sheets in the Santa Catalina forerange (Figs. 5 and 6); the Youtcy granite, an irregular stocklike mass that may be the east end of the Wilderness pluton in the Redington Pass area, between the Santa Catalina and Rincon Mountains; the Espiritu Canyon granite, an east-northeast–elongated pluton having diffuse contacts with the Wrong Mountain granite 2 km northeast of Mica Mountain in the Tanque Verde Mountains; and the Wrong Mountain granite, a batholith-sized laccolithic mass widespread throughout the Rincon and Tanque Verde Mountains. Previous nomenclature is summarized in Appendix 1. The Wilderness-Youtcy plutons are coextensive with Tolman's original batholith (Moore and others, 1949).

Wilderness-type plutons occupy a distinct mineralogic nitch in the Arizona Phanerozoic magmatic framework. For this reason, we affix the term "granite" to plutons of this suite to emphasize their unique mineralogy, even though some phases are technically quartz monzonites (Peterson, 1961) or monzogranites (Streckeisen, 1976). All Wilderness suite plutons contain muscovite with biotite and garnet as common accessories. Average modes for the Wilderness granite, the best-documented pluton of the suite, are listed in Table 4. The data are listed in ascending structural order so that pegmatites at the top of the mountain represent the top of the ~4.5-km-thick section of granitic rocks, much of which contains low-angle mylonitic deformation. The data seem to suggest several mineralogic changes within the Wilderness pluton. Structurally low sills are characterized by more biotite relative to muscovite and less K-feldspar relative to plagioclase feldspar than structurally higher phases. Garnet is locally present throughout the pluton but is evidently more abundant in upper structural levels, particularly in the Lemmon Rock leucogranite. A biotite-rich phase of the Wilderness granite is present south of Mount Bigelow 8 km east-southeast of Mount Lemmon.

Crosscutting relationships indicate that Wilderness suite granites occupy an intermediate stage in evolution of the Late Cretaceous–middle Tertiary batholith. Contact relationships were previously discussed between plutons of the Leatherwood suite and Wilderness suite. These indicate overwhelmingly that Wilderness suite granites intrude and are younger than Laramide Leatherwood suite plutons.

Contact relationships between different muscovite granite phases are commonly gradational (Shakel, 1978; Thorman and Drewes, 1978). Undeformed phases of two-mica Wilderness granite pass gradationally upward into Lemmon Rock leucogranite 2 km south of Mount Lemmon (Shakel, 1978). The Lemmon Rock leucogranite is composed of a complex of alaskites, pegmatites, and aplites. Several generations of dikes may be present in a single outcrop. In the northwest Tortolita Mountains, pegmatite phases of the Derrio Canyon granite extend some 10 km eastward from the main mass in the northwestern Tortolita Mountains. Many pegmatites become more granitic in texture and imperceptibly grade into the main Derro Canyon mass to the west. A few pegmatites cut this mass, but most exhibit a gradational relationship.

Wilderness suite plutons are clearly older than middle Tertiary quartz monzonites (Catalina suite). In the central Tortolita Mountains, pegmatite apophyses of the Derrio Canyon granite are truncated by Catalina quartz monzonite. In Cargodera Canyon in the western Santa Catalina Mountains, the foliated Lemmon Rock leucogranite phase of the Wilderness granite is intruded by and occurs as inclusions in the border phase of the Catalina quartz monzonite (see subsequent discussion in Catalina suite section). In the eastern Rincon Mountains, apophyses from the southern mass of Happy Valley quartz monzonite intrude Wrong Mountain granite, according to the mapping of Drewes (1975).

In summary, field relationships suggest approximate contemporaneity between various muscovite granite phases. Contact relationships clearly indicate that the muscovite granite suite intrudes and is younger than the Laramide Leatherwood suite but is intruded by and is therefore older than the middle Tertiary quartz monzonites (Catalina suite).

Most of our Rb-Sr geochemistry (Fig. 17) is on the Wilderness granite and related Lemmon Rock leucogranite and the Derrio Canyon granite. H. Drewes and R. F. Marvin of the U.S. Geological Survey have provided us with one analysis from the Wrong Mountain granite in the Rincon Mountains (1978, written commun.). The Rb and Sr abundances (Fig. 17) define a field for the Wilderness suite granites and their mylonitic equivalents that is offset from all other igneous rock types, a relationship consistent with the suite's unique mineralogy. Pegmatites with low Sr abundances form a distinct field separate from main phases of the Wilderness suite. We interpret these pegmatites to represent a low Sr residuum formed late in the differentiation history of the Wilderness sequence.

Isotopic data for Wilderness suite rocks (Fig. 18) show much scatter that we interpret as being mainly due to large variation in $^{87}Sr/^{86}Sr$ initial ratios and contamination by nearby highly radiogenic host rocks. Samples whose analyses scatter along the 1.44-b.y. reference isochron (Fig. 18) are generally in geologic settings in which they could have been easily contaminated by highly radiogenic Sr from nearby 1.45-b.y.-old Oracle Granite (see Damon and others, 1980). Some samples of pegmatite from the Santa Catalina forerange are anomalously radiogenic (Damon and others, 1980) and could represent radiogenically disturbed pegmatites of original Precambrian ancestry or small pockets of metamorphically differentiated material "sweated out" from Precambrian protoliths during Wilderness intrusion.

An outstanding example of the contamination phenomenon in Wilderness suite rocks occurs in the Lemmon Rock leucogranite phase of the Wilderness granite intrusion east and southeast of Mount

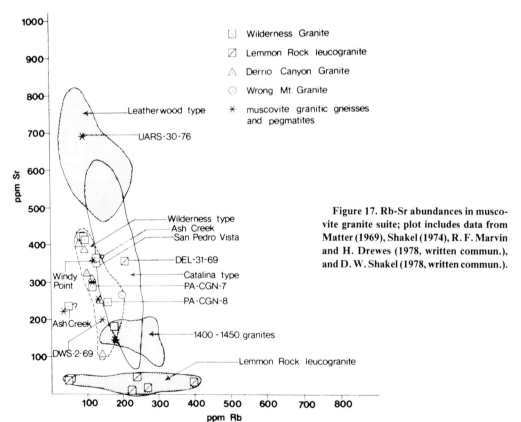

Figure 17. Rb-Sr abundances in muscovite granite suite; plot includes data from Matter (1969), Shakel (1974), R. F. Marvin and H. Drewes (1978, written commun.), and D. W. Shakel (1978, written commun.).

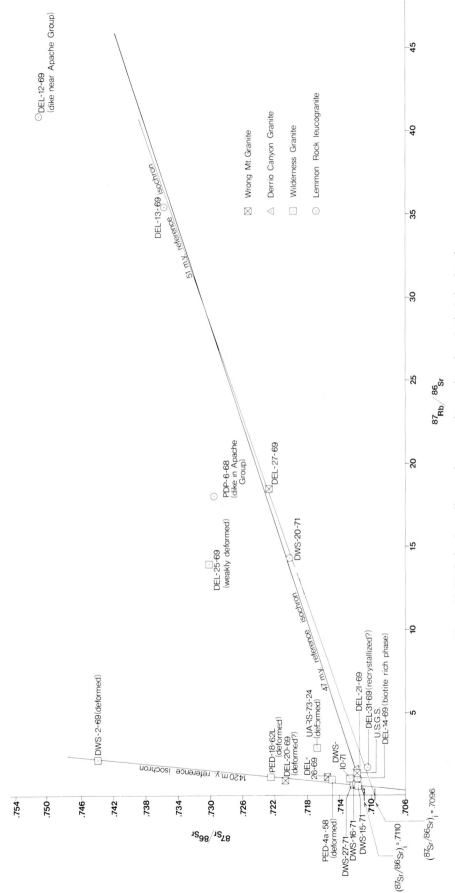

Figure 18. Rb-Sr isochron diagram for muscovite granite suite; plot includes data from Shakel (1974), D. W. Shakel (1978, written commun.), and R. F. Marvin and H. Drewes (1978, written commun.). The 47-m.y. reference isochron is constructed through non-deformed rocks that are sufficient distances from older highly radiogenic host rocks (Damon and others, 1980), and assumes Wilderness granite and Lemmon Rock leucogranite are comagmatic.

Lemmon. A sample of Lemmon Rock leucogranite (DEL-10-69) that contains numerous 1.44-b.y.-old Oracle Granite inclusions about 4 km southeast of Mount Lemmon is anomalously radiogenic ($^{87}Sr/^{86}Sr$ = 0.7772 based on a 49-m.y. assumed age). The Precambrian Oracle Granite host rock in this area (DEL-29-69) is radiogenic enough (0.7936 at 49 m.y. B.P.) to have been a likely source of radiogenic Sr contamination. Also, pegmatite members of Lemmon Rock leucogranite that are near or in Precambrian Apache Group strata (samples PDP-6-68 and DEL-12-69) contain anomalous amounts of radiogenic Sr and plot above the 47-m.y. reference isochron for Wilderness granite. In contrast, samples of Lemmon Rock leucogranite that intrude Leatherwood quartz diorite about 4 km east of Mount Lemmon near Summerhaven (samples DEL-13-69 and DWS-20-71) plot on the 51-m.y. reference isochron (Fig. 18) for the Lemmon Rock leucogranite. Importantly, the Leatherwood quartz diorite is comparatively nonradiogenic (at 49 m.y. measured $^{87}Sr/^{86}Sr$ values for five samples range from 0.7081 to 0.7087). Thus, the only apparent way to explain the large variation in measured ratios from the Lemmon Rock leucogranite in the Mount Lemmon area is by the variable addition of radiogenic Sr to Lemmon Rock leucogranite magma during its intrusion presumably 44 to 50 m.y. ago. Precambrian Oracle Granite and Apache Group host rocks provided ready sources of radiogenic Sr that was mobilized by hydrothermal metasomatism during emplacement of the water-rich Lemmon Rock leucogranite assemblage of aplites, pegmatites, and alaskites.

Samples of Wilderness suite rocks that were *not* collected near highly radiogenic host rocks and that have only weak or no mylonitic fabric define a 47-m.y. reference isochron which has an initial $^{87}Sr/^{86}Sr$ value of 0.7110. This isochron is based on the assumption that the Wilderness granite and Lemmon Rock leucogranite are comagmatic. The fact that one Derrio Canyon granite analysis and two Wrong Mountain granite analyses fall on the same isochron suggest that these plutons may be comagmatic with Wilderness granite and ~47 m.y. old.

In addition to the Rb-Sr isotope data which indicate a 47-m.y.-age assignment, numerous other radiometric data are concordant with the Rb-Sr data. The first of these data are in the form of K-Ar ages on coarse muscovite (Table 1). Coarse muscovite from the Lemmon Rock leucogranite yielded a 47.9-m.y. K-Ar age. Damon and others (1963) interpreted this age as an incomplete degassing of Ar due to the large dimensions of the mica books. The age in this context was a minimum or inherited age for the pegmatite, which could have had an unknown older age. Within the current radiometric context, however, the 48-m.y. age may date the emplacement of the Lemmon Rock leucogranite. More K-Ar support is found in the Redington Pass area. Marvin and others (1978) reported a 45.8-m.y. K-Ar age for muscovite in a sample of Wrong Mountain granite.

Discordant and relatively old K-Ar ages have been known from the two-mica granitic rocks since the work of Damon and others (1963; Table 1). A coarse muscovite from the Wrong Mountain granite yielded a 33.5-m.y. age. Muscovite from deformed Wilderness granite yielded K-Ar ages of 30.2 m.y. on muscovite versus 25.4 m.y. for biotite (Damon and others, 1963). Rb-Sr dates on the same sample reported by Livingston and others (1967) yielded 36.9 m.y. for muscovite as opposed to 25.4 m.y. for biotite. Another muscovite from deformed Wilderness granite at the same location yielded a 33-m.y. K-Ar age (Catanzaro and Kulp, 1964). Coarse-grained muscovite gave a 31.9-m.y. K-Ar age for a light band (Wilderness equivalent) in the mylonitic forerange gneisses as compared to 26.0 m.y. for fine-grained muscovite and 25.6 m.y. for fine-grained biotite from another sample at the same location. Recently, discordant K-Ar ages from the Wrong Mountain granite in Redington Pass were reported by Marvin and others (1978). Here, biotite gave an age of 21.9 m.y., and muscovite yielded a strongly discordant age of 35.1 m.y.

All of the relatively old and/or discordant ages reported from the Santa Catalina–Rincon–Tortolita complex are on coarse muscovites from Wilderness-type granitic material. Significantly, none of these ages is older than 50 m.y. The ages are distinctly older than the majority of K-Ar and fission-track ages in the complex, which are 20 to 30 m.y. (Table 1). We view these post–45-m.y. and pre–30-m.y.

muscovite ages as reduced ages that probably reflect the relatively retentive nature of coarse muscovites (Damon and others, 1963; Mauger and others, 1968). As a rule, these muscovites have had their ages reduced less than those for finer-grained micas. This is in opposition to the interpretation of Banks (1977 and this volume) that these ages represent contamination of the muscovites by excess Ar during what he considers a middle Tertiary event of batholithic intrusion.

As pointed out by Damon (1970), because all minerals have a measurable solubility for Ar in the presence of an external Ar pressure, excess environmental Ar can be detected by analyzing minerals in which K is a minor or trace component. For such minerals, the presence of significant amounts of excess environmental Ar yields highly discordant ages. Livingston and others (1967) analyzed plagioclase from mylonitic Wilderness granite and concluded that the plagioclase contained no more than 1.5×10^{-11} mol/g of excess Ar relative to muscovite and biotite. This amount of excess Ar would increase the apparent age of the muscovite or biotite by no more than 1.2 m.y. However, in order to place closer limits on the amount of excess environmental Ar, we have analyzed garnet containing only 0.04% K (Damon and others, 1980). This garnet contained only 3.16×10^{-12} mol/g of nonatmospheric ^{40}Ar of which no more than one-half could be excess environmental Ar. This amount would increase the age of muscovite or biotite by only about 0.1 m.y. Therefore, we conclude that the Santa Catalina part of the complex was open to the escape of Ar during the final elevated thermal history. Consequently, the nonatmospheric ^{40}Ar contained within K-bearing minerals of the complex was generated since that time or inherited from older minerals only partially degassed during the middle Tertiary thermal event that preceded final middle Tertiary cooling.

More convincing evidence for the Eocene igneous event is found in U-Th-Pb data recently reported from the Wilderness granite and associated aplite, alaskite, and pegmatite bodies. U-Th-Pb ages of 44 to 47 m.y. on "igneous-looking" monazite from a sample of Wilderness granite in the Santa Catalina main range were reported by Shakel and others (1977). Monazites from foliated two-mica mylonitic gneiss at the west end of the Santa Catalina forerange also yielded 44- to 47-m.y. U-Th-Pb ages (Shakel and others, 1977). It is interesting that the lower-intercept age reported by Catanzaro and Kulp (1964) for the highly discordant zircons from garnet-bearing two-mica mylonite gneiss (deformed Wilderness granite at Windy Point in the Santa Catalina main range) was 50 m.y. Catanzaro and Kulp (1964) emphasized the upper-intercept age of 1,660 m.y. and suggested that the 50-m.y. lower intercept was too old to represent the metamorphic event, which they believed was dated by K-Ar as 30 to 40 m.y. B.P. They thought that middle Tertiary episodic Pb loss from a Precambrian protolith might explain the data. In the context of our work, the 50-m.y. lower intercept probably dates the emplacement of the Wilderness granite. Catanzaro and Kulp included a picture of the zircons they dated. In the picture, there appear to be two populations present. About 10% of the zircons are stubby, opaque, and cloudy and perhaps could be older "inherited" zircons. The other 90% are clear, elongate, and euhedral zircons that might be younger igneous zircons formed during Eocene Wilderness granite crystallization. It is interesting that the actual analysis plots on the discordia line close to the 50-m.y. intercept about nine-tenths of the way down the discordia line toward the 50-m.y. lower intercept. If the source of the zircon contamination was Oracle Granite (as might be predicted by studies of Shakel and others, 1977; Shakel, 1978; and Rb-Sr data in Fig. 18), a discordia cord upper intercept at 1.45 b.y. would require a lower intercept *slightly* younger than 50 m.y. This younger lower-intercept age (interpreted by us to be age of emplacement) would be approximately concordant with our 47-m.y. Rb-Sr reference isochron.

In summary, field relationships require that Wilderness suite granites postdate Late Cretaceous–early Tertiary Leatherwood suite plutons but predate middle Tertiary Catalina suite plutons. Isotopic data from different methods are impressively concordant at 44 to 50 m.y. These include an 11-point whole-rock Rb-Sr isochron, two K-Ar dates of coarse muscovite, two U-Th-Pb dates on monazite, and one zircon sample dated by the U-Pb discordia method. Most of the dates are from the Wilderness granite, Lemmon Rock leucogranite, and related pegmatite phases in the Santa Catalina forerange.

However, one sample from the Derro Canyon granite and two from Wrong Mountain granite fall on our Rb-Sr reference isochron; this suggests that these plutons correlate with Wilderness granite in age as well as mineralogy and relative position within the intrusive sequence of the Santa Catalina–Rincon–Tortolita crystalline complex.

Middle Tertiary Quartz Monzonite (Catalina) Suite. Emplacement of one granite and three quartz monzonite plutons (designated the Catalina suite) in late Oligocene–early Miocene time marked the final event in the growth of the batholith component of the Santa Catalina–Rincon–Tortolita crystalline complex. By far, the greatest volume of magma emplaced during this time was that in the northwestern Santa Catalina Mountains and southern Tortolita Mountains where three coalescing plutons—the Reef of Rock granite and the Catalina and Tortolita quartz monzonites—form a small batholithic mass. The Catalina quartz monzonite forms a large, half-circle–shaped pluton in the northwestern Santa Catalina Mountains and an east-northeast–trending dikelike mass in the south-central Tortolita Mountains. The Tortolita quartz monzonite occupies a rectangular area with an east-northeast–trending long axis in the southern Tortolita Mountains and occurs as north-northwest–striking dikes in the western Santa Catalina Mountains. The Reef of Rock granite forms a jagged north-northeast–trending spine that borders the southeast margin of the Catalina quartz monzonite in the central Santa Catalina Mountains (Suemnicht, 1977). This granite is one of the youngest plutons of the Santa Catalina–Rincon–Tortolita crystalline complex. A fourth pluton, the Happy Valley quartz monzonite, is provisionally placed in the Catalina suite pending further work and appears as two stocklike masses on the eastern slopes of the Rincon Mountains. Together, Catalina suite plutons form about 15% of the batholithic rocks exposed in the complex.

Catalina suite plutons are mineralogically distinct from Wilderness and Leatherwood suites. Average modal analyses are reported in Table 5. Catalina suite plutons contain more quartz and more K-feldspar, less (but more sodic) plagioclase, and much less biotite than Leatherwood suite plutons. Both suites contain hornblende and sphene as important accessories. Catalina suite intrusions have similar quartz content, slightly more K-feldspar, slightly less (but more calcic) plagioclase, and similar amounts of biotite compared with Wilderness type plutons. Catalina plutons (with the exception of the northern mass of Happy Valley quartz monzonite) contain no muscovite compared with 1% to 7% in Wilderness plutons. Catalina suite plutons contain hornblende and sphene as common accessories and *no* garnet. The reverse is true for Wilderness suite plutons.

Upon casual inspection, Catalina quartz monzonite, the largest pluton of the Catalina suite, may easily be confused with Precambrian Oracle Granite. This confusion resulted in an incorrect age assignment for this rock initially by Tolman (1914, unpublished manuscript). This error was continued in many subsequent publications including the recent 1:500,000-scale Arizona State geologic map (Wilson and others, 1969). The mineralogic contrast between the two plutons was first observed by Wallace (1954), who assigned an older Precambrian age to the Oracle Granite on the basis of its intrusive contacts into Pinal Schist of older Precambrian age. A fossil horn coral found in a limestone inclusion by McCullough (1963) and Rb-Sr isotope data published by Hoelle (1976) established a Phanerozoic age for this intrusion. Subsequent isotopic results (Damon and others, 1963; Creasey and others, 1977; Shakel and others, 1977; this paper) firmly establish a late Oligocene age for this pluton. Quartz, plagioclase, K-feldspar, and biotite contents for Oracle Granite and Catalina quartz monzonite are very similar. The important mineralogic difference is in the accessory minerals. Catalina quartz monzonite contains hornblende and sphene in essentially all samples, but these minerals are absent from Oracle Granite.

Published modes of Catalina quartz monzonite are all from that part of the pluton in the western Santa Catalina Mountains (Table 5). The presumed analogue in the Tortolita Mountains has about 25% to 30% quartz, 30% to 35% K-feldspar, 25% to 30% plagioclase (An_{25-35}), 6% to 8% biotite, 1% hornblende, and 0.5% to 1% sphene. Fine- to medium-grained nonporphyritic border and coarser-grained porphyritic main phases of Catalina quartz monzonite are present in both the Santa Catalina and Tortolita Mountains. The border phase (Table 5) is commonly present along outer margins of the Catalina intrusion and surrounds many of the larger inclusions within the Catalina pluton.

Within the small batholith of coalescing Catalina suite plutons in the western Santa Catalina and Tortolita Mountains, there is a compositional variation between plutons (Table 5). Catalina quartz monzonite contains less quartz, less (but more calcic) plagioclase, the same amount of K-feldspar, and more mafic minerals than Tortolita quartz monzonite. In turn, Reef of Rock granite contains more quartz, more K-feldspar, and less plagioclase than Tortolita quartz monzonite. Both plutons intrude Catalina quartz monzonite. Their overlapping isotopic ages (see below) suggest a differentiation continuum.

As discussed earlier, the Catalina quartz monzonite intrudes and postdates the Leatherwood quartz diorite (Suemnicht, 1977). The nature of the southern contact of the Catalina quartz monzonite with phases of the Wilderness granite has been much debated. All previously published opinions (McCullough, 1963; Creasey and others, 1977; Banks, 1977 and this volume) regard the contact as some type of metamorphic front. A new interpretation proposed in this paper is that the contact represents an intrusive contact of Catalina quartz monzonite into Lemmon Rock leucogranite border phase of the Wilderness granite. Much of the contact is occupied by a steeply inclined migmatite zone that contains an interleaved assemblage of metasedimentary rocks (quartzites, calc-silicate skarns, siliceous gneiss) and variably foliated intrusive rocks including Leatherwood quartz diorite, Lemmon Rock leucogranite, and Catalina quartz monzonite. The migmatite—which is interpreted by us to represent a screen of metamorphosed and highly *injected,* pre-Catalina intrusive and metasedimentary rocks—thins to the west. Here a fine- to medium-grained border phase of the Catalina quartz monzonite, similar to that described along the eastern margin of the pluton by Hoelle (1976) and Suemnicht (1977), sharply intrudes and contains inclusions of foliated Lemmon Rock leucogranite. This relationship persists for at least a 2-km length of Catalina-Wilderness contact. Both plutons contain numerous metasedimentary inclusions that locally obscure the intrusive relationships along the contact. In contrast to the conclusions of other workers (McCullough, 1963; Banks, this volume), thin sections and field observations along the Catalina-Wilderness contact in Cargodera Canyon indicate that foliated fabric in the metasedimentary inclusions, Catalina quartz monzonite, and Lemmon Rock leucogranite is mostly nonmylonitic. Our observations do not support the contentions of Creasey and others (1977) and Banks (1977 and this volume) that Cargodera Canyon represents a transition zone ("gneiss front") between mylonitic and nonmylonitic parts of a *single* pluton. In the Tortolita Mountains, Catalina quartz monzonite postdates the Leatherwood and Wilderness suites, as discussed above.

The youngest major pluton in the crystalline core of the complex is the Tortolita quartz monzonite. It is easily distinguished from the Catalina quartz monzonite by its finer grain size and nonporphyritic hypidiomorphic-granular texture. Wherever the two plutons are in contact, the Tortolita quartz monzonite clearly cuts the Catalina quartz monzonite (McCullough, 1963; Banks, 1976). In lower Cargodera Canyon, two large north-northwest–striking dikes of Tortolita quartz monzonite clearly crosscut and contain inclusions of Catalina quartz monzonite and Wilderness granite. These dikes were correlated by Banks (1976) with the main Tortolita pluton in the southern Tortolita Mountains 8 km to the west-northwest. We concur with Banks's correlation. In the southern Tortolita Mountains, the main pluton of Tortolita quartz monzonite occupies the entire southern third (about 70 km²) of the mountain range. The contact between the Tortolita and Catalina plutons in the south-central Tortolitas was mapped by Budden (1975) as a mixed zone with many different phases present. More recent mapping by one of us (Keith) has indicated that the zone represents inclusions of quartz diorite that have been engulfed and injected by Catalina quartz monzonite and subsequently intruded by sheets of Tortolita quartz monzonite from the south (Fig. 4). Keith believes that the presence of K-feldspar porphyroblasts in and near the edges of the dioritic inclusions and the occurrence of biotitic "patches" in the Catalina pluton suggest K metasomatism during intrusion of Catalina quartz monzonite. The Tortolita quartz monzonite and older rocks are locally intruded by pegmatite, lamprophyre, and granodiorite dikes which volumetrically are insignificant. At the southeast corner of the main Tortolita pluton, intrusions of Tortolita quartz monzonite into Catalina quartz monzonite have been recognized by Budden (1975) and Banks (1976, 1977, and this volume).

Another small pluton, the Reef of Rock granite, is exposed in the Santa Catalina Mountains, 2 km north of Mount Lemmon (Suemnicht, 1977). The granite intrudes and contains inclusions of Leatherwood quartz diorite and Catalina quartz monzonite (Suemnicht, 1977). Possible equivalents of the Reef of Rock granite intrude Tortolita quartz monzonite in the Tortolita Mountains where they have been mapped as a phase of the Tortolita pluton.

A possible fourth pluton of the Catalina suite is the Happy Valley quartz monzonite of the eastern Rincon Mountains (Drewes, 1974). Drewes (1974) mapped projections of the southernmost mass of Happy Valley quartz monzonite cutting Wrong Mountain granite (a Wilderness suite pluton). The northernmost mass contains muscovite (Miles, 1965) and may be a Wilderness suite pluton.

Plutons of the Catalina suite are the youngest set of intrusions in the batholithic sequence and mark its final consolidation. Two of the *youngest* plutons are intruded by northwest-striking dike swarms. Early Miocene K-Ar ages have been obtained from some of these dikes (Banks and others, 1978). The eastern part of the Catalina quartz monzonite is crosscut by northwest-striking rhyolite porphyry dikes. Tortolita quartz monzonite is crosscut by northwest-striking quartz latite and granodiorite dikes (Fig. 2). No dikes are known to crosscut the Happy Valley or Reef of Rock intrusions.

We have Rb-Sr geochemistry on two plutons of the Catalina suite, the Catalina and Tortolita quartz monzonites. Rb-Sr abundances for the two plutons (Fig. 19) do not overlap but are aligned along a similar trend, which is permissive of a differentiation continuum.

Isotopic analyses have been determined for main-phase quartz monzonites and correlative dike and aplite samples. We believe that data for both the Catalina and Tortolita plutons (including their dikes and aplites) are best explained by adherence to a reference isochron of $\sim$26 m.y. (Fig. 20). The slope of the isochron is to a large degree governed by analyses of aplites, but we feel confident that the aplites sampled are comagmatic with the enclosing plutons. Samples that plot well above the reference isochron were collected near to and could have been easily contaminated by highly radiogenic wall rocks or inclusions. Rb-Sr abundances of dark inclusions (Fig. 21) in these young quartz monzonites suggest that a majority of the dark inclusions are from the Leatherwood suite of rocks.

The late Oligocene age suggested by the 26-m.y. Rb-Sr isochron is supported by abundant isotopic data. Single biotite K-Ar ages of 25.6 m.y. in the Santa Catalina Mountains and 28.0 m.y. in the Tortolita Mountains (Damon and others, 1963) for the Catalina pluton are semiconcordant with a single K-Ar biotite age of 23.8 m.y. and concordant K-Ar biotite-hornblende pairs of 23.9 m.y. (biotite) and 23.7 m.y. (hornblende), 24.7 m.y. (biotite) and 27.9 m.y. (hornblende), and 21.1 m.y. (biotite) and 21.6 m.y. (hornblende) for the Catalina quartz monzonite reported by Creasey and others (1977). The last concordant pair is from the Catalina quartz monzonite in the Tortolita Mountains. Similarly, biotite K-Ar ages from the Tortolita pluton of 24.5 m.y. in the Santa Catalina Mountains (Damon and others, 1963) and 22.7 m.y. in the Tortolita Mountains (Creasey and others, 1977) strongly overlap with those of the Catalina intrusion and suggest temporal equivalence. The notion of temporal equivalence is further supported by numerous fission-track ages of 30.0, 28.3, and 28.0 m.y. on sphene; 28.9, 27.1, and 25.9 m.y. on zircon; and 23.5, 20.8, and 21.7 m.y. on apatite for the Catalina quartz monzonite and 18.5 and 17.0 m.y. on apatite from the Tortolita quartz monzonite reported by Creasey and others (1977). With the exception of apatite, which represents final cooling of Catalina and Tortolita intrusions, the fission-track ages are all concordant with the K-Ar ages.

Any doubts about the age of the Catalina pluton were removed by the 27-m.y. U-Pb concordant zircon age reported by Shakel and others (1977). The U-Pb data showed that Rb-Sr "isochrons" published earlier by Shakel and others (1972) and Hoelle (1976) probably involved erroneous assumptions regarding sample selection (see comments in Table 1).

In summary, field relationships indicate that plutons of the Catalina suite are younger than rocks of the Late Cretaceous–early Tertiary Leatherwood suite and Eocene Wilderness suite. Concordant K-Ar biotite ages and hornblende-biotite pairs, sphene and zircon fission-track ages, a U-Pb zircon age, and a poorly constrained Rb-Sr whole-rock isochron require a middle Tertiary age for quartz monzonites of the Catalina suite.

IMPLICATIONS OF PLUTONIC EPISODES FOR MYLONITIC DEFORMATION

Previous sections have detailed the emplacement of three suites of plutons from 75 to 20 m.y. ago. Plutons of *each* suite have been deformed to varying degrees by distinctive, gently inclined mylonitic foliation with conspicuous lineation that plunges east-northeast and west-southwest. At least three episodes or events of mylonitization (and probably more) are recorded in relationships where undeformed parts of younger plutons cut deformed parts of older plutons.

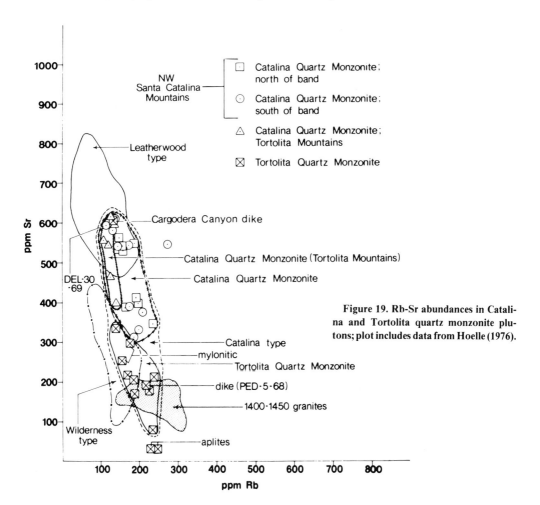

Figure 19. Rb-Sr abundances in Catalina and Tortolita quartz monzonite plutons; plot includes data from Hoelle (1976).

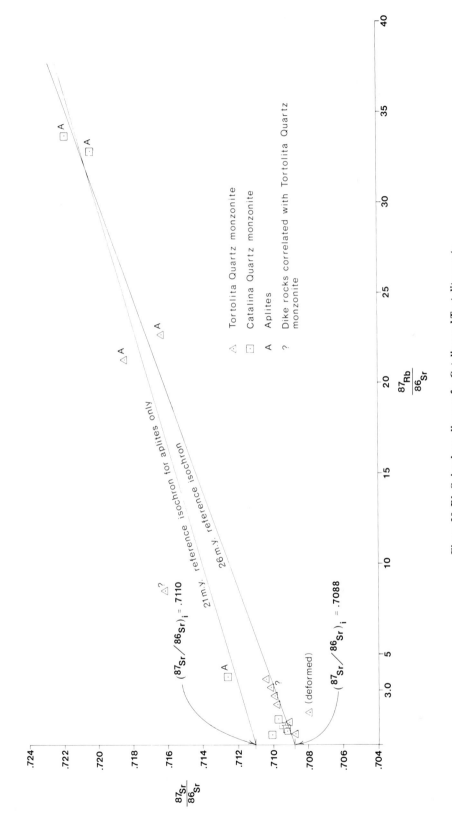

Figure 20. Rb-Sr isochron diagram for Catalina and Tortolita quartz monzonite plutons; plot includes data from Hoelle (1976); 26-m.y. reference isochron assumes Catalina and Tortolita quartz monzonite plutons are comagmatic.

The oldest mylonitic event is post–75 m.y. B.P. because it deforms Leatherwood quartz diorite. This fabric must be pre–50 m.y. because mylonitic foliation in Leatherwood is intruded in many places by dated 44- to 50-m.y.-old undeformed muscovite pegmatites east of Mount Lemmon (Hanson, 1966). In a well-exposed roadcut along the new paved-highway access to the Mount Lemmon ski area 2.5 km east-northeast of Mount Lemmon, several instructive relationships may be observed between mylonitic events in the Leatherwood quartz diorite and muscovite pegmatites of the Lemmon Rock leucogranite. Here, an older, coarser-grained mylonitic foliation is cut at a low angle by a younger, fine-grained, more intense mylonitic foliation. Two generations of pegmatites are present. The older generation consists of shallowly inclined sheets about 0.1 to 0.2 m thick which cut the older, coarser-grained mylonitic foliation but are conspicuously boudined where they cross the younger, fine-grained mylonitic foliation. The two foliation events and the shallowly inclined pegmatites are clearly crosscut by large, steeply inclined, undeformed pegmatite dikes that constitute 90% of the pegmatite exposed in the roadcut.

Similar relationships occur 20 km to the east-southeast of Mount Lemmon (Wilson, 1977; Ted Theodore, 1979, oral commun.) where undeformed pegmatite apophyses of the Wilderness pluton discordantly cut mylonite schist derived from Leatherwood. In addition, equigranular main-phase Wilderness granite may intrude mylonitic Leatherwood quartz diorite east of Green Mountain, about 12 km east-southeast of Mount Lemmon (Pilkington, 1962). The southern part of the Leatherwood pluton is continuously deformed along its east-southeast–trending margin for more than 20 km. As

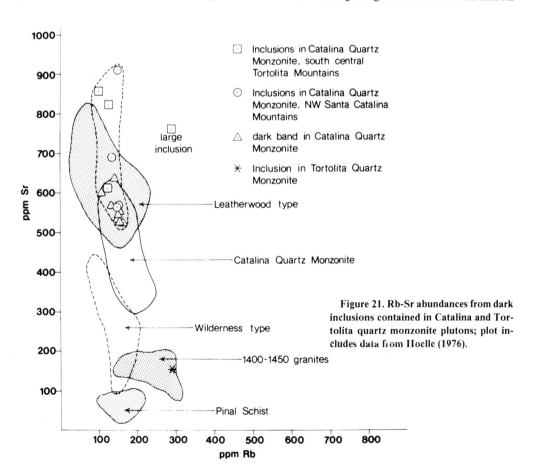

Figure 21. Rb-Sr abundances from dark inclusions contained in Catalina and Tortolita quartz monzonite plutons; plot includes data from Hoelle (1976).

such, mylonitic fabric in the Leatherwood constitutes a major pre–50 m.y. B.P. and post–75 m.y. B.P. mylonitic event in the Santa Catalina–Rincon–Tortolita crystalline complex.

Orientation of mineral lineation in mylonitic Leatherwood quartz diorite is distinct from that in main-range exposures of mylonitic Wilderness granite. Based on 116 measurements, mean lineation in Leatherwood quartz diorite exposures along the Oracle road 3 to 6 km northeast of Mount Lemmon is 16° N85°E (Tom Heidrick, 1979, written commun.). Lineation on low-angle mylonitic surfaces in the deformed Wilderness intrusion in the Windy Point and Spencer Peak areas about 5 and 15 km, respectively, southeast of Mount Lemmon is more northeasterly (about N35°E to N65°E). The difference in lineation orientation suggests that mylonitic foliation in the Leatherwood and Wilderness intrusions may have formed during two distinct episodes of cataclasis.

The next major mylonitic episode deformed the structurally lower parts of the Wilderness granite and its wall rocks in the Santa Catalina forerange. Analogous deformation may be represented by widespread mylonitic fabric in the mineralogically similar Wrong Mountain granite (Drewes, 1977) of the Rincon Mountains. This mylonitization deformed and therefore postdated the 44- to 50-m.y.-old muscovite granites but predated their cooling between 31 and 25 m.y. B.P. as defined by K-Ar and fission-track ages. Keith infers that relationships between deformed and undeformed pegmatites in the Santa Catalina forerange suggest that the mylonitic fabric exposed there formed during the emplacement of Wilderness equivalent pegmatites. Mylonitization in the forerange must have been completely over by the time of intrusion of an undeformed 21-m.y.-old trachyte dike (Shakel, 1974; Damon and others, 1980).

One of the above two episodes may be widespread throughout the northwestern Tortolita Mountains where the Chirreon Wash (Leatherwood suite) and Derrio Canyon (Wilderness suite) intrusions are strongly mylonitized—particularly so in the area of the Derrio Canyon granite sill sequence (Figs. 4 and 7). At least part of this deformation predated emplacement of the Catalina pluton as evidenced by large, strongly deformed and lineated inclusions of quartzite, Pinal Schist, stretched-pebble metaconglomerate, and Oracle Granite in relatively less-deformed Catalina quartz monzonite. Presence of mylonitic inclusions of Leatherwood quartz diorite in Catalina quartz monzonite in the Santa Catalina Mountains (Suemnicht, 1977) also indicates a mylonitic event that predated emplacement of the Catalina quartz monzonite.

In the southwestern Tortolita Mountains, the youngest episode of mylonitization clearly affects to varying degrees the Catalina quartz monzonite and the entire western half of the Tortolita quartz monzonite. This episode must postdate the two ~26-m.y.-old plutons but predate cooling of the mylonitic rocks between 17 and 20 m.y. B.P. as defined by fission-track apatite ages. An additional minimum age for this episode is provided by a series of northwest-trending dikes which discordantly intrude mylonitic foliation. One of these dikes which cuts mylonitic Chirreon Wash granodiorite has yielded a 24-m.y. K-Ar biotite date (Table 1). Thus, a significant mylonitic episode is bracketed very close to 25 m.y. B.P. In the South Mountains near Phoenix, Reynolds and Rehrig (this volume) have documented almost exactly the same age for mylonitic fabric that deforms a pluton which resembles phases of Tortolita quartz monzonite in rock type, texture, and style of mylonitic deformation.

In the Tortolita Mountains, several *events* of mylonitization are recognizable within the youngest *episode* of mylonitization. For example, near the mouth of Wild Burro Canyon, strongly mylonitized inclusions of Oracle Granite that were deformed during an earlier episode are included in the coarsely porphyritic phase of Catalina quartz monzonite. This phase in turn contains a younger, much weaker mylonitic foliation. These are both crosscut by low-dipping sheets of the granitic phase of Tortolita quartz monzonite which has been strongly mylonitized by a still-younger mylonitic event. This youngest mylonitic event may be related to a widespread set of shallow-dipping, relatively wide-spaced shears that cut older, more steeply inclined foliations (some of which are mylonitic) in the Chirreon and Catalina intrusions.

SUMMARY AND CONCLUSIONS

Various rocks within the Santa Catalina–Rincon–Tortolita crystalline complex can be correlated with rocks inside and locally outside of the complex by utilizing field relationships and lithologic, trace-element, and isotopic analyses. Deformed Precambrian, Paleozoic, Mesozoic, and Cenozoic rocks all occur within the complex, but its geology is *dominated* by a series of Late Cretaceous(?) through middle Tertiary plutonic and deformational episodes. Major plutonism was apparently episodic and produced three distinct ages and suites of intrusions. The Laramide (75 to 64 m.y. B.P.) quartz diorite and granodiorite (Leatherwood suite) were emplaced earliest. These were followed by Eocene (44 to 50 m.y. B.P.) muscovite granite, pegmatite, and alaskite (Wilderness suite). Finally, quartz monzonites (Catalina suite) were intruded in middle Tertiary time (27 to 25 m.y. B.P.). At least three episodes of mylonitic deformation occurred between 75 and 20 m.y. B.P. Although the plutonism was episodic, it cannot be definitively demonstrated that the mylonitic deformation was also episodic rather than part of a prolonged continuum. However, the close spatial and temporal association of mylonitization with plutonism suggests a genetic relationship between the two and therefore supports episodicity of mylonitization.

ACKNOWLEDGMENTS

This paper benefited greatly from conversations and communication with many geologists who unfortunately cannot all be mentioned. Discussions with colleagues D. W. Shakel, G. H. Davis, N. G. Banks, W. A. Rehrig, H. W. Peirce, P. J. Coney, E. J. McCullough, E. B. Mayo, S. C. Creasey, C. H. Thorman, T. G. Theodore, H. Drewes, P. Meyers, R. T. Budden, G. Suemnicht, P. Anderson, J. Hoelle, and C. W. Kiven particularly contributed to the development of concepts presented herein. We are especially grateful to E. J. McCullough, P. J. Coney, and G. H. Davis for their advice and encouragement. We also express appreciation to numerous students who have provided laboratory assistance during the past 18 yr. The quality of the manuscript was greatly improved by thoughtful reviews of N. G. Banks, S. C. Creasey, R. L. Armstrong, R. E. Zartmann, and M. Silberman. Unpublished K-Ar and Rb-Sr data which greatly influenced many of our geochronologic interpretations were generously provided by S. C. Creasey, R. F. Marvin, and H. Drewes of the U.S. Geological Survey, D. W. Shakel of Pima College, P. Anderson of the University of Arizona, and S. M. Soderman of the Oracle Ridge Mining Company. T. L. Heidrick of Gulf Minerals, Inc., kindly provided chemical data for the Leatherwood quartz diorite. Rb-Sr and K-Ar isotopic work performed at the Laboratory of Isotope Geochemistry, University of Arizona, was funded by National Science Foundation Grant EAR78-11535, Atomic Energy Commission contract AT(11-1)-689, and the state of Arizona. We thank J. LaVoie and R. DuPont of the Bureau of Geology and Mineral Technology and S. Harralson and B. Bible for their assistance in preparing this manuscript.

APPENDIX 1. NOMENCLATURE FOR PLUTONIC ROCKS IN THE SANTA CATALINA-RINCON-TORTOLITA CRYSTALLINE COMPLEX

INTRODUCTION

This appendix compares our preferred terminology* for what we believe are, in essence, plutonic rocks in the Santa Catalina-Rincon-Tortolita crystalline complex with nomenclature of other past and current workers. The discussion is arranged from the oldest to youngest rock unit. The number in parentheses that follows certain rock names corresponds to the plutons numbered in Figure 3. The subsections entitled "Location" designate the geographic area (also refer to Fig. 3) where *we* are using the preferred name. The subsections entitled "Comments" present historical notes and commentary about previous or current usages by other workers for the rock within the geographic area previously outlined.

The interested reader will quickly notice that all of our nomenclature lacks the term "gneiss." Although we recognize the presence of widespread areas of mylonite gneiss within the crystalline core of the Santa Catalina-Rincon-Tortolita complex, we are confident that in most cases, the igneous protolith that has been texturally modified by the mylonitization event(s) can be recognized. Future mapping in the complex should attempt to identify protoliths as closely as possible. Areas of mylonite overprint (textural modification of the protolith) should also be identified. The Mount Lemmon quadrangle mapped by Banks (1976) and Figure 3 of this paper are two approaches to the overprint problem. Banks (1976) enclosed the protolith symbol in parentheses to indicate areas where the protolith was strongly mylonitized. Figure 3 shows mylonitic or deformed areas by aligned linear patterns and areas of nonmylonitic or nondeformed igneous rocks by random patterns. These techniques avoid the impression imparted by previous maps (for example, see Wilson and others, 1969) that areas mapped as "gneiss" represent a unique rock unit. This in turn leads to the false generalization that the "gneiss" units are different from adjacent rocks and have a different origin. The emerging view instigated by Creasey and Theodore (1975), Banks (1976), Creasey and others (1977), and this paper is that the gneisses for the most part represent areas where the plutonic terrane is merely deformed. Thus, contacts between areas of mylonite gneiss and nonmylonitic counterparts are gradational as depicted in Figure 3.

MIDDLE PROTEROZOIC (1.7 TO 1.4 B.Y. B.P.) PLUTONIC ROCKS

Madera Diorite

Location. Exposures in Pinal Mountains and northern Dripping Spring Mountains, 60 km north of the Santa Catalina-Rincon-Tortolita complex.

Comments. Named by Ransome (1903) for exposures near Mount Madera in the Pinal Mountains. Intrusion in northern Tortolita Mountains, 6 km north of the crystalline core of the Santa Catalina-Rincon-Tortolita complex was questionably correlated with Madera Diorite by Banks and others (1978). Mylonitized diorites in the Santa Catalina forerange may correlate with Madera Diorite.

Johnny Lyon Granodiorite*

Location. Large half-moon-shaped exposure in Johnny Lyon Hills 20 km east of the Rincon Mountain part of the complex. Also widely exposed in the eastern Rincon Mountains in and east of Happy Valley according to Silver (1978).

Comments. Named by Silver (1955) for large exposure in western Johnny Lyon Hills and formally adopted by Cooper and Silver (1964). Silver (1978) indicated that exposures in the Happy Valley area of the eastern Rincon Mountains mapped by Drewes (1974) as Rincon Valley Granodiorite, are instead Johnny Lyon equivalent and recommended that the term "Rincon Valley Granodiorite" be discontinued for rocks in this area. Catanzaro and Kulp (1964) suggested that highly discordant zircons extracted from granitic gneiss at Windy Point in the Santa Catalina main range (viewed by us as mylonitically deformed Wilderness granite) were the same age (about 1,625 m.y.) as those extracted from Johnny Lyon Granodiorite by Silver and Deutsch (1963). We place more significance on the lower intercept age of 50 m.y. as the age of intrusion for Wilderness granite and consider the discordance to be due to contamination by older zircons of a possible 1,450-m.y. age (see discussion in text).

---

*The rock names that we prefer for current map usage within the complex are denoted by an asterisk the first time they are mentioned.

Plutons between 1,400 and 1,450 M.Y. Old

Oracle Granite*

Location. Large west-northwest–elongated exposure at Oracle, Arizona, and numerous scattered exposures north of the complex. Numerous mylonitic equivalents within the Santa Catalina and Tortolita parts of the complex are correlated with Oracle Granite by Tolman (1914, unpublished manuscript), Creasey and others (1977), Shakel and others (1977), and this paper.

Comments. Originally named by Tolman (1914, unpublished manuscript, as referred to by Moore and Tolman in their 1938 unpublished manuscript) for exposures at Oracle, Arizona. (The 1938 unpublished manuscript by B. N. Moore and C. F. Tolman is untitled and contains 131 single-spaced pages in elite-sized type. The manuscript is available for inspection at the Arizona Bureau of Geology and Mineral Technology, 845 N. Park Avenue, Tucson, Arizona 85719.) First published reference is Peterson (1938). Name has been subsequently retained for rock in type area by all subsequent workers (for example, Banerjee, 1957; Creasey, 1967). Mylonitic equivalents have recently been mapped in the Santa Catalina forerange by Creasey and Theodore (1975), Banks (1976), Creasey and others (1977). This correlation was confirmed by U-Pb studies reported by Shakel and others (1977). Previous workers had included these mylonitic gneisses as dark bands in the Catalina gneiss or forerange banded gneiss complex and assigned various names to some of the larger bands (for example, Peterson, 1968; Shakel, 1974).

Ruin Granite

Location. Widespread exposures north and northwest of Globe, Arizona, 100 km north of the complex.

Comments. Named by Ransome (1903) for exposures in Ruin Basin 20 km northwest of Globe. Oracle Granite and Ruin Granite, if not the same rock, are the same age (Livingston and others, 1967; Silver, 1968).

Continental Granodiorite*

Location. Large north-trending exposure on west side of northern Santa Rita Mountains 20 km south of the complex. Mapped as a widespread unit by Drewes (1974, 1977) throughout Rincon and Tanque Verde parts of the complex.

Comments. Named by Drewes (1968) for exposures in the northern Santa Rita Mountains where the rock is predominantly porphyritic biotite granodiorite and quartz monzonite with small areas of hornblende-bearing granodiorite and quartz monzonite. Drewes assigned an imprecise age of 1,450 to 1,700 m.y. to the Continental Granodiorite in its type area in the northern Santa Rita Mountains. This age was based on what he regarded as equivocal ages from discordant K-Ar, Rb-Sr, and Pb-α data ranging from 55 m.y. to 1,450 m.y. on coarsely porphyritic samples (see Table 1). Drewes (1976b) regarded all five ages as reduced ages and assigned a possible true age of 1,600 to 1,700 m.y. Field and U-Pb studies reported by Silver (1978) revealed that at least two intrusive phases are present, "one 1,430 m.y. old and the other more than 1,600 m.y. old" (Silver, 1978, p. 162). Most of the Continental Granodiorite of Drewes is coarsely porphyritic. This fact combined with two imprecise 1,360- and 1,450-m.y. Pb-α ages reported by Drewes (1968) and the 1,430-m.y. U-Pb age reported by Silver (1968) indicate that most of the Continental Granodiorite in the type area is probably of the 1,400- to 1,450-m.y.-old pluton generation which has extensively intruded the Johnny Lyon Granodiorite and/or rocks of similar age (1,600 m.y.). For the above reasons, we suggest that the term "Continental Granodiorite" be restricted in the Santa Rita, northern Empire, and Rincon Mountains to coarsely porphyritic biotite quartz monzonite and its mylonitic equivalents that can be shown to have a 1,400- to 1,450-m.y. ancestry. "Continental Granodiorite" in this sense is correlative to "Oracle Granite" of the Santa Catalina and Tortolita parts of the complex.

Rincon Valley Granodiorite*

Location. Several large outcrop areas mapped by Drewes (1974, 1977) on both east and west sides of the Rincon Mountains, adjacent to but not within the crystalline core of the complex. As used by us, "Rincon Valley Granodiorite" is restricted to exposures in Rincon Valley on the southwest side of the Rincon Mountains.

Comment. Exposures in Rincon Valley were originally referred under the term "Rincon granite" by Moore and Tolman (1938, unpub. manuscript). Referred to as Rincon Valley Granite by Moore and others (1949) and Acker (1958),who assigned a Cretaceous-Tertiary (Laramide) age to the rock. Renamed Rincon Valley Granodiorite, assigned a middle Proterozoic age on the basis of K-Ar radiometric data, and extended to outcrops in the eastern Rincon Mountains by Drewes (1974, 1977). Exposures on the east side of the Rincon Mountains are probably Johnny Lyon Granodiorite as suggested by Silver (1978). Exposures in Rincon Valley are also petrographically similar to Johnny Lyon Granodiorite, but it is not clear whether they are from the same pluton. Hence, we suggest "Rincon Valley Granodiorite" be limited to Rincon Valley exposures pending further work. The coarsely porphyritic quartz monzonite, which was mapped by Drewes (1977) as Rincon Valley Granodiorite(?) northwest of Saguaro

National Monument in the Tanque Verde Mountains, mineralogically and texturally resembles the 1,400- to 1,450-m.y.-old granitic suite. Three K-Ar ages of 1,380, 1,390, and 1,400 m.y. and one 1,400-m.y. Rb-Sr model age are essentially concordant and are alternatively interpreted by us as an emplacement age. Also, the single Rb-Sr analysis reported by Marvin and Cole (1978) plots very near the composite 1,380-m.y. Rb-Sr isochron for Ruin Granite and Oracle Granite of the 1,400- to 1,450-m.y.-old granitic clan (Fig. 14).

POST–PALEOZOIC BATHOLITHIC ROCKS

Laramide Quartz Diorite–Granodiorite (Leatherwood) Suite

Rice Peak granodiorite porphyry* (1)

Location. Two stocklike masses intruding Apache Group rocks in the northern Santa Catalina Mountains between the Mogul and Geesman faults.

Comments. Described under unnamed "andesite" by Moore and Tolman (1938, unpub. manuscript) and Moore and others (1949). Eastmost mass shown as Laramide volcanics on the "Arizona Highway Geologic Map" compiled by Cooley (1967). Mapped as "meta-diorite" by Wallace (1954) and as "granodiorite porphyry" by Creasey (1967) and Suemnicht (1977). Granodiorite porphyry at San Manuel mapped as same rock by Creasey (1967). Waag (1968) and Creasey and Theodore (1975) suggested that the rocks we call Rice Peak granodiorite and Leatherwood quartz diorite may be closely related if not the same rock.

Leatherwood quartz diorite* (2)

Location. Subcircular stocklike mass in central Santa Catalina Mountains continuous with and gradational into several mylonitized sills in the northeast Santa Catalina Mountains. May be present in Rincon Mountains although it is not recognized by current mapping (Drewes, 1974, 1977).

Comment. Originally named by Tolman in an unpublished 1914 manuscript for exposures near the Leatherwood Mine group in the north-central Santa Catalina Mountains. Referred to as unnamed "quartz diorite and related schistose rock" by Moore and others (1949) and is accompanied by map symbology "Tld" and "Tldp." Name was first used in print by Bromfield (1952) and subsequently used by Peirce (1958), DuBois (1959a, 1959b), Wood (1963), Hanson (1966), Braun (1969), Matter (1969), Banks (1976), Creasey and others (1977), Suemnicht (1977), and Banks (this volume). Included within "TKm" unit of Cooley (1967) (see comment under Wilderness granite for more discussion). Pilkington (1962) and Waag (1968) mapped the schistose rock as quartz latite porphyry which Waag (1968) correlated with Leatherwood quartz diorite. Creasey and Theodore (1975) mapped the unit as unnamed quartz diorite but acknowledged the local use of the term "Leatherwood."

Chirreon Wash granodiorite* (3)

Location. East-elongated intrusion in the north-central Tortolita Mountains.

Comments. Mapped as a composite granodiorite, quartz diorite, monzodiorite, and diorite intrusion under the heading "quartz monzonite of Samaniego Ridge and related rocks" by Banks and others (1977). Named granodiorite of Chirreon Wash by Banks and others (1978) and adopted here as Chirreon Wash granodiorite.

Eocene Muscovite Granite (Wilderness) Suite

Derrio Canyon granite* (4) and Related Pegmatites

Location. Sheetlike mass of muscovite-garnet granite and related pegmatite dikes and sheets in the northwest to north-central Tortolita Mountains.

Comments. Main mass mapped as mylonitic "quartz monzonite of Samaniego Ridge" by Banks and others (1977). Pegmatite, aplite, and alaskite apophyses which extend eastward into and intrude Chirreon Wash granodiorite mapped under the heading "quartz monzonite of Samaniego Ridge" as a "pegmatite complex" by Banks and others (1977). Banks (this volume) refers to the body that is co-extensive with our "Derrio Canyon granite" as Cottonwood Canyon gneiss and infers that it was metamorphically derived from the Chirreon Wash granodiorite.

Wilderness granite* and Related Lemmon Rock leucogranite* (5)

Location. Large sill-like or laccolithic mass which extends westward from upper Sabino Canyon (4 km south of Mount Lemmon) through the Wilderness of Rocks, 3 km south of Mount Lemmon to Romero Pass and Romero Canyon. Mylonitized varieties extend southwesterly and south into the Santa Catalina forerange where they are a main component of the forerange mylonitic gneiss complex. A related assemblage of pegmatites, aplites, and alaskites (Lemmon Rock leucogranite) occurs along the north margin of the pluton. Wilderness granite includes biotite-rich phases (for example, a biotite-rich phase in the Spencer Campground area about 5 km east-southeast of Mount Lemmon). Wilderness pluton extends eastward from upper Sabino Canyon through Molino Canyon (9 km southeast of Mount Lemmon) to the Bellota Ranch region (15 km southeast of Mount Lemmon). East of Bellota Ranch, muscovite granite outcrops that are continuous with Wilderness exposures referred to as Youtcy granite (see below) may in fact be the eastward extension of the Wilderness intrusion.

Comments. The area for which we are using the term "Wilderness Granite" contains a subarea (Santa Catalina main range) composed mainly of granitic rock referred under the term "Catalina batholith" by Moore and Tolman (1938, unpub. manuscript). Mylonitic varieties in the Santa Catalina forerange were referred under the term "Catalina gneiss." The contact between the two units was originally mapped by Tolman (1911 and 1912) and has appeared on most subsequent pre-1976 regional maps (beginning with the 1924 Arizona state geologic map by Darton and others) as a west-northwest–trending, solid-line contact. Interestingly, the map contained in the 1949 open-file report by Moore and others carries the notation "scratch contact" for the west-northwest–trending contact. The text contains the admonition, "it should be noted that in general there is no distinct boundary between the two kinds of rocks" (p. 12). Moore and others (1949) included all exposures in the Santa Catalina main range and forerange under the umbrella term "Santa Catalina granitic complex," which consisted of three subdivisions: (1) a gneissic variety in the forerange south of the west-northwest–trending contact which contained a leucocratic, medium- to coarse-grained rock resembling a sheared granite (mostly our deformed Wilderness granite), and a porphyritic, mesocratic, coarse-grained rock resembling a sheared granite (mostly the deformed Oracle Granite of Creasey and others, 1977; Banks, 1976 and this volume; this paper); (2) a granitic gneiss in the Santa Catalina main range (equivalent to mylonitic Wilderness granite); and (3) a muscovite granite in the Santa Catalina main range (equivalent to nonmylonitic, undeformed Wilderness granite). Peirce (1958) mapped the granitic gneisses of the main range near Mount Lemmon as unnamed remobilized Precambrian granite. DuBois (1959a, 1959b) included all gneissic varieties of the Santa Catalina main range and forerange under the term "Catalina gneiss." This usage was adopted by Pilkington (1962), McCullough (1963), Mayo (1964), Hanson (1966), Pashley (1966), Peterson (1968), Waag (1968), Matter (1969), Sherwonit (1974), and Shakel (1972, 1974). As did Bromfield (1952), these workers informally recognized textural differences between the forerange gneisses and main range gneisses by informal terms such as forerange banded gneiss and main range granitic gneiss. Some of the larger leucocratic layers of the forerange gneisses were given various names by Peterson (1958) and Shakel (1974). All of the above workers believed that the gneisses were derived metamorphically from a sedimentary protolith generally thought to be older Precambrian in age.

"Catalina gneiss" nomenclature as originally envisioned by DuBois (1959a, 1959b) reached its maximum usage on the 1:1,000,000 "Arizona Highway Geologic Map" compiled by Cooley (1967). On this map all gneissic rocks in the Santa Catalina main range and forerange are designated "TKm," which is captioned "mainly schist and gneiss, including the Catalina Gneiss." This map does not show the Tolman contact discussed above. The "TKm" unit of Cooley (1967) includes *all* of the following units shown in Figure 3 of this paper: Leatherwood quartz diorite, Wilderness granite; Wrong Mountain granite, and mylonitic Oracle Granite and Continental Granodiorite.

Comparatively undeformed varieties of muscovite granite in the Santa Catalina main range were renamed "Wilderness granite" for exposures in the Wilderness of Rocks 3 km south of Mount Lemmon by Doug Shakel around 1972 (1975, oral commun.). Name first appeared in print in Budden (1975). Budden's Wilderness granite is co-extensive with the "TKg" unit outlined on the 1969 Arizona state geologic map by Wilson and others. Unit was redesignated "quartz monzonite of Samaniego Ridge" by Creasey and Theodore (1975) and Banks (1976, 1977). These workers dissolved Tolman's long-standing contact and mapped large sill-like apophyses of deformed Samaniego quartz monzonite into the Santa Catalina forerange. Rocks mapped as quartz monzonite of Samaniego Ridge in the Santa Catalina main range and forerange were regarded by Creasey and others (1977) as a metamorphic differentiate of a porphyritic sphene-bearing hornblende-biotite quartz monzonite on Samaniego Ridge in the northwest Santa Catalina Mountains, 5 km northwest of Mount Lemmon (see discussion under Catalina quartz monzonite). The term "Wilderness granite" has been continued by Shakel and others (1977), Shakel (1978), and this paper because of textural and mineralogic differences between rocks on Samaniego Ridge and in the Wilderness of Rocks. Alaskite and pegmatite unit has been informally referred to as "Lemmon Rock intrusive" by Hoelle (1976) and "Lemmon Rock leucogranite" by Shakel (1978). Banks (this volume) has named the area co-extensive with the "Santa Catalina granitic complex" of Moore and others (1949) the "gneiss of Windy Point," which includes all of our Wilderness and Wrong Mountain granite.

Espiritu Canyon granite* (6) and Associated Aplites

Location. East-northeast–elongated pluton exposed on the east end of the Tanque Verde Mountains.

Comments. Named by Thorman and Drewes (1978) for exposures 3 km northeast of Mica Mountain in the

Tanque Verde–Rincon Mountains and assigned a Tertiary age. Southwest one-half of pluton mapped by Drewes (1977) as Wrong Mountain Quartz Monzonite and assigned an older Precambrian age. Thorman and Drewes (1978) view Espiritu Canyon pluton as a Tertiary remobilization of Precambrian Wrong Mountain Quartz Monzonite. Redesignated as quartz monzonite of Samaniego Ridge by Banks (this volume), who shows it to be undeformed and have gradational contacts with gneiss of Windy Point (Wrong Mountain quartz monzonite of Drewes, 1977). In our view, the Espiritu Canyon pluton may be an undeformed variety of Wrong Mountain granite. The two bodies may be part of a single Eocene muscovite granite pluton.

Wrong Mountain granite* (7)

Location. Large sheetlike laccolithic mass widespread throughout the Tanque Verde and Rincon Mountains.

Comments. Named Wrong Mountain Quartz Monzonite by Drewes (1977) for exposures at Wrong Mountain 3 km south of Rincon Peak in the Rincon Mountains and assigned an older Precambrian age. It is a major map unit of Drewes (1974, 1977) and Thorman and Drewes (1978) throughout the Rincon and Tanque Verde Mountains. Considered by Drewes (1977) as the leucocratic component of formational rank in his proposed "Santa Catalina Group" terminology, which is equivalent to Catalina gneiss terminology of DuBois (1959a, 1959b) for the Santa Catalina Mountains. If the Wrong Mountain mass is instead an Eocene pluton as proposed here, the Santa Catalina Group terminology as proposed by Drewes (1977) no longer has any useful meaning and should be discontinued. The entire unit has been included under "gneiss of Windy Point" nomenclature of Banks (this volume).

Youtcy granite* (8)

Location. As provisionally used by us, "Youtcy granite" applies to an igneous body of undetermined shape and size underlying the Youtcy Ranch area in Redington Pass. This body may be an undeformed extension of "Wilderness granite" farther west.

Comments. Name originally given by Tolman and Moore (1938, unpub. manuscript) to nongneissic exposures of muscovite granite around Youtcy Ranch in Redington Pass. Included as nongneissic component of the Santa Catalina granitic complex by Moore and others (1949). Shown on the map in Bromfield (1952) as "Catalina granite." An extension of the Youtcy mass in the Korn Kob mine area north of Redington Pass was mapped as "Catalina granite" by Wilson (1977). Entire area assigned to the Youtcy granite was mapped as quartz monzonite of Samaniego Ridge by Creasey and Theodore (1975) and Creasey and others (1977). Redesignated "quartz monzonite of Youtcy Ranch" by Thorman and Drewes (1978), who showed continuous exposure into but no contact with "quartz monzonite of Samaniego Ridge west and southwest of the Youtcy ranch. Thorman and Drewes (1978) considered the Youtcy pluton to be a remobilized phase of the "quartz monzonite of Samaniego Ridge" (in this area, equivalent to our Wilderness granite).

Middle Tertiary Quartz Monzonite (Catalina) suite

Catalina quartz monzonite* (9)

Location. Large half-circle–shaped exposure in the northwest Santa Catalina Mountains. Large east-northeast–trending elongate mass in the south-central Tortolita Mountains.

Comments. Originally mapped as Oracle Granite of older Precambrian age by Tolman [unpub. 1914 manuscript, cited in 1938 unpub. manuscript by Moore and Tolman (see comments on Oracle Granite)]. Similar treatment and citation is given by Moore and others (1949) and Bromfield (1952). Renamed Samaniego granite by Wallace (1954) on the basis of mineralogic differences of rocks exposed in the vicinity of Samaniego Ridge 5 km northwest of Mount Lemmon from Oracle Granite in the type area. Wallace retained the Precambrian age. Renamed Catalina granite by McCullough (1963), who assigned a post-Paleozoic age to the body based on a rugose coral in a carbonate inclusion. Eastern contact of Catalina quartz monzonite with Oracle Granite was first recognized by Hoelle about 1974 and more precisely located by the mapping of Banks (1976) and Suemnicht (1977). Appeared as Precambrian granite on the Arizona state geologic map by Wilson and others (1969), which used Tolman's mapping. "Catalina granite" nomenclature continued by Shakel and others (1972), Budden (1975), and Hoelle (1976), who all assigned a Late Cretaceous age to the body on the basis of improperly interpreted Rb-Sr isotopic data (see Table 1). Shown as a Laramide intrusion ("Li") on the "Arizona Highway Geologic Map" compiled by Cooley (1967). Renamed quartz monzonite of Samaniego Ridge by Banks (1976). Samaniego term was extended by Creasey and Theodore (1975), Banks (1976), Creasey and others (1977), Banks and others (1978) to include many presumed Phanerozoic plutonic bodies in the Tortolita, Durham-Suizo, southwest Tortilla, and Santa Catalina ranges. "Catalina granite" term was continued for the body in the northwest Santa Catalina Mountains by Suemnicht (1977), Shakel and others (1977), and Shakel (1978). Banks (this volume) has discontinued the Samaniego nomenclature for plutons in the Tortolita Mountains. As used by Banks (this volume), "quartz monzonite of Samaniego Ridge" is restricted to exposures in northwest Santa Catalina Mountains (equivalent to our Catalina quartz monzonite) and exposures around Youtcy Ranch in Redington Pass and Espiritu Canyon (in our view probable Wilderness and Wrong Mountain equivalents, respectively) northeast of Mica Mountain in the Tanque Verde Mountains. The east-northeast–trending mass in the south-central Tortolita Mountains, which is correlated by us with the Catalina quartz monzonite body in the northwest Santa Catalina Mountains, is designated by Banks (this volume) as "quartz monzonite of Wild Burro Canyon." On the basis of impressively similar texture, modal mineralogy, Rb-Sr trace-element data, and Sr isotopes of Tortolita exposures with the Catalina quartz monzonite in the northwest Santa Catalina Mountains, we believe that the Wild Burro Canyon term seems inappropriate.

Tortolita quartz monzonite* (10)

Location. Large east-northeast–elongated pluton in the southern Tortolita Mountains and large northwest-trending dikes in the western Santa Catalina Mountains.

Comments. Originally named Tortolita granite by Moore and Tolman (1938, unpub. manuscript) for exposures in the southwest Tortolita Mountains. Included by Moore and others (1949) as a pluton member of the "Santa Catalina granitic complex." Mapped by Budden (1975) as Tortolita granodiorite and later as quartz monzonite of the Tortolita Mountains by Banks and others (1976), Creasey and others (1977), and Banks and others (1977). Dike exposures in Santa Catalina Mountains originally mapped as "Cargadero [Cargodera] Canyon granite" by McCullough (1963) and later as quartz monzonite of the Tortolita Mountains by Banks (1976).

Reef of Rock granite* (11)

Location. North-northeast–trending dikelike mass of leucocratic granite along Reef of Rock ridge 2 km north of Mount Lemmon. Scattered dikes of biotite granite which intrude Tortolita quartz monzonite (but not differentiated from Tortolita quartz monzonite during reconnaissance mapping) in the southwest Tortolita Mountains may equate with Reef of Rock granite.

Comments. Exposures at Reef of Rock originally mapped and named "Mount Lemmon granite" by Moore and Tolman (1938, unpub. manuscript), who indicated the body intruded Oracle Granite (Catalina quartz monzonite) and Leatherwood quartz diorite. Included by Moore and others (1949) as a nongneissic granite within the "Santa Catalina granitic complex." Shown as "Catalina granite" on the maps of Bromfield (1952), Peirce (1958), Hanson (1966), and Waag (1968). Shown on the 1924 (Darton and others) and 1969 (Wilson and others) Arizona state maps as "Kg" and "TKg," respectively. Renamed "Reef granite" by Budden (1975). Included within the quartz monzonite of Samaniego Ridge unit of Banks (1976), Creasey and others (1977), and Banks (this volume). Renamed "Reef of Rock granite" by Suemnicht (1977), who showed the rock to be a distinct intrusive phase into Catalina quartz monzonite and Leatherwood quartz diorite.

Happy Valley quartz monzonite (12)

Location. Two stocklike masses within the eastern Rincon Mountains.

Comments. Named "Happy Valley quartz monzonite" by Miles (1965) for exposures near Barney Ranch south of Lechequilla Peak (8 km east of Mica Mountain) in the eastern Rincon Mountains. Exposures near Barney Ranch referred to as granodiorite of Happy Valley by Drewes (1974, 1977), who also mapped a large mass to the south in the Little Rincon Mountains (8 km east of Rincon Peak) under the same name. Northern mass is mineralogically similar to (Table 5) and may belong to the muscovite granite (Wilderness) suite.

REFERENCES CITED

Acker, C. J., 1958, Geologic interpretations of a siliceous breccia in the Colossal Cave area, Pima County, Arizona [M.S. thesis]: Tucson, University of Arizona, 50 p.

Banerjee, A. R., 1975, Structure and petrology of the Oracle Granite, Pinal County, Arizona [Ph.D. dissert.]: Tuscon, University of Arizona, 112 p.

Banks, N. G., 1976, Reconnaissance geologic map of the Mount Lemmon quadrangle, Arizona: U.S. Geological Survey Map MF-747.

—— 1977, Geologic setting and interpretation of a zone of middle Tertiary igneous-metamorphic complexes in south-central Arizona: U.S. Geological Survey Open-File Report 77-376.

—— 1980, Geology of a zone of metamorphic core complexes in southeastern Arizona: Geological Society of America Memoir 153 (this volume).

Banks, N. G., and others, 1972, Chronology of intrusion and ore deposition at Ray, Arizona: Part I, K-Ar ages: Economic Geology, v. 67, p. 864–878.

Banks, N. G. and others, 1977, Reconnaissance geologic map of the TortoMountains quadrangle: U.S. Geological Survey Miscellaneous Field Studies Map MF-864.

—— 1978, Radiometric and chemical data for rocks of the Tortolita Mountains 15' quadrangle, Pinal County, Arizona: Isochron/West, no. 22, p. 17–21.

Blake, W. P., 1908a, Geological sketch of the region of Tucson, Arizona, in MacDougal, D. T., ed., Botanical features of North American deserts: Carnegie Institution of Washington Publication 99, p. 45–68.

—— 1908b, Note upon the Santa Catalina gneiss, Arizona: Science, n.s., v. 28, p. 379–380.

Braun, E. R., 1969, Geology and ore deposits of the Marble Peak area, Santa Catalina Mountains, Pima County, Arizona [M.S. thesis]: Tucson, University of Arizona, 75 p.

Bromfield, C. S., 1950, Geology of the Maudina Mine area, northern Santa Catalina Mountains, Pinal County, Arizona [M.S. thesis]: Tucson, University of Arizona, 63 p.

—— 1952, Some geologic features of the Santa Catalina Mountains, in Guidebook for field trip excursions in southern Arizona: Arizona Geological Society Digest, v. 5, p. 51–55.

Bryant, D. L., 1968, Diagnostic characteristics of the Paleozoic formations of southeastern Arizona: Arizona Geological Society Guidebook III, p. 33–49.

Budden, R. T., 1975, The Tortolita–Santa Catalina Mountain complex [M.S. thesis]: Tucson, University of Arizona, 133 p.

Catanzaro, E. J., and Kulp, J. L., 1964, Discordant zircons from the Little Belt (Montana), Beartooth (Montana), and Santa Catalina (Arizona) Mountains: Geochimica et Cosmochimica Acta, v. 28, p. 87–124.

Coney, P. J., 1979, Tertiary evolution of Cordilleran metamorphic core complexes, in Armentrout, J. W., and others, eds., Cenozoic paleogeography of western United States: Society of Economic Paleontologists and Mineralogists, Pacific Section Symposium III, p. 15–28.

—— 1980, Cordilleran metamorphic core complexes: An overview: Geological Society of America Memoir 153 (this volume).

Cooley, M. E., 1967, Arizona highway geologic map: Arizona Geological Society, scale 1:1,000,000.

Cooper, J. R., and Silver, L. T., 1964, Geology and ore deposits of the Dragoon quadrangle, Cochise County, Arizona: U.S. Geological Survey Professional Paper 416, 196 p.

Creasey, S. C., 1965, Isotopic age of fresh and altered igneous rocks associated with copper deposits, southeastern Arizona [abs.]: Geological Society of America Special Paper 87, p. 39.

—— 1967, General geology of the Mammoth quadrangle, Pima County, Arizona: U.S. Geological Survey Bulletin 1218, 94 p.

Creasey, S. C., and Kistler, R. W., 1962, Age of some copper-bearing porphyries and other igneous rocks in southeastern Arizona, in Geological Survey research, 1962: U.S. Geological Survey Professional Paper 450-D, p. D1–D5.

Creasey, S. C., and Theodore, T. G., 1975, Preliminary reconnaissance geologic map of the Bellota Ranch quadrangle, Pima County, Arizona: U.S. Geological Survey Open-File Report 75-295.

Creasey, S. C., and others, 1977, Middle Tertiary plutonism in the Santa Catalina and Tortolita Mountains, Arizona: U.S. Geological Survey Journal of Research, v. 5, p. 705–717.

Damon, P. E., 1968, Application of the potassium-argon method to the dating of igneous and metamorphic rocks within the Basin-Ranges of the Southwest: Arizona Geological Society Guidebook III, p. 7–20.

—— 1970, A theory of "real" K-Ar clocks: Eclogae Geologicae Helvetiae, v. 63, p. 69–76.

Damon, P. E., and Giletti, B. J., 1961, The age of basement rocks of the Colorado Plateau and adjacent areas, in Kulp, J. L., ed., Geochronology of rock systems: New York Academy of Science Annals, v. 91, p. 443–453.

Damon, P. E., and Mauger, R. M., 1966, Epeirogeny-orogeny viewed from the Basin and Range Province: Society of Metallogenetic Engineers Transactions, v. 235, p. 99–112.

Damon, P. E., Livingston, D. E., and Erickson, R. C., 1962, New K-Ar dates for the Precambrian of Pinal, Gila, Yavapai and Coconino Counties, Arizona: New Mexico Geological Society, 13th Field Conference Guidebook, p. 56–57.

Damon, P. E., Erickson, R. C., and Livingston, D. E., 1963, K-Ar dating of Basin and Range uplift, Catalina Mountains, Arizona: Nuclear Geophysics, National Academy of Sciences/National Research Council Publication 1075, p. 113–121.

Damon, P. E., and others, 1969, Correlation and chronology of ore deposits and volcanic rocks, in Annual progress report no. 1969 C00-689 to U.S. Atomic Energy Commission: Tucson, Geochronology Labs., University of Arizona, 90 p.

—— 1980, New Rb-Sr and K-Ar data for the Santa Catalina–Rincon–Tortolita metamorphic core complex: Isochron/West (in press).

Darton, N. H., Lausen, C., and Wilson, E. D., 1924, Geologic map of the State of Arizona: Arizona Bureau of Mines and U.S. Geological Survey, scale 1:500,000.

Davis, G. A., and others, 1980, Mylonitization and detachment faulting in the Whipple-Buckskin-Rawhide Mountains terrane, southeast California and western Arizona: Geological Society of America Memoir 153 (this volume).

Davis, G. H., 1975, Gravity-induced folding off a gneiss dome complex, Rincon Mountains, Arizona: Geological Society of America Bulletin, v. 86, p. 979–990.

———1977a, Characteristics of metamorphic core complexes, southern Arizona: Geological Society of America Abstracts with Programs, v. 9, p. 944.

———1977b, Reply to Discussion on Gravity-induced folding off a gneiss dome complex, Rincon Mountains, Arizona: Geological Society of America Bulletin, v. 88, p. 1212–1216.

———1978, Third day, road log from Tucson to Colossal Cave and Saguaro National Monument: New Mexico Geological Society, 29th Field Conference Guidebook, p. 77–87.

———1979, Laramide folding and faulting in southeastern Arizona: American Journal of Science, v. 279, p. 543–569.

———1980, Structural characteristics of metamorphic core complexes, southern Arizona: Geological Society of America Memoir 153 (this volume).

Davis, G. H., and Coney, P. J., 1979, Geological development of Cordilleran metamorphic core complexes: Geology, v. 7, p. 120–124.

Davis, G. H., and Frost, E. G., 1976, Internal structure and mechanism of emplacement of a small gravity-glide sheet, Saguaro National Monument (east), Tucson, Arizona: Arizona Geological Society Digest, v. 10, p. 287–304.

Davis, G. H., and others, 1975, Origin of lineation in the Catalina-Rincon-Tortolita gneiss complex, Arizona: Geological Society of America Abstracts with Programs, v. 7, p. 602.

Drewes, H., 1968, New and revised stratigraphic names in the Santa Rita Mountains of southeastern Arizona: U.S. Geological Survey Bulletin 1274-C, p. 1–15.

———1974, Geologic map and sections of the Happy Valley quadrangle, Cochise County, Arizona: U.S. Geological Survey Map I-832, scale 1:48,000.

———1976a, Laramide tectonics from Paradise to Hells Gate, southeastern Arizona: Arizona Geological Society Digest, v. 10, p. 151–167.

———1976b, Plutonic rocks of the Santa Rita Mountains, southeast of Tucson, Arizona: U.S. Geological Survey Professional Paper 915, 76 p.

———1977, Geologic map and sections of the Rincon Valley quadrangle, Pima County, Arizona: U.S. Geological Survey Map I-998, scale 1:48,000.

Drewes, H., and Thorman, C. H., 1978, New evidence for multiphase development of the Rincon metamorphic core complex east of Tucson, Arizona: Geological Society of America Abstracts with Programs, v. 10, p. 103.

DuBois, R. L., 1959a, Geology of the Santa Catalina Mountains: Arizona Geological Society Guidebook II, p. 106–116.

———1959b, Petrography and structure of a part of the gneissic complex of the Santa Catalina Mountains: Arizona Geological Society Guidebook II, p. 117–127.

Eberly, L. D., and Stanley, T. B., 1978, Cenozoic stratigraphy and geologic history of southwestern Arizona: Geological Society of America Bulletin, v. 89, p. 921–940.

Erickson, R. C., 1962, Petrology and structure of an exposure of the Pinal Schist, Santa Catalina Mountains, Arizona [M.S. thesis]: Tucson, University of Arizona, 71 p.

Frost, E. G., 1977, Mid-Tertiary, gravity-induced deformation in Happy Valley, Pima and Cochise Counties, Arizona [M.S. thesis]: Tucson, University of Arizona, 86 p.

Giletti, B. J., and Damon, P. E., 1961, Rubidium-strontium ages of some basement rocks from Arizona and northwestern Mexico: Geological Society of America Bulletin, v. 72, p. 639–644.

Hanson, H. S., 1966, Petrography and structure of the Leatherwood quartz diorite, Santa Catalina Mountains, Pima County, Arizona [Ph.D. dissert.]: Tucson, University of Arizona, 104 p.

Hayes, P. T., 1970, Cretaceous paleogeography of southeastern Arizona and adjacent areas: U.S. Geological Survey Professional Paper 658-B, 42 p.

Hayes, P. T., and Drewes, H., 1968, Mesozoic sedimentary and volcanic rocks of southeastern Arizona, in Titley, S. R., ed., Southern Arizona guidebook III: Tucson, Arizona Geological Society, p. 49–58.

———1978, Mesozoic depositional history of southeastern Arizona: New Mexico Geological Society, 29th Field Conference Guidebook, p. 201–208.

Hernon, R. M., 1932, Pegmatite rocks of the Catalina-Rincon Mountains, Arizona [M.S. thesis]: Tucson, University of Arizona, 65 p.

Higgins, M. E., 1971, Cataclastic rocks: U.S. Geological Survey Professional Paper 687, 97 p.

Hoelle, J. L., 1976, Structural and geochemical analysis of the Catalina granite, Santa Catalina Mountains, Arizona [M.S. thesis]: Tucson, University of Arizona, 79 p.

Johnson, J. P., 1972, K-Ar dates on intrusive rocks and alteration associated with the Lakeshore porphyry copper deposit, Pinal County, Arizona: Isochron/West, no. 4, p. 29–30.

Keith, S. B., 1978, Paleosubduction geometries inferred from Cretaceous and Tertiary magmatic patterns in southwestern North America: Geology, v. 6, p. 515–521.

Livingston, D. E., 1969, Geochronology of older Precambrian rocks in Gila County, Arizona [Ph.D. dissert.]: Tucson, University of Arizona, 224 p.

Livingston, D.E., and Damon, P. E., 1968, The ages of

stratified Precambrian rock sequences in central Arizona and northern Sonora: Canadian Journal of Earth Sciences, v. 5, p. 763–772.

Livingston, D. E., and others, 1967, Argon 40 in cogenetic feldspar-mica mineral assemblages: Journal of Geophysical Research, v. 72, p. 1362–1375.

Marvin, R. F., and Cole, J. L., 1978, Radiometric ages: Compilation A, U.S. Geological Survey: Isochron/West, no. 22, p. 3–14.

Marvin, R. F., and others, 1973, Radiometric ages of igneous rocks from Pima, Santa Cruz, and Cochise Counties, southeastern Arizona: U.S. Geological Survey Bulletin 1379, 27 p.

Marvin, R. F., Naeser, C. W., and Mehnert, H. H., 1978, Tabulation of radiometric ages—including unpublished K-Ar and fission track ages—for rocks in southeastern Arizona and southwestern New Mexico: New Mexico Geological Society, 29th Field Conference Guidebook, p. 243–257.

Matter, P., 1969, Petrochemical variations across some Arizona pegmatites and their enclosing rocks [Ph.D. dissert.]: Tucson, University of Arizona, 173 p.

Mauger, R. L., Damon, P. E., and Livingston, D. E., 1968, Cenozoic argon ages on metamorphic rocks from the Basin and Range Province: American Journal of Science, v. 266, p. 579–589.

Mayo, E. B., 1964, Folds in gneiss beyond North Campbell Avenue, Tucson, Arizona: Arizona Geological Society Digest, v. 7, p. 123–145.

McCullough, E. J., Jr., 1963, A structural study of the Pusch Ridge–Romero Canyon area, Santa Catalina Mountains, Arizona [Ph.D. dissert.]: Tucson, University of Arizona, 67 p.

Miles, C. H., 1965, Metamorphism and hydrothermal alteration in the Lecheguilla Peak area of the Rincon Mountains, Cochise County, Arizona [Ph.D. dissert.]: Tucson, University of Arizona, 98 p.

Moore, B. N., and others, 1949, Geology of the Tucson quadrangle, Arizona: U.S. Geological Survey Open-File Report, 20 p.

Pashley, E. F., 1966, Structure and stratigraphy of the central, northern, and eastern parts of the Tucson Basin, Arizona [Ph.D. dissert.]: Tucson, University of Arizona, 273 p.

Peirce, F. L., 1958, Structure and petrology of part of the Santa Catalina Mountains [Ph.D. dissert.]: Tucson, University of Arizona, 86 p.

Peirce, H. W., 1976, Elements of Paleozoic tectonics in Arizona: Arizona Geological Society Digest, v. 10, p. 37–58.

Peterson, D. W., 1961, AGI data sheet 22: Geotimes, v. 5, no. 6, p. 30–36.

Peterson, N. C., 1938, Geology and ore deposits of the Mammoth mining camp area, Pinal County, Arizona: University of Arizona, Arizona Bureau of Mines Bulletin 144, 63 p.

Peterson, R. C., 1968, A structural study of the east end of the Catalina Forerange, Pima County, Arizona [Ph.D. dissert.]: Tucson, University of Arizona, 105 p.

Pilkington, H. D., 1962, Structure and petrology of a part of the east flank of the Santa Catalina Mountains, Pima County, Arizona [Ph.D. dissert.]: Tucson, University of Arizona, 120 p.

Ransome, F. L., 1903, Geology and ore deposits of the Globe-Miami district, Arizona: U.S. Geological Survey Professional Paper 342, 151 p.

—— 1916, Some Paleozoic sections of Arizona and their correlation: U.S. Geological Survey Professional Paper 98-K, p. 144–145.

Rehrig, W. A., and Reynolds, S. J., 1980, Geologic and geochronologic reconnaissance of a northwest-trending zone of metamorphic core complexes in southern and western Arizona: Geological Society of America Memoir 153 (this volume).

Reynolds, S. J., and Rehrig, W. A., 1980, Mid-Tertiary plutonism and mylonitization, South Mountains, central Arizona: Geological Society of America Memoir 153 (this volume).

Rose, A. W., and Cook, D. R., 1965, Radioactive age dates of porphyry copper deposits in the western United States [abs.]: Geological Society of America Special Paper 87, p. 139.

Scarborough, R. B., and Peirce, H. W., 1978, Late Cenozoic basins of Arizona: New Mexico Geological Society, 29th Field Conference Guidebook, p. 253–259.

Shafiqullah, M., Damon, P. E., and Peirce, H. W., 1976, Late Cenozoic tectonic development of Arizona Basin and Range Province [abs.]: International Geological Congress, 25th, Sydney, v. 1, p. 99.

Shafiqullah, M., and others, 1978, Mid-Tertiary magmatism in southeastern Arizona: New Mexico Geological Society, 29th Field Conference Guidebook, p. 231–242.

Shakel, D. W., 1972, "Older" Precambrian gneisses in southern Arizona: International Geological Congress, 24th, Montreal, section 1, p. 278–287.

—— 1974, The geology of layered gneisses in part of the Santa Catalina forerange, Pima County, Arizona [M.S. thesis]: Tucson, University of Arizona, 233 p.

—— 1978, Supplemental road log number 2: Santa Catalina Mountains via Catalina Highway: New Mexico Geological Society, 29th Field Conference Guidebook, p. 105–111.

Shakel, D. W., Livingston, D. E., and Pushkar, P. D., 1972, Geochronology of crystalline rocks in the Santa Catalina Mountains near Tucson, Arizona: A progress report: Geological Society of America Abstracts with Programs, v. 4, p. 408.

Shakel, D. W., Silver, L. T., and Damon, P. E., 1977, Observations on the history of the gneissic core complex, Santa Catalina Mountains, southern Arizona: Geological Society of America Abstracts with Programs, v. 9, p. 1169–1170.

Sherwonit, W. E., 1974, A petrographic study of the Catalina gneiss in the forerange of the Santa Catalina Mountains [M.S. thesis]: Tucson, University of Arizona, 165 p.

Shride, A. F., 1967, Younger Precambrian geology in southern Arizona: U.S. Geological Survey Professional Paper 566, 89 p.

Silver, L. T., 1955, The structure and petrology of the Johnny Lyon Hills area, Cochise County, Arizona [Ph.D. dissert.]: Pasadena, California Institute of Technology, 407 p.

—— 1968, Precambrian batholiths of Arizona [abs.]: Geological Society of America Special Paper 121, p. 558–559.

—— 1978, Precambrian formations and Precambrian history in Cochise County, southeastern Arizona: New Mexico Geological Society, 29th Field Conference Guidebook, p. 157–164.

Silver, L. T., and Deutsch, S., 1963, Uranium-lead isotopic variations in zircons: A case study: Journal of Geology, v. 71, p. 721–758.

Streckeisen, A., 1976, To each plutonic rock its proper name: Earth-Science Reviews, v. 12, p. 1–33.

Suemnicht, G. A., 1977, The geology of the Cãnada del Oro headwaters, Santa Catalina Mountains, Arizona [M.S. thesis]: Tucson, University of Arizona, 108 p.

Thorman, C. H., 1977, Discussion *on* Gravity induced folding off a gneiss dome complex, Rincon Mountains, Arizona: Geological Society of America Bulletin, v. 88, p. 1211–1212.

Thorman, C. H., and Drewes, H., 1978, Mineral resources of the Rincon Wilderness study area, Pima County, Arizona: U.S. Geological Survey Open-File Report 78-596, 58 p.

Titley, S. R., 1976, Evidence for a Mesozoic linear tectonic pattern in southeastern Arizona: Arizona Geological Society Digest, v. 10, p. 71–102.

Waag, C. J., 1968, Structural geology of the Mount Bigelow–Bear Wallow–Mount Lemmon area, Santa Catalina Mountains, Arizona [Ph.D. dissert.]: Tucson, University of Arizona, 133 p.

Wallace, R. M., 1954, Structures of the northern end of the Santa Catalina Mountains, Arizona [Ph.D. dissert.]: Tucson, University of Arizona, 45 p.

Wilson, E. D., Moore, R. T., and Cooper, J. R., 1969, Geologic map of Arizona: Arizona Bureau of Mines and U.S. Geological Survey, scale 1:500,000.

Wilson, J. R., 1977, Geology, alteration, and mineralization of the Korn Kob Mine area, Pima County, Arizona [M.S. thesis]: Tucson, University of Arizona, 103 p.

Wood, M. M., 1963, Metamorphic effects of the Leatherwood quartz diorite, Santa Catalina Mountains, Pima County, Arizona [M.S. thesis]: Tucson, University of Arizona, 68 p.

MANUSCRIPT RECEIVED BY THE SOCIETY JUNE 21, 1979

MANUSCRIPT ACCEPTED AUGUST 7, 1979

Printed in U.S.A.

Distribution and U-Pb isotope ages of some lineated plutons, northwestern Mexico

T. H. ANDERSON
Department of Geology and Planetary Sciences
University of Pittsburgh
Pittsburgh, Pennsylvania 15260

L. T. SILVER
Division of Geological and Planetary Sciences
California Institute of Technology
Pasadena, California 91109

G. A. SALAS
Departamento de Geologia
Universidad de Sonora
Hermosillo, Sonora, Mexico

ABSTRACT

Characteristically lineated and foliated rocks of middle to late Mesozoic(?) age crop out in ranges in north-central Sonora throughout an area of 15,000 km², between lat 31°30′N and 30°30′N. The terrane consists predominantly of layered sedimentary, volcanic, and volcaniclastic units, commonly metamorphosed to greenschist facies. Associated plutons, which are unambiguously intrusive, consist mainly of biotite- or biotite-muscovite–bearing granite. The layered suite and intrusive rocks are distinguished by subhorizontal, penetrative lineation, commonly defined by smeared mineral grains, which consistently trend northeast. Foliation is also predominantly low dipping. South of lat 30°30′N, sporadic occurrences of lineated granitic and metamorphic rocks suggest the existence of this deformational fabric at least to Sierra Mazatan, which lies east of Hermosillo near lat 29°00′N.

Cogenetic suites of zircon from one undeformed and three deformed plutons yield U-Pb ages from 75 to 55 m.y. These intrusive bodies are elements of a widespread, time-transgressive, late Mesozoic magmatic suite. They are not known to be affected by folds and faults commonly related to the Laramide orogeny. The apparent ages of the plutons are interpreted to be crystallization ages and therefore indicate that lineation and foliation formed, at least locally, later than 55 m.y. ago.

Outcrops of distinctively deformed rocks appear to crudely define a north-trending belt. Rocks outside of this general zone are composed of sedimentary, volcanic, and volcaniclastic rocks of

Precambrian and Mesozoic age which locally have been strongly folded and metamorphosed to greenschist or higher facies. However, postdeformational pegmatites and intrusive rocks indicate pre-Tertiary minimum ages for related episodes of deformation and metamorphism.

INTRODUCTION AND OBJECTIVE

In Sonora, Arizona, and southeastern California, widespread regions are underlain by terranes composed of igneous and metamorphic rocks. Radiometric ages and structural relations in these terranes indicate a complex sequence of magmatic, metamorphic, and deformational events between 1.8 b.y. and latest Cretaceous time. The Tertiary period was generally characterized by extensive volcanism and high-angle faulting, with less conspicuous plutonism. Recent geologic and geochronologic studies of metamorphic core complexes that crop out in Utah and Arizona have suggested the existence of a surprising and perplexing history of mid-Tertiary sedimentation, magmatism, metamorphism, and deformation (Compton and others, 1977; Banks, 1976; Banks and others, 1977; Creasey and others, 1976). These events and processes are characterized throughout extensive regions by distinctive similarities in style and timing, although they are not obviously related to any typical sequence associated with orogenic belts formed along continental margins.

For example, among core complexes south of the Snake River Plain, Crittenden and others (1978) pointed out that existing data suggest that (1) core complexes were unroofed by erosion and gravitational gliding during late Tertiary time, and (2) extensive areas are underlain by penetrative, unidirectional lineated gneiss (which also is commonly characterized by subhorizontal foliation).

Based on their studies in southern Arizona, G. H. Davis (this volume) and Davis and Coney (1979) suggest that some structural elements such as core, carapace, and decollement zone can be recognized in many of the core complexes of southern Arizona. According to these authors, these tectonic features developed in response to regional extension, and they emphasize the role of the major unconformity between Precambrian crystalline basement and overlying cover rocks in the structural evolution of the complexes.

Recent evidence supports the conclusion that among the crystalline core complexes discussed herein the development of penetrative fabrics, plutonic activity, metamorphism, uplift, and accompanying denudation occurred during early and middle Tertiary time. Timing of plutonism and of associated deformation and metamorphism is basic to understanding the evolution of core complexes. Shakel and others (1977), who reported results of K-Ar, Rb-Sr, and U-Pb isotopic studies of core rocks from the Santa Catalina Mountains in Arizona, concluded that the final "gneiss-forming event" in this range probably occurred about 45 m.y. ago and was followed by a middle Tertiary thermal event with plutonism. Identification and description of the young plutonism had been reported by Creasey and others (1976).

Eberly and Stanley (1978) concluded on the basis of their seismic and stratigraphic data, gathered from terrigenous deposits in southwestern Arizona, that middle Tertiary orogeny, whose final, waning stages occurred 17 to 20 m.y. ago, resulted in the development of a profound regional unconformity which has been identified throughout southwestern Arizona.

Considerable data now indicate the existence of a regionally correlatable suite of distinctive tectonic elements whose formation was initiated after the beginning of Tertiary time and was completed in many places about 20 m.y. ago. Reconnaissance geologic investigations indicate that such elements are present in Sonora, Mexico, in the form of abundant, strikingly lineated and foliated layered metamorphic rocks and orthogneisses.

Reconnaissance and more detailed studies indicate that (1) exposed crystalline rocks—distinguished by the presence of mylonitic orthogneiss with low-dipping foliation and associated unidirectional lineation, and similarly deformed greenschist-grade Phanerozoic strata—underlie, at the very least, 2,000 km² and perhaps as much as 15,000 km², mainly in north-central Sonora; (2) in central Sonora, mylonitic gneiss crops out sporadically within restricted areas of at least two ranges, where it is transitional into undeformed plutonic rocks of late Cretaceous–early Tertiary age; and (3) clastic beds of presumed Tertiary age, which consistently crop out along the flanks of the uplifted core complexes and in places occupy inter-range basins, record the history of uplift and have the potential to contribute much toward our understanding of the evolution of the complexes. Apparent ages, derived from U-Pb isotopic ratios in cogenetic suites of zircons from some of the mylonitic orthogneisses, fall within the interval of 78 to 57 m.y. We interpret these apparent ages to be maximum ages for the development of the penetrative fabric recorded by the plutonic rocks in northern and central Sonora, respectively.

The objectives of this paper are to identify some distinctively lineated and foliated rocks in Sonora, to report preliminary U-Pb isotopic age data and geologic observations pertaining to these rocks, and to provide enough information to make comparisons to similar terranes in other areas.

REGIONAL SETTING

The distribution of lineated rocks shown by Figure 1 encompasses those areas considered to have acquired their distinctive penetrative fabric during the first half of the Tertiary. Most of these rocks can be designated as metamorphic core complexes.

Metamorphic and plutonic rocks whose fabrics were formed during Precambrian or Mesozoic events are abundant throughout Sonora, and care must be taken to distinguish among various generations of metamorphic core complexes.

Precambrian basement within Sonora comprises two mutually exclusive terranes, each composed of elements of former orogenic belts. These belts, previously recognized in the southwestern United States, consist of chronologically distinctive volcanic, plutonic, and metamorphic rock suites whose interpreted U-Pb isotopic ages on zircons range from ~1.8 to 1.7 b.y. and ~1.7 to 1.6 b.y. (Anderson and Silver, in prep.). The older terrane, which crops out in western and central Sonora, is dominated by plutonic rocks, schist, and feldspathic gneiss commonly of upper amphibolite facies. U-Pb isotope ages of zircons from paragneisses and pegmatites that formed during the culmination of metamorphism indicate an age of about 1,650 m.y. for metamorphism. In northeastern Sonora, volcanic and sedimentary beds that accumulated about 1.7 b.y. ago are commonly strongly folded but record only greenschist-grade metamorphism. Younger, anorogenic plutonic suites consist of ubiquitous porphyritic quartz monzonite ($1,450 \pm 25$ m.y.) (Anderson and Silver, 1977) and rare, distinctive, red, granophyric granite ($1,100 \pm 15$ m.y.) (Anderson and Silver, in prep.).

Throughout Paleozoic time this terrane acted as a stable block upon which hundreds to thousands of metres of late Precambrian and early Paleozoic carbonate and clastic rocks accumulated. During much of Jurassic time, renewed volcanism and plutonism in northern Sonora resulted in the development of a superposed orogenic belt. A fault of late Jurassic age may have caused severe disruption of Precambrian trends along a zone called the Mojave-Sonora megashear (Silver and Anderson, 1974). After a brief lull at the end of Jurassic time, widespread volcanism and plutonism further modified the old basement. Late Cretaceous plutonic rocks that belong to this suite are progressively younger from west to east and crop out at many places throughout the state (Anderson and Silver, 1974).

In places, layered sequences of Jurassic and Cretaceous age are strongly deformed and metamorphosed. Although it is clear that deformation and metamorphism of Tertiary age affected rocks throughout considerable areas, locally the radiometric results from posttectonic, undeformed plutons of Cretaceous and Jurassic age that cut metamorphic rocks indicate the existence of tectonic and thermal events of Mesozoic age.

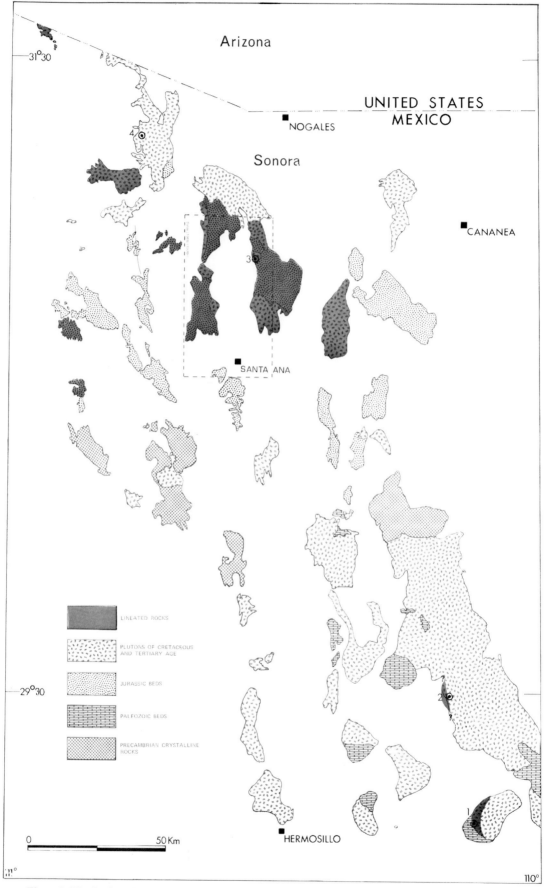

Figure 1. Distribution of pre-Cretaceous rocks and associated, younger plutons, north-central Sonora.

GEOCHRONOLOGICAL STUDIES

Introduction

Geologic setting, rock description, and analytical results are provided for four samples, three of which are weakly to strongly foliated plutons and one of which is from a large, undeformed pluton. This pluton, which is surrounded almost completely by ranges composed of metamorphic core complex rocks, has somehow escaped the effects of Tertiary deformation and metamorphism.

From south to north these samples are (1) Sierra Mazatan porphyritic granodiorite gneiss, (2) weakly foliated Puerto del Sol prophyritic granite, (3) Sierra Guacomea granite gneiss, and (4) Rancho Los Alamos granodiorite.

Sierra Mazatan

Sierra Mazatan lies 70 km east of Hermosillo, the capital city of Sonora (Fig. 1, locality 1). It is a roughly circular range with a diameter of 12 km; the crest of the range is marked by knobs of granite that rise a few tens of metres above a rolling surface of low relief which in places is dissected by shallow stream valleys. As seen on aerial and orbiter photographs, Sierra Mazatan is one of the most distinctive and easily recognized physiographic features in northwestern Mexico because it rises nearly 1,000 m above the surrounding terrane and its circular flanks which outline the crest are emphasized by a cover of dark oak trees above 600 m elevation, in contrast to the lighter shades of brown and yellow of the adjacent desert (Fig. 2).

Sierra Mazatan exhibits a strikingly smooth profile that is broken along the western flank by jutting outcrops of carbonate beds of Paleozoic age. These beds are structurally the highest elements of the range. They rest upon a strongly lineated surface developed upon a thin layer of mylonitic gneiss which is transitional into somewhat less strongly sheared porphyritic granodiorite gneiss from which the analyzed sample was collected (Table 1). In thin section this rock is characterized by classic mortar texture in which porphyroclasts of feldspar are set in a fine-grained aggregate of quartz (Fig. 3). The porphyroclasts are from 1 mm to a few centimetres in diameter and display undulose extinction, deformation twinning, and abundant fractures. The nature of the porphyroclasts indicates that they are pretectonic and have persisted as relicts from primary igneous texture. Fractures that traverse the larger crystals are filled with granulated feldspar, and foliation defined by quartz aggregates wraps around the large feldspars. Quartz is abundant and its grain size, texture, and degree of strain are commonly a function of proximity to grain boundaries. Biotite, commonly partially altered to chlorite, and rare muscovite occur as (1) fine-grained aggregates that wrap around coarser feldspar crystals; (2) bent and twisted flakes where preserved from the effects of high strain; and (3) lenticular and ribbonlike aggregates of fine-grained, parallel flakes and prisms distributed among the well-oriented, strongly granulated recrystallized quartz that forms the matrix. Less abundant constituents such as sphene, apatite, zircon, and opaque minerals are commonly associated with the biotite. Of these, the opaque minerals in places show distinct granulation, whereas other minor components surrounded by more easily deformed mica seem to have escaped much of the deformation.

To the east, porphyritic granodiorite gneiss interfingers with and becomes subordinate to a sequence of layered quartzo-feldspathic, biotite-rich, and locally quartzitic gneiss. Concordant layers of pegmatitic leucogranite are common. This zone of heterogeneous gneisses appears to be restricted to an interval of a few kilometres. Within this zone, the rocks are generally characterized by low-dipping foliation and penetrative lineation which commonly plunges less than 20°. Impressive swarms of aplitic and pegmatitic veins locally cut the layered country rock, and although they are not transposed into concordance with compositional layering, they are consistently slightly foliated. At other places

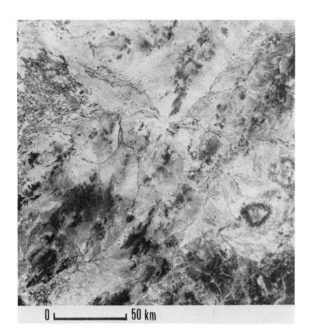

Figure 2. Skylab 4 photograph of part of central Sonora. North is toward the upper right-hand corner. North-trending basins and ranges are obvious physiographic elements. Sierra Mazatan is the distinctive, crudely circular feature not far from the lower right corner. Small, dark spots slightly west of the Sierra are composed of carbonate beds of Paleozoic age. The geometric pattern near the upper left corner is the result of cultivation near the mouth of Rio Sonora. Hermosillo is the dark circular area near the center of the photograph. Rio Sonora runs through Hermosillo, and its trace can be discerned extending east across the region to a point near the center of the right margin. This point is just west of the sample locality for Puerto del Sol porphyritic granite.

migmatitic gneisses record extensive flowage, spectacular boudinage, and folds and faults. Folded layers of lineated mylonitic rock are not uncommon.

Higher on the western flank of the range, homogeneous porphyritic biotite granodiorite gneiss again predominates, but foliation and lineation become less distinct as the top of the range is approached. Most of the sierra is composed of typical porphyritic biotite granodiorite, which in places is only slightly foliated. The rock is cut by numerous small slickensided surfaces, but it is not deformed to any unusual degree. Reconnaissance traverses down the eastern and northern flanks of the range suggest that deformed rocks are absent in those areas.

Puerto del Sol

The village of Puerto del Sol lies ~75 km northeast of Hermosillo, on the south bank of Rio Sonora, a few kilometres west of where the river debouches from a canyon that cuts across one of the largest exposures of granite in Sonora. Although little mapping has been carried out in this area, granitic rocks appear to underlie a complex of generally north-trending ranges which compose an area ~100 km long and ~25 km wide. This is a minimum area because it does not take into account the probable existence of granite under pediments or valleys. The batholith is approximately bisected by Rio Sonora, and the walls of its canyon reveal granite as the predominant rock type.

The granite is characteristically light colored in fresh samples, with pink feldspar phenocrysts commonly 1 cm or more in long diameter. Accompanying gray quartz occurs as ovoid or irregular eyes as much as 1 cm in length, as well as smaller anhedral crystals in a matrix that includes abundant feldspar and biotite with less common muscovite. Cross-cutting pegmatites are common and may contain muscovite, biotite, and garnet in addition to quartz and feldspar. Mafic dikes cut the silicic rocks. Thin sections from the sample locality (Fig. 1, locality 2) 5 km east of Puerto del Sol reveal quartz that is distinctly undulose and feldspar crystals that exhibit bent or wedge-shaped polysynthetic twin lamellae (Fig. 4). These features suggest strain, although foliation is at best weakly developed in outcrop. However, in road cuts between the village of Puerto del Sol and our sample locality, crude foliation is defined by aligned biotite flakes and scattered larger feldspar crystals whose long axes are

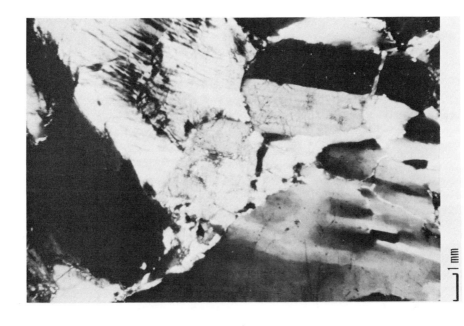

Figure 3. Photomicrograph of mylonitic gneiss from the sample locality of Sierra Mazatan porphyritic granodiorite gneiss. See text for description.

Figure 4. Photomicrograph of Puerto del Sol porphyritic granite whose mineral constituents record strain effects.

TABLE 1. MINERALOGICAL COMPOSITION OF DATED PLUTONS

	Sierra Mazatan porphyritic granodiorite gneiss (%)	Puerto del Sol porphyritic granite (%)	Sierra Guacomea granite gneiss (%)	Rancho Los Alamos granodiorite (%)
Plagioclase	47.0	37.0	31.9	36.5
Orthoclase (microcline)	18.0	29.5	21.0	18.4
Quartz	26.7	26.8	37.4	34.2
Biotite	6.8	5.1	8.3	6.7
Muscovite	0.3	0.5	..	..
Hornblende	..	..	..	2.6
Chlorite	..	..	..	0.5
Sphene	<0.1	0.1	0.2	0.4
Opaque	1.3	0.9	1.2	0.5
Apatite	<0.1	0.2	<0.1	0.1
Zircon	<0.1	<0.1	<0.1	<0.1

Note: Two thin sections were counted from each sample locality; 1,000 points per section.

Figure 5. Outcrop of Sierra Guacomea granite gneiss which shows well-developed, subhorizontal foliation.

aligned. Locally, quartz has been more strongly deformed, and the elongate grains define a distinct low-dipping foliation. Reconnaissance traverses in this area indicate that foliated granite is probably not widespread within the batholith and that the west to east transition from distinctly foliated to undeformed granite on the western flank is not mirrored by exposures on the eastern side. However, the age and the style of the limited deformation as recorded east of Puerto del Sol suggest correlation with regional events, which resulted in the formation of more extensively developed core complexes.

Sierra Guacomea

In north-central Sonora, strongly deformed, layered rocks of at least low- to medium-grade metamorphic facies and associated silicic plutons underlie major north-trending sierras. Initial recon-

Figure 6. Photomicrograph of Sierra Guacomea granite gneiss which shows moderately developed mortar texture.

naissance field mapping of a widespread area (Fig. 1) by Salas (1968 [1970]) revealed the presence of an extensive suite of greenschist-facies rocks composed mainly of phyllite with subordinate marble, metaconglomerate, quartzite, and metagraywacke. In places, mineral assemblages characteristic of the almandine-amphibolite facies occur. Salas (1968 [1970]) concluded that the layered rocks were composed of two suites, both of which were of Precambrian age. His conclusion was based on the fact that metamorphism recorded in this region is comparable to that reported from areas of Precambrian crystalline rocks that crop out farther west, south of Caborca. Our radiometric data now indicate that one of the strongly deformed plutons considered to be Precambrian is actually of latest Cretaceous age. To date, no paleontologic or radiometric studies that bear directly on the age of any of the *layered* rocks have been conducted. Regional geologic studies do not support a unique Precambrian age assignment for these strata, and younger ages cannot be discounted.

In Sierra Guacomea (Fig. 1, locality 3) we collected a sample from an intrusive mass whose crudely circular outcrop pattern is about 3.5 km in diameter. Salas (1968 [1970]) described this as a leucogranite. Near our sample locality within the southwestern margin of the mass, the strongly foliated rock ranges in composition from granite to granodiorite (Fig. 5, Table 1). It is not clear whether the modal variations are primary igneous features or the result of metamorphic processes. In outcrop, the gneiss appears generally leucocratic and has been cut by numerous pegmatites that have been deformed with the granite, as have associated country rocks composed of porphyritic and aplitic gneisses. Black, largely aphanitic rock layers concordantly interfingered with the foliated granite are probably zones of mylonite.

The foliation at the sample locality dips consistently within 10° of horizontal, and lineation as reported by Salas (1968 [1970]) ubiquitously trends northeast.

In thin section the effects of cataclasis are clearly evident (Fig. 6). The overprint is not as strong as that observed in the thin sections from Sierra Mazatan, but distinctly stronger than the effects observed in rocks near Puerto del Sol. Quartz is strongly undulose, and granular aggregates of this mineral form ragged mosaics between many grains; however, relict grains as much as several millimetres in diameter are common. Predominantly xenoblastic plagioclase with ragged grain edges is commonly undulose with bent and wedge-shaped deformation twins. Patches of microcline have formed in strongly undulose orthoclase. Biotite exists as subhedral flakes or as trains of fine-grained flakes, depending upon its position with respect to highly strained zones.

Los Alamos Granodiorite

Los Alamos granodiorite is a dated sample representative of an undeformed plutonic mass whose relevance to the metamorphic core complex problems exists in the fact that it lies within a large area predominantly underlain by characteristically foliated and lineated crystalline rocks, yet it has escaped deformation on a mesoscopic scale.

The pluton as shown on photogeologic maps (United Nations, 1969) crops out over an area of at least 300 km^2 and forms low hills with rare higher peaks in an elongate north-trending range more than

TABLE 2. ISOTOPIC DATA FOR ZIRCONS FROM SOME LINEATED PLUTONS OF LATE CRETACEOUS-EARLY TERTIARY AGE

Sample	Observed ratios			Radiogenic Pb* Atom (%)			Concentrations (ppm)	
	206/204	207/204	208/204	206	207	208	rad$_{Pb}$	U
Sierra Guacomea granite gneiss								
Fraction 1	1877.2	105.72	159.67	89.77	4.35	5.88	27.849	2457.6
Fraction 2	1175.0	70.45	150.66	*87.35	4.14	8.51	28.899	2620.6
Sierra Mazatan porphyritic granodiorite gneiss								
Fraction 1	577.4	43.01	96.17	*86.73	4.25	9.02	12.763	1396.2
Fraction 2	882.8	57.88	115.61	87.82	4.30	7.89	14.229	1571.8
Puerto del Sol porphyritic granite								
Fraction 1	1550.9	100.15	171.46	87.54	4.83	7.63	11.957	1241.8
Fraction 2	1526.7	99.15	184.01	86.79	4.81	8.40	14.972	1576.1
Rancho Los Alamos granodiorite								
Fraction 1	1712.8	97.18	243.21	85.52	4.12	10.36	16.141	1411.0
Fraction 2	2386.4	128.74	388.07	83.64	4.00	12.36	22.441	1888.3

* Leads are corrected with a lead of isotopic composition: $^{206}Pb/^{204}Pb = 18.6$; $^{207}Pb/^{204}Pb = 15.6$; $^{208}Pb/^{204}Pb = 38.0$; except those marked by an asterisk*, which were corrected with a lead of $^{206}Pb/^{204}Pb = 18.2$.

30 km long and 10 km wide (Fig. 1, locality 4). In hand specimen, the granodiorite is commonly distinguished by thick books of biotite which, where weathered out, result in a deeply pitted surface texture (Table 1). Thin sections from our sample locality reveal stubby laths of subhedral plagioclase that range in length from 1 to 2 mm to several millimetres in long diameter, anhedral to subhedral biotite and hornblende, and euhedral sphene, set in a matrix of anhedral quartz and potassium feldspar. Granulation along grain boundaries was not observed. The only deformation effects are distinctly undulose quartz and rare, bent or kinked flakes of biotite.

ANALYTICAL RESULTS AND GEOCHRONOLOGIC INTERPRETATION

Results of isotopic analyses for two cogenetic fractions from each of the plutons described in the preceding paragraphs are presented in Tables 2 and 3. The behavior of isotope systems for these zircon fractions is analogous to older suites of zircon as described by Banks and Silver (1966). The presence of Precambrian basement and possible contributions of old radiogenic lead to the zircon suites of younger magmatic generations is recorded by the ratios of $^{207}Pb/^{235}U$ and $^{207}Pb/^{206}Pb$, which are sensitive monitors of the amount of inherited lead. Corrections for this component can be made simply by plotting the lower intersection of a chord, defined by each individual fraction and an assumed old component (for this region, 1.8 b.y.), with the Concordia curve. The resulting $^{206}Pb/^{238}U$ ratios can then be interpreted as per Banks and Silver (1966).

Sierra Mazatan porphyritic granodiorite (58 ± 3 m.y.) and Puerto del Sol porphyritic granite (57 ± 3 m.y.) yield essentially synchronous ages. These ages and the widespread, although presently unmapped, outcrops of granite that extend at least from Sierra Mazatan (Fig. 1) northward about 100 km, in a belt at least 20 km wide, suggest the presence of a major batholith.

Sierra Guacomea granite gneiss (78 ± 3 m.y.) and Rancho Los Alamos granodiorite (74 ± 2 m.y.) are in striking contrast, not so much in age but with respect to the varying degree of deformation recorded by each. As noted above, Rancho Los Alamos granodiorite is almost an island surrounded by lineated and foliated rocks.

TABLE 3. ATOM RATIOS AND APPARENT AGES FOR ZIRCONS FROM SOME LINEATED PLUTONS OF LATE CRETACEOUS-EARLY TERTIARY AGE

Sample	Atom ratios			Apparent ages (m.y.)			Interpreted age (m.y.)
	$^{206}Pb/^{238}U$	$^{207}Pb/^{235}U$	$^{207}Pb/^{206}Pb$	$^{206}Pb/^{238}U$	$^{207}Pb/^{235}U$	$^{207}Pb/^{206}Pb$	
Sierra Guacomea granite gneiss							
Fraction 1	0.01184	0.07913	0.04849	76	78	124	
Fraction 2	0.01121	0.07324	0.04742	72	72	70	78 ± 3
Sierra Mazatan porphyritic granodiorite gneiss							
Fraction 1	0.00922	0.06232	0.04902	59	62	149	
Fraction 2	0.00925	0.06237	0.04892	59	61	144	58 ± 3
Puerto del Sol porphyritic granite							
Fraction 1	0.00980	0.07460	0.05518	63	73	420	
Fraction 2	0.00959	0.07325	0.05540	62	72	428	57 ± 3
Rancho Los Alamos granodiorite							
Fraction 1	0.01138	0.07553	0.04815	73	74	107	
Fraction 2	0.01156	0.07614	0.04778	74	75	88	74 ± 2

Note: U^{238}/U^{235} = 137.88. U^{235}, λ = 0.984850 × 10^{-9}/yr. U^{238}, λ = 0.155125 × 10^{-9}/yr.

CONCLUSIONS

Some plutons with crystallization ages of between 78 and 57 m.y. record deformation intensity that varies greatly but style that is strikingly consistent. The age and character of deformation suggest its affinity with correlatable events in southern Arizona and perhaps areas farther north. We conclude that the unique characteristics of age, geography, and deformational style, which distinguish this early to middle Tertiary tectonic event, intermittently persist more than 250 km into Sonora, close to the apparent southernmost extent of the Basin and Range province.

The effects of tectonic denudation that are strikingly displayed and that appear to play a conspicuous role in the evolution of core complexes in Arizona (Davis and Coney, 1979) are not strongly manifested in northern Sonora. The major unconformity that separates Precambrian basement from overlying cover in southern Arizona and that is a locus for decollement (Davis, this volume) is not known to crop out in north-central Sonora. Furthermore, geochronologic reconnaissance (Anderson and Silver, unpub. data) has not resulted in the identification of Precambrian crystalline rocks in this area. Throughout this region, lineated, foliated rocks predominate in contrast to their sporadic distribution in areas to the north and south. The widespread development of these deformed rocks appears to have been enhanced by the complete absence or, if not absence, the lack of coherent Precambrian basement.

Although the limited data presented herein indicate the presence of rocks that were penetratively deformed during post-Cretaceous time as far south as Hermosillo, much additional information pertaining to structural trends, style of deformation, areal distribution, relationship of lineated and foliated rocks to older transected belts, and timing of plutonism, metamorphism, and deformation must be gathered before the formation and evolution of this enigmatic terrane can be understood.

ACKNOWLEDGMENTS

Our investigations have benefited from logistical support at times from the Instituto de Geología, Universidad Nacional Autónoma de México; Universidad de Sonora; and Consejo de Recursos Naturales no Renovables. Conversations with Jaime Roldan Q., Claude Rangin, George H. Davis, Peter Coney, and Doug Shakel have contributed to our perspective of regional geology.

Mineral separations by O. Shields and Jamie Alvarez; microscope work by F. Corona; and ever friendly, helpful contributions of chemistry and mass spectrometry by Gerri Silver and Maria Pearson are most kindly acknowledged.

This work was supported by NSF Grants GA-15989 and EAR 74-00155 A01 (formerly GA-40858) awarded to California Institute of Technology and EAR 76-84167 awarded to the University of Arizona.

APPENDIX 1. SAMPLE LOCALITIES

Sierra Mazatan porphyritic granodiorite gneiss was collected from the northwestern flank of Sierra Mazatan from outcrops that form low hills between Rancho El Parian and Rancho La Feliciana, a short distance southeast upstream of the corral at El Garambullal, where rock is exposed at the mouth of Canada Los Pedernales (Cetenal topographic map, scale 1:50,000, series Mazatan sheet H12 D43).

Puerto del Sol porphyritic granite was collected from blasted outcrops on the north side of the highway (Sonora Route 21) between Ures and Mazocahui, 5 km east of the turnoff to the Village of Puerto del Sol (Cetenal topographic map, scale 1:50,000, series Puerto del Sol sheet H12 D33).

Sierra Guacomea granite gneiss forms a small pluton on the western margin of Sierra Guacomea. The dated sample was collected about 0.5 km northeast of Rancho San Antonio from a stream-transported boulder derived from outcrops of similar lithology within the mass (reconnaissance geologic map and section of the Santa Ana region, Sonora, scale 1:50,000; Salas, 1970).

Rancho Los Alamos granodiorite was collected from stream-polished outcrops just below a small concrete dam at Rancho Los Alamos. This ranch lies about 6 km (map distance) northwest of Rancho Las Chollas on a road that connects Rancho Las Chollas to Rancho Luisillo. Cerro del Chile is a distinctive hill that lies about 4 km south-southwest of Rancho Los Alamos (geological survey map of the northern part of the state of Sonora, Mexico, scale 1:250,000, Consejo de Recursos Naturales no Renovables; United Nations, 1969).

APPENDIX 2. ZIRCON CONCENTRATES

Sierra Mazatan Porphyritic Granodiorite Gneiss

Most crystals are transparent, colorless, euhedral to subhedral individuals with length-to-width ratios that range from 1:1 to 5.5:1, with an average ratio between 2:1 and 3.5:1. Anhedral and broken grains are not uncommon, but tetragonal dipyramidal crystals are characteristic. The outlines of many of the well-formed crystals are modified by the presence of pits or scalloped edges. Rod- and needle-shaped inclusions as well as irregularly shaped black grains, opaque, finely disseminated particles, and very round to irregularly rounded forms can be discerned in almost every grain. Distinct, fine zoning is not uncommon. A few grains show elongate tubes parallel to the long axis of the crystal.

Puerto del Sol Porphyritic Granite

The suite is characterized by elongate, transparent, colorless, euhedral to subhedral individuals with length-to-width ratios that range from ~6:1 to 1:1, with an average ratio between 4:1 and 5:1. Many grains have tetragonal dipyramidal forms. Crystals that are less well formed may be strongly modified by pits, small facets, and broken or scalloped edges. Almost every grain contains a few cavities in the form of tubes, spheres, or pits or inclusions such as slender rods of apatite(?), zircon-shaped grains, fine opaque matter, or anhedral, black(?) opaque grains. Zoned crystals are rare. A few grains appear to be very faint pink or brown.

Sierra-Guacomea Granite Gneiss

The population is predominantly composed of transparent, colorless subhedral or euhedral individuals with length-to-width ratios between 6:1 and 1:1, with an average ratio of ~3:1. Commonly, grains contain abundant growth cavities in the form of tubes, spheres or pits, or inclusions such as slender rods of apatite, zircon-shaped grains, fine opaque matter, or anhedral black(?) opaque grains. Many crystals are characterized by simple tetragonal dipyramidal forms, but grains that are less well formed and that are strongly modified by numerous pits, small facets, and broken or scalloped edges are abundant. Individuals that record distinct fine zoning are scattered throughout the suite. A few grains show pale orangish-brown staining, and rare grains appear to have very pale brown interiors.

Rancho Los Alamos Granodiorite

Individuals that comprise this suite are the least well formed of the four populations. Most crystals are transparent, colorless, subhedral to euhedral individuals with length-to-width ratios that range from 4:1 to 1:1, with an average ratio of about 2:1. The suite is characterized by tetragonal pyramidal forms, most of which are strongly modified by broken and scalloped edges as well as numerous small facets and pits, which give the grains a corroded appearance. Needle- and rod-shaped inclusions are common but subordinate to cavities in the form of tubes, spheres, or pits and fine, opaque matter. Crystals that show zoning are rare, as are faintly colored brown grains.

Analytical Procedure

Fractions of the total zircon population were selected for isotopic analysis on the basis of their uranium content, with the objective of obtaining the greatest range in the degree of discordance. The zircon fractions were fused with about six times their weight of purified, sodium-tetraborate flux for varying time periods, depending upon the sample weight. The analytical procedure for extracting lead is the same as that described by Silver and others (1963). The few modifications made of the proven techniques include the substitution of 10-ml pyrex beakers for teflon during the final digestion of the lead dithizonate and a change in loading technique from running PbS on a tantalum filament to $PbNO_3$ in a silica gel-phosphoric acid medium on a rhenium filament (modified from the procedure of Cameron and others, 1969). Uranium was extracted by normal ion exchange techniques.

Isotopic analyses were made on 12-in., single-focusing, solid-source mass spectrometers constructed at the California Institute of Technology. Data obtained on the electron multiplier and simple collector have been corrected for mass discrimination and for very slight nonlinearity in the shunt factors. Mass spectrometer reproducibility is generally $\pm 1\%$ for radiogenic $^{206}Pb/^{204}Pb$ ratios, and $\pm 0.3\%$ for the $^{206}Pb/^{207}Pb$, $^{206}Pb/^{208}Pb$, and $^{235}U/^{238}U$ ratios. Mass spectrometer precision (1σ) for $^{206}Pb/^{207}Pb$ averages about $\pm 0.2\%$, as does precision for $^{206}Pb/^{208}Pb$ ratios. Uncertainty for $^{206}Pb/^{204}Pb$ is about $\pm 0.5\%$.

The common lead blank contributed by reagents and glassware is slightly less than those determined by Silver and others (1963). The use of the silica gel-phosphoric acid loading technique enables us to run 1-μg quantities of lead that permit significant reduction of sample size. This combined with the smaller amount of flux used (6 rather than 10 times the weight of sample) can reduce the blank significantly. Contribution of common lead from the rest of the procedure has also been reduced so that the range is about 0.06 to 0.04 μg.

REFERENCES CITED

Anderson, T. H., and Silver, L. T., 1974, Late Cretaceous plutonism in Sonora, Mexico and its relationship to circum-Pacific magmatism [abs.]: Geological Society of America Abstracts with programs, v. 6, no. 6, p. 484.
——1977, U-Pb isotope ages of granitic plutons near Cananea, Sonora: Economic Geology, v. 72, p. 827–836.
Banks, N. G., 1976, Reconnaissance geologic map of the Mount Lemmon quadrangle, Arizona: U.S. Geological Survey Miscellaneous Field Studies Map MF-747.
Banks, N. G., and others, 1977, Reconnaissance geologic map of the Tortolita Mountains quadrangle, Arizona: U.S. Geological Survey Miscellaneous Field Studies Map MF-864, with text.
Banks, Philip O., and Silver, Leon T., 1966, Evaluation of the decay constant of Uranium 235 from lead isotope ratios: Journal of Geophysical Research, v. 71, p. 4037–4046.
Cameron, A. E., Smith, D. H., and Walker, R. L., 1969, Mass spectrometry of nanogram-size samples of lead: Analytical Chemistry, v. 41, p. 525–526.

Compton, Robert T., and others, 1977, Oligocene and Miocene metamorphism, folding, and low-angle faulting in northwestern Utah: Geological Society of America Bulletin, v. 88, p. 1237–1250.
Creasey, S. C., and others, 1976, Middle Tertiary plutonism in the Santa Catalina and Tortolita Mountains, Arizona: U.S. Geological Survey Open-File Report 76-262, 20 p.
Crittenden, Max, Jr., Coney, Peter J., and Davis, George, conveners, 1978, Tectonic significance of metamorphic core complexes in the North American Cordillera: Penrose Conference Report, Geology, v. 6, no. 2, p. 79–80.
Davis, G. H., 1980, Structural characteristics of metamorphic core complexes, southern Arizona: Geological Society of America Memoir 153 (this volume).
Davis, George H., and Coney, Peter J., 1979, Geologic development of the Cordilleran metamorphic core complexes: Geology, v. 7, no. 3, p. 120–124.
Eberly, L. D., and Stanley, T. B., Jr., 1978, Cenozoic stratigraphy and geologic history of southwestern Arizona: Geological Society of America Bulletin, v. 89, p. 921–940.

Salas, Guillermo A., 1968 [1970], Areal geology and petrology of the igneous rocks of the Santa Ana region, northwest Sonora: Boletin Sociedad Geologica Mexicana, v. 31, p. 11–63.

Shakel, Douglas W., Silver, Leon T., and Damon, Paul E., 1977, Observations on the history of the gneissic core complex, Santa Catalina Mtns., Southern Arizona [abs.]: Geological Society of America Abstracts with Programs, v. 9, no. 7, p. 1169–1170.

Silver, L. T., and others, 1963, Precambrian age determinations in the western San Gabriel Mountains, California: Journal of Geology, v. 71, p. 196–214.

Silver, Leon T., and Anderson, Thomas H., 1974, Possible left-lateral early to middle Mesozoic disruption of the southwestern North American craton margin [abs.]: Geological Society of America Abstracts with Programs, v. 6, p. 955.

United Nations, 1969, Survey of metallic mineral deposits, Mexico, 1969, Final Report: New York, United Nations Development Program, 72 p.

MANUSCRIPT RECEIVED BY THE SOCIETY JUNE 21, 1979
MANUSCRIPT ACCEPTED AUGUST 7, 1979
CALTECH CONTRIBUTION #3208

Printed in U.S.A.

PART 3

EASTERN GREAT BASIN

Geological Society of America
Memoir 153
1980

Transition from infrastructure to suprastructure in the northern Ruby Mountains, Nevada

Arthur W. Snoke
Department of Geology
University of South Carolina
Columbia, South Carolina 29208

ABSTRACT

In the northern Ruby Mountains, Nevada, the transition from high-grade, migmatitic infrastructure to the nonmetamorphosed but allochthonous rocks of the suprastructure is a zone of intense ductile deformation and high strain. This zone, which is disharmonic to the terranes both structurally above and below, is characterized by amphibolite-facies metamorphism, polyphase deformation, mylonitic textures, a west-northwest lineation, and tabular masses of orthogneiss. Major map-scale folds are overturned westward opposite the common direction of overturning in the infrastructure. Many of the large folds in the transition zone (*Abscherungszone* after the concepts of Haller, 1956, 1971) have evolved into ductile faults (tectonic slides) and form complex braided systems. Tectonic slices of low-grade metasedimentary rocks (greenschist facies?) mark the boundary between the top of the *Abscherungszone* and an overlying low-angle fault complex of unmetamorphosed upper Paleozoic and locally Tertiary rocks. These upper Paleozoic rocks represent attenuated fragments of the suprastructure which was structurally above the metamorphic complex during middle Mesozoic regional metamorphism and deformation. The Tertiary sedimentary and volcanic rocks are Miocene (K-Ar age on sanidine from a rhyolite flow), and conglomeratic facies contain clasts derived from the underlying metamorphic complex. Therefore, the low-angle fault complex is a Neogene brittle phenomenon apparently superimposed on an older Mesozoic "stockwerk" terrane. Despite these relations, a tectonic continuum involving a diachronistic transition from ductile to brittle deformation is an alternative explanation; however, such a tectonic regime must have spanned an interval from middle Mesozoic to Miocene.

INTRODUCTION

At scattered localities in the hinterland province of the Sevier orogenic belt, eastern Great Basin, metamorphic complexes are overlain by allochthonous slices of unmetamorphosed miogeoclinal sedimentary rocks (Misch, 1960; Fig. 1). Recent stratigraphic and structural studies in the metamorphic complexes indicate that these terranes include Precambrian, lower, middle, and in a few cases even upper Paleozoic rocks. The stratigraphic ascent of the metamorphic front was apparently widely variable throughout the region, for in some ranges only late Precambrian rocks are metamorphosed (for example, Schell Creek Range, Misch and Hazzard, 1962), while in other areas (for example, northwest Utah and adjacent Idaho) Carboniferous and Permian rocks have been recrystallized during metamorphism (Armstrong, 1968; Compton, 1972, 1975; Compton and others, 1977; Todd, 1973a, 1973b).

Misch (1960) considered the metamorphic rocks to have been deformed and recrystallized during the middle Mesozoic and referred to the tectonic contact with the overlying nonmetamorphic strata as a decollement. Many subsequent workers (for example, Snelson, 1957; Nelson, 1966, 1969; Thorman, 1970; Cebull, 1970) supported the decollement model and argued for widespread Mesozoic compres-

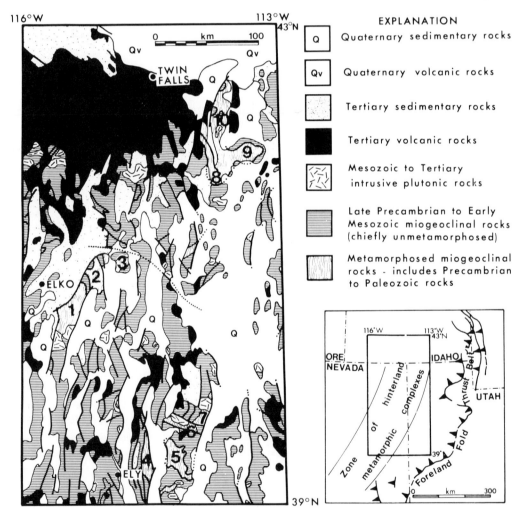

Figure 1. Regional geologic map of the eastern Great Basin showing the distribution of metamorphic core complexes (after King and Beikman, 1974). Key to the numbered localities: 1, Ruby Mountains; 2, East Humboldt Range; 3, Wood Hills; 4, Schell Creek Range; 5, Northern Snake Range; 6, Kern Mountains; 7, Deep Creek Range; 8, Grouse Creek Mountains; 9, Raft River Mountains; 10, Albion Range.

sional thrusting throughout the eastern Great Basin. Young (1960) was one of the first workers to present an alternative model based on gravitational sliding related to extensional tectonics. Later workers (Drewes, 1967; Moores and others, 1968) expanded this interpretation. Armstrong and Hansen (1966) applied the "stockwerk" tectonic model to the hinterland province as originally developed by Wegmann (1935). They interpreted the "decollement" to be a Mesozoic *"Abscherungszone"* or a zone of disharmonic movement and detachment between mobile infrastructure and rigid suprastructure. Willden and others (1967) first, and subsequently Armstrong (1972) suggested that many of the supposed Mesozoic thrust faults were actually Tertiary features or in a few cases Mesozoic faults reactivated during the Tertiary. Armstrong (1972) argued that the younger-over-older tectonic style of the eastern Great Basin is related to local gravitational sliding along Tertiary normal faults that flatten with depth and is not a manifestation of compressional tectonics or crustal shortening. Other workers (for example, Coney, 1974; Snoke, 1975) have supported this contention. Tertiary deformation within the metamorphic complex has been documented in Nevada (Willden and others, 1967) and northwest Utah (Todd, 1973a, 1973b; Compton and others, 1977), and Tertiary K-Ar metamorphic mineral ages have been reported in several other areas throughout the province (for example, Thorman, 1965; Armstrong and Hansen, 1966; Kistler and Willden, 1969b; Lee and others, 1970; Snoke, 1975). These Tertiary K-Ar ages are often interpreted to be cooling ages related to deep burial and rapid uplift, or ages reset during basin-range magmatism.

From the foregoing summary, it is obvious that the structural history of the region is complex and diverse in character. Furthermore, no study has documented, in detail, the transition from high-grade mobile infrastructure to allochthonous suprastructure (the nonmetamorphic cover rocks). The northern Ruby Mountains, Nevada (Fig. 2), offer an exceptional opportunity to examine this transition zone. Structural analysis of the mobile core terrane is available (Howard, 1966), and the stratigraphy and structure of correlative but autochthonous, nonmetamorphic lithologic units are well known in the southern Ruby Mountains (Sharp, 1942; Willden and Kistler, 1967, 1969; Hose and Blake, 1976). Furthermore, geochronologic studies have dated some of the events in the range (Willden and others, 1967; Willden and Kistler, 1969; Kistler and Willden, 1969b; Kistler and O'Neil, 1975). Therefore, the purpose of this report is to summarize the results of recent field studies in the Ruby Mountains metamorphic complex and the allochthonous nonmetamorphic cover rocks; a major goal of this analysis was to integrate the structural features into a consistent sequence as well as to account for contrasting structural style.

TECTONIC FRAMEWORK

The northern Ruby Mountains can be conveniently divided into three distinct structural terranes: (1) a high-grade metamorphic core characterized by large-scale recumbent folding and widespread migmatization (Howard, 1966, and this volume); (2) an overlying zone of intense strain manifested by a prominent flattening foliation and associated elongation lineation; and (3) a structurally higher low-angle fault complex (a brittle phenomenon) whose base is typically marked by tectonic slivers of low-grade metasedimentary rocks but which chiefly consists of unmetamorphosed Paleozoic and locally Tertiary rocks. Although each of these terranes has a distinct structural style, some transitional features between them can be recognized. Furthermore, taken together, they provide a framework for the recognition of contrasting tectonic levels that may characterize wider areas in this segment of the Cordilleran orogen.

Following the suggestion of Armstrong and Hansen (1966), one can equate these three structural terranes to the various tectonic tiers within Wegmann's (1935) "stockwerk" model (deep-seated to shallow): infrastructure, transition zone (called *"Abscherungszone"* by Haller, 1956), and suprastruc-

Figure 2. Generalized geologic map of the Ruby Mountains, Nevada (after Howard and others, 1979). The areas shown in Figures 4 and 5 are outlined.

ture. In this report, I use these terms in a descriptive sense to classify areas of relatively unified and distinct structural style that evolved under contrasting physical conditions. However, I emphasize that the structural evolution of the Ruby Mountains was a polyphase process that involved the superposition of both Mesozoic and Cenozoic tectonic events.

Perhaps one of the most perplexing problems in the Ruby Mountains is the age of regional metamorphism. Although the data are equivocal, a Mesozoic age as originally proposed by Misch (1960) and Misch and Hazzard (1962) seems probable. Data from the Ruby Mountains that support this hypothesis are (1) Cretaceous Rb-Sr whole-rock age determination from the postkinematic Dawley Canyon two-mica granite (Willden and Kistler, 1969); (2) Jurassic Rb-Sr whole-rock age determinations from pegmatitic granite of the migmatite complex (Willden and Kistler, 1969); and (3) a clustering of minimum ages in the Mesozoic of discordant zircons from orthogneisses of the infrastructural core as well as the structurally overlying *Abscherungszone* (T. W. Stern, 1978, written commun.).

ROCK UNITS

Metamorphic Complex

Howard (1966, 1971) recognized that the high-grade metamorphic complex of the Ruby Mountains could be divided into mappable stratigraphic units. Although large parts of the complex consist of more than 50% granitoid rock, layers and lenses of metasedimentary rocks are sufficiently abundant to consistently identify each unit. Furthermore, in adjoining areas, where granitic rocks are less abundant, the stratigraphic sequence and character can be developed in considerable detail (Fig. 3). In the areas shown in Figures 4 and 5, large parts of the stratigraphic section are either attenuated or inverted so that at no single locality can one identify a complete, right-side-up section. However, Howard (1966) has found a convenient reference section near the mouth of Lamoille Canyon (Fig. 2) that appears to be right-side-up and to contain all the major metasedimentary units found in the northern Ruby Mountains. I accept this section as the "standard" stratigraphic datum for the metamorphic complex. In the following paragraphs and in Figure 3, the units, modified from Howard (1971), are described as characterized in the northern Ruby Mountains.

Late Precambrian-Cambrian Impure Metaquartzite and Schist. The oldest unit recognizable is a widespread impure metaquartzite-schist unit (Kistler and Willden, 1969a) tentatively correlated with the late Precambrian McCoy Creek Group of Misch and Hazzard (1962) and the Lower Cambrian Prospect Mountain Quartzite. In the high-grade core mapped by Howard (1966), massive brown-weathering metaquartzite is characteristic and may nearly exclusively correlate with the Prospect Mountain Quartzite (Howard, 1971). North of Soldier Creek Canyon the analogous unit is more heterogeneous and may therefore include a substantial part of the McCoy Creek Group. In the infrastructure terrane (for example, Hidden Lakes uplift, Fig. 4), the metaquartzite is coarse grained and blocky (Howard, 1971). The metaquartzite characteristically contains subordinate mica and feldspar; accessory minerals include sillimanite, garnet, opaque oxides, and zircon. Scarce calcareous metaquartzite contains calc-silicate minerals such as hornblende, diopside, and garnet. Interlayered metapelite consists of the assemblage quartz + biotite + muscovite + plagioclase + sillimanite + opaque oxides ± garnet. In the thinned *Abscherungszone* above the infrastructural core terrane, the metaquartzite becomes progressively flaggy and mylonitic. The rock is strongly foliated and is pervasively lineated. The lineation resembles slickensides but is penetrative. Primary sedimentary textures and structures have not been recognized in either the massive or flaggy varieties of this unit.

Impure Calcite Marble and Calc-Silicate Rocks. A widespread but heterogeneous unit in the

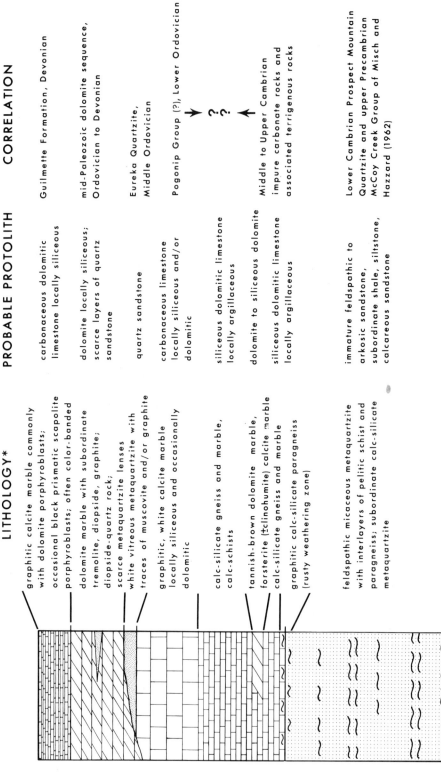

Figure 3. Generalized stratigraphic column of the metasedimentary rocks of the Ruby Mountains meta-morphic complex (modified from Howard, 1971).

northern Ruby Mountains is characterized by impure calcite marble and calc-silicate rocks. The calcite marble varies from massive but foliated graphite-bearing white marble to calc-schists with abundant diopside, tremolite-actinolite, phlogopitic biotite, quartz, and feldspar. Metadolomite is not an abundant component of the sequence; however, a distinctive brown-weathering, massive dolomite marble forms a discontinuous layer a few hundred feet above the top of the Hidden Lakes uplift (Fig. 4). Howard (1971) mapped a similar brown metadolomite south of my area and correlated it with a widespread horizon apparently characteristic of the base of the miogeoclinal Cambrian carbonate sequence.

The calc-silicate rocks are characterized by diopside-rich segregation bands, but invariably show tremendous modal as well as textural variations. The mineral assemblages of some representative calc-silicate rocks and calc-schists are summarized in Table 1. The most common calc-silicate rock is thinly layered; resistant calc-silicate–rich layers alternate with less resistant carbonate-rich layers. Although this compositional layering probably reflects general original lithologic and chemical variations in the protolith, the present uniform regularity of the layers is more a manifestation of metamorphic segregation and tectonic transposition. These calc-silicate rocks represent siliceous dolomitic limestone apparently locally argillaceous, for biotite-rich layers are common.

The contact between the calcite marble–calc-silicate rock unit and the presumably older impure metaquartzite-schist sequence is often characterized by the presence of graphitic paragneiss (Howard, 1971).

Blue-gray Calcite Marble. In the Secret Creek Gorge area (Fig. 5), thin tectonic slivers of low-grade metasedimentary rocks occur immediately below the low-angle fault complex that emplaces unmetamorphosed upper Paleozoic and Tertiary rocks onto the metamorphic substratum. Three distinct lithologies occur as tectonic slices: blue-gray, fine-grained calcite marble; white metaquartzite; and fine-grained metadolomite. The white metaquartzite and the metadolomite appear to be metamorphosed Eureka Quartzite and a middle Paleozoic dolomite, respectively. The stratigraphic position of the blue-gray marble is uncertain, but it may be metamorphosed Cambrian Windfall Formation or perhaps Ordovician Pogonip Group. Regardless of the precise correlation, the unit probably is stratigraphically below the Eureka Quartzite, for a remarkable aspect of these tectonic slices is that, despite extreme attenuation and faulting, their apparent proper stratigraphic order is maintained.

Textural variations of the marble are extreme. Some rocks are barely recrystallized though intensely brecciated, whereas other rocks within the same tectonic slice are strongly folded tectonites with well-developed axial plane schistosity. Color banding consisting of an alternation of bluish-gray and white marble is especially characteristic of the tectonite.

Eureka Metaquartzite. Lenses of white, nearly pure metaquartzite occur at numerous localities throughout the northern Ruby Mountains. The metaquartzite varies from massive to strongly foliated and lineated quartz mylonite. Scarce impurities include white mica, graphite, and feldspar. Howard (1971) correlated this unit with the Middle Ordovician Eureka Quartzite, and Thorman (1970) has described similar metamorphosed Eureka Quartzite in the Wood Hills (Fig. 1).

Metadolomite. Massive white metadolomite, which overlies the Eureka metaquartzite, or the impure calcite marble–calc-silicate rock unit when the metaquartzite is missing, is the metamorphic equivalent of the middle Paleozoic dolomite interval characteristic of the eastern Great Basin miogeocline. Metadolomite which has undergone amphibolite facies metamorphism is massive and coarse grained. Visible layering is scarce. The rock is typically almost free of impurities except for minor graphite. However, tremolite- and diopside-bearing metadolomite, as well as diopside-quartz rock, is present in the unit. South of Wilson Creek, lenses of white metaquartzite (metamorphosed Simonson sandstone?) occur in the metadolomite (Fig. 4).

The metadolomite unit also occurs as low-grade (greenschist facies?) tectonic slices under the nonmetamorphic allochthonous complex in the Secret Creek Gorge area (Fig. 5). These metadolomitic

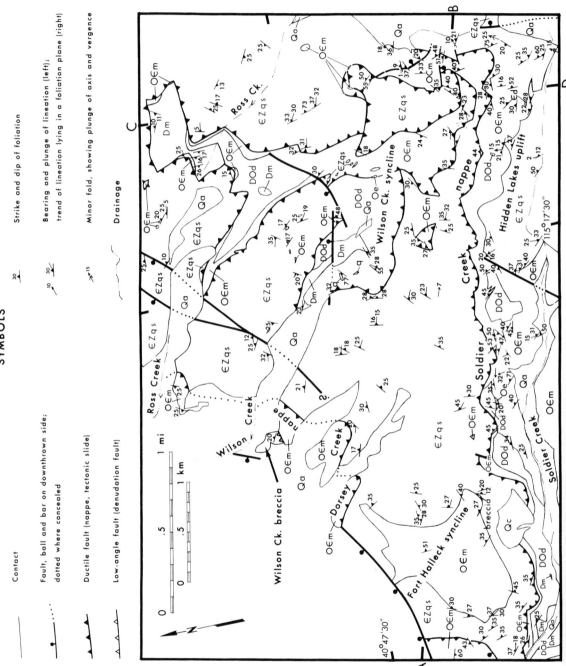

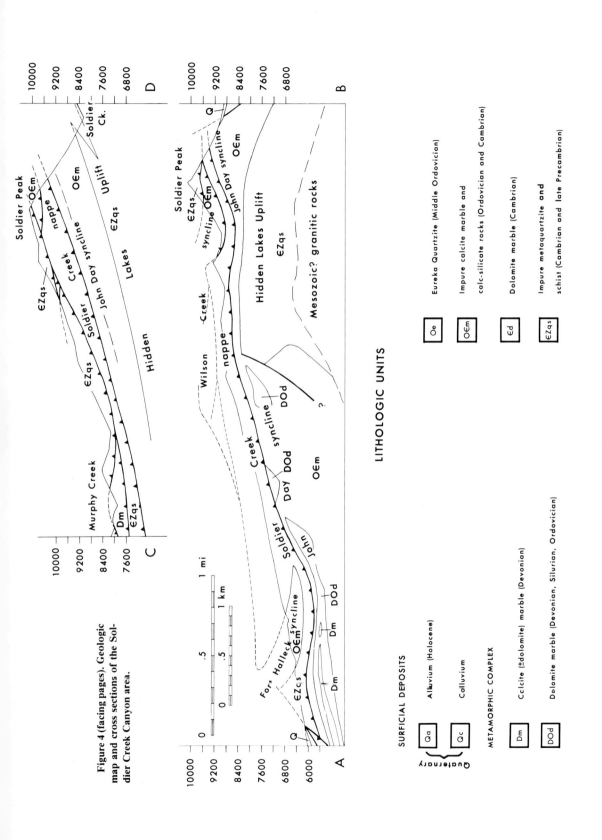

Figure 4 (facing pages). Geologic map and cross sections of the Soldier Creek Canyon area.

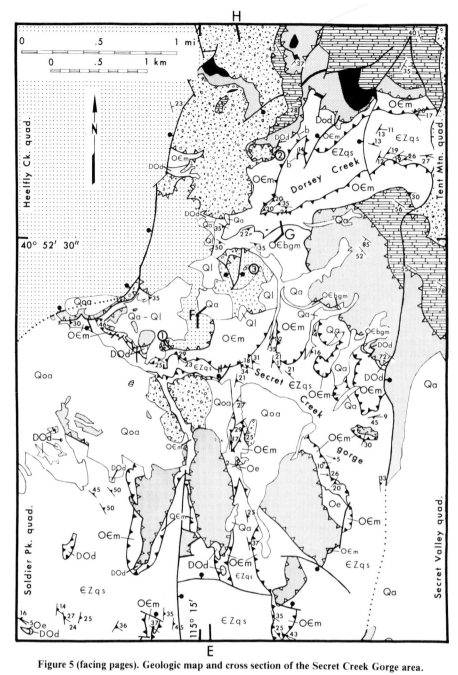

Figure 5 (facing pages). Geologic map and cross section of the Secret Creek Gorge area.

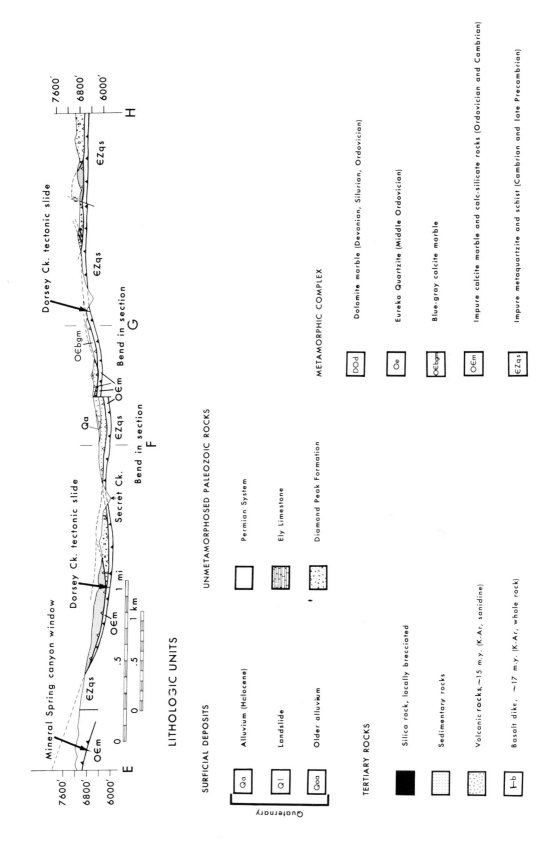

TABLE 1. REPRESENTATIVE MINERAL ASSEMBLAGES OF IMPURE MARBLES AND CALC-SILICATE ROCKS

	1	2	3	4	5	6	7	8	9	10	11	12	13	14	15
Sample no.	8-4*	934-T2*	172-10	934G*	197-4*	170-58	189-17	188-15B	B5	143-3	170-2	170-31X	170-41A	170-47	170-49
Calcite	X	X	-	X	X	X	X	X	-	0	X	X	X	X	X
Dolomite	-	X	-	-	-	-	-	-	-	-	-	-	-	X	-
Quartz	-	-	X	X	-	-	X	-	X	-	X	-	X	-	-
Plagioclase	X	-	X	X	X	X	X	X	X	X	X	X	X	-	-
K-feldspar	-	-	-	-	-	-	-	0	X	X	X	X	X	-	-
Diopside	-	-	-	X	X	X	X	X	X	X	X	X	X	-	-
Tremolite-Actinolite	X	X	X	X	0	X	X	-	X	X	X	X	-	0	X
Biotite	-	-	X	-	X	X	-	-	X	X	X	X	0	-	-
Phlogopite	X	X	-	-	-	-	-	-	-	-	-	-	-	-	X
Scapolite	-	-	-	-	X	X	X	-	-	-	X	X	X	-	X
Humite group	-	Xch	-	-	-	-	-	-	-	-	-	-	-	Xs	Xcl
Forsterite	-	-	-	-	-	-	-	-	-	-	-	-	-	-	Xs
Zoisite	-	-	-	-	-	-	-	X	-	-	-	-	-	-	-
Chlorite	0r	-	-	-	Xr	-	-	0r	-	0r	-	0r	-	-	0r
Sphene	0	X	0	0	X	0	0	X	0	X	0	0r	X	-	-
Apatite	-	-	X	-	-	0	0	0	0	0	0	0	0	-	0
Graphite	0	-	-	-	-	0	0	-	X	-	-	-	-	-	0
Opaque oxides-Hematite	0	0	-	0	0	0	0	-	0	0	0	0	0	0	0
Zircon	-	-	0	-	0	-	0	-	0	0	-	0	-	-	-

Note:

X = mineral present in amounts >1.0%.
0 = mineral present in amounts <1.0%.
- = mineral not present.
* = sample from Dorsey Creek nappe.
ch = chondrodite.
cl = clinohumite.
s = locally altered to serpentine-group minerals.
r = mineral species is a retrograde replacement after a prograde mineral.

rocks are massive but typically intensely fractured and fine grained. They weather whitish but vary from dark to very light gray on freshly broken surfaces. Occasionally the low-grade metadolomite is finely laminated, which appears to reflect both original composition as well as penetrative deformation.

Calcite-Dolomite Marble. Overlying the metadolomite and apparently gradational with it is a calcite marble commonly with subordinate dolomite (Howard, 1966). The marbles characteristic of this unit are graphitic, giving them a typical gray color. Black scapolite porphyroblasts, colored by disseminated carbonaceous inclusions, are especially distinctive; other accessory minerals include white mica, phlogopite, tremolite, and opaque oxides. This unit based on its composition and apparent stratigraphic position best correlates with the Devonian Guilmette Formation (Howard, 1971).

Granitic Rocks. Granitic rocks are conspicuous throughout the metamorphic complex, and locally account for more than 50% of the metamorphic terrane. The most widespread rock is leucocratic, often pegmatitic, consisting dominantly of K-feldspar, sodic plagioclase, and quartz; modal variations span the granite to trondhjemite fields. Muscovite and biotite are commonly present, while garnet, sillimanite, and zircon are occasional accessory minerals. The rock forms dikes, sills, and irregular masses in the metasedimentary sequence, and the emplacement history ranges from the early synkinematic stage to postkinematic phases. This type of granitoid rock is apparently widespread in infrastructural terranes (Haller, 1971); Griffin (1974) referred to analogous granitic rocks in the southern Appalachian Inner Piedmont as "felsic mobilizate" and that name is adopted here.

Concordant layers of mafic to felsic gneiss are locally abundant in the metamorphic complex. The available field, petrographic, and geochemical data (Snoke, unpub. data) suggest that these gneisses are chiefly of igneous origin. The composition of the orthogneisses is variable, ranging from quartz gabbro to leucogranite; however, biotite ($\pm$ muscovite) granite (IUGS classification) is perhaps the most abundant variety. The more mafic gneisses contain hornblende and biotite, whereas biotite is the characteristic mafic mineral in the granodiorite to granitic gneisses. Some extremely leucocratic orthogneisses (color index <5) are garnet-bearing, two-mica granites.

Zircons extracted from granitic gneisses found in the impure metaquartzite-schist unit and the presumably younger impure marble–calc-silicate rock sequence have yielded strikingly discordant U-Pb isotopic data, indicating a complex thermal history. However, $^{207}Pb/^{206}Pb$ ratios in both samples exceed 1 b.y.; this fact strongly suggests the presence of a Precambrian Pb component in these gneisses (T. W. Stern, 1978, written commun.). These geochronological studies are not complete and several interpretations are feasible. However, if the gneisses are truly igneous and the tentative stratigraphic correlation is correct, the gneisses may represent anatexites remobilized from a Precambrian terrane during Mesozoic(?) regional metamorphism. Compton and others (1977) have described what may be an analogous situation in the Grouse Creek–Raft River Mountains, northwest Utah, where remobilized Precambrian rocks have intruded younger rocks but have retained their Precambrian isotopic signature.

Mafic and Ultramafic Rocks. Hornblende-rich mafic rocks are a distinctive but volumetrically insignificant magmatic element of the metamorphic complex. These rocks are late kinematic to postkinematic and typically occur as small dikes, sills, or irregular mases in the metasedimentary rocks. The hornblende-rich rocks commonly vary from quartz diorite to gabbro, but at one locality feldspathic websterite is associated with rocks of this assemblage.

Unmetamorphosed Paleozoic Rocks

The unmetamorphosed Paleozoic rocks that are part of the low-angle fault complex exposed in the Secret Creek Gorge area (Fig. 5) have been divided into three broad units: Diamond Peak Formation, Ely Limestone, and Permian System (undifferentiated). Detailed lithologic and stratigraphic data are available for these units from other areas where structural complexity is significantly less and rock exposures are superior (Berge, 1960; Steele, 1960; Brew, 1971; Marcantel, 1975). Therefore, the lithologic descriptions of these units are concise and paleontological data are summarized in Appendix 1.

Diamond Peak Formation. A distinctive orangish-yellow to red color typically characterizes exposures of the Diamond Peak Formation. Grit and pebble conglomerate are common lithologies, while tan-weathering sandstone and gray micrite limestone are subordinate components in this formation. No attempt has been made to separate the Chainman Shale from the Diamond Peak Formation; therefore, several areas of poorly exposed, fine-grained detrital rocks (shale and siltstone) have been included with the Diamond Peak Formation. The contact with the overlying Ely Limestone is preserved in several of the low-angle fault slices and is clearly gradational. Conglomerates of the Diamond Peak type form distinct interlayers within the basal Ely Limestone. The Diamond Peak Formation is of Mississippian to Early Pennsylvanian age and it, along with the Chainman Shale, is commonly considered to be part of a clastic wedge derived from the Antler orogenic belt of central Nevada.

Ely Limestone. Rocks mapped as Ely Limestone are massive to slabby, medium gray limestone often chiefly composed of carbonate mud and organic detritus. Even the most fossiliferous limestones appear to have a mud-supported texture. Chert, white to dark gray, commonly forms nodular, somewhat irregular masses in the limestone. Fusulinid-rich calc-arenite is a subordinate lithology.

Contacts between the Ely Limestone and the other rock units are commonly faults (Fig. 5); however, as previously indicated, a gradational contact between the Mississippian Diamond Peak Formation and the overlying Ely Limestone is preserved locally. Contacts between the Ely Limestone and younger rocks (that is, the Permian System) are always faulted in the Secret Creek Gorge area. These faults include high-angle as well as low-angle faults; the low-angle faults include both older-over-younger and younger-over-older types.

Permian System. Permian rocks are an important component of the low-angle fault complex exposed in the Secret Creek Gorge area and apparently include elements of the Ferguson Mountain and Pequop Formations. Limited exposures and structural complexities have necessitated that all the rocks assigned to the Permian be mapped as an undifferentiated unit, although paleontological data (see App. 1) clearly indicate the presence of several distinct time-stratigraphic units. Massive gray silty limestone is perhaps the most common lithology, although many of the Permian rocks contain much terrigenous material and are best classified as calcareous sandstone. In general, the abundance of terrigenous debris in the Permian rocks is the most distinctive aspect of the unit.

Marcantel (1975) has recognized distinctive facies characteristic of Pennsylvanian-Permian stratigraphy in Nevada and has developed various tectono-stratigraphic models to explain these variations. I have recognized several of Marcantel's facies in the Secret Creek Gorge area: fusulinid-biomicrite, brachiopod-bryozoan biomicrite, crinoid-foraminifer biosparite, coral biolithite, conglomerate, and very fine sand. However, in that these Permian rocks are part of a low-angle fault complex, many original depositional textures have been partially obliterated by brecciation, silicification, and recrystallization.

Tertiary Rocks

Tertiary rocks are widespread near the western entrance to Secret Creek Gorge, and several small klippen of Tertiary strata rest directly on upper Paleozoic and metamorphic rocks north of the gorge (Fig. 5). These rocks were mapped by Sharp (1939a) as part of his Miocene Humboldt Formation. Later Snelson (1957) restudied the Cenozoic rocks on the flanks of the East Humboldt Range and northern Ruby Mountains and separated the Tertiary rocks into four informal units. Snelson's (1957)

Clover Valley unit is exclusively exposed in my study area, although basalt dikes that may be temporally equivalent to his Warm Springs unit intrude the metamorphic rocks.

The Tertiary sedimentary rocks vary in color from shades of red and lavender to pale green; texture ranges from siltstone to conglomerate. The finer-grained rocks are commonly tuffaceous, whereas the conglomeratic deposits contain a variety of clasts including reworked Tertiary sedimentary rocks, volcanic rocks, various unmetamorphosed Paleozoic units, and scarce metamorphic rock fragments. Some of the metamorphic rock fragments contain the conspicuous lineation characteristic of the *Abscherungszone* and thereby indicate that the Ruby Mountains–East Humboldt Range metamorphic complex was exposed during the deposition of these Tertiary rocks. This observation puts definite constraints on the time of metamorphism and chronology of low-angle faulting.

Volcanic rocks are a subordinate component in the Tertiary sequence of the Secret Creek gorge area, but both Sharp (1939a) and Snelson (1957) indicate that volcanic rocks are more common along the western flank of the East Humboldt Range in Starr Valley. I have not examined these rocks, but lithologic descriptions (Sharp, 1939a; Snelson, 1957) suggest a similarity to the volcanic rocks of the Secret Creek Gorge area. The volcanic rocks in my study area (Fig. 5) vary from dense black vitrophyre to pink porphyritic rhyolite to a vuggy rhyolite partially silicified. Sanidine and quartz phenocrysts are common in all varieties, and the vitrophyre also contains green pyroxene (hedenbergitic) phenocrysts. The volcanic rocks appear to be local flows intercalated with the sedimentary rock sequence. Immediately north of Secret Creek Gorge (Fig. 5, locality 3), the vitrophyric rocks occur at the base of a small patch of volcanic rocks which in turn is part of a klippe of Tertiary rocks overlying the metamorphic substratum. Sanidine from the vitrophyre yielded a K-Ar date of 15.0 ± 1.5 m.y. (E. H. McKee, 1978, written commun.). Although the exact stratigraphic position of the volcanic rocks within the Tertiary section is uncertain, my data coupled with Snelson's (1957) stratigraphy imply that the Tertiary rocks along the flanks of the northern Ruby Mountains and East Humboldt Range are probably Miocene or younger.

STRUCTURAL FEATURES

The structural transition from the migmatitic core of the Ruby Mountains metamorphic complex to the nonmetamorphic but allochthonous upper Paleozoic cover rocks records a diachronous history of polyphase deformation. The purpose of the following structural analysis is to describe and classify the various deformation elements manifested in the rocks of this transition zone (that is, the *Abscherungszone,* after the concept of Haller, 1956). The structural style of the core has been documented in detail by Howard (1966, 1968), and his nomenclature has generally been retained to facilitate a continuity between our studies.

The change from the core terrane into the *Abscherungszone* is best exposed in the Soldier Creek Canyon area (Fig. 4). In this area a series of remarkably appressed recumbent folds passes into a braided family of ductile faults (that is, tectonic slides of Fleuty, 1964). The boundary between the *Abscherungszone* and the allochthonous cover rocks is well exposed in Secret Creek Gorge (Fig. 5) where tectonic slivers of low-grade metasedimentary rocks delineate the base of a low-angle fault complex. A thin but extensive sheet of metacarbonate and associated rocks (Dorsey Creek nappe) is the most prominent structural feature at the top of the *Abscherungszone*. The nonmetamorphic rocks of the suprastructure are dramatically attenuated into fault-bounded slivers which generally emplace younger rocks on older. Field relations supplemented by geochronological studies (in progress with E. H. McKee and T. W. Stern) indicate that these structural relations are the product of tectonic overprinting of contrasting ages. Therefore, although a well-defined transition zone between infrastructure and suprastructure is preserved in the Ruby Mountains, some aspects of this structural and metamorphic transition have been masked by younger deformations. The *Abscherungszone* as developed in the northern Ruby Mountains is a zone of intense strain where the dynamic effects of regional metamorphism were prominent, as compared to the infrastructure where continual heating outlasted deformation, or the suprastructure where brittle deformation dominated.

Fold Geometry and Succession

The oldest penetrative deformation in the range is characterized by scarce flattened isoclines and a widespread northwest to east-west elongation lineation. This F_1 fold system in part merges with but is distinctly overprinted by a variety of synmetamorphic to late(?) metamorphic fold phases [Howard's (1966) F_2 system] which vary in style and orientation. Although some of these fold phases unquestionably postdate the F_1 system, the precise interrelations between them are uncertain, and they are given geographic names so as to provide type areas as well as to be free of chronologic restriction. A comparison of style of the F_1 fold system and the so-called F_2 system is given in Table 2, together with an overview of postmetamorphic fold characteristics.

To detect variations in structural style and to isolate relatively homogeneous domains, the study area has been divided into subareas (Fig. 6). Structural data collected throughout the northern Ruby Mountains have been plotted on a series of equal-area stereographic projections (Fig. 7); only representative structural readings are shown on the detailed geologic maps (Figs. 4, 5). Synoptic diagrams (Fig. 8) have also been prepared to provide an overview of the study area.

F_1 **Fold System.** Folds of this generation form tight, cylindrical isoclines that are often difficult to detect unless the examined rock surface is crudely perpendicular to the fold axis. These folds are classic slip folds in that original bedding (S_0) acts as a passive element only to outline the shape of the fold (Fig. 9A, 9B). Although mesoscopic F_1 folds occur nearly throughout the study area, no macroscopic F_1 folds have been found. Furthermore, the map pattern of the lithologic units is not controlled by this early deformation phase.

A well-developed penetrative foliation (S_1) parallels the axial surface of these folds and is the dominant planar structural element in the northern Ruby Mountains (Fig. 8A). Throughout much of the study area, this surface is a mylonitic flattening foliation that developed during a ductile deformation whose strains are still well preserved in the rocks. Commonly within this foliation is an elongation lineation defined by rodded and streaked-out mineral grains. At many localities in the metasedimentary sequence, this lineation is the product of an intersection between original bedding (S_0) and the penetrative foliation (S_1). In other rocks, especially the relatively homogeneous orthogneisses, the lineation is not an intersection phenomenon but is defined solely by the streaking of mineral grains in the plane of foliation. This conspicuous L_1 lineation shows an amazing uniformity, trending approximately west-northwest throughout the entire study area (Fig. 8B). The L_1 lineation is commonly parallel to F_1 fold axes, but several examples have been found where L_1 and adjacent F_1 fold axes are not precisely coincident. This lineation, which is often down-the-dip of foliation, is similar to so-called transverse linear structures well-documented in other orogens such as the Caledonides of Scandinavia and Scotland as well as the southern Appalachians of the United States (Kvale, 1953; Bryant and Reed, 1969; Olesen and Sorensen, 1972). Controversy has long existed as to whether these structures parallel the a or the b tectonic axis.

To further describe the fabric elements associated with the F_1 fold system, the quartz microfabric of two impure metaquartzites were studied (Fig. 10). Quartz constitutes over 80% of the analyzed rocks, but the modal abundances of subordinate mineral phases are variable between the specimens. In 15-2E, lensoidal muscovite porphyroclasts elongated parallel to L_1 are widespread with streamlined feldspar microaugen slightly less common. In 140-15E, feldspar is the common varietal mineral, while biotite and scarce muscovite constitute only a small percentage of the rock. The character of the quartz grains

TABLE 2. MESOSCOPIC STRUCTURAL ELEMENTS OF THE NORTHERN RUBY MOUNTAINS METAMORPHIC COMPLEX

	F_1 fold system (synmetamorphic)	F_2 fold system (synmetamorphic to late metamorphic)				F_3 fold system (postmetamorphic)
		Secret Creek Gorge generation	Sharps Creek generation	Soldier Creek Canyon generation	Hidden Lakes uplift generation	
Fold style	Small-scale cylindrical, tight isoclines; slip folds	Flexural slip to flexural flow to passive, locally disharmonic	Flexural slip to flexural flow, locally disharmonic	Flexural slip to flexural flow, locally disharmonic	Cylindrical, tight isoclines; slip folds	Kink folds; concentric, open folds; scarce fold mullions (Dorsey Creek nappe)
Surfaces folded	S_0 (bedding) and compositional layering	S_0, compositional layering, S_1 foliation	S_0, compositional layering, S_1 foliation	S_0, compositional layering, S_1 foliation	S_0, compositional layering	S_0, compositional layering S_1 foliation
Planar structures	S_1, the dominant metamorphic foliation of the Abscherungszone; commonly a mylonitic flattening fabric	S_2, secondary foliation locally present in the crestal areas of F_2 folds	S_2, locally present in the crestal areas of F_2 folds; crenulation cleavage in pelitic layers	S_2 foliation parallels axial surface of F_2 folds, often weak but locally strongly developed	S_2, penetrative foliation, parallels axial surface of F_2 folds	Local kink bands but characteristically no penetrative foliation
Linear structures	L_1, elongation lineation; intersection between S_0 and S_1 in the metasedimentary rocks (L_{0x1}); elongation and streaking of mineral grains in orthogneisses	L_{0x2}, L_{1x2} lineations along hinge lines of F_2 folds	L_{0x2}, L_{1x2} lineations along hinge lines of F_2 folds; small-scale crenulations in pelitic layers	L_{0x2}, L_{1x2} lineations along hinge lines of F_2 folds; scarce folded early L_2 lineations by later F_2 folds	L_{0x2} lineation, strong orientation of synkinematic mineral phases parallel to F_2 folds	L_{1x3}, crinkling of micaceous folia, and kinking of mylonitic S_1 foliation
Orientation	Strong west-southwest trend throughout the Abscherungszone, no consistent vergence	North-northwest to northeast trend with consistent westward vergence	Northwest trend, south-west vergence is common	North-northwest to northeast trend, eastward vergence is common	North-northwest to north-northeast trend, no consistent vergence	Insufficient data but apparently widely variable

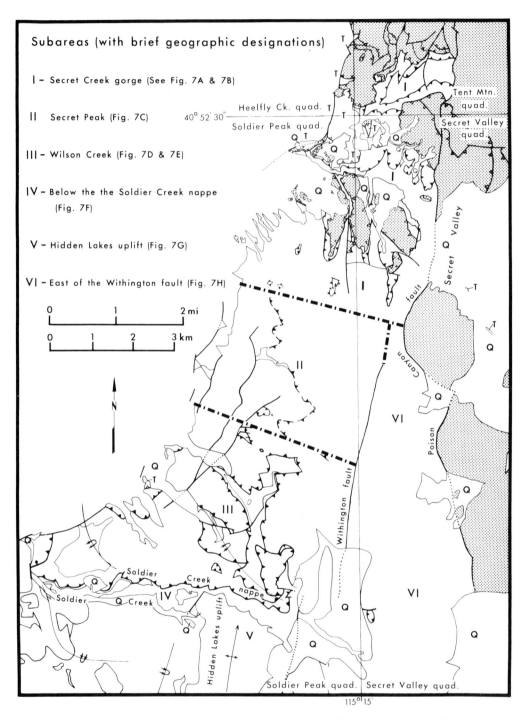

Figure 6. Structural map of the northern Ruby Mountains showing division into subareas. Stippled pattern indicates unmetamorphosed upper Paleozoic sedimentary rocks; T = Tertiary rocks; Q = Quaternary deposits; otherwise symbols are the same as used in Figures 4 and 5. Geologic data south of Soldier Creek after Howard (1966).

also contrasts between the rocks. In 15-2E, quartz occurs either as (1) strained ribbonlike grains exhibiting widespread subgrain formation (maximum length is parallel to L_1 and commonly exceeds several millimetres) or (2) considerably smaller (typically < 0.25 mm), inequant strained to strain-free grains flattened in the foliation plane. The small grains appear to represent syntectonic recrystallization of the ribbonlike grains. In 140-15E, the quartz grains are lensoidal, intensely strained, and complexly intergrown with adjacent grains. Along the sutured boundaries of these large quartz grains (commonly the maximum dimension perpendicular to L_1 is several millimetres) are scattered recrystallized quartz grains. Although the details of the microfabric diagrams differ (double maximum versus single maximum), the symmetry of the quartz c-axis fabric in both rocks is essentially orthorhombic. This aspect combined with the mylonitic foliation, elongation lineation, and tight isoclinal character of the F_1 folds is a characteristic apparently universally common in quartz-rich rocks that have undergone intense ductile deformation (Christie, 1963; Dalziel and Bailey, 1968; Ross, 1973). Flattening perpendicular to the mylonitic foliation seems certain; however, there has been continuing controversy concerning whether the deformation history involved progressive pure shear (for example, Johnson, 1967) or progressive simple shear (for example, Escher and Watterson, 1974). Although symmetry arguments based on the orthorhombic microfabric of the quartz c axes would favor progressive pure shear as the dominant deformational mechanism, apparently nearly analogous quartz microfabrics are also found in tectonites deformed by progressive simple shear (Hara and others, 1973). Furthermore, the monoclinic quartz fabrics determined by X-ray analysis by Riekels and Baker (1977) of Christie's (1963) specimen 62 cast doubt on the detailed orthorhombic nature of quartzite mylonites.

Nevertheless, taking the progressive simple shear and pure shear mechanism a step further, at least two tectonic models are consistent with the data: (1) deep-seated ductile deformation involving uniform translation (that is, regional thrusting) of rock material parallel to the L_1 lineation or (2) vertical uplift (that is, diapiric?) with flattening parallel to S_1 and the axis of maximum extension parallel to L_1. At present, my data are not sufficient to differentiate between these models; furthermore, it is conceivable, if not likely, that elements of both models were operative during the deformation. For example, Bell (1978) has argued that the evolution of the Woodroffe thrust, a ductile mylonite zone, initially involved inhomogeneous bulk flattening with a progressive increase in the inhomogeneous simple-shear component leading eventually to brittle deformation and detachment. Such a model involving a diachronous deformation path seems most consistent with the structural history of the northern Ruby Mountains as it is now understood.

As previously emphasized, the origin of elongation lineations is a recurring enigma characteristic of the crystalline core of many orogens. The azimuths of these lineations are characteristically remarkably uniform over large regions, and although the best known examples are transverse to the orogenic belt, elongation lineations parallel to the orogenic belt are also known (for example, southern Appalachians–Kiokee belt, Secor and Snoke, 1978). It is tempting to suggest that this fabric may be a good approximation of the flow lines characteristic of the crystalline core. If so, the fabric is a powerful tool for outlining terranes of uniform tectonic transport as well as for recognizing significant variations in movement direction along the length of the orogen.

In summary, the parallelism of S_1 to the axial surface of F_1 mesoscopic folds and the common alignment of synkinematic prismatic minerals parallel to F_1 fold axes indicate that the development of the F_1 fold system was coincident with the peak of regional metamorphism. Local nonparallelism between L_1 and F_1 fold axes is interpreted to indicate that initial folding somewhat preceded the maximum ductile deformation (that is, the mylonization) characteristic of the *Abscherungszone*. Total strain associated with this first-phase deformation was apparently high, but the presence of synkinematic sillimanite also indicates high-temperature conditions during the mylonization. The rocks of the *Abscherungszone,* therefore, have undergone amphibolite-facies mylonization and are another example of a growing list of high-grade mylonite zones common in many orogenic belts (Theodore, 1970).

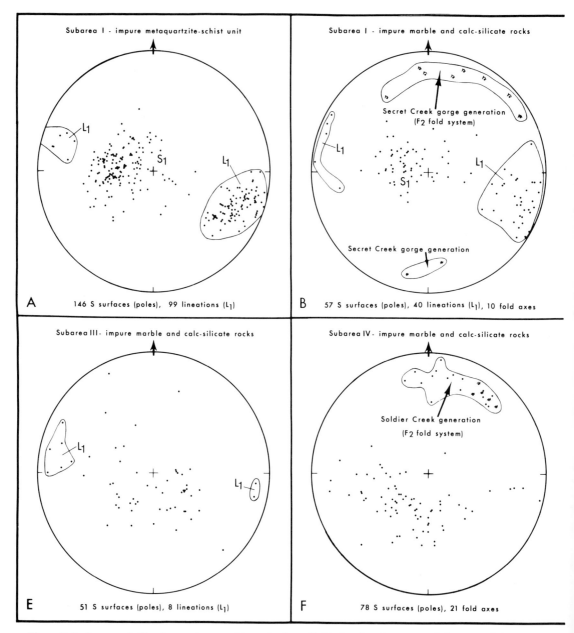

Figure 7 (facing pages). Equal-area stereographic projections (A-H) showing the structural elements of the various subareas.

Although later deformations deform the mylonitic foliation and associated L_1 lineation, the texture and microfabric of the rocks suggest that much of the strain experienced during the first-phase deformation is retained. In contrast, the rocks of the high-temperature core are characterized by fold trends nearly at right angles to the F_1 system, and textures generally indicate that recrystallization outlasted deformation. If F_1 was ever present in these rocks, it has been totally obliterated during later high-temperature deformation.

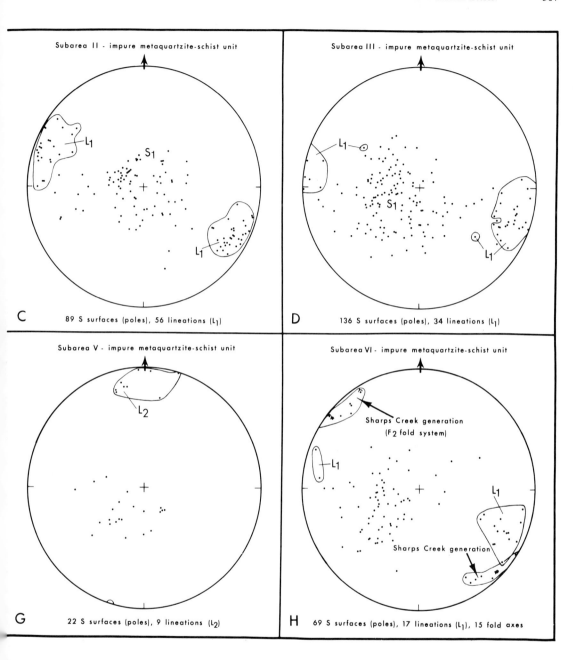

F_2 Fold System. In contrast to the F_1 fold system, the F_2 system is characterized by considerable variation in style and orientation. Folds of this system apparently formed during changing or at least contrasting physical conditions. Some F_2 folds are characterized by foliations that parallel their axial surface, whereas others fold an earlier flattening fabric (S_1 and L_1). In the migmatitic core, Howard (1966) has mapped large F_2 folds that control the distribution of lithologic units. The most extensive and largest F_2 fold is the southeastward overturned Lamoille Canyon nappe, which is well exposed in

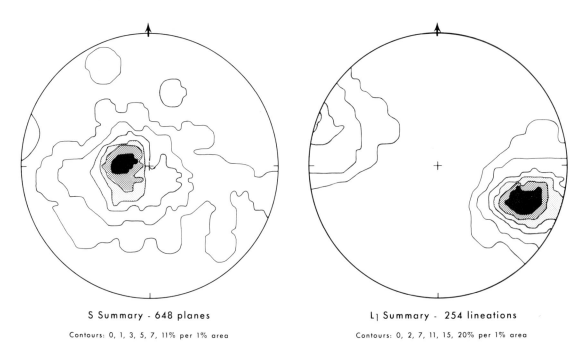

Figure 8. Synoptic equal-area stereographic projections (contoured) showing S and L_1 data for the study area.

numerous glaciated canyons south of my study area. In Figure 2, the distribution of the Lower Cambrian to upper Precambrian impure metaquartzite-schist unit north and south of Lamoille Canyon serves to outline the trend and shape of this major structure of the infrastructural core.

As previously emphasized, the following "fold generations" represent groups of folds that share similarity in style and orientation. In that none of these "fold generations" occur in overlapping domains, I am uncertain of their relative chronology; however, I judge that they are relatively coeval, and apparent variations are related to contrasting physical conditions such as flow direction, ductility, lithology, and so forth.

Secret Creek Gorge Generation. Spectacular examples of folds of this group are well developed in the thin plate of metacarbonate and associated rocks (Dorsey Creek nappe) that occurs both north and south of Secret Creek Gorge. These folds deform both S_0 and S_1 and locally rotate L_1 lineations. The folds are flexural-flow in style and are sometimes intensely disharmonic, but they consistently show a westward vergence (Fig. 11). Local reorientation of mica platelets and $L_{1\times 2}$ lineations are common near the crestal areas of the folds, but petrographic data imply that the folds are late metamorphic and therefore postdate the main phase of penetrative S-surface development. On the other hand, the consistency of the vergence despite variations in fold-axis trends suggests that the structures may have developed during ductile faulting. The available structural data on these folds (Fig. 7B) imply a slip line directed westward roughly along a west-northwest trend. This azimuth is parallel to the early elongation lineation (L_1), and Howard (1966, 1968) has argued that both the synmetamorphic and late metamorphic structures in the northern Ruby Mountains maintained a unique flow line and are part of a continuum of deformation. My data substantiate this interpretation; but it should be emphasized that the direction of overturning (that is, tectonic transport) in part of the infrastructural core is west to east, while an opposite sense of movement, east to west, is characteristic of the overlying tectonically flattened and thinned *Abscherungszone.*

Figure 9. F_1 folds in the impure metaquartzite-schist unit. (A) Tight isoclines in impure metaquartzite. The hammer handle is parallel to the L_1 lineation. The folds are located on the north side of Secret Peak. (B) Sawed slab of impure metaquartzite showing an F_1 isocline. Note that the S_1 foliation is parallel to the axial surface of the fold and transects the hinge of the fold. Thin feldspathic veinlets transect the S_1 foliation.

Finally, although distinctly less abundant, folds of the Secret Creek Gorge generation are also present in the impure metaquartzite-schist unit immediately below the Dorsey Creek nappe. These folds are best developed in the pelitic-rich horizons of the sequence and vary in style from flexural-flow folds to crenulation structures (Fig. 12). Wherever vergence is determinable, it is consistently to the west.

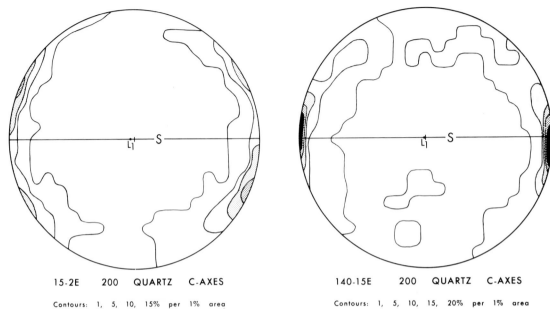

Figure 10. Orientation diagrams for quartz [0001] in impure metaquartzites. The analyzed sections were cut perpendicular to foliation (S_1) and nearly perpendicular to the L_1 lineation.

Soldier Creek Canyon Generation. The folds of the Soldier Creek Canyon generation trend north-northwest to northeast and are consistently overturned to the east (Fig. 7F). They are particularly widespread in the impure calcite marble–calc-silicate rock sequence along the lower limb of the John Day recumbent syncline (Fig. 4). The character of these folds is variable, many are flexural-flow folds, some of them disharmonic, with thickened crests and thinned limbs (Fig. 13). Typically these folds lack a pervasive axial plane foliation, although reorientation of mica platelets is common near the crest and intersection lineations paralleling the fold axes are widespread. Other folds of this generation are passive with a penetrative foliation parallel to the axial surface of the fold. These style variations suggest that the folds that I have grouped together in this generation actually span the interval synmetamorphic to late metamorphic. In fact, at one locality early synmetamorphic lineations are folded by nearly coaxial younger folds. I do not interpret this refolding as a separate event, but envision it as a product of a somewhat unsteady flow gradient in the manner proposed by Wynne-Edwards (1963). The similarity in orientation and consistency of vergence despite style variations suggest that the folds grouped as the Soldier Creek Canyon generation developed in response to a common stress system. Equally important is the fact that the folds of this generation parallel the prominent synmetamorphic folds of the infrastructure (Hidden Lakes uplift generation; see below). The Soldier Creek Canyon generation folds, therefore, constitute a link between the mobile infrastructure and the overlying *Abscherungszone*. However, it is important to note that despite similarity in orientation, the vergence of the major macroscopic structure of the *Abscherungszone,* the Soldier Creek nappe, is opposite to the direction of overfolding characteristic of the Soldier Creek Canyon fold generation. In summary, the metacarbonate rocks that contain Soldier Creek Canyon generation folds occupy the boundary between the infrastructure and the overlying *Abscherungszone*; the remarkably flattened and attenuated John Day syncline (Fig. 4) appears to be the macroscopic structure that connects these contrasting tectonic levels.

Hidden Lakes Uplift Generation. Within the impure metaquartzites of the Hidden Lakes uplift, slip folds with approximately north-trending axes are locally common (Figs. 4, 7G). These folds are

Figure 11. F₂ folds of the Secret Creek Gorge generation in rocks of the Dorsey Creek nappe. The main lithologies are m, impure calcite marble with pelitic interlayers; gn, biotite augen gneiss; and qtz, flaggy white metaquartzite. This exposure is part of a cascade of folds uniformly overturned to the west. Note the disharmonic nature of folding as controlled by contrasting layer competency. Both S_1 foliation and L_1 lineation are deformed by these folds, and F_1 slip folds are present in individual marble layers.

remarkably similar in style to the tight, cylindrical isoclines of the F_1 system, but their orientation is nearly normal to the west-northwest trend of the F_1 system. Lineation within the Hidden Lakes uplift also trends crudely north, and the L_1 trend is totally absent. Furthermore, the metaquartzites of the Hidden Lakes uplift are massive and generally lack the flaggy aspect characteristic of rocks containing the F_1 fabric elements.

The strong axial plane foliation associated with folds of the Hidden Lakes uplift and the alignment of prismatic minerals, such as sillimanite and hornblende parallel to fold axes, indicate that these

Figure 12. F_1 folds of the Secret Creek Gorge generation in interlayered metapelite and pegmatitic granite (that is, felsic mobilizate) of the impure metaquartzite-schist unit. The granitoid layers are characterized by a mylonitic fabric which is folded by the F_2 folds. The vergence of the folds is westward. These folds are exposed in a roadcut along Nevada 11 at the east end of Secret Creek Gorge.

structures developed during an intense period of regional metamorphism. Therefore, what are the temporal relations between the Hidden Lakes uplift generation and the F_1 fold system? The available data do not clearly resolve these relations, and I see three possible interpretations:

1. The F_1 fold system and the Hidden Lakes fold generation developed concurrently but represent fabric elements of contrasting tectonic levels (that is, the *Abscherungszone* and the mobile infrastructural core, respectively).

2. The F_1 fold system of the *Abscherungszone* represents folds of the Hidden Lakes uplift generation that have been rotated parallel to the direction of maximum elongation during tectonic flattening and attenuation (analogous to the model proposed by Sanderson, 1973).

3. The Hidden Lakes uplift folds postdate the F_1 fold system of the *Abscherungszone* and represent a deeper manifestation of a north-trending system that included the Secret Creek Gorge and Soldier Creek Canyon generations. This interpretation would imply that metamorphism and deformation continued longer in the hot infrastructural core than in the structurally overlying *Abscherungszone*.

Although the available data are far from convincing, I tentatively favor the third hypothesis. As noted by Price and Mountjoy (1970) and Price (1972), with time any rheologic zone related to regional metamorphism must advance and recede through the whole rock mass. The result is that brittle zones will become more ductile during the advance of the metamorphic front and ductile zones will become more brittle during its retreat. These observations imply that diachronism is an inherent property of any metamorphic core complex. Furthermore, it is therefore predictable that the mobile migmatitic infrastructure will remain hotter and deform ductilely throughout a longer interval of time than the mantling *Abscherungszone*.

Sharps Creek Generation. The metamorphic rocks east of the Withington fault in the northern Ruby Mountains are predominantly the impure metaquartzite-schist unit and associated granitic rocks. Within this sequence at several localities, well-developed northwest-trending late metamorphic folds have been found. These folds are nearly coaxial with local L_1 lineation trends, but they display a more

Figure 13. F_2 fold of the Soldier Creek Canyon generation in calc-silicate gneiss. This fold is reclined with vergence to the east. Note the thickened crest and thinned limbs; a weak foliation paralleling the axial surface is defined by oriented biotite flakes in the crestal areas of the fold. This fold occurs approximately 0.9 km south-southeast of Soldier Peak on the north wall of Soldier Creek Canyon above the Hidden Lakes uplift.

northward azimuth and clearly deform the S_1 foliation (Fig. 7H). Nevertheless, these folds may actually merge with the F_1 system and represent a continuum of deformation as the peak of metamorphism waned. Southwestern vergence is characteristic at several localities, but the sample is insufficient to unequivocally document a systematic direction.

These folds are invariably asymmetric and commonly overturned; they are often tight and locally disharmonic (Fig. 14A). A weak axial plane foliation is occasionally developed, and crenulation of pelitic layers is common (Fig. 14B). These aspects suggest that the folds developed during the waning of metamorphism when the rocks were still ductile but after the main-phase folding. Massive pegmatitic granite dikes cut these folds, implying that the regional migmatization characteristic of the northern Ruby Mountains had not completely subsided.

Nappes (Tectonic Slides) Formed by Ductile Faulting

The outcrop pattern of the metasedimentary rock units in the northern Ruby Mountains is essentially governed by the geometry of several major low-angle ductile faults (that is, tectonic slides). I use the term "tectonic slide" as defined by Fleuty (1964) for a fault that developed "in close causal connection with folding." These features are, therefore, not thrust faults in the strict sense, for they evolved in a continuum from plastic attenuation associated with recumbent folding to brittle fracture manifested as detachment and displacement. The structural relations associated with these features display a hybrid character in that they reflect the early ductile deformation as well as later brittle phenomena. For example, occasionally it is difficult to precisely locate the base of a nappe in a sequence of strongly foliated metamorphic rocks; however, along strike the basal contact of the same slice may be characterized by tectonic breccia and cataclastic rocks. Moreover, it should be emphasized that development of these nappes in the northern Ruby Mountains appears to postdate the F_1 fold system, for structural elements such as S_1 and L_1 are commonly truncated or deformed at the basal

Figure 14. F_2 folds of the Sharps Creek generation in the impure metaquartzite-schist unit. (A) The folds are overturned to the southwest and trend N32°W–5°NW. These folds deform S_1 and L_1 and L_1 but are intruded by massive, unfoliated pegmatite dikes. (B) Although these disharmonic folds clearly fold an earlier schistosity, reorientation of biotite is common in impure metaquartzite layers, and a crenulation cleavage is well developed in the pelitic layers. Note that thin layers of felsic mobilizate have participated in this folding.

contacts. On the other hand, the nappes are clearly older than the low-angle fault complex exposed in the Secret Creek Gorge area which emplaces unmetamorphosed Paleozoic and Tertiary rocks onto the metamorphic substratum.

A large and well-exposed tectonic element in the northern Ruby Mountains is the Soldier Creek nappe (Fig. 4). This feature was originally mapped by Howard (1966), and I have subsequently studied it in detail. The base of the nappe is exposed on the north wall of Soldier Creek Canyon, and structurally above it is a complex system of braided tectonic slices and several highly appressed, faulted recumbent synclines (here designated the Fort Halleck and Wilson Creek synclines; see Fig. 4).

A variety of field relations suggests that the Soldier Creek nappe first developed as a fold, then became detached from the immediately underlying metacarbonate rocks during its final emplacement. The critical data are (1) rock units are locally truncated at the basal contact (Fig. 4), (2) structural attitudes in rocks both above and below the boundary are truncated at the contact, (3) mylonite and locally tectonic breccia occur along the contact, and (4) retrogression is common along the contact. In detail, however, the base of the nappe is not always defined by the contact between the overlying impure metaquartzite-schist unit and metacarbonate rocks. Locally the fault occurs within the metaquartzite-schist sequence, and a conformable unfaulted contact exists between the lithologic units. Along such a conformable contact, characteristic graphitic paragneiss may be found. These relations suggest that although detachment occurred, the zone of brittle fracture did not always precisely follow the lithologic boundary between rock units.

Another important aspect of the Soldier Creek nappe is that the basal mylonite horizon which is present locally has been strongly folded with vergence to the east in at least one locality. Although an interpretation of this feature is very tentative, the folding may represent a form of backsliding and readjustment after the emplacement of the nappe. The facing of the recumbent synclines within the nappe suggests east to west tectonic transport; however, the magnitude of this transport is unknown.

The recumbent synclines above the Soldier Creek nappe are cored by metacarbonate rocks, and detachment and faulting are typically apparent along contacts between the competent metaquartzite envelope and the core rocks (Fig. 4). The concept of faulting synchronous with or at least associated with folding is best documented in the Wilson Creek syncline where the metaquartzite-schist envelope truncates various lithologic units of the core sequence. The Wilson Creek syncline is well exposed between Wilson Creek and Murphy Creek (Fig. 4), but windows exposing its metacarbonate core are found along Mineral Spring Canyon and near Granite Spring about 3 km south of Secret Creek Gorge (Fig. 5). The recumbent syncline, therefore, appears to be a very flat and elongated feature. Just east of Soldier Peak, the metacarbonate rocks that form the core of the syncline are completely squeezed out, and rocks of the flanking metaquartzite-schist envelope are juxtaposed. The outcrop pattern along Wilson Creek suggests that a similar phenomenon may happen in that area. Therefore, the metacarbonate core is essentially a large "tectonic fish" between competent plates of the older metaquartzite-schist unit.

Lithologic contacts that outline the Fort Halleck fold are generally comformable and appear unfaulted. However, spectacular tectonic breccia characterized by clasts of pegmatitic granite and graphitic paragneiss in a gray quartzose matrix is developed near the hinge of the Fort Halleck syncline. The geometry of the fold appears similar to the Wilson Creek structure; namely, near the range front the underlying and overlying rocks of the metaquartzite-schist unit progressively approach each other with the elimination of the intervening metacarbonate rocks.

In addition to these large tectonic "fish" which appear to be detached synclinal fragments, a multitude of considerably smaller "fish" are found in the terrane between Soldier Creek Canyon and Secret Creek Gorge. These tectonic inclusions commonly appear to define distinct horizons which may represent zones of maximum transposition and detachment. A particularly spectacular example is exposed on the ridge north of Wilson Creek. At least seven distinct metacarbonate "fish" define a tectonic zone in the metaquartzite-schist unit (Fig. 4). The rocks that constitute these tectonic "fish" are essentially ductile breccias characterized by a strongly foliated carbonate matrix containing detached and rotated fragments of more competent rock. Small-scale F_2 system folds are abundant and appear to trend crudely north to northeast; vergence where determinable appears to be westward.

The Dorsey Creek nappe is a thin allochthonous plate of metacarbonate and associated rocks that occurs at the top of the *Abscherungszone*. Commonly, tectonic slivers of low-grade metasedimentary rocks occur along the boundary between the Dorsey Creek plate and the nonmetamorphic but allochthonous rocks of the suprastructure. The Dorsey Creek nappe is well exposed in the Secret Creek Gorge area (Fig. 5), and erosional remnants can be traced south as far as the mouth of Soldier Creek Canyon (Fig. 4). Mapping by Snelson (1957) indicates that an analogous tectonic plate (his Horse Creek thrust sheet) is present in the Secret Hills and may extend throughout the southern East Humboldt Range (Fig. 1).

The metacarbonate rocks within the plate are chiefly impure calcite marbles often containing diopside, tremolite-actinolite, biotite, quartz, and feldspar. Graphite-bearing calcite marble and dolomite-calcite marble as well as calc-silicate rock are distinctly subordinate metacarbonate lithologies within this plate. Furthermore, an unusual aspect of this sequence is the presence of terrigenous rocks including white metaquartzite, gray impure metaquartzite, and laminated metasiltstone. Other lithologies in this sequence are gneisses (mafic to leucocratic), amphibolite, and pegmatitic granite. All the rocks within the plate are intensely deformed, and mylonitic textures are widespread. The metacarbonate rocks display spectacular evidence of ductile flowage; competent layers within these rocks have been boudined so completely that the rock is a tectonic breccia (Fig. 15). These mylonitic metacarbonate rocks are analogous to the marble tectonite that Misch (1960) and Coney (1974) described from beneath the Snake Range decollement in east-central Nevada.

Secret Creek Low-Angle Fault Complex

Both north and south of Secret Creek is a complex of low-angle faults that emplace unmetamorphosed rock onto the metamorphic substratum. Sharp (1939b) recognized that the contact between the fossiliferous upper Paleozoic rocks and the metamorphic sequence was a major fault and considered it to be an early Tertiary thrust. However, Snelson (1957) made the first detailed study of the structural relations of the area. He considered the low-angle faults to be part of a complex thrust zone perhaps analogous to the Snake Range decollement as originally described by Hazzard and others (1953). Later, Misch (1960) cited the Secret Creek Gorge locality as well as numerous other examples as evidence for widespread decollement-type thrusting throughout eastern Nevada and western Utah.

A unique character of the Secret Creek fault complex as well as many similar structures in the region is that younger rocks are commonly emplaced on older rocks. This aspect has been emphasized by many workers (for example, see Young, 1960; Drewes, 1967; Nelson, 1966, 1969; Moores and others, 1968; Thorman, 1970) and is apparently a widespread structural style in the Basin and Range province from Arizona (Shackelford, 1976; Davis and others, 1977) to southern Idaho (Armstrong, 1968; Compton and others, 1977). The evolution of thought concerning the origin of these features has been unusually complex and varied. The regional decollement model originally developed by Misch and his students has in recent years been challenged by various workers who favor gravitational gliding mechanisms (see, for example, Moores and others, 1968; Armstrong, 1972; Hose and Daneš, 1973; Coney, 1974). The details of the various models differ, but the role of gravity rather than compressive stresses is emphasized. My mapping in the Secret Creek Gorge area does not make it possible to differentiate unequivocally between a gravity and a compressional model, but it does suggest a chronology in the history of these faults that imposes some constraints on mechanism.

Figure 15. Mylonitic metacarbonate breccias of the Dorsey Creek nappe (Secret Creek Gorge area). (A) Note strong fluxion structure with rotated clasts, of competent rock. (B) F_2 folds of the Secret Creek Gorge generation in mylonitic impure marble. These folds have been dismembered during extreme boudinage.

The following list summarizes the character of the low-angle faults of the Secret Creek Gorge area, including some chronologic restrictions:

1. The unmetamorphosed rocks of the upper plate are commonly brecciated and complexly faulted, but no mesoscopic folds were observed.

2. Bedding within the upper plate commonly is steep and intersects the basal fault surface at a high angle.

3. The fault surface has been gently folded on a northwest-trending axis.

4. Older-over-younger faults occur as well as younger-over-older structure (see north of Dorsey Creek, Fig. 5).

5. The basal detachment fault is invariably delineated by tectonic slices of low-grade metasedimentary rocks.

6. High-angle normal faults related to the evolution of the present-day basin and range topography truncate the low-angle faults.

7. Hydrothermal silicification is a common phenomenon along fault planes within the upper plate.

8. The metamorphic rocks of the lower plate are commonly retrograded or locally brecciated. However, the ductile mylonitic texture of these metamorphic rocks predates the brittle cataclasis characteristic of the fault zone.

9. No rock units of the lower plate ever are found structurally above the nonmetamorphic strata.

10. Tertiary sedimentary and volcanic rocks (Miocene) are involved in the low-angle fault complex. Basalt dikes (dated at approximately 17 m.y., E. H. McKee, 1978, written commun.) are truncated by a low-angle fault.

The involvement of Tertiary rocks in the low-angle fault complex exposed in the Secret Creek Gorge area supplements the earlier discovery by Willden and others (1967) of post–36-m.y.-old low-angle faulting in the southern Ruby Mountains. Structural relations between the Tertiary and older rocks are summarized in Figure 5. Although poor exposure of the Tertiary rocks commonly makes contact relations with the underlying rock units difficult to analyze, at two localities north of Secret Creek the relations are unusually well exposed (see Fig. 5, localities 1, 2). At locality 1, along Nevada 11, green Tertiary sandstone and siltstone structurally overlies a thin wedge of Diamond Peak Formation–Chainman Shale(?). Bedding in both the overlying Tertiary rocks and the underlying Paleozoic rocks is truncated at the fault contact. The overlying Tertiary rocks have been brecciated while the underlying Diamond Peak–Chainman lithology has been intensely dislocated and has the appearance of so-called broken formation. Bedding is partially transposed and boudins of competent grit and pebble conglomerate occur within a silty shale matrix. Along strike, the Diamond Peak–Chainman unit wedges out and the Tertiary rocks rest directly on mylonitic metacarbonate rocks of the Dorsey Creek nappe. An important aspect of this low-angle fault is that the rocks of the underlying plate (that is, the Diamond Peak–Chainman unit) are intensely deformed; this fact suggests that surficial sliding is an unlikely mechanism.

At locality 2, approximately 2.2 km north of Secret Creek, a small patch of Tertiary volcanic rock partially rests on both unmetamorphosed upper Paleozoic rocks and the metamorphic substratum. The volcanic rocks that constitute this exposure have been profoundly brecciated and silicified, but are lithologically similar to Tertiary volcanic rocks exposed at locality 3, approximately 1.1 km due south (Fig. 5). The rhyolitic rocks at locality 3 rest depositionally on Tertiary sedimentary rocks with a basal vitrophyre at the lower contact. I believe, therefore, that the Tertiary volcanic rocks at locality 2 are allochthonous and have been faulted onto the older rocks. The volcanic rocks at locality 3 are also allochthonous; however, the low-angle fault occurs below the volcanic rock–sedimentary rock contact which at this locality is clearly stratigraphic. Sanidine from the vitrophyre at locality 3 has yielded a K-Ar age of 15.0 ± 1.5 m.y. (E. H. McKee, 1978, written commun.), which requires a post–middle Miocene age for the postulated low-angle faults separating the Tertiary strata from the older rocks. In support of these data, a basalt dike, located about 0.3 km northeast of locality 2, yields a whole-rock K-Ar age of approximately 17 m.y. and is truncated by a low-angle fault. At this locality (see Fig. 5), the upper plate rocks are Permian. In summary, the similarity in structural style between these low-angle faults that involve Tertiary rocks and the other low-angle faults of the Secret Creek complex that emplace unmetamorphosed Paleozoic rocks onto metamorphic rocks suggests an analogous age as well as mechanism.

Wilson Creek Breccia

South of the mouth of Wilson Creek, a small breccia sheet is structurally above impure metacarbonate and associated granitic rocks (Fig. 4). As previously concluded, I believe these metacarbonate rocks could be erosional remnants of the Dorsey Creek nappe; therefore, the breccia occupies a position analogous to the unmetamorphosed Paleozoic rocks of the Secret Creek Gorge area. Willden and Kistler (1969) have mapped a similar breccia sheet near the range front in the southern Ruby Mountains, and the Cedar Mountain klippe which structurally overlies the Tertiary Harrison Pass pluton is perhaps also analogous (Fig. 2).

The breccia near Wilson Creek is characterized by a tannish-brown siliceous matrix containing angular clasts (commonly 1 to 5 cm maximum dimensions) of white vein quartz, quartzite, metaquartzite, pegmatitic granite, fine-grained hydrothermal(?) silica rock, and rebrecciated composite fragments of matrix and clasts. Commonly the breccia is massive, locally vuggy, but remarkable foliated zones are also present. The clasts within these foliated zones are flattened and elongated much like pebbles in a metaconglomerate. At one spectacular exposure, the breccia has been strongly folded into a westward vergent recumbent fold. The character of the breccia suggests nearly concurrent ductile and brittle deformation and therefore implies a mechanism capable of producing a spectrum of deformational features. Further conclusions are speculative, but I am impressed that the above relations are completely compatible with the denudation fault model of Armstrong (1972). The Wilson Creek breccia may be the sole of a deep-seated denudation fault along which a large block of suprastructure was emplaced onto metamorphic rocks. Nearly continuous brecciation coupled with local frictional shear heating would facilitate the development of contrasting deformational styles along the glide block interface. Therefore, I argue, that the Wilson Creek breccia is another manifestation of the low-angle faulting that characterizes the boundary between suprastructure and the metamorphic substratum in the northern Ruby Mountains.

SYNTHESIS

Chronology of Events

The preceding discussions have documented the variety of deformational styles characteristic of the northern Ruby Mountains. In Figure 16, I have attempted to integrate these events into a consistent chronology. Nevertheless, a major question remains; namely, whether or not there was a continuum from the ductilely deformed metamorphic terrane to the overlying brittle low-angle fault complex. Certainly the presence of metamorphic rock clasts in Tertiary conglomerates and the subsequent involvement of Tertiary rocks in low-angle faults in the Secret Creek Gorge area indicate late faulting that considerably postdates metamorphism. On the other hand, these data do not preclude the possibility that some low-angle faulting in the suprastructure could merge with the deformation of the metamorphic complex. The widespread occurrence of tectonic slides (that is, low-angle ductile faults) in the *Abscherungszone* argues for a transition in deformational style from the ductile to the brittle

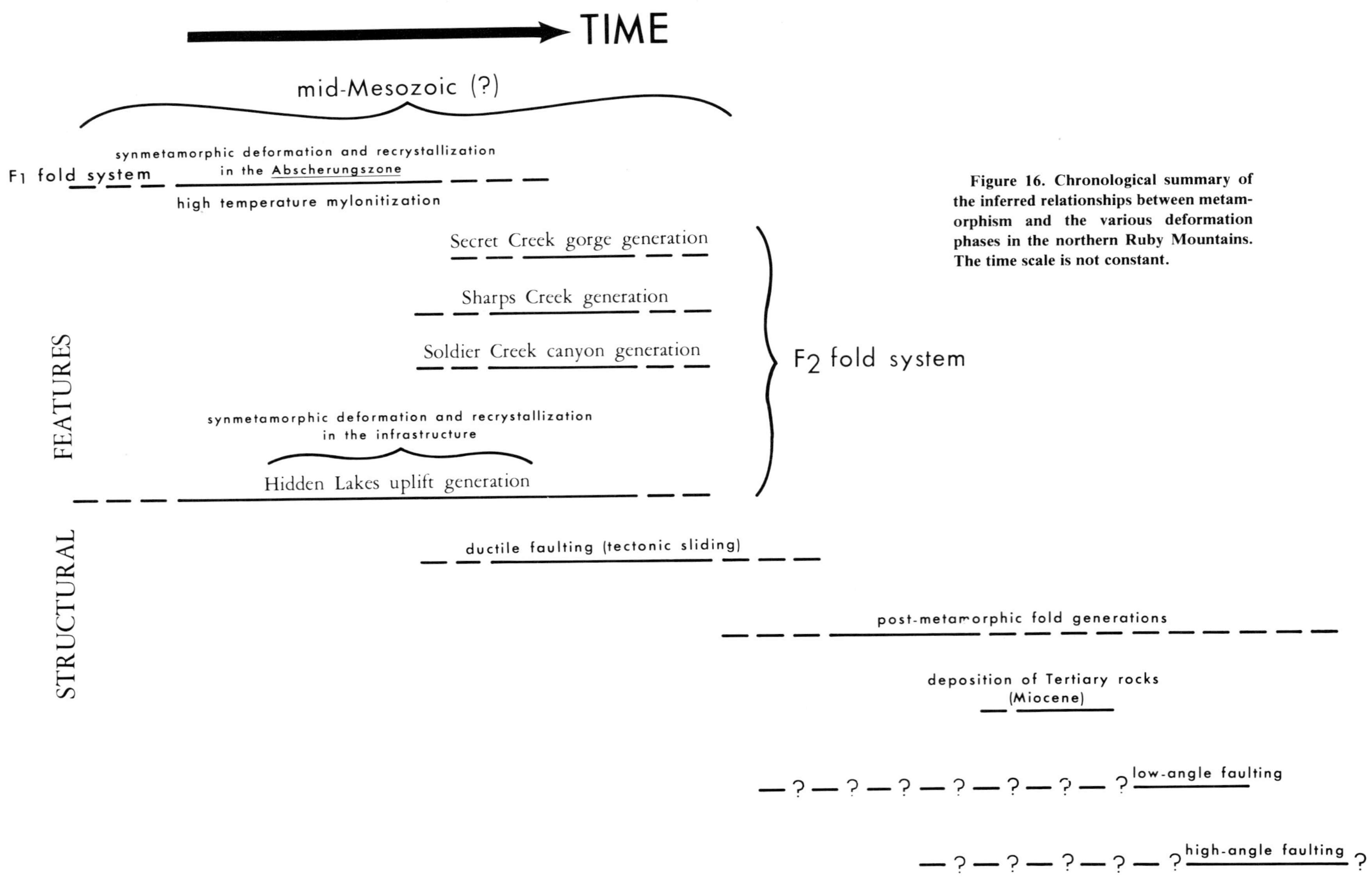

Figure 16. Chronological summary of the inferred relationships between metamorphism and the various deformation phases in the northern Ruby Mountains. The time scale is not constant.

→ TIME

mid-Mesozoic (?)

STRUCTURAL FEATURES

F₁ fold system
- synmetamorphic deformation and recrystallization in the Abscherungszone
- high temperature mylonitization

F₂ fold system
- Secret Creek gorge generation
- Sharps Creek generation
- Soldier Creek canyon generation
- synmetamorphic deformation and recrystallization in the infrastructure
- Hidden Lakes uplift generation

- ductile faulting (tectonic sliding)
- post-metamorphic fold generations
- deposition of Tertiary rocks (Miocene)
- ?—?—?—?—?—?—? low-angle faulting
- ?—?—?—?—?—?—? high-angle faulting

regime. Also the Wilson Creek breccia, locally folded, suggests an overlap between ductile and brittle phenomena. The crux of the problem is that although tectonic overprinting is clear (Mesozoic metamorphism and deformation and Neogene low-angle faulting), the extent to which this overprinting may have masked earlier deformational structures in the suprastructure is unclear. At present, I see no straightforward method to solve this problem in the Ruby Mountains.

Evaluation of the Infrastructure-Suprastructure Concept

A steep metamorphic gradient between infrastructure and suprastructure has been an important element in the "stockwerk" tectonic model since Wegmann's (1935) classic account. This thermal gradient, also manifested in contrasting rock ductility, is part of the *Abscherungszone* concept developed by Haller (1956, 1971). In the Ruby Mountains, the *Abscherungszone* is characterized by mylonitic textures, elongation lineations, polyphase deformation, low-angle ductile faults, and concordant bodies of orthogneiss. A steep metamorphic gradient can be inferred with the structurally overlying but nonmetamorphic upper Paleozoic rocks. However, the only evidences of the transition from high- to low-grade metamorphic rocks are the tectonic slices of blue-gray calcite marble, white metaquartzite, and metadolomite immediately below the low-angle fault complex. These rocks, however, do not contain index minerals suitable for the identification of metamorphic grade, although textural characteristics suggest greenschist-facies(?) conditions. Therefore, in the Secret Creek Gorge area, tectonic attenuation has made a detailed study of the transition from metamorphic to nonmetamorphic strata impossible. In the Secret Hills at the end of North Ruby Valley, units analogous to the tectonic slices under the low-angle fault complex are more extensively exposed and thicker. Furthermore, the same units extend into the southeastern East Humboldt Range. Therefore, these areas which have only been mapped in reconnaissance (Snelson, 1957; Hope, 1972, written commun.; Howard and others, 1979) offer the possibility of documenting the steep metamorphic gradient between infrastructure and suprastructure.

Metamorphism and associated polyphase folding in the Ruby Mountains were part of one protracted deformation event. The F_1 fold system, especially characteristic of the *Abscherungszone,* often has synkinematic sillimanite or hornblende prisms parallel to fold axes in rocks of the appropriate protolith. Furthermore, the regional S_1 schistosity is parallel to the axial surfaces of all observed F_1 folds. In a few isolated localities, F_1 fold axes and the L_1 lineation are not parallel. This fact suggests that F_1 folding locally somewhat preceded the maximum development of the elongation lineation. Nevertheless, the bulk of the data suggest that the F_1 system developed roughly synchronously with the peak of regional metamorphism.

In the Hidden Lakes uplift, the trend of the predominant fold generation is at a high angle to the F_1 system. These F_2 system folds are somewhat similar in style to F_1 folds and therefore appear to be synchronous with regional metamorphism. How these conflicting observations can be reconciled is uncertain, but I favor a model that allows the rocks of the mobile core to continue to recrystallize and flow after the metamorphic front has subsided from the transition zone between the infrastructure and suprastructure.

The previous description appears somewhat different from the relations between metamorphism and deformation recently summarized by Compton and others (1977) for the Raft River–Grouse Creek complex in northwest Utah. That metamorphic core complex is in general not a migmatite terrane, and sillimanite is only present locally. On the other hand, distinct metamorphic mineralogic and textural changes are detectable in both the autochthonous and allochthonous rocks. Furthermore, postmetamorphic—or at least late metamorphic—low-angle faults have juxtaposed rocks of contrasting metamorphic grade and relative reconstitution. An important aspect of the Raft River–Grouse Creek complex is that shear strain in the metamorphic complex appears to decrease downward such that the deepest exposures of Precambrian granite in the autochthon are nearly undeformed. This interpretation is, of course, considerably different from the infrastructure-suprastructure concept that employs a mobile migmatitic core, often involving basement, as a major active force in the deformation.

Regional Tectonic Overview

Recent geologic mapping and reconnaissance along the length of the Cordillera indicate that a zone of metamorphic core complexes can be traced from Alaska to Sonora (Crittenden and others, 1978). This zone of intense metamorphism, polyphase folding, and igneous activity is a distinct tectonic element west of the foreland fold-thrust belt but often east of the magmatic arc terrane of extensive granitoid plutons. However, some geologists (for example, Price and Mountjoy, 1970) have previously recognized the possible tectonic significance of the zone of metamorphic infrastructure and have attempted to relate metamorphism and gravitational spreading in the core complexes to deformation along the margin of the orogen. This hypothesis is attractive, for it eliminates the mechanical problem of transmitting compressive stresses from an east-dipping subduction zone across the orogen to the foreland terrane. However, a major problem in this argument is the growing evidence of significant Tertiary deformation and metamorphism in some metamorphic core complexes (for example, Compton and others, 1977—northwest Utah; Rehrig and Reynolds, 1977—Arizona). These new data suggest that these metamorphic terranes were actually deformed later than the external foreland fold-thrust belt, and any simplistic cause-and-effect relationship is obscure. The problem of the timing of the metamorphism in the core terranes is far from resolved, and in light of the extent of the metamorphic complexes, a variable chronology is almost certain.

Furthermore, as pointed out by Campbell (1973), a consequence of the Price-Mountjoy model is that either the metamorphic core complexes are allochthonous relative to a deeper basement layer or substantial shortening of basement occurs in the core terranes. Campbell (1973) cites evidence for a relatively autochthonous history for the core terrane of the Canadian Cordillera and suggests that relations between the Omineca crystalline belt and the foreland fold-thrust zone are best explained by underthrusting of the craton into a mobile high-temperature zone. According to this mechanism, the sedimentary cover overlying the craton would peel off along a decollement forming a stack of east-directed thrust sheets. The down-flowing cratonic crust would thicken the core zone and cause major vertical uplift.

The concept of cratonic underthrusting was earlier suggested by Misch (1960, 1971) as a mechanism to relate the deformation in the hinterland province of the Sevier orogenic belt (that is, the eastern Great Basin of Utah and Nevada) and the external foreland fold-thrust belt. Misch's concept, as originally envisioned, involved the westward underthrusting of "stress-transmitting and non-yielding" basement beneath "stress-absorbing" strata of the "geosyncline." Moores (1970, Fig. 4g) recast the cratonic underthrusting hypothesis as an example of intracontinental westward-dipping subduction coeval with eastward-dipping oceanic subduction along the western margin of North America. According to Moores (1970), the zone of metamorphic infrastructure developed between two oppositely dipping subduction zones. Recently Burchfiel and Davis (1975) have argued that magmatic heating of the leading edge of the North American plate during the evolution of the Mesozoic volcano-plutonic arc generated a profound ductility contrast between the western margin and the eastern portion of the plate. The result of this ductility contrast was manifested in intraplate yielding such that the more rigid eastern portion of continental lithosphere moved beneath the ductile marginal zone. The zone of highest ductility contrast became the locus of intracontinental thrusting which according to Burchfiel and Davis (1975) shifted eastward in time along much of the length of the orogen. Furthermore, the partial subduction of continental lithosphere coupled with the back-arc high-heat flux served to localize the zone of metamorphic infrastructure (Dickinson, 1976).

The suggestion by Fox and others (1977) that the overriding of an oceanic rise by the North American plate (see also Kistler and others, 1971) is another attempt to integrate the Cordilleran metamorphic core complexes with the plate tectonic paradigm. According to their model, the overridden rise system causes widespread partial melting of mantle and sialic crust producing a broad zone of magmatism and regional metamorphism. The culmination of this orogenic cycle involved the mobilization of the infrastructure by diapiric rise and the collapse of a dense mafic residuum depleted of its low-temperature melting fraction. A somewhat similar scenario could be devised using a thermal diapir of mobilized mantle material generated in the back-arc region according to the model proposed by Karig (1971) for modern marginal basins. The diapir would be initiated by frictional shear heating of asthenosphere during deep subduction. Scholz and others (1971) presented geologic and geophysical evidence that a subduction-related diapir existed beneath the continental lithosphere of the western United States during the Cenozoic.

In summary, a variety of tectonic models is now available to evaluate the metamorphic complexes and their relationships with adjacent terranes. However, before meaningful conclusions can emerge, I am impressed that numerous unresolved problems hamper these analyses. I perceive some of the more important questions are as follows:

1. Are the individual metamorphic complexes relatively independent phenomena or are they part of a continuous metamorphic terrane that underlies this segment of the Cordillera?

2. How does the deformation chronology, in particular the age of metamorphism, vary along the belt?

3. What is the role of Precambrian basement in the evolution of these metamorphic complexes? For example, does significant basement shortening occur in the complexes? Is basement reactivation a widespread response, and what are its manifestations?

4. What is the regional significance of the tranverse elongation lineation and mylonitic fabric often common in parts of the metamorphic complexes? Do these fabrics reflect regional stretching of the continental lithosphere?

5. What is the role of tectonic diachronism in the history of the metamorphic core complexes? Can a systematic cycle be developed that accounts for the common observation of an early ductile deformation superimposed by later brittle features (that is, low-angle faults in the hinterland of the Sevier orogenic belt)? Are these deformation features part of a continuum or examples of overlapping tectonism?

The belt of metamorphic core complexes of western North America, therefore, is an important petrotectonic element of Andean or cordilleran-type orogens. A thorough understanding of the evolution of these Cordilleran core complexes could provide important insights into the development of metamorphic terranes in other orogens of a reputed more complex history (for example, the Piedmont province of the southern Appalachians).

APPENDIX 1. SUMMARY OF FOSSIL ASSEMBLAGES IN PENNSYLVANIAN-PERMIAN ROCKS FROM THE SECRET CREEK GORGE AREA, NORTHERN RUBY MOUNTAINS, NEVADA

The following identifications were made by various members of the Branch of Paleontology and Stratigraphy, U.S. Geological Survey. This report is divided into three divisions based on the kinds of fossils studied: foraminifera (Raymond C. Douglass), brachipods (Bruce Wardlaw), bryozoa (Olgerts L. Karklins).

FORAMINIFERA - Report prepared by Raymond C. Douglass

f24703 (954-2) Elko County, Nevada   L. Permian
    NW cor. sec. 30, T. 35 N., R. 60 E., Tent Mtn. 7-1/2 min. quad. Silty limestone with some bryozoans, small textularid forams and PARAFUSULINA? sp. possibly closer to MONODIEXODINA sp. Similar forms occur in the latest Wolfcampian to early Leonardian.

f24704 (954-3) - Locality same as 954-2   L. Permian
    Calcareous fine sandstone with bryozoa, brachiopods, and PARAFUSULINA? sp. suggesting an early Leonardian age.

f24705 (957-6) Elko County, Nevada   L. Permian
    SE cor. sec. 30, T. 35 N., R. 60 E., Tent Mtn. 7-1/2 min. quad. Crinoidal limestone with bryozoans and brachiopods, small textularid and miliolid forams, and the fusulinids, PSEUDOFUSULINA sp. and PARAFUSULINA sp. of Leonardian age.

f24706 (932-1) Elko County, Nevada   L. Permian
    SW cor. sec. 32, T. 35 N., R. 60 E., Secret Valley 7-1/2 min. quad. Fine silty limestone with scattered crinoid columnals and fusulinids including a staffellid and SCHWAGERINA sp. of probable Leonardian age but possibly late Wolfcampian.

f24707 (114-17) Elko County, Nevada   L. Permian
    SE cor. sec. 32, T. 34 N., R. 60 E., Secret Valley 7-1/2 min. quad. Sandy limestone with some bryozoans and abundant forams including the smaller forms BRADYINA sp. and CLIMACAMMINA sp. and the fusulinids SCHUBERTELLA sp., SCHWAGERINA sp., PSEUDOFUSULINA? sp. and rare MONODIEXODINA sp. This is a common Riepe Spring (Wolfcampian) assemblage.

f24708 (114-17B) Locality same as 114-17   L. Permian
    Calcareous sandstone with scattered crinoidal debris, the small foram BRADYINA sp. and PSEUDOFUSULINA sp. (Wolfcampian age).

f24709 (114-18) Locality same as 114-17   L. Permian
Two lithologies are included in this sample. One is a
yellow tan calcareous sand with a few crinoid columnals
and abundant large specimens of MONODIEXODINA sp. The
other is a dark gray silty-sandy limestone principally
composed of fusulinids. The smaller forams include BRADYINA
sp. and CLIMACAMMINA sp. and the fusulinids include SCHUBERTELLA
sp., PSEUDOSCHWAGERINA sp. and SCHWAGERINA sp. This is a
common Riepe Spring fauna.

f24710 (135-1) Elko County, Nevada   L. Permian
SE cor. sec. 19, T. 25 N., R. 60 E., Tent Mtn. 7-1/2 min. quad.
Crinoidal limestone and calcareous grit to sand of this
sample has some scattered bryozoans and many fragments and
larger eroded specimens of fusulinids. PSEUDOFUSULINA sp.
can be recognized, but an age closer than Early Permian was
not determined.

f24711 (955-12A) Elko County, Nevada   L. Permian
NE cor. sec. 30, T. 35 N., R. 60 E., Tent Mtn. 7-1/2 min. quad.
Fine calcareous sand with over half the rock made up of
the small fusulinid EOPARAFUSULINA sp. This fusulinid is reported from California and Alaska, but I have not seen it
previously in Nevada. It is probably an offshoot from one
of the forms listed as MONODIEXODINA sp. in this report
(Wolfcampian).

f24712 (955-12B) Locality same as 955-12A   L. Permian
Similar to 955-12A but with less matrix and the fusulinids
randomly oriented. EOPARAFUSULINA sp. (Wolfcampian).

f24713 (955-14A) Locality same as 955-12A   L. Permian
Dark silty limestone with abundant smaller forams in the
matrix and abundant large fusulinids. BRADYINA sp.,
CLIMACAMMINA sp., and many unidentified texturlarids, and
the fusulinids SCHUBERTELLA? sp., PSEUDOFUSULINA sp. and
PSEUDOSCHWAGERINA sp. This is a typical Riepe Spring fauna
(Wolfcampian).

f24714 (955-14B) Locality same as 955-12A   L. Permian
The sample is not significantly different from 955-14A.
The fauna is similar but with fewer of the large
PSEUDOSCHWAGERINA sp. specimens (Wolfcampian).

f24715 (932-4) Elko County, Nevada   L. Permian
NW cor. sec. 32, T. 35 N., R. 60 E., Secret Valley 7-1/2 min.
quad. Silty limestone essentially composed of a large
PARAFUSULINA sp. of early Leonardian age.

f24716 (955-29A) Elko County, Nevada   L. Permian
NE cor. sec. 30, T. 35 N., R. 60 E., Tent Mtn. 7-1/2 min.
quad. Fine yellow-tan calcareous sandstone with abundant
large MONODIEXODINA sp. and a few specimens of SCHWAGERINA
sp. (Wolfcampian).

f24717 (955-29B) Locality same as 955-29A   L. Permian
   Fine silty-sand with some textularids and with the fusulinids
   SCHWAGERINA sp. and MONODIEXODINA sp. (Wolfcampian).

f24718 (932-6) Elko County, Nevada   L. Permian
   NW cor. sec. 32, T. 35 N., R. 60 E., Tent Mtn. 7-1/2 min.
   quad. Dark silty limestone with scattered crinoid columnals
   and common large specimens of PARAFUSULINA sp. of Leonardian
   age.

f24719 (826-14) Elko County, Nevada   L. Permian
   SW cor. sec. 6, T. 34 N., R. 60 E., Soldier Peak 7-1/2 min.
   quad. Clastic limestone composed largely of crinoidal debris
   and fusulinids in a dark silty matrix. The small foram
   CLIMACAMMINA can be recognized along with the fusulinids
   SCHUBERTELLA sp., SCHWAGERINA sp. and a possible PARAFUSULINA
   sp. or MONODIEXODINA sp. The age is uncertain but probably
   represents latest Wolfcampian or early Leonardian.

f24720 (955-28) Elko County, Nevada   M. Pennsylvanian
   NE cor. sec. 30, T. 35 N., R. 60 E., Tent Mtn. 7-1/2 min. quad.
   Gray limestone with a matrix including some silt, but largely
   made up of smaller forams. These include BRADYINA sp.,
   CLIMACAMMINA sp., TETRATAXIS sp. and other unidentified
   forms. Fusulinids are common and include MILLERELLA sp.,
   PSEUDOSTAFFELLA sp., and PROFUSULINELLA sp. This fauna is
   characteristic of the Atokan (or Derryan) part of the Middle
   Pennsylvanian.

f24721 (955-21) Elko County, Nevada   L. Permian
   NW cor. sec. 29, T. 35 N., R. 60 E., Tent Mtn. 7-1/2 min. quad.
   Fine clastic limestone matrix with abundant CLIMACAMMINA
   sp. in fragments of shell and crinoid columnals. Large
   fusulinids are PSEUDOFUSULINA sp. and possibly PARAFUSULINA
   sp. suggesting late Wolfcampian to early Leonardian age.

Suggested lithostratigraphic assignments:

   Pequop Formation (or Arcturus) - 932-4, 932-6, 957-6.

   Intermediate (Ferguson Mountain Formation)- 954-2, 954-3, 955-21,
      826-14, 135-1, 932-1.

   Riepe Spring Limestone (or Ferguson Mountain Formation)- 114-17A, B;
      114-18; 955-12A, B; 955-14A, B; 955-29A, B.

   Ely Limestone - 955-28.

BRACHIOPODS - Report prepared by Bruce Wardlaw

26349-PC (114-14) Elko County, Nevada   M. Pennsylvanian
NE cor. sec. 32, T. 35 N., R. 60 E., Secret Valley 7-1/2 min. quad.
The fauna consists of:

KOSLOWSKIA sp.
DESMOINESIA sp.
BUXTONIA sp.
ANTIQUATONIA sp.
small Pontisiid Rhynchonellid resembling ACOLOSIA
CLEIOTHYRIDINA sp.
COMPOSITA SUBTILITA (Hall)
HUSTEDIA MISERI GIBBOSA Lane
CRURITHYRIS PLANOCONVEXA (Shumard)
NEOSPIRIFER sp.
DIELASMA sp.
BEECHERIA sp.
indeterminate Notothyridid Terebratulid brachiopod
small, high spired gastropods
Bryozoans
This is a well silicified fauna of Desmoinesian age (Pennsylvanian).

26350-PC (933-1) Elko County, Nevada   M. Pennsylvanian
NW cor. sec. 32, T. 35 N., R. 60 E., Tent Mtn. 7-1/2 min. quad.
The fauna consists of:

RHIPIDOMELLA sp.
HETERALOSIA sp.
DESMOINESIA sp.
LINOPRODUCTUS sp.
small Rhynchonellids, similar to those in 114-14
CLEIOTHYRIDINA of. C. ORBICULARIS (McChesney)
HUSTEDIA of. H. MISERI GIBBOSA Lane
NEOSPIRIFER sp.
RETICULARINA sp.
BEECHERIA sp.
CRANAENA sp.
Trilobite
Corals
small, high spired gastropods
Bryozoans

This is a coarsely silicified fauna of Desmoinesian age.

26351-PC (955-27) Elko County, Nevada   M. Pennsylvanian
NE cor. sec. 30, T. 35 N., R. 60 E., Tent Mtn. 7-1/2 min. quad.
The fauna consists of:

DESMOINESIA?
NEOSPIRIFER sp.

This fauna indicates a Desmoinesian age. The NEOSPIRIFER is the same as in the other collections.

26352-PC (956-14) Elko County, Nevada   M. Pennsylvanian
   NE cor. sec. 29, T. 35 N., R. 60 E., Tent Mtn. 7-1/2 min. quad.
   The fauna consists of:

   LISSOCHONETES?
   KOSLOWSKIA sp.
   CLEIOTHYRIDINA sp.
   COMPOSITA sp.
   NEOSPIRIFER sp.
   RETICULARIINA sp.
   Crinoid debris

   The fauna is partially silicified and occurred in a very sandy limestone. This is a typical representative of the NEOSPIRIFER-KOSLOWSKIA assemblage of the middle Pennsylvanian generally occurring in shallow water. The age is Desmoinesian.

26353-PC (956-18) Elko County, Nevada   M. Pennsylvanian
   SW cor. sec. 20, T. 35 N., R. 60 E., Tent Mtn. 7-1/2 min. quad.
   The fauna consists of:

   DERBYIA sp.
   ANTIQUATONIA sp.
   LINOPRODUCTUS sp.
   COMPOSITA sp.
   HUSTEDIA sp.
   RETICULARIINA?
   indeterminate Terebratulid brachiopod
   Corals
   Bryozoans

   This fauna is partially silicified. It is dominated by large Productids. This represents a different biofacies than most of the other collections (except 956-14, which is also different) and probably represents some minor change in lithofacies, perhaps higher mud content. The age is Desmoinesian.

955-16 (location same as 956-18)   M. Pennsylvanian
   The fauna consists of:

   DESMOINESIA?
   HUSTEDIA sp.
   Crinoid debris
   Bryozoans

   This collection can only be dated as probably Pennsylvanian but its relation to the other collections from the same area indicate a Desmoinesian age.

E-1-- Elko County, Nevada   M. Pennsylvanian
   NE cor. sec. 1, T. 34 N., R. 59 E., Soldier Peak 7-1/2 min. quad.
   The fauna consists of:

RHIPIDOMELLA sp.
RETICULARIINA?
HUSTEDIA sp.
Crinoid debris

Though from a small sample and poorly preserved, this fauna indicates a Pennsylvanian age and appears to be related to the other samples which are more easily datable.

All the collections are probably Desmoinesian, equivalent to the upper part of the Ely Limestone.

BRYOZOA - Report prepared by Olgerts L. Karklins

26349-PC (114-14) Elko County, Nevada
NE cor. sec. 32, T. 35 N., R. 60 E., Secret Valley 7-1/2 min. quad.
The fauna identified:

FENESTELLA sp., three specimens
Rhobdomesids, several specimens

Typical upper Paleozoic taxa; specimens too fragmented for further determination.

RHOMBOTRYPELLA sp., 6 specimens

Genus ranges from the lower Pennsylvanian through the Permian. Although it is common in the upper Paleozoic strata of western North America, its species remain undescribed at present.

26350-PC (933-1) Elko County, Nevada
NW cor. sec. 32, T. 35 N., R. 60 E., Tent Mtn. 7-1/2 min. quad.
The fauna identified:

ASCOPORA sp., five specimens

Probable new species. Genus apparently is common in the late Paleozoic strata of western North America, but its species remain undescribed. Probable range of genus: Carboniferous - middle Permian.

TABULIPORA sp., one specimen

Typical upper Paleozoic genus. Species not identifiable because the skeletal microstructure of the colony is almost completely replaced by silica.

26352-PC (956-14) Elko County, Nevada
NE cor. sec. 29, T. 35 N., R. 60 E., Tent Mtn. 7-1/2 min. quad.
The fauna identified:

RHOMBOPORA sp., one specimen

This form is similar to RHOMBOPORA LEPIDODENDROIDES Meek, which is common in strata of Carboniferous age.

RHOMBOPORA? sp., four specimens

It can also be RHOMBOTRYPELLA sp., but is different from RHOMBOTRYPELLA in coll. 26349-PC of this report. There are some RHOMBOPORA spp. that appear to be intermediate morphologically between RHOMBOPORA and certain RHOMBOTRYPELLA forms.

26353-PC (956-18) Elko County, Nevada
SW cor. sec. 20, T. 35 N., R. 60 E., Tent Mtn. 7-1/2 min. quad.
The fauna identified:

PRISMOPORA? sp., 4 specimens
RHOMBOPORA sp., 5 specimens

Taxa poorly preserved; PRISMOPORA is a common upper Paleozoic genus.

ACKNOWLEDGMENTS

My studies in the Ruby Mountains–East Humboldt Range, Nevada, began while I was a National Research Council–U.S. Geological Survey Research Associate (1972 to 1974) at the U.S. Geological Survey, Menlo Park, California. The U.S. Geological Survey provided stipend and field and technical support as well as an exciting atmosphere for geologic discussion. M. D. Crittenden, Jr., and K. A. Howard introduced me to the geology of northeast Nevada and have continued to provide valuable insight into regional and local geologic problems. Many other geologists have helped me with discussion, criticism, and advice; I particularly appreciate the aid of the following: J. M. Christie, R. R. Compton, R. W. Kistler, E. H. McKee, D. Miller, R. A. Schweickert, T. W. Stern, and V. R. Todd. Acknowledgment is also made to the donors of The Petroleum Research Fund, administered by the American Chemical Society, for partial support of this research (PRF #8831-G2). Numerous ranchers in the area provided access to and across their property, and my family and I greatly appreciated the hospitality of Sim and Dee Duval and Steve and Mavis Wright.

REFERENCES CITED

Armstrong, R. L., 1968, Mantled gneiss domes in the Albion Range, southern Idaho: Geological Society of America Bulletin, v. 79, p. 1295-1314.

——1972, Low-angle (denudation) faults, hinterland of the Sevier orogenic belt, eastern Nevada and western Utah: Geological Society of America Bulletin, v. 83, p. 1729-1754.

Armstrong, R. L., and Hansen, E., 1966, Cordilleran infrastructure in the eastern Great Basin: American Journal of Science, v. 264, p. 112-127.

Bell, T. H., 1978, Progressive deformation and reorientation of fold axes in a ductile mylonite zone: The Woodroffe thrust: Tectonophysics, v. 44, p. 285-320.

Berge, J. S., 1960, Stratigraphy of the Ferguson Mountain area, Elko County, Nevada: Brigham Young University Research Studies Geology Series, v. 7, no. 5, 63 p.

Brew, D. A., 1971, Mississippian stratigraphy of the Diamond Peak area, Eureka County, Nevada: U.S. Geological Survey Professional Paper 661, 84 p.

Bryant, B., and Reed, J. C., Jr., 1969, Significance of lineation and minor folds near major thrust faults in the southern Appalachians and the British and Norwegian Caledonides: Geological Magazine, v. 106, p. 412-429.

Burchfiel, B. C., and Davis, G. A., 1975, Nature and controls of Cordilleran orogenesis, western United States: Extensions of an earlier synthesis: American Journal of Science, v. 275, p. 363-396.

Campbell, R. B., 1973, Structural cross-section and tectonic model of the southeastern Canadian Cordillera: Canadian Journal of Earth Sciences, v. 10, p. 1607-1620.

Cebull, S. E., 1970, Bedrock geology and orogenic succession in southern Grant Range, Nye County, Nevada: American Association of Petroleum Geologists Bulletin, v. 54, p. 1828-1842.

Christie, J. M., 1963, The Moine thrust zone in the Assynt region, northwest Scotland: University of California Publications in Geological Sciences, v. 40, p. 345-440.

Compton, R. R., 1972, Geologic map of the Yost quadrangle, Box Elder County, Utah, and Cassia County, Idaho: U.S. Geological Survey Miscellaneous Geologic Investigations Map I-672.

——1975, Geologic map of the Park Valley quadrangle, Box Elder County, Utah, and Cassia County, Idaho: U.S. Geological Survey Miscellaneous Geologic Investigations Map I-873.

Compton, R. R., and others, 1977, Oligocene and Miocene metamorphism, folding, and low-angle faulting in northwestern Utah: Geological Society of America Bulletin, v. 88, p. 1237-1250.

Coney, P. J., 1974, Structural analysis of the Snake Range "decollement," east-central Nevada: Geological Society of America Bulletin, v. 85, p. 973-978.

Crittenden, M. D., Jr., Coney, P. J., and Davis, G., 1978, Tectonic significance of metamorphic core complexes in the North American Cordillera: Geology, v. 6, p. 79-80.

Dalziel, I.W.D., and Bailey, S. W., 1968, Deformed garnets in a mylonitic rock from the Grenville Front and their tectonic significance: American Journal of Science, v. 266, p. 542-562.

Davis, G. A., and others, 1977, Enigmatic Miocene low-angle faulting, southeastern California and west-central Arizona—suprastructural tectonics?: Geological Society of America Abstracts with Programs, v. 9, no. 7, p. 943-944.

Dickinson, W. R., 1976, Sedimentary basins developed during evolution of Mesozoic-Cenozoic arc-trench systems in western North America: Canadian Journal of Earth Sciences, v. 13, p. 1268-1287.

Drewes, H., 1967, Geology of the Connors Pass quadrangle, Schell Creek Range, east-central Nevada: U.S. Geological Survey Professional Paper 557, 93 p.

Escher, A., and Watterson, J., 1974, Stretching fabrics, folds and crustal shortening: Tectonophysics, v. 22, p. 223-231.

Fleuty, M. J., 1964, Tectonic slides: Geological Magazine, v. 101, p. 452-456.

Fox, K. F., Jr., Rinehart, C. D., and Engels, J. C., 1977, Plutonism and orogeny in north-central Washington—Timing and regional context: U.S. Geological Survey Professional Paper 989, 27 p.

Griffin, V. S., Jr., 1974, Analysis of the Piedmont in northwest South Carolina: Geological Society of America Bulletin, v. 85, p. 1123-1138.

Haller, J., 1956, Probleme der Tiefentektonik: Bauformen in Migmatit-Stockwerk der ostgronlandisccchen Kaledoniden: Geologisches Rundschau, v. 45, p. 159-167.

——1971, Geology of the East Greenland Caledonides: New York, Interscience Publishers, 413 p.

Hara, I., Takeda, K., and Kimura, T., 1973, Preferred lattice orientation of quartz in shear deformation: Journal of Science of the Hiroshima University, ser. C, v. 7, p. 1-10.

Hazzard, J. C., and others, 1953, Large-scale thrusting in northern Snake Range, White Pine County, northeastern Nevada [abs.]: Geological Society of America Bulletin, v. 64, p. 1507-1508.

Hose, R. K., and Danes, Z. F., 1973, Development of late Mesozoic to early Cenozoic structures of the eastern Great Basin, in DeJong, K. A., and Scholten, R., eds., Gravity and tectonics: New York, John Wiley & Sons, Inc., p. 429-441.

Hose, R. K., and Blake, M. C., Jr., 1976, Geology and mineral resources of White Pine County, Nevada; Part I, Geology: Nevada Bureau of Mines and Geology Bulletin 85, p. 1–35.

Howard, K. A., 1966, Structure of the metamorphic rocks of the northern Ruby Mountains, Nevada [Ph.D. thesis]: New Haven, Connecticut, Yale University, 170 p.

——1968, Flow direction in triclinic folded rocks: American Journal of Science, v. 266, p. 758–765.

——1971, Paleozoic metasediments in the northern Ruby Mountains, Nevada: Geological Society of America Bulletin, v. 82, p. 259–264.

——1980, Metamorphic infrastructure in the northern Ruby Mountains, Nevada: Geological Society of America Memoir 153 (this volume).

Howard, K. A., and others, 1979, Geologic map of the Ruby Mountains, Nevada, 1:125,000: U.S. Geological Survey Miscellaneous Investigations Series Map I-1136.

Johnson, M.R.W., 1967, Mylonite zones and mylonite banding: Nature, v. 213, p. 246–247.

Karig, D. E., 1971, Origin and development of marginal basins in the western Pacific: Journal of Geophysical Research, v. 76, p. 2542–2561.

King, P. B., and Beikman, H. M., 1974, Geologic map of the United States (exclusive of Alaska and Hawaii): U.S. Geological Survey, scale 1:2,500,000.

Kistler, R. W., and Willden, R., 1969a, Precambrian-Cambrian boundary in the Ruby Mountains, Nevada [abs.]: Geological Society of America Abstracts with Programs, v. 1, no. 4, p. 32.

——1969b, Age of thrusting in the Ruby Mountains, Nevada [abs.]: Geological Society of America Abstracts with Programs, v. 1, no. 5, p. 40–41.

Kistler, R. W., Evernden, J. F., and Shaw, H. R., 1971, Sierra Nevada plutonic cycle: Part I, Origin of composite granitic batholiths: Geological Society of America Bulletin, v. 82, p. 853–868.

Kistler, R. W., and O'Neil, J. R., 1975, Fossil thermal gradients in crystalline rocks of the Ruby Mountains, Nevada, as indicated by radiogenic and stable isotopes [abs.]: Geological Society of America Abstracts with Programs, v. 7, p. 334.

Kvale, A., 1953, Linear structures and their relation to movement in the Caledonides of Scandinavia and Scotland: Quarterly Journal of Geological Society of London, v. 109, p. 51–74.

Lee, D. E., and others, 1970, Modification of K/Ar ages by Tertiary thrusting in the Snake Range, White Pine County, Nevada, in Geological Survey Research, 1970: U.S. Geological Survey Professional Paper 700-D, p. D92–D102.

Marcantel, J., 1975, Late Pennsylvanian and Early Permian sedimentation in northeast Nevada: American Association of Petroleum Geologists Bulletin, v. 59, p. 2079–2098.

Misch, P., 1960, Regional structural reconnaissance in central-northeast Nevada and some adjacent areas: Observations and interpretations, in Geology of east-central Nevada: Intermountain Association of Petroleum Geologists 11th Annual Field Conference Guidebook, p. 17–42.

——1971, Geotectonic implications of Mesozoic decollement thrusting in parts of eastern Great Basin: Geological Society of America Abstracts with Programs, v. 3, p. 164–166.

Misch, P., and Hazzard, J. C., 1962, Stratigraphy and metamorphism of late Precambrian rocks in central northeastern Nevada and adjacent Utah: American Association of Petroleum Geologists Bulletin, v. 46, p. 289–343.

Moores, E. M., 1970, Ultramafics and orogeny, with models of the U.S. Cordillera and the Tethys: Nature, v. 228, p. 837–842.

Moores, E. M., Scott, R. B., and Lumsden, W. W., 1968, Tertiary tectonics of the White Pine–Grant Range region, east-central Nevada, and some regional implications: Geological Society of America Bulletin, v. 79, p. 1703–1726.

Nelson, R. B., 1966, Structural development of northernmost Snake Range, Kern Mountains, and Deep Creek Range, Nevada and Utah: American Association of Petroleum Geologists Bulletin, v. 50, p. 921–951.

——1969, Relation and history of structures in a sedimentary succession with deeper metamorphic structures, eastern Great Basin: American Association of Petroleum Geologists Bulletin, v. 53, p. 307–339.

Olesen, N. O., and Sorensen, K., 1972, Caledonian fold- and fabric-elements: A model: International Geological Congress, 24th, Montreal, sec. 3, p. 533–544.

Price, R. A., 1972, The distinction between displacement and distortion in flow, and the origin of diachronism in tectonic overprinting in orogenic belts: International Geological Congress, 24th, Montreal, sec. 3, p. 545–551.

Price, R. A., and Mountjoy, E. W., 1970, Geologic structure of the Canadian Rocky Mountains between Bow and Athabasca Rivers—A progress report, in Wheeler, J. O., ed., Structure of the Southern Canadian Cordillera: Geological Association of Canada Special Paper no. 6, p. 8–25.

Rehrig, W. A., and Reynolds, S. J., 1977, A northwest zone of metamorphic core complexes in Arizona [abs.]: Geological Society of America Abstracts with Programs, v. 9, no. 7, p. 1139.

Riekels, L. M., and Baker, D. W., 1977, The origin of the double maximum pattern of optic axes in quartzite mylonite: Journal of Geology, v. 85, p. 1–14.

Ross, J. V., 1973, Mylonitic rocks and flattened garnets

in the southern Okanagan of British Columbia: Canadian Journal of Earth Sciences, v. 10, p. 1–17.

Sanderson, D. J., 1973, The development of fold axes oblique to the regional trend: Tectonophysics, v. 16, p. 55–70.

Scholz, C. H., Barazangi, M., and Sbar, M. L., 1971, Late Cenozoic evolution of the Great Basin, western United States, as an ensialic interarc basin: Geological Society of America Bulletin, v. 82, p. 2979–2990.

Secor, D. T., Jr., and Snoke, A. W., 1978, Stratigraphy, structure, and plutonism in the central South Carolina Piedmont, in Snoke, A. W., ed., Geological investigations of the eastern Piedmont, southern Appalachians: South Carolina Geological Survey, South Carolina State Development Board, Carolina Geological Society, Guidebook, p. 65–123.

Shackelford, T. J., 1976, Juxtaposition of contrasting structural and lithologic terranes along a major Miocene gravity detachment surface, Rawhide Mountains, Arizona [abs.]: Geological Society of America Abstracts with Programs, v. 8, no. 6, p. 1099.

Sharp, R. P., 1939a, The Miocene Humboldt Formation in northeastern Nevada: Journal of Geology, v. 47, p. 133–160.

—— 1939b, Basin-range structure of the Ruby–East Humboldt Range, northeastern Nevada: Geological Society of America Bulletin, v. 50, p. 881–920.

—— 1942, Stratigraphy and structure of the southern Ruby Mountains, Nevada: Geological Society of America Bulletin, v. 53, p. 647–690.

Snelson, S., 1957, The geology of the northern Ruby Mountains and the East Humboldt Range, Elko County, Nevada [Ph.D. thesis]: Seattle, Washington, 214 p.

Snoke, A. W., 1975, A structural and geochronological puzzle: Secret Creek Gorge area, northern Ruby Mountains, Nevada [abs.]: Geological Society of America Abstracts with Programs, v. 7, p. 1278–1279.

Steele, G., 1960, Pennsylvanian-Permian stratigraphy of east-central Nevada and adjacent Utah, in Geology of east central Nevada: Intermountain Association of Petroleum Geologists, 11th Annual Field Conference, Guidebook, p. 91–113.

Theodore, T. G., 1970, Petrogenesis of mylonites of high metamorphic grade in the Peninsular Ranges of southern California: Geological Society of America Bulletin, v. 81, p. 435–450.

Thorman, C. H., 1965, Mid-Tertiary K-Ar dates from late Mesozoic metamorphosed rocks, Wood Hills and Ruby–East Humboldt Range, Elko County, Nevada: [abs.]: Geological Society of America Special Paper 87, p. 234–235.

—— 1970, Metamorphosed and nonmetamorphosed Paleozoic rocks in the Wood Hills and Pequop Mountains, northeast Nevada: Geological Society of America Bulletin, v. 81, p. 2417–2448.

Todd, V. R., 1973a, Structure and petrology of metamorphosed rocks in central Grouse Creek Mountains, Box Elder County, Utah [Ph.D. thesis]: Stanford, California, Stanford University, 316 p.

—— 1973b, Tectonic mobilization of Precambrian gneiss during Tertiary metamorphism and thrusting, Grouse Creek Mountains, northwestern Utah [abs]: Geological Society of American Abstracts with Programs, v. 5, p. 116.

Wegmann, C. E., 1935, Zur Deutung der Migmatite: Geologisches Rundsahau, v. 26, p. 303–350.

Willden, R., Thomas, H. H., and Stern, T. W., 1967, Oligocene or younger thrust faulting in the Ruby Mountains, northeastern Nevada: Geological Society of America Bulletin, v. 78, p. 1345–1358.

Willden, R., and Kistler, R. W., 1967, Ordovician tectonism in the Ruby Mountains, Elko County, Nevada, in Geological Survey Research, 1967: U.S. Geological Survey Professional Paper 575-D, p. D64–D75.

—— 1969, Geologic map of the Jiggs quadrangle, Elko County, Nevada: U.S. Geological Survey Map GQ-859.

Wynne-Edwards, H. R., 1963, Flow folding: American Journal of Science, v. 261, p. 793–814.

Young, J. C., 1960, Structure and stratigraphy in north-central Nevada: Intermountain Association of Petroleum Geologists, 11th Annual Field Conference Guidebook, p. 158–172.

MANUSCRIPT RECEIVED BY THE SOCIETY JUNE 21, 1979

MANUSCRIPT ACCEPTED JUNE 26, 1979

Printed in U.S.A.

Geological Society of America
Memoir 153
1980

Metamorphic infrastructure in the northern Ruby Mountains, Nevada

KEITH A. HOWARD
U.S. Geological Survey
345 Middlefield Road
Menlo Park, California 94025

ABSTRACT

The metamorphic complex of the northern Ruby Mountains in northeastern Nevada exposes Paleozoic strata that are metamorphosed to sillimanite grade, migmatized, and recumbently folded. Nappes are variously overturned to the east, north, south, and west. The deeper part of this metamorphic infrastructure is a migmatitic zone pervaded by pegmatitic two-mica granite. A structurally higher transition zone underwent extreme tectonic flattening and some thrusting as the mobile infrastructure rose buoyantly against more rigid suprastructure. Relief by flow and stretching to the west-northwest and east-southeast resulted in a regionally constant lineation in this transition zone. The age of metamorphism is uncertain but may be Jurassic.

INTRODUCTION

The metamorphic complex of the northern Ruby Mountains lies in the Cenozoic Basin and Range province, in the western hinterland of the Mesozoic Sevier orogenic belt and east of the Paleozoic Antler orogenic belt (Fig. 1). Precambrian(?) to Devonian carbonate and quartzite units in the complex are metamorphosed, intimately injected by granite, and recumbently folded. Extensive migmatization and the ductile style of deformation support the concept that a mobile metamorphic infrastructure developed in this area of the Cordilleran miogeocline that contrasts with more brittlely deformed suprastructural rocks higher in the pile, as originally suggested by Armstrong and Hansen (1966; see also Howard, 1971; Snoke, 1974, 1975, and this volume). Equivalent strata that are relatively little metamorphosed are exposed in nearby areas. This paper briefly summarizes the tectonic style of the infrastructure in the northern Ruby Mountains as determined from detailed structural data and analysis (Howard, 1966; Smith and Howard, 1977; K. A. Howard, unpub. mapping, 1966 through 1977).

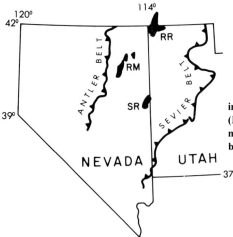

Figure 1. Location of major metamorphic complexes in Nevada and Utah: Ruby Mountains (RM), Raft River (RR), and Snake Range (SR) terranes. Sawteeth indicate major thrust faults of the Antler and Sevier orogenic belts.

SETTING

The metamorphic complex (Fig. 2) is exposed in the northern Ruby Mountains and adjacent East Humboldt Range, which are horsts, and in the uparched Wood Hills, the Paleozoic rocks of which were described by Thorman (1970). Rocks in the migmatitic cores of the Ruby and East Humboldt Ranges are metamorphosed to upper amphibolite facies in which sillimanite–muscovite–K-feldspar is a common assemblage. They are flanked along the west by a *transition zone,* largely in lower amphibolite facies, that is characterized by extreme attenuation of stratigraphic units and by blastomylonitic fabric with a regional west-northwest–trending streaky lineation.

A generalized geologic map (Fig. 3) shows rock units of the area of this report. The part of the Ruby Mountains north of this area is a transition zone containing low-angle ductile faults overlain by fault sheets of upper Paleozoic sedimentary rocks (Snelson, 1957; Snoke, 1974, 1975, and this volume). South of the study area, granitic rocks in an anticlinal high largely separate the metamorphic complex from Paleozoic sedimentary rocks in the southern Ruby Mountains (Willden and Kistler, 1969). A generalized geologic map and sections of the Ruby Mountains by Howard and others (1979) show relations among these areas.

The metasedimentary stratigraphic units of the northern Ruby Mountains are summarized in Table 1. Correlations of these units with strata elsewhere in eastern Nevada are based on lithology and stratigraphic sequence (Howard, 1971), as no fossils have been found in the northern Ruby Mountains. Fossils have been found, however, in similar metasedimentary rocks in the Wood Hills (Thorman, 1965, 1970).

The absence of kyanite in the Ruby Mountains, together with other characteristics of the mineral assemblages (Table 1; Howard, 1966), indicates that amphibolite-facies metamorphism was of low-pressure or low-pressure–intermediate type (Miyashiro, 1961).

GRANITIC ROCKS

The complex as exposed in the northern Ruby Mountains is a migmatite terrane in which granitic rocks, intimately intermixed with the metasedimentary rocks, account for about half the exposed volume, yet do not obliterate the stratigraphic sequence and structure. Metasedimentary rocks in the highest levels of the complex are intruded only by sparse pods of pegmatite, much as in the nearby

Wood Hills (Thorman, 1970). The proportion of granitic material increases downward until gneissic pegmatitic granite occupies the lowest structural levels with few or no metasedimentary relics. The predominant granitic rock is leucocratic two-mica pegmatitic granite (varying to trondhjemite) that occurs in sills, dikes, and irregular bodies (Table 2). Other rock types include biotite adamellite, two-mica adamellite, and biotite granodiorite. Garnet is common in these rocks, many of which are gneissic. Metasedimentary relics interleaved with and enveloped by granitic rocks show a coherent ghost stratigraphy and structure which demonstrate that emplacement of the granitic rocks was largely

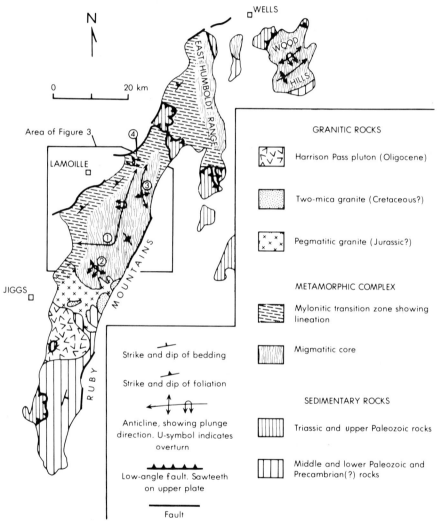

Figure 2. Sketch map of the Ruby Mountains, East Humboldt Range, and Wood Hills (from Thorman, 1970; Snelson, 1957; Stewart and Carlson, 1977; Howard and others, 1979). In the Wood Hills, the metamorphic complex consists of folded metasedimentary rocks. In the Ruby Mountains and East Humboldt Range, the metamorphic complex consists of a mylonitic transition zone with west-northwest lineation and an underlying migmatitic core in which folds and lineations are variably oriented. Numbered anticlines are (1) Lamoille Canyon nappe, (2) King Peak nappe, (3) Hidden Lakes uplift, and (4) Soldier Creek nappe. Lower and middle Paleozoic strata in the southern Ruby Mountains are largely separated from the metamorphic complex by granitic plutons of Jurassic(?), Cretaceous(?), and Oligocene ages. Klippen of upper Paleozoic strata overlie the complex. Cenozoic sedimentary deposits (unpatterned) flank the ranges.

passive. Intrusion of the granitic bodies spanned the period of penetrative deformation as indicated by all gradations from folded gneissic sills to undeformed crosscutting dikes.

The pegmatitic granite is extensively exposed (Fig. 2) just south of the area of this study. Willden and Kistler (1969) and Howard and others (1979) have assigned it a Jurassic age (160 m.y.) there, and they have reported that it is intruded by two-mica granite of Cretaceous age (82 m.y. old) and by the undeformed granodiorite pluton of Harrison Pass of Oligocene age (38 m.y. old). The Oligocene age (Willden and others, 1967) is well established by several radiometric techniques; the Cretaceous and Jurassic ages are based solely on Rb-Sr data, which possibly may be subject to alternative interpretations.

RECUMBENT FOLDS

The dominant structures in the part of the range studied (Fig. 3) are large recumbent folds (Figs. 4, 5). Foliation parallels the axial surfaces of these folds and dips gently over large areas. The folds vary widely in trend and vergence but generally plunge northward, with the result that higher structural levels are exposed in that direction. The principal anticlinal folds are the Lamoille Canyon nappe (overturned southeastward, maximum crest-to-trough amplitude 9 km), the King Peak nappe (overturned northeastward, maximum amplitude 6 km), the Hidden Lakes uplift (amplitude >1 km), and the Soldier Creek nappe (overturned westward, maximum amplitude 4 km). These large folds are paralleled by small parasitic folds and by mineral lineations.

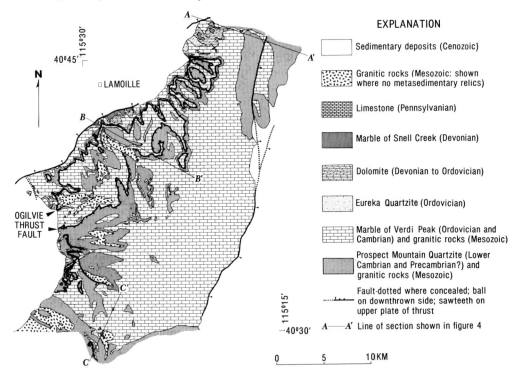

Figure 3. Generalized geologic map of report area, northern Ruby Mountains (located in Fig. 2). Granitic rocks are abundant but are shown only where metasedimentary relics are lacking. A thrust fault (Ogilvie thrust) exposed in the Lamoille Canyon nappe predates metamorphism and folding.

TABLE 1. STRATIGRAPHIC UNITS OF NORTHERN RUBY MOUNTAINS

Rock	Metamorphic minerals*	Tectonic thickness (m)	Inferred age
Marble of Snell Creek Banded gray and white limestone marble with a few dolomitic laminae. Possible crinoid stems. Tentatively correlated with Devonian Guilmette Formation	Calcite, dolomite	70	Devonian
Dolomite Massive white dolomite marble	Dolomite, tremolite	50-150	Devonian, Silurian and Ordovician
Eureka Quartzite White metaquartzite	Quartz, diopside	0-70	Orodovician
Marble of Verdi Peak Impure limestone, marble, calc-silicate rock, some schist and gneiss. Brown dolomite marble in lower part. Base commonly graphitic	Calcite, diopside, phlogopite, plagioclase, scapolite, actinolite, idocrase, garnet, sphene, K-feldspar, epidote, forsterite in calc-silicate rocks. Plagioclase, hornblende, biotite, quartz, muscovite, sillimanite in schists	50-3,000	Ordovician and Cambrian
Prospect Mountain Quartzite (unrestricted) Brown foliated metaquartzite, feldspathic and micaceous. Minor schist and calc-silicate	Quartz, biotite, muscovite, K-feldspar, plagioclase, sillimanite	50-1,000	Early Cambrian and Precambrian(?)

*Coexisting mineral assemblages are given in Howard (1966).

The tightly appressed Lamoille Canyon nappe (Fig. 4, B-B') is well exposed in many deep glaciated canyons. It folds the premetamorphic Ogilvie thrust fault (Smith and Howard, 1977), and Prospect Mountain Quartzite (Precambrian? and Lower Cambrian) and marble of Verdi Peak (Cambrian and Ordovician), which had originally been duplicated by that thrust. Two bodies of orthogneiss occupy the core of the nappe beneath the level of the folded thrust: biotite granodiorite gneiss of Seitz Canyon is enveloped by the Prospect Mountain Quartzite, and two-mica garnet adamellite gneiss of Thorpe Creek is enveloped by marble (Fig. 4). The nappe arches gently parallel to its north-trending axis so that the nose bends down eastward. This bending, together with coaxial eddylike and refolded small folds, suggests that the nappe began to roll over on itself as it grew. To the south the nappe hinge bends abruptly to the west, the nappe decreases in amplitude, and becomes locally detached. In its southwesternmost exposures, in the transition zone, the nappe becomes a thrust sheet with a displacement of only 1 km (Smith and Howard, 1977).

The King Peak nappe displaces the Prospect Mountain Quartzite northeastward over marble of

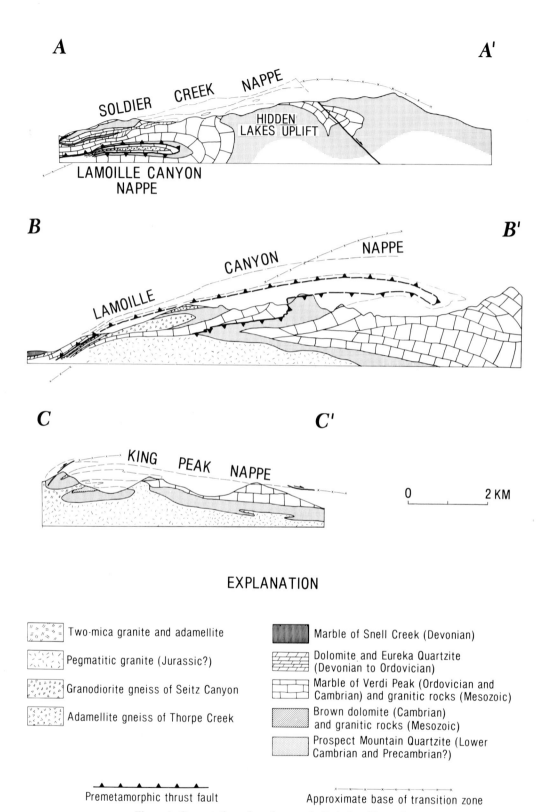

Figure 4. Cross sections. Locations are indicated on Figure 3.

Figure 5. Cartooned oblique view of relations among large folds in the northern two-thirds of the area shown in Figure 3. Surfaces shown are top of Prospect Mountain Quartzite (lined) and top of Eureka Quartzite (dotted). Post-metamorphic faulting of the Hidden Lakes uplift and Soldier Creek nappe been restored.

Verdi Peak (Fig. 4, C–C'). The south end of the area mapped in Figure 3 (Rattlesnake Canyon) exposes three, perhaps four, synclinal roots beneath the nappe; at least one of these is partly detached by thrusting. Quartzite in the nappe is exposed continuously for 1 km in front (north) of its root(s), and outliers are as far as 6 km north and a little lower than the root. Those beyond 4 km rest in thrust contact on younger but higher grade marbles, a relation suggesting that the nappe slid down on them from a higher position of lower grade. A plexus of late metamorphic thrust slivers lies at the west foot of the range, where the King Peak nappe, if it projects westward, opposes the southward overturned Lamoille Canyon nappe (Smith and Howard, 1977).

The Hidden Lakes uplift is a doubly plunging, mostly upright anticline entirely within the sillimanite-zone core of the complex (Fig. 5). Foliation in the Prospect Mountain Quartzite and the overlying marble of Verdi Peak wraps around the uplift, and locally small folds are coaxially refolded. In the northern part, the uplift overhangs westward so that the opposing Lamoille Canyon nappe nestles against it (Fig. 4, A–A'). Southward, the overhang disappears as the uplift diverges from the nappe, and the synclinal complex of high-grade rocks between them becomes many kilometres broad. Marble of Verdi Peak within the synclinal complex is thus greatly thickened. Variously oriented small folds and other fabric elements suggest variable flow directions, perhaps related to buoyant(?) rising of quartzite in the uplift and sinking of marble in the synclinal complex. Variable flow directions and a supple style of deformation thus characterize the high-grade core of the metamorphic infrastructure. The rocks in this core are coarsely crystalline and annealed (Fig. 6, right), which indicates that heating kept pace with, and outlasted, deformation.

The Soldier Creek nappe is rooted in a steeply dipping, migmatitic zone east of the Hidden Lakes uplift (Fig. 4, A–A') and is folded some 4 km westward down over the uplift and the Lamoille Canyon nappe. It is a tight fold that inverts rocks ranging in age from Precambrian(?) and Cambrian (Prospect Mountain Quartzite) to Devonian (marble of Snell Creek, probably equivalent to the Guillmette Formation). Most of the nappe is in the transition zone, and Snoke (1974) has found evidence that the nappe is locally detached. The underlying synclinal hinges of marble of Verdi Peak, Eureka Quartzite, dolomite, and marble of Snell Creek show a complex pattern of extraordinary fold disharmony and variation in trend (Fig. 5). At the level of the Ordovician Eureka Quartzite, the nappe decreases in amplitude to 0.5 km and the hingeline changes direction 180°. Despite this disharmony, all fold hinges in the nappe and underlying syncline share a component of vergence to the west-northwest. This sense of overturn is opposite that of the underlying Lamoille Canyon nappe (Fig. 4, A–A', B–B') although it matches the sense shown by many small folds on the upper limb of that nappe; their reverse sense recalls spruce-tree folds of New England (Skehan, 1961; Thompson and Rosenfeld, 1951).

TABLE 2. CHARACTERISTICS OF GRANITIC ROCKS IN THE NORTHERN RUBY MOUNTAINS

Rock	Description
Pegmatitic granite	Leucocratic. Ranges in composition from granite to trondhjemite. Biotite, muscovite, some garnet, sillimanite. By far the most voluminous granitic rock in the complex. Sills, dikes, and irregular bodies from lit-par-lit size to more than 200 m thick. Proportion increases downward, reaching 100% in some deep parts of the complex. Ranges from synkinematic (gneiss) to late kinematic (dikes). Metasedimentary relics intermixed with pegmatitic granite show ghost stratigraphy. Pegmatitic granite south of study area is assigned a Jurassic age by Willden and Kistler (1969) on the basis of Rb-Sr dating
Adamellite gneiss of Thorpe Creek	Muscovite-biotite-garnet gneiss. Early synkinematic. Forms a sill-like body averaging 60 m thick in core and upper limb of Lamoille Canyon nappe
Granodiorite gneiss of Seitz Canyon	Biotite granodiorite gneiss. Synkinematic. Forms core of Lamoille Canyon nappe. Contains metasedimentary relics showing a ghost stratigraphy
Biotite adamellite	Forms irregular bodies of mylonite gneiss in transition zone (early synkinematic) and foliated sill-like to irregular bodies in sillimanite zone (late synkinematic) and dikes (postkinematic)
Two-mica granite and adamellite	Some contain garnet. Small irregular bodies that in the sillimanite zone may postdate most pegmatitic granite. Biotite granite mylonite gneiss in the transition zone (early synkinematic) may be related

TRANSITION ZONE

Along the west flank of the range, the Lamoille Canyon and Soldier Creek nappes are both so highly attenuated that the stratigraphic sequence is thinned to as little as one-fifteenth to one-twentieth of the original thickness. This extreme thinning occurred during metamorphism and partly outpaced crystallization, as indicated by blastomylonitic textures in augen gneisses, flaggy quartzite (Fig. 6, left), and marble. Quartzofeldspathic rocks have a pervasive foliation and streaky lineation (Fig. 7). The transition zone is composed of rocks showing these features characteristic of dominant strain. The lineation is formed by highly elongate quartz and other minerals; crystalline quartz and mica mortar typically surround more resistant feldspar augen. The lineation has a constant west-northwest orientation, perpendicular to the main trend of the Lamoille Canyon nappe.

The transition zone encompasses parts of several folds, not all at the same structural level (Figs. 2, 4). For example, the streaky lineation occurs in the nose and part of the root zone of the Soldier Creek nappe in the north, in both limbs but not the nose of the lower Lamoille Canyon nappe, and in pegmatitic granite below the mapped nappes in the southwest corner of the area of the geologic map (Fig. 3). Despite their mylonitic character, rocks of the transition zone deformed at high temperature and lack effects of retrograde metamorphism. The isograd defined by the first appearance of sillimanite in micaceous quartzite generally lies near the base of the transition zone and cuts obliquely across structures, which suggests that high temperatures continued to the end of deformation.

Figure 6. Comparison of Prospect Mountain Quartzite in (above left) transition zone, where rock is tectonically thinned, flaggy, and blastomylonitic, and in (above right) migmatitic core, where rock is higher grade, coarsely crystalline, annealed, and not greatly thinned. Both specimens are cut normal to foliation.

Figure 7. Character of streaky west-northwest–trending lineation of the transition zone, here seen on the foliation surface of a granite gneiss.

The blastomylonitic fabric that characterizes the transition zone is believed to result from the extraordinary tectonic flattening and stretching. This fabric system is mesoscopically orthorhombic in symmetry; the few small folds that parallel the streaky lineation are tightly appressed in the plane of the subhorizontal foliation and show no consistent vergence. The symmetry is compatible with an origin for the foliation and lineation by orthorhombic thinning and stretching. The strain must have been broadly synchronous with nappe folding, for the stretching lineation is commonly deformed by folds parasitic to the Lamoille Canyon nappe, yet an unfolded aplite dike bears the lineation, and both limbs of the nappe have been flattened and thinned (Fig. 4).

Calc-silicate marbles in the transition zone typically show folds parasitic to the nappes, even though some on the upper limb of the Lamoille Canyon nappe show a reverse or spruce-tree sense of vergence. Slip line of flow, as determined by Hansen's (1971) method using vergence separation arcs, in the Lamoille Canyon nappe near and in the transition zone strike west-northwest (Howard, 1966). This orientation obtains even in the southern part of the nappe, where the nappe and its most prominent parasitic folds and lineation veer obliquely to or subparallel to the flow line (Howard, 1968). The separation arcs containing the flow lines, taken cumulatively, show that slip flow to the west-northwest and east-southeast paralleled the direction of stretching shown by the streaky lineation (Fig. 8).

Tectonic transport was west-northwest in the Soldier Creek nappe, as indicated by the map pattern of its hinge lines (Fig. 5). Slickensides on small thrust faults between the Lamoille Canyon and King Peak nappes parallel the streaky lineation and further indicate tectonic transport parallel to the lineation. Rocks in the transition zone apparently squeezed outward through both thinning and shear to the east-southeast and west-northwest. This is a pattern of extending flow (Price, 1972). It suggests gravitational compression with a single direction of relief that is perpendicular to Cordilleran trends.

DISCUSSION

The variability in fold styles and trends and their independence at different structural levels are striking. Yet the orientation of flow and extension within the transition zone was surprisingly uniform.

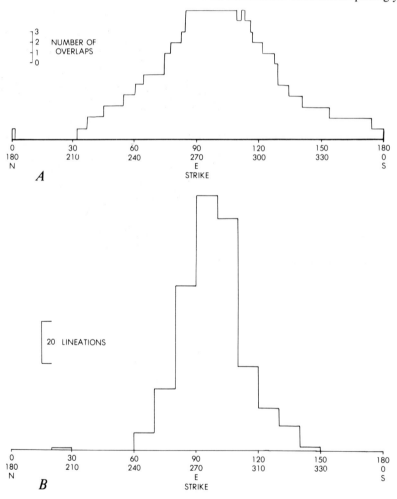

Figure 8. The streaky west-northwest lineation of the transition zone parallels the slip directions from vergence separation arcs of folds. (A) Azimuthal frequency of 13 fold-separation arcs graphed as the number of overlapping arcs. (B) Azimuthal frequency of streaky lineation.

The infrastructure core, which had high temperatures and abundant granitic material, had low strength that allowed large-scale rock flow. Its deformation was probably driven by thermal and gravitational instabilities; the opposing overturns of the nappes argue against lateral compression. Crustal stretching cannot be ruled out, for volumes of metasedimentary rocks in known folds are less than original stratigraphic thicknesses would allow. Structural disharmony probably developed as the infrastructure rose buoyantly against a more rigid suprastructure (Fig. 9). Outward flow at this level resulted in development of the transition zone, thereby flattening folds that originated below. Some of these movements outpaced metamorphic crystallization, which resulted in detached folds. The shear and extension direction is remarkably constant. West-northwest lineations in transition-zone rocks can be recognized along the Ruby Mountains and East Humboldt Range for 80 km across the strike of the lineation (Fig. 2A). Similar fabric and lineation with virtually the same orientation occur in transition-zone rocks in the northern Snake Range and Kern Mountains 150 km southeast (Fig. 1; Nelson, 1969). The orientation of extending flow, then, may have been constant over a large region of northeastern Nevada.

Gravitational flattening coupled with horizontal tectonic transport also characterizes the Raft River–Grouse Creek Mountains metamorphic terrane 200 km northeast (Fig. 1; Compton and others, 1977). There, deformation decreased downward into Precambrian basement, whereas in the Ruby Mountains terrane, migmatite and infrastructural mobility increased downward.

The regional distribution of Phanerozoic metamorphic rocks in the Great Basin is patchy; the Ruby Mountains contain the largest exposure of migmatite (Misch and Hazzard, 1962; Armstrong and Hansen, 1966; Compton and others, 1977). Does this mean that the Ruby Mountains terrane is an isolated thermal high or part of a much broader infrastructure at depth like that now exposed in the Omineca crystalline terrane of British Columbia (Wheeler, 1970)? One clue may come through studies in progress by Anita Harris of regional distribution of low-grade metamorphism as indicated by conodont color alteration (Epstein and others, 1977). Other clues may come through mapping of

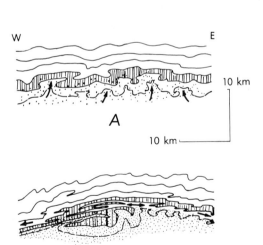

Figure 9. A possible model of infrastructure development, cartooned by east-west cross sections. (A) Diapirs form as migmatite front rises. (B) Mobile infrastructure presses buoyantly against suprastructure and causes outward flow and attenuation of nappes. (C) Cooling transforms infrastructure into new rigid basement, against which detachment faults bottom.

regional metamorphic terranes along strike in Idaho and in southeast California and Arizona. Resolution of the timing of metamorphism is of course critical.

The age of metamorphism and infrastructure development in the Ruby Mountains is unsettled, despite extensive geochronologic study (Kistler and Willden, 1969; Willden and Kistler, 1969; Kistler and O'Neil, 1975; Snoke, 1975 and this volume). Tertiary K-Ar ages have been interpreted to be reset, not original (Armstrong and Hansen, 1966; Thorman, 1966; Mauger and others, 1968; Kistler and O'Neil, 1975). Sr and Pb isotopic data are not yet published, and interpretation of these data is complicated by apparent mixing. Metamorphism must be younger than the Devonian strata involved and is probably younger than the youngest (Triassic) conformable deposits of the Cordilleran miogeocline. If confirmed, Jurassic age for the pegmatitic granite assigned by Willden and Kistler (1969) on the basis of Rb-Sr data should date the main stage of intrusion and metamorphism. Westward overturning of the Soldier Creek nappe, and of higher folds to the north (Snoke, 1974), possibly is related to westward-directed folds and thrusts 20 to 40 km to the west in the Carlin-Pinon Range area, which Smith and Ketner (1977) date as latest Jurassic or Early Cretaceous. Table 3 summarizes the sequence of events as known from the Ruby Mountains. Continuing geochronologic study by Kistler and by Snoke and coworkers may eventually resolve the age of the metamorphic complex. The age is the key to how the complex is related to other metamorphic complexes and how or whether it is related to Sevier thrusting, basin-range extension, or other tectonic events.

TABLE 3. SUMMARY OF GEOLOGIC HISTORY OF THE RUBY MOUNTAINS

PALEOZOIC	MESOZOIC(?)	CENOZOIC
Miogeoclinal sedimentation (Precambrian? through at least Devonian-regionally to Triassic)	Metamorphism	Granodiorite pluton (Harrison Pass-Oligocene**, *)
	Blastomylonitic thinning	Uplift and cooling (Oligocene and Miocene†)
	Nappe folding Thrusting	
	2-mica adamellite	Unroofing (late Miocene§)
Westward thrusting	Granodiorite	
	Pegmatitic granite (Jurassic?*)	Basalt dikes (Miocene#)
	Biotite adamellite	Detachment faulting (Neogene#)
	2-mica granite (Cretaceous?*)	
		Block faulting (Miocene§ to Holocene)

Note: Infrastructure development in the Mesozoic(?) is shown by the relative sequence of events interpreted from geologic relations.

*Willden and Kistler (1969); R. W. Kistler (1975, personal commun.).
†Kistler and O'Neil (1975).
§Smith and Ketner (1977); Sharp (1939).
#Snoke (this volume).
**Willden and others (1967).

REFERENCES CITED

Armstrong, R. L., and Hansen, Edward, 1966, Cordilleran infrastructure in the eastern Great Basin: American Journal of Science, v. 264, p. 112–127.

Compton, R. R., and others, 1977, Oligocene and Miocene metamorphism, folding, and low-angle faulting in northwestern Utah: Geological Society of America Bulletin, v. 88, p. 1237–1250.

Epstein, A. G., Epstein, J. B., and Harris, L. D., 1977,

Conodont color alteration—An index to organic metamorphism: U.S. Geological Survey Professional Paper 995, 27 p.

Hansen, Edward, 1971, Strain facies: New York, Springer-Verlag, 207 p.

Howard, K. A., 1966, Structure of the metamorphic rocks of the northern Ruby Mountains, Nevada [Ph.D. thesis]: New Haven, Conn., Yale University, 170 p.

——1968, Flow direction in triclinic folded rocks: American Journal of Science, v. 266, p. 758–765.

——1971, Paleozoic metasediments in the northern Ruby Mountains, Nevada: Geological Society of America Bulletin, v. 82, p. 259–264.

Howard, K. A., and others, 1979, Geologic map of the Ruby Mountains, Nevada: U.S. Geological Survey Map I-1136, scale 1:125,000.

Kistler, R. W., and O'Neil, J. R., 1975, Fossil thermal gradients in crystalline rocks of the Ruby Mountains, Nevada, as indicated by radiogenic and stable isotopes: Geological Society of America Abstracts with Programs, v. 7, p. 334.

Kistler, R. W., and Willden, Ronald, 1969, Age of thrusting in the Ruby Mountains, Nevada: Geological of America Abstracts with Programs for 1969, Part 5, p. 40.

Mauger, R. L., Damon, P. E., and Livingston, D. E., 1968, Cenozoic argon ages on metamorphic rocks from the Basin and Range province: American Journal of Science, v. 266, p. 579–589.

Misch, Peter, and Hazzard, J. C., 1962, Stratigraphy and metamorphism of late Precambrian rocks in central northeastern Nevada and adjacent Utah: American Association of Petroleum Geologists Bulletin, v. 46, p. 289–343.

Miyashiro, Akiho, 1961, Evolution of metamorphic belts: Journal of Petrology, v. 2, p. 277–311.

Nelson, R. B., 1969, Relation and history of structures in a sedimentary succession with deeper metamorphic structures, eastern Great Basin: American Association of Petroleum Geologists Bulletin, v. 53, p. 307–339.

Price, R. A., 1972, The distinction between displacement and distortion in flow, and the origin of diachronism in tectonic overprinting in orogenic belts: International Geological Congress, 24th, Montreal, Proceedings, Section 3, p. 545–551.

Sharp, R. P., 1939, Basin-range structure of the Ruby–East Humboldt Range, northeastern Nevada: Geological Society of America Bulletin, v. 50, p. 881–920.

Skehan, J. W., 1961, The Green Mountain anticlinorium in the vicinity of Wilmington and Woodford, Vermont: Vermont Geological Survey Bulletin 17, 159 p.

Smith, J. F., and Howard, K. A., 1977, Geologic map of the Lee 15-minute quadrangle, Elko County, Nevada: U.S. Geological Survey Map GQ-1393, scale 1:62,500.

Smith, J. F., and Ketner, K. B., 1977, Tectonic events since early Paleozoic in the Carlin-Pinon Range area, Nevada: U.S. Geological Survey Professional Paper 867-C, 18 p.

Snelson, Sigmund, 1957, The geology of the northern Ruby Mountains and the East Humboldt Range, Elko County, Nevada [Ph.D. thesis]: Seattle, University of Washington, 268 p.

Snoke, A. W., 1974, The transition from infrastructure to suprastructure in the northern Ruby Mountains, Nevada: Geological Society of America Abstracts with Programs, v. 6, p. 258.

——1975, A structural and geochronological puzzle: Secret Creek gorge area, northern Ruby Mountains, Nevada: Geological Society of America Abstracts with Programs, v. 7, p. 1278–1279.

——1980, The transition from infrastructure to suprastructure in the northern Ruby Mountains, Nevada (this volume).

Stewart, J. H., and Carlson, J. E., 1977, Million-scale geologic map of Nevada: Nevada Bureau of Mines and Geology Map 57.

Thompson, J. B., and Rosenfeld, J. R., 1951, Tectonics of a mantled gneiss dome in southeastern Vermont: Geological Society of America Bulletin, v. 62, p. 1484–1485.

Thorman, C. H., 1965, Biotitized graptolites from northeastern Nevada: American Association of Petroleum Geologists Bulletin, v. 49, p. 610–613.

——1966, Mid-Tertiary K-Ar dates from late Mesozoic metamorphosed rocks, Wood Hills and Ruby–East Humboldt Range, Elko County, Nevada: Geological Society of America Special Paper 87, p. 234–235.

——1970, Metamorphosed and nonmetamorphosed Paleozoic rocks in the Wood Hills and Pequop Mountains, northeast Nevada: Geological Society of America Bulletin, v. 81, p. 2417–2448.

Wheeler, J. O., 1970, Summary and discussion, *in* Structure of the southern Canadian Cordillera: Geological Association of Canada Special Paper 6, p. 155–166.

Willden, Ronald, and Kistler, R. W., 1969, Geologic map of the Jiggs quadrangle, Elko County, Nevada: U.S. Geological Survey Map GQ-859, scale 1:62,500.

Willden, Ronald, Thomas, H. H., and Stern, T. W., 1967, Oligocene or younger thrust faulting in the Ruby Mountains, northeastern Nevada: Geological Society of America Bulletin, v. 78, p. 1345–1358.

Manuscript Received by the Society June 21, 1979

Manuscript Accepted August 7, 1979

Printed in U.S.A.

Structure and petrology of a Tertiary gneiss complex in northwestern Utah

VICTORIA R. TODD
U.S. Geological Survey
La Jolla, California 92093

ABSTRACT

A gneiss complex in the Grouse Creek Mountains, northwestern Utah, consists of 2.5-b.y.-old adamellite that was remobilized and intruded younger Precambrian and Paleozoic cover rocks 25 m.y. ago during an extended period of synkinematic metamorphism. One or more structural domes formed in the region during metamorphism. The remains of a middle Tertiary culmination in the central Grouse Creek Mountains is an asymmetrical welt of schistose Precambrian adamellite that protrudes into greatly attenuated autochthonous and allochthonous cover rocks. Fine- to medium-grained gneiss and schist of the upper 200 m of the welt, or dome, grade upward to retrograde phengite-quartz-albite schist that intruded the lowermost beds of upper Precambrian(?) quartzite. The age of the schistose Precambrian adamellite and its geographic continuity with less-metamorphosed Precambrian adamellite lying unconformably beneath the quartzite indicate that remobilization of the former was post-Precambrian. Simultaneously, cover rocks were thinned to one-fifth their original thickness by metamorphic flattening, extension, and low-angle faulting.

A 25-m.y.-old pluton underlies the area of remobilization. Its outer shell bears metamorphic folds related to a second deformation. These folds also predominate in a metasomatic aureole in the Precambrian adamellite surrounding the Tertiary pluton and in the upper, schistose part of the gneiss dome. Thus, development of the dome was closely related to the second deformation and to the intrusion 25 m.y. ago.

Late-metamorphic folds and low-angle faults chiefly affected allochthonous cover rocks, and postmetamorphic movements carried parts of these rocks over 11-m.y.-old sedimentary rocks in the adjoining basin. Isotopic data on autochthonous rocks suggest that metamorphic temperatures persisted into Miocene time and provided the potential for continuing uplift and basinward shedding of sheets which ended after 11 m.y. ago.

INTRODUCTION

A gneiss complex in the central Grouse Creek Mountains, northwestern Utah, consists of remobilized 2.5-b.y.-old adamellite that diapirically intruded autochthonous and allochthonous upper Precambrian and Paleozoic cover rocks about 25 m.y. ago. It is part of a Tertiary metamorphic and igneous complex in the Raft River and Grouse Creek Mountains which has been described by Compton and others (1977) (Fig. 1). Age relations between metamorphic minerals and superimposed fold sets in the gneiss complex and surrounding rocks indicate that its rise was related to intrusion of adamellite (rock names according to the usage of Williams and others, 1954, p. 121) magma during regional synkinematic metamorphism in early Miocene time.

The Grouse Creek Mountains gneiss complex shares many features in common with gneiss-dome terranes in western North America (Armstrong, 1968; Davis, 1975; Fox and others, 1976; McMillan, 1970; Reesor, 1965; Reesor and Moore, 1971). These features include remobilization of basement rocks and their intrusion into younger mantling rocks, low- to high-grade metamorphism with a steep metamorphic gradient, penetrative deformation and polyphase recumbent folding, low-angle faulting, and igneous intrusion with formation of abundant leucocratic rock. The Grouse Creek gneiss complex has, in addition, features that make it a particularly useful example of gneiss-dome dynamics: it is small in area (150 km^2), transitions from one structural element to another are well defined, the rocks that compose the complex can be traced into less-modified rocks in the surrounding region, and events in its development can be tied to radiometric ages.

METAMORPHIC, STRUCTURAL, AND INTRUSIVE FRAMEWORK OF THE GNEISS COMPLEX

Rocks of the Raft River and Grouse Creek Mountains underwent upper amphibolite–facies metamorphism accompanied by multiple deformation in middle Tertiary time (Compton and others, 1977). Figure 1 shows the distribution of autochthonous, allochthonous, and intrusive rocks within this area. Figure 2 is a simplified structural map and cross section of the central Grouse Creek Mountains showing remnants of the three thin allochthonous sheets that once covered an autochthon composed chiefly of Precambrian adamellite. The uppermost allochthon is not metamorphosed, and the metamorphic grades attained in the underlying autochthon and two allochthons differ markedly. This mismatch exists throughout the Raft River–Grouse Creek region. For example, the highest-grade part of the lower sheet currently rests upon the lowest-grade part of the autochthon; this situation indicates considerable late-metamorphic or postmetamorphic transport on low-angle faults (Compton and others, 1977). Mineral assemblages of increasing and decreasing metamorphic grade are listed in Table 1 for each structural unit present in the central Grouse Creek Mountains. Strain outlasted recrystallization in most rocks.

Low-Angle Faults

The three principal low-angle faults have divided a possibly complete—although for the eastern Great Basin anomalously thin—upper Precambrian(?) through lower Mesozoic sedimentary sequence into three superimposed, nearly horizontal allochthonous sheets, each emplaced upon older rocks. In the underlying autochthon, the lowest unit of this sequence [upper Precambrian(?) Elba Quartzite] rests nonconformably upon 2.5-b.y.-old adamellite throughout the region (Armstrong, 1968; Compton, 1972, 1975). The age of the Elba Quartzite and its overlying units remains uncertain. If Crittenden is correct in correlating the Elba Quartzite with a formation in the Wasatch Range, then the Elba may be as old as 1.6 to 1.8 b.y. (Crittenden, 1979). The lowest fault lies within a schist unit above the Elba Quartzite, and the lowermost sheet consists of fault-bounded slices of a folded sequence of upper Precambrian(?) through Lower Ordovician stratified rocks. Large recumbent isoclinal folds were broken by low-angle faults nearly parallel to axial planes of the folds. The faults are localized in metamorphosed shale and limestone, and individual units are tectonically thinned by an order of

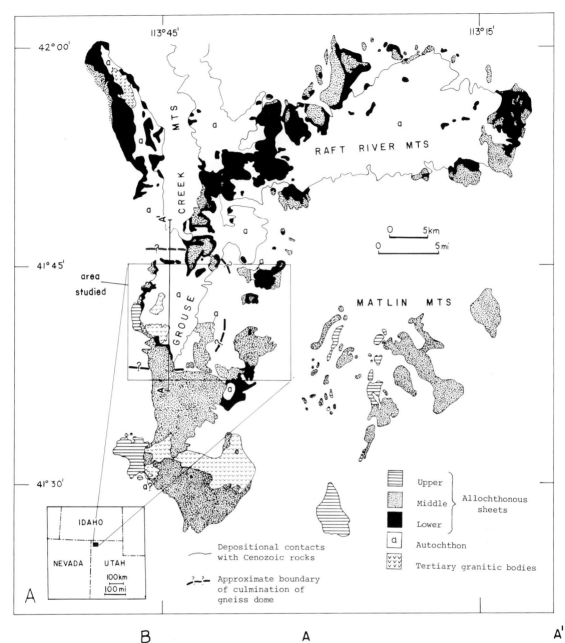
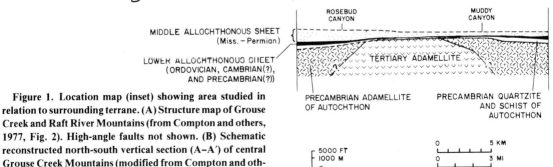

Figure 1. Location map (inset) showing area studied in relation to surrounding terrane. (A) Structure map of Grouse Creek and Raft River Mountains (from Compton and others, 1977, Fig. 2). High-angle faults not shown. (B) Schematic reconstructed north-south vertical section (A–A′) of central Grouse Creek Mountains (modified from Compton and others, 1977, Fig. 4). Horizontal and vertical scales equal; section somewhat enlarged relative to map.

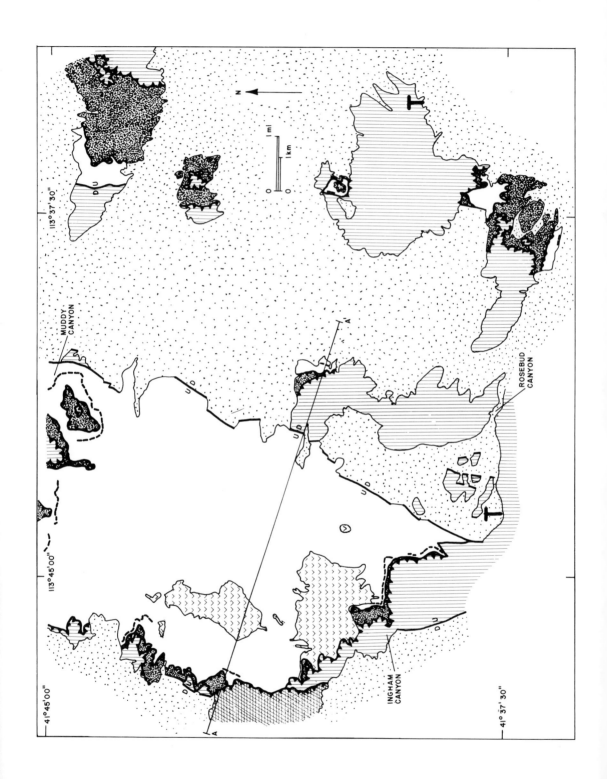

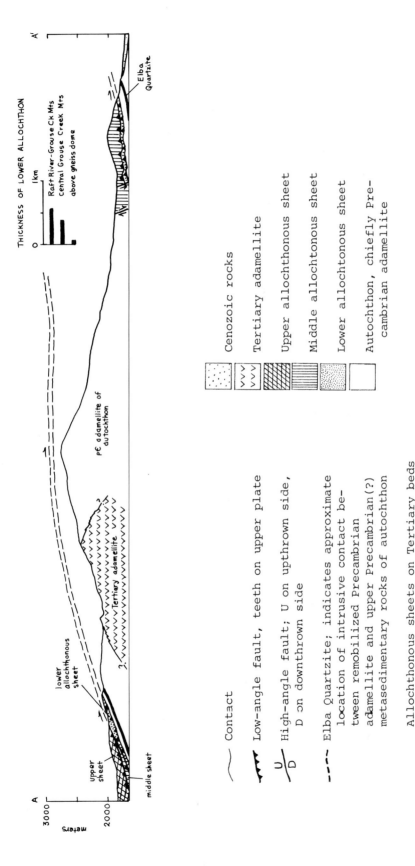

Figure 2 (facing pages). Generalized structure map and vertical cross section of central Grouse Creek Mountains. Cross section A-A' has equal horizontal and vertical scales and is slightly enlarged relative to map. Thicknesses of rock units in lower sheet are maximal for region after metamoprhism (from Compton and others, 1977, and Todd, 1973).

TABLE 1. METAMORPHIC MINERAL ASSEMBLAGES IN ROCKS IN CENTRAL GROUSE CREEK MOUNTAINS

Structural unit	Original rock	Metamorphic minerals
Upper allochthon		Unmetamorphosed
Middle allochthon	Silty limestone	Calcite, dolomite, quartz, K-feldspar, plagioclase (feldspars detrital)
	Shale	White mica, quartz, biotite, chloritoid, graphite [chlorite]
	Calcareous impure sandstone	Quartz, calcite, white mica (locally chlorite and actinolite)
Lower allochthon	Impure limestone	Calcite, dolomite, quartz, white mica
	Shale	Biotite, quartz, garnet, oligoclase, graphite [chlorite, white mica]
	Feldspathic sandstone	Quartz, white mica (locally K-feldspar, plagioclase, garnet, biotite) [chlorite]
Autochthon	Adamellite	Oligoclase, orthoclase, quartz, biotite [phengite, quartz, albite, epidote, sphene]
	Gabbro	Green hornblende, plagioclase, quartz, garnet, ilmenite [epidote, sphene, biotite, quartz]
	Shale	Biotite, quartz, oligoclase, garnet (altered metasomatically to assemblages with kyanite, staurolite, sillimanite) [chlorite*, sphene, white mica*, andalusite, apatite*]
	Impure limestone (as sparse tectonic inclusions in upper part of adamellite)	Calcite, quartz, calcic plagioclase, clinozoisite, garnet, biotite, sphene

Note: Minerals listed in order of decreasing abundance. Assemblages listed are prograde; brackets indicate retrograde assemblage. Most minerals retain preferential alignment with axes of first folds but have recrystallized to some degree parallel to second (northwest) fold axes. Minerals that grew, or recrystallized almost completely, parallel to second (northwest) folds are underlined. Local posttectonic recrystallization indicated by asterisk.

magnitude (Fig. 3). A single large recumbent anticline appears to form the lowermost allochthonous sheet over much of the central Grouse Creek Mountains (Todd, 1973). Satellitic folds on the limbs of this fold trend northwest; those on the upper limb are overturned to the southwest, whereas those on the lower (overturned) limb are overturned to the northeast.

The middle low-angle fault is exposed for tens of kilometres across the Grouse Creek and Raft River Mountains (Compton and others, 1977). In the central Grouse Creek Mountains, it has emplaced Upper Mississippian through Permian rocks onto Lower Ordovician rocks and, locally, onto Precambrian rocks (Fig. 2). In the northern part of the area, metamorphosed black shale of the Upper Mississippian Chainman Formation is preserved as a series of tectonic slivers at the base of the middle sheet; south of Ingham Canyon, a relatively thick section of the Upper Mississippian Diamond Peak Formation forms the lower part of the sheet. The fault zone locally extends downward to include Upper Cambrian(?) and Lower Ordovician units that occur as slices along the fault, everywhere in normal stratigraphic order beneath the Chainman shale. The presence of Silurian, Devonian, and Lower Mississippian rocks in the southern Grouse Creek Mountains in fault-bounded slices (Stanford University Department of Geology, 1973, 1976, unpub. data) suggests that Upper Ordovician through Lower Mississippian units probably have been removed by faulting in the central Grouse Creek Mountains. Well-cemented breccia occurs locally above the fault plane, and in one locality, Precambrian adamellite has been converted to phyllonite below the middle fault. In some places, rocks above the fault are tightly folded on northwest-trending axes with southwestward overturn.

In the eastern part of the Grouse Creek Mountains where the middle allochthonous sheet is thickest, it consists of a sequence of large, north-northeast–trending, nearly recumbent isoclinal folds with eastward overturn (Fig. 4). These folds are interpreted as late metamorphic because associated recrystallization and flow within layers are minor. However, the folds must have formed under moderate load because they are accompanied by parallel crenulations that have overprinted earlier (first and second) metamorphic folds. The folds are cut by younger-on-older low-angle faults and by older-on-younger imbrications with eastward displacement (Fig. 4). In the eastern part of the Grouse Creek Mountains, the middle sheet rests on undated tuffaceous beds in two localities (Fig. 2).

The structurally highest low-angle fault is exposed in the central and southern Grouse Creek Mountains. It carried Upper Permian rocks onto Mississippian through Pennsylvanian-Permian rocks and, where the latter are tectonically cut out, onto the Precambrian autochthon (Fig. 2). This large stratigraphic displacement is accompanied by an abrupt metamorphic break—rocks above the fault are neither metamorphosed nor penetratively folded, although they are brecciated, attenuated, and broken by tear faults.

Rocks identical to those of the middle and upper allochthons of the central Grouse Creek Mountains are interlayered with a sequence of 11-m.y.-old tuffaceous sedimentary rocks and fanglomerate (K-Ar dating of volcanic plagioclase by E. H. McKee, U.S. Geological Survey; fission-track dating of volcanic zircon by C. W. Naeser, U.S. Geological Survey) 5 km to the east in the Matlin Mountains (Fig. 1) (Todd, 1975). Upper Paleozoic and lower Mesozoic rocks moved over tilted or folded Tertiary rocks repeatedly; this resulted in the stacking up of thin, westward-dipping slices.

Folds

The distribution of metamorphic folds in relation to the three allochthonous sheets and the autochthon is shown in Figure 5. Fold axes and mineral lineation are parallel. Deformed pebbles and fossils are elongated in the direction of fold axes, less extended perpendicular to fold axes, and greatly flattened in the horizontal plane (Fig. 6). Table 1 indicates age relationships between growth of metamorphic minerals and episodes of folding.

Folds with wavelengths larger than 5 cm and as large as several metres occur sporadically in the

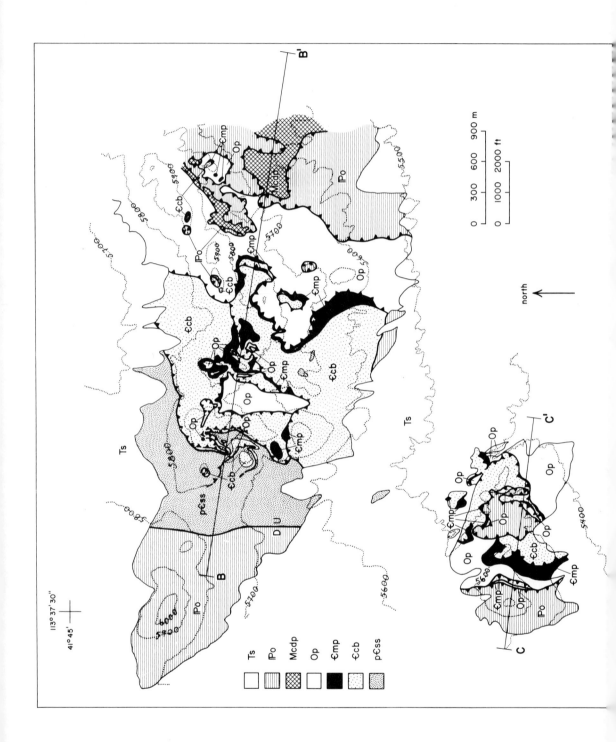

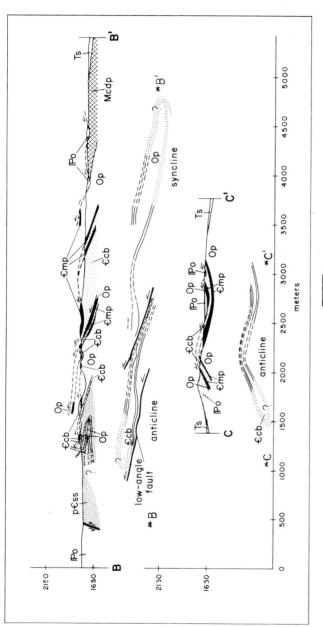

Figure 3. Geologic map and vertical cross sections of an area in the northeast corner of Figure 2, showing faulted recumbent folds in lower allochthon. Contours (dotted) are from Grouse Creek 4 NE and Grouse Creek 4 NW 7½-minute quadrangles. Map area includes secs. 19–21 and 28–33, R. 15 W., T. 12 N.; secs. 4–6, R. 15 W., T. 11 N.; secs. 24, 25, and 36, R. 16 W., T. 12 N.; and sec. 1, R. 16 W., T. 11 N. Informal names from Compton (1972). Cross sections B–B′ and C–C′ have equal horizontal and vertical scales. Arrows indicate low-angle faults, not necessarily actual direction of tectonic transport. Interpretive sections *B–*B′ and *C–*C′ are designed to indicate structure inferred by completing folds above ground.

358   V. R. TODD

lower and middle allochthons, but the upper part of the gneiss complex is pervasively folded (Fig. 7). Early folds were isoclinal, recumbent to strongly overturned, and tightly appressed; later folds were kink or chevron folds with moderate overturning. The geometry of rock layers suggests that folding began by a flexural mechanism but continued, with increasing metamorphism, by flow within and across layers. Metamorphic foliation and axial planes of large folds typically diverge by no more than 15° to 20°, and small folds lie within or on foliation surfaces: these relationships indicate that folds and foliation formed approximately simultaneously; both foliation and axial planes are nearly horizontal when the effects of postmetamorphic folding and high-angle faulting are removed.

Two sets of pervasive metamorphic folds are present in the autochthon and lower and middle

Figure 4. Late-metamorphic, north-northeast–trending recumbent folds in middle allochthon. (A) View northeast from upper Rosebud Canyon; structurally higher antiform (light dashed lines) is thin-bedded Lower Permian metasandstone; structurally lower anticline is Upper Pennsylvanian–lowermost Permian metaquartzite with core of pre–Upper Pennsylvanian marble and is overturned to east; folded strata broken by westward-dipping low-angle fault (heavy line). (B) View to south along axis of satellitic, north-northeast–trending recumbent isocline in metaconglomerate (hammer for scale); first-stage pebble lineation is folded around hinge of late-metamorphic fold, which verges northwestward.

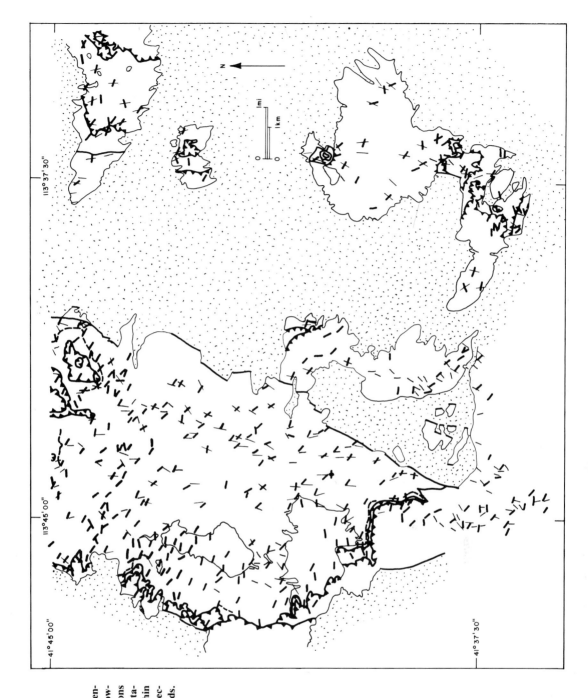

Figure 5. Structure map of central Grouse Creek Mountains showing axial trends of two generations of small (wavelength <5 cm) metamorphic folds and lineations. Thin lines = first folds; heavy lines = second folds and scattered third folds. Base is map of Figure 2.

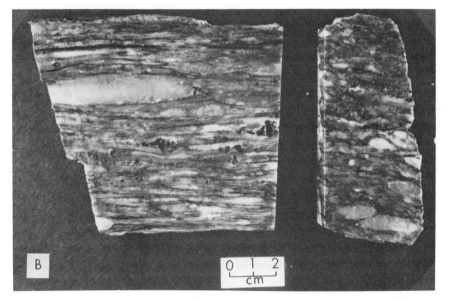

Figure 6 (facing pages). Character of lineation in rocks of central Grouse Creek Mountains. (A) N2° E-trending (first) folds (lower left to upper right) crumpled by N55° W-trending (second) crenulation (vertical, and parallel to pencil, upper right) on quartzite bed in middle allochthon. (B) Pebbles elongated parallel to northwest-trending (second) folds in metaconglomerate of middle allochthon. Both surfaces are cut perpendicular to foliation. Left, surface cut parallel to lineation. Right, surface cut perpendicular to lineation. (C) Approximate foliation surface in muscovitic Precambrian adamellite showing coarse, north-trending (first) quartzofeldspathic and micaceous segregations, parallel to watch band. (D) Closeup of foliation surface in Precambrian adamellite. Early, north-trending lineation = vertical light (quartzofeldspathic) and dark (micaceous) segregations. Faint second, northwest-trending lineation composed of single mica grains indicated by arrow.

allochthonous sheets (Fig. 5 and Table 2). Folds of the first set are most variable in axial trend (they strike from north-northwest to northeast) and are overturned to the northwest. The second set refolds the first and strikes from northwest to west; where direction of overturning can be determined, it is most often southwestward in the central Grouse Creek Mountains. Sparse data suggest a change in direction of overturning of second-stage folds from northeastward in the lower allochthon to southwestward in the middle allochthon.

Locally in the lower and middle allochthonous sheets, a third set of metamorphic folds refolds first-

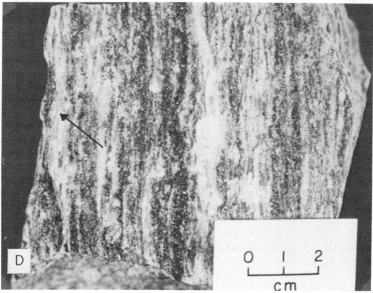

and second-stage folds. The third set trends east ±10° with local northward overturn. Second folds in the gneiss complex adjacent to outcrops of Tertiary adamellite and in the upper part of the complex have this trend; rarely, all three sets are present. The dominant trend of lineation, microfolds, and aligned minerals in the recrystallized outer part of the Tertiary stock is also approximately east. These folds are interpreted as a local late stage of the second folding.

A fourth, late-metamorphic set of folds trends north-northeast to northeast and is overturned to the east and southeast. Muscovite and chlorite grew parallel to axial planes of sparse overturned kink and chevron folds of this trend in the lower allochthonous sheet. Folds of this trend are ubiquitous in the middle sheet where they range from microfolds that crumple first and second folds to recumbent folds that have amplitudes of several kilometres and that locally have broken into upthrusts (Fig. 4). The middle low-angle fault was folded during this folding.

Postmetamorphic folds consist of north-trending, upright, tightly appressed folds that have wavelengths from 1 to 10 m and that locally have been refolded by northwest- to east-trending folds of similar size.

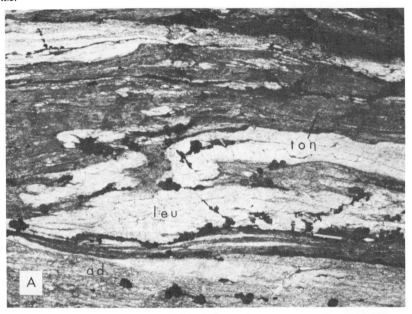

Figure 7 (facing pages). Second folds in upper part of gneiss dome. (A) Partly detached recumbent folds; ad = adamellite, leu = leucoadamellite, ton = tonalite. Pencil, lower right, parallel to foliation. (B) Almost completely detached recumbent folds in adamellite (ad) and tonalite (ton). Pencil, center, parallel to foliation. (C) Closeup of deformed adamellite (ad) and tonalite (ton).

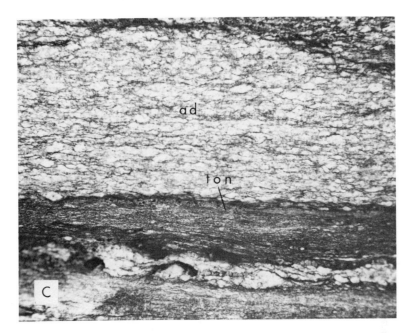

TABLE 2. SEQUENCE AND DISTRIBUTION OF METAMORPHIC FOLDS, AND RELATIONS TO METAMORPHIC GRADE

	First folds	Second folds	Third folds	Fourth folds
Strike of fold axes and associated lineation	N.20°W. to north to N.45°E.	West to N.45°W.	East-west ± 10 degrees	N.20°E. to N.45°E.
Preferred direction of overturn of folds	West to northwest	South to southwest	North	East and southeast
Stage of metamorphism during folding	Prograde	Prograde to retrograde	Retrograde	Retrograde
Character of folds	Isoclinal, recumbent	Isoclinal, recumbent	Recumbent; overturned kink or chevron	Overturned kink or chevron folds; large recumbent folds
Distribution of folds and lineation	Pervasive in autochthon, and in lower and middle allochthons	Pervasive in autochthon, and in lower and middle allochthons; outer part of Tertiary pluton	Sporadic in lower and middle allochthons; locally, refold first and second folds in autochthon	Pervasive in middle allochthon; sporadic in lower allochthon

Normal Faults

The Grouse Creek Mountains have been raised relative to flanking basins and tilted westward along a series of en echelon, north-northeast–trending normal faults that dip about 70° eastward on the eastern flank of the range (Compton, 1972; Todd, 1973). These faults cut all metamorphic and postmetamorphic structures. To the east of the main ridge, high-angle faults of similar trend have blocked out horsts from which Tertiary fanglomerate is being stripped. In the Matlin Mountains 10 to 15 km east of the Grouse Creek Mountains, high-angle faults cut low-angle faults that have emplaced upper Paleozoic rocks on 11-m.y.-old tuffaceous beds. Gravity studies of the Grouse Creek Mountains and surrounding areas indicate that Grouse Creek Valley, which flanks the mountains on the west, is a graben with alluvium as much as 1,830 m thick (Cook and others, 1964). On the west side of the range, north-trending faults have downdropped Tertiary rocks and Pliocene-Pleistocene gravels against crystalline rocks (Fig. 2). Lineaments and aligned springs occur in the alluvium. This places a probable maximum age limit of late Pliocene on basin-range faulting in this area.

Intrusive Rocks

Adamellite dated by the Rb-Sr whole-rock method as 24.9 ± 0.5 m.y. old (Compton and others, 1977) crops out in two stocks on the west side of the central Grouse Creek Mountains, and related alaskite forms a small outcrop on the east side of the range (Fig. 2). A metasomatic aureole in the Precambrian adamellite indicates that a more extensive Tertiary pluton underlies most of the gneiss complex. The interior parts of the Tertiary stocks are massive, but the margins were moderately to strongly deformed and moderately recrystallized during the second period of folding.

Tertiary adamellite is a medium-grained, hypidiomorphic-equigranular rock in which inclusions are rare. Locally, coarse-grained rock in the interior is nearly porphyritic (2- to 4-mm-long white K-feldspar phenocrysts). Euhedral phenocrysts, delicate oscillatory zoning in both feldspars, and synneusis aggregates in plagioclase imply magmatic origin. The average rock is adamellite with 5.5% biotite and 0.5% muscovite, the latter both primary and deuteric. Accessory minerals are monazite, allanite, zircon, apatite, and an opaque oxide; a late alaskite phase of the pluton carries abundant muscovite and purplish-pink garnet. Dilative alaskite and aplite dikes cut the pluton, but are noticeably more abundant in the surrounding Precambrian adamellite and lower and middle allochthonous sheets where their orientation is predominantly north-trending, or approximately perpendicular to the second lineation. Alaskite dikes range in thickness from 1 to 100 m and are massive or layered parallel to their walls. Some show a west-northwest lineation.

Tertiary adamellite contains rotated inclusions of Precambrian adamellite near the contact. Numerous large dikes and small bodies of Tertiary adamellite occur in the Precambrian adamellite in the area between the two stocks. The contact between Precambrian and Tertiary adamellites is broadly concordant with low-dipping foliation in both rocks, but foliation appears to transect the intrusive contact locally. No chilled margins were seen in the younger pluton. The Tertiary pluton intrudes the lower and middle allochthons concordantly with low-dipping unit contacts, metamorphic foliation, and low-angle faults (Fig. 2).

Where it is in contact with Precambrian adamellite, the Tertiary pluton has a moderate to strong, nearly horizontal foliation consisting of the planar alignment of biotite grains, biotite aggregates, and mafic schlieren. Subparallel jointing gives this rock a platy to slabby appearance. West-northwest–trending to east-trending lineation marked by elongate biotite aggregates and slightly elongate quartz and feldspar groups lies within this foliation. Some outcrops also show 2.5- to 5-mm-wavelength crenulations parallel to mineral lineation. These structures are parallel to foliation and lineation in the surrounding metamorphic rocks. Thin sections from rocks of this zone show evidence of variable cataclasis and solid-state recrystallization (Fig. 8B); feldspars and quartz have moderately preferred crystal fabrics. Deformed rock grades inward to faintly foliated and lineated rock and finally to massive adamellite in the interior of the pluton (Fig. 8A).

Tertiary adamellite is strongly gneissic through a 400-m thickness beneath the middle allochthon, and the rock has a marked 1- to 2-mm-wavelength, west-northwest-trending crenulation and a parallel mineral lineation. All quartz except quartz inclusions in feldspar phenocrysts has broken and recrystallized to 0.5-cm-wide irregular lenticles whose subgrains have strong c-axis fabric and marked undulose extinction. Plagioclase grains retain subhedral shapes, but much of the biotite forms platy, recrystallized aggregates and fine-grained, chloritized, and muscovitized folia that wrap around feldspar. The matrix of comminuted quartz in this rock averages less than 0.1 mm in grain size. The pluton was mylonitized throughout its upper few metres beneath the middle allochthon (Fig. 8C); at about the same time, it thermally metamorphosed carbonate rocks in that allochthon. At the base of the middle allochthon, dolomite produced during contact metamorphism is locally tectonitic and folded on west-northwest-trending axes that are parallel to crenulation and mineral-streaking lineation in the underlying mylonitized Tertiary adamellite. Overturning of these folds is toward the southwest.

Summary

Important age relationships among structural, metamorphic, and intrusive events in the central Grouse Creek Mountains are as follows: (1) The lower and middle allochthonous sheets were emplaced mainly before or early in metamorphism and folding because both first and second folds (north- and northwest-striking axes, respectively) are present in both sheets and in the autochthon. (2) Metamorphic folding was going on 24.9 m.y. ago when Tertiary adamellite was intruded because the outer part of the pluton shows west-northwest-trending second (or third) lineation and because contact metamorphic minerals in carbonate rocks of the middle sheet above the stock grew parallel to axes of west-northwest-trending folds. (3) Some movement on the middle low-angle fault took place at this time because the upper part of the Tertiary stock and the skarn at the base of the middle sheet were mylonitized and folded on northwest-trending axes with southwestward overturn. The fact that the thermal aureole in the middle sheet is not displaced from the stock suggests that postintrusive movement on this segment of the fault was not large. (4) Because carbonate rocks of the upper allochthonous sheet are neither thermally nor synkinematically metamorphosed, this sheet must have been emplaced later than 24.9 m.y. ago. (5) Postmetamorphic movements have displaced parts of the middle and upper allochthons onto 11-m.y.-old Tertiary beds over an east-west distance of 10 to 15 km on the east side of the range. Parts of the upper sheet also rest on Tertiary beds in the southern Grouse Creek Mountains (Stanford University Department of Geology, 1973, 1976, unpub. data).

DEVELOPMENT OF THE GNEISS COMPLEX

On the basis of regional metamorphic and structural relationships, Compton and others (1977) proposed that during Tertiary time the Raft River–Grouse Creek Mountains region was the site of widespread heating and uplift leading to the formation of a large dome whose shape and culmination changed with time. Opposed directions of metamorphic flowage during the second deformation suggest that the culmination was in the vicinity of Muddy Canyon in middle Tertiary time (Fig. 2). Thus, the gneissic complex of the central Grouse Creek Mountains was produced by a culmination within the broader region of heating, uplift, and deformation.

The central Grouse Creek Mountains culmination is defined by diapiric rise and intrusion of schistose to gneissose, reconstituted Precambrian adamellite into younger cover rocks and by the

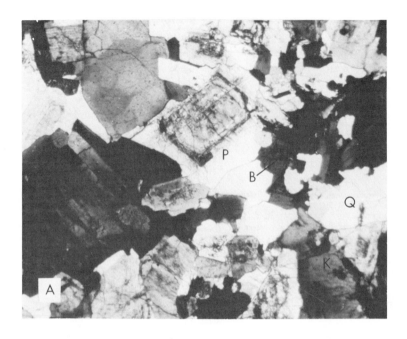

Figure 8 (facing pages). Photomicrographs of Tertiary adamellite showing progression from relatively undeformed through mylonitized adamellite. (A) Weakly foliated and lineated adamellite with unstrained quartz (Q), subhedral plagioclase (P), K-feldspar (K), and large biotite (B) grains. (B) Strongly lineated gneiss less than 400 m below middle allochthon showing recrystallized quartz aggregates (Q), broken large K-feldspar (K), and comminuted biotite (B). (C) Mylonite about 3 m below middle allochthon. Only large feldspar grains are intact. Scale is same in all views.

extreme attenuation of these cover rocks. Schistose and gneissose Precambrian adamellite and tectonically thinned cover rocks occur throughout the region, but only in the central Grouse Creek Mountains does Precambrian adamellite appear to intrude its cover. The diapiric nature of the intrusion is indicated by (1) increasing extension and metamorphic flow upward in the Precambrian adamellite and overlying autochthonous and allochthonous metasedimentary rocks and (2) isotopic data (Compton and others, 1977) which indicate that the Precambrian adamellite remained solid during intrusion, although it was rendered plastic by heat and metasomatizing fluids. The marginally foliated and lineated middle Tertiary pluton lies directly within the culmination—that is, the area of remobilization and maximum attenuation (Fig. 1B).

The east-west extent of the intrusive part of the culmination is not known because it has been overlapped by Tertiary and Quaternary sediments and broken by normal faults. When the effects of basin-range faulting and tilting have been removed, gneissic foliation and concordant low-angle faults and unit contacts are essentially horizontal in the central Grouse Creek Mountains but to the north and south probably plunge gently northward and southward, respectively (Fig. 1). The north-south cross section (Fig. 1B) shows that the Precambrian adamellite forms a gentle asymmetrical convexity beneath the metasedimentary rocks. Thus, although the gneissic rocks of the central Grouse Creek Mountains may represent only a spatially and temporally transient culmination within a broader uplift which persisted for tens of millions of years, there is evidence for the existence of at least a gentle domal structure in the area. For brevity, the central Grouse Creek Mountains gneissic complex will be referred to as the gneiss dome in the discussion to follow.

Protolith

Virtually continuous exposure of the autochthon from Grouse Creek Mountains on the west to the eastern end of Raft River Mountains permits identification of the protolith of the complexly folded, locally diverse rocks that compose the gneiss dome in the central Grouse Creek Mountains. The gneiss dome formed in part of an adamellite batholith roughly 2.5 b.y. old which underlies northwestern Utah and south-central Idaho (Armstrong, 1968; Compton and others, 1977). Relict igneous textures are seen in the deepest exposures of this body in the Grouse Creek Mountains (about 900 m), and gneissic adamellite can be traced continuously northeastward into slightly metamorphosed adamellite of comparable age in the eastern Raft River Mountains (Compton, 1972, 1975). Although the adamellite of the eastern Raft River Mountains yielded a whole-rock Rb-Sr isochron of 2,180 ± 190 m.y., it was thought that this is a minimum age due to postcrystallization metamorphism and/or weathering (Compton and others, 1977).

On the basis of well-preserved igneous and sedimentary textures, Compton and Stanford University field classes were able to determine the Precambrian history of the relatively low grade metamorphic rocks of the autochthon in the eastern Raft River Mountains. Flat-lying semipelitic sedimentary rocks were intruded sequentially by basaltic sills, trondhjemite, and adamellite; these reacted locally with the sedimentary rocks across a distance of 10 m (Compton and others, 1977). The more strongly metamorphosed adamellite in the central Grouse Creek Mountains contains numerous stratiform inclusions of semipelitic schist and amphibolite that are as much as 10 m thick and 1.5 km long and contain isoclinally folded bodies of gneissic trondhjemite. The surrounding adamellite has been contaminated to granodiorite and tonalite near them. This rock association appears to be equivalent to the less-metamorphosed igneous and sedimentary rocks of the eastern Raft River Mountains.

In the central Grouse Creek Mountains, all of these rocks occur in flat-lying, schlierically layered and isoclinally folded complexes, but much of the compositional heterogeneity probably is a remnant of the older Precambrian terrane. The Tertiary metamorphic folds and lineations seen in adamellite and older basement rocks of the eastern Raft River Mountains can be traced continuously into those of the higher-grade, more-deformed rocks of the Grouse Creek Mountains. Thus, the striking gneissic fabric of the 2.5-b.y.-old adamellite and its inclusions of older Precambrian basement in the Grouse Creek Mountains is post-Precambrian and, as suggested by radiometric data, in large part or entirely of Tertiary age (Compton and others, 1977).

Structural Relationship of Gneiss Dome to Surrounding Rocks

Within the area of maximum resurgence, the Precambrian adamellite is converted to mylonite gneiss (as used by Higgins, 1971, p. 11) with a pronounced low-dipping foliation consisting of flattened mineral grains, mineral aggregates, and schlieren (Fig. 7). Deformation was most intense in the upper part of the adamellite, and the intensity increases upward toward the intrusive contact between adamellite and autochthonous upper Precambrian(?) rocks and the lowermost allochthonous sheet. At outcrop scale, this is shown by a gradual upward decrease in grain size and at microscopic scale by an increase in the number of broken and milled grains; strain twins in plagioclase; bent, kinked, and undulose grains; and the disappearance of relict grains retaining an igneous texture. The coherence of broken phenocrysts and the range in grain size from large, relict phenocrysts to fine-grained recrystallized aggregates suggest that the gneissic texture formed by slow, concurrent strain and recrystallization.

Recrystallization of strained phenocrysts to aggregates of smaller grains and replacement of both feldspars by mica were also important in the development of fine- to medium-grained rock in the upper 100 to 200 m of the gneiss dome. In the uppermost 2 to 3 m, this rock grades into phengitic mica–quartz–albite schist that is interlayered with the lower beds of the Elba Quartzite (Fig. 9). Stratiform

Figure 9 (facing page). Contact relations between upper part of gneiss dome and Elba Quartzite. (A) Pods of reconstituted Precambrian adamellite, now phengite-quartz-albite schist (ad), in base of Precambrian Elba Quartzite (between adamellite pods). Hammer for scale. (B) Closeup of phengite-quartz-albite schist showing isoclinally folded white quartz segregations. Hammer for scale.

metaquartzite inclusions with relict bedding are found as much as 450 m below the projected contact of adamellite with the Elba Quartzite. These relationships suggest a metamorphosed intrusive contact, but when they are combined with the 2.5-b.y. age of the adamellite protolith and the fact that it can be traced into virtually undeformed adamellite in the eastern Raft River Mountains, they clearly demonstrate that the gneiss was remobilized and invaded its cover after Precambrian time.

The Elba Quartzite thins abruptly in the Muddy Canyon area where the intrusive contact is well exposed. About 2.5 km north of Muddy Canyon, the quartzite is about 215 m thick and lies with angular unconformity on Precambrian adamellite. The quartzite thins to 6 m in the north wall of Muddy Canyon and, to the west, is cut out completely. Precambrian adamellite is there in contact with a younger schist unit. Contorted and granitized schist inclusions are especially abundant in the upper part of the gneiss dome in this area. Some of these inclusions may be primary, that is, inclusions of basement older than 2.5 b.y., but the clean white quartzite and calc-silicate inclusions in the upper part of the gneiss dome probably represent upper Precambrian(?) and lower Paleozoic units that were tectonically engulfed during rise of the dome. These rock types are not found elsewhere in the area in basement older than 2.5 b.y.

The interface of the gneiss dome with overlying rocks has aspects of both a low-angle fault and a passive intrusive contact. The interfingering contact between comminuted, recrystallized, remobilized adamellite and the Elba Quartzite may have formed initially by stoping and later have broken into a low-angle fault that rose across units into the lowermost allochthon. Retention of Rb and Sr ratios indicative of a 2.5-b.y. age in all gneiss samples except those from the uppermost 2 to 3 m (Compton and others, 1977) argues against mobilization of Precambrian adamellite by melting.

The lower allochthonous sheet is also anomalously thin above the gneiss dome, in part owing to attenuation by movement on the middle fault (Fig. 1). Typically, units remain in normal stratigraphic position, and normal vertical lithologic changes and gradations are preserved; these circumstances suggest that thinning was caused by metamorphic flow (extension and flattening) in addition to low-angle faulting.

Two factors suggest a genetic relationship between emplacement of the Tertiary pluton and rise of the gneiss dome. First, the pluton directly underlies the area where Precambrian adamellite has intruded its cover (Fig. 1). Second, the linear structures (small folds and mineral alignments) that predominate in the upper part of the dome have the same trend as those in Precambrian adamellite adjacent to the two stocks in which the underlying pluton is exposed and in the outer part of the stocks themselves (Fig. 5). Throughout most of the Precambrian adamellite, the chief linear structures are north-trending rod-shaped aggregates of recrystallized quartz, feldspar, and biotite, which are crossed by smaller northwest-trending crenulations and small mica grains. These northwest-trending (second) crenulations are parallel to axes of the larger, isoclinal folds described previously. Whereas the north-trending (first) lineation lies within the nearly horizontal foliation, the northwest-trending (second) lineation is formed by the intersection of this foliation with a poorly developed, steeply dipping, secondary mica foliation (Fig. 10). In the upper part of the gneiss dome and in gneiss adjacent to the Tertiary stocks, quartz, feldspar, and mica have recrystallized along the northwest trend to form prominent rodding that locally obliterates the earlier north-trending structures (Fig. 11A). Axial planes of northwest-trending minor folds in these rocks are nearly vertical, that is, the secondary mica foliation becomes dominant in these areas. This intensification of structures related to the northwest-trending (second) folding suggests that these parts of the Precambrian adamellite recrystallized more completely during the second folding than the rest of the adamellite, possibly because heat, fluids, and the upward buoyant pressure of Tertiary intrusion were localized in these areas. The upper part of the gneiss dome and gneiss adjacent to the Tertiary stocks are also places where metasomatic reconstituting reactions have been most complete. Thus, the stocks and the underlying pluton have a contact

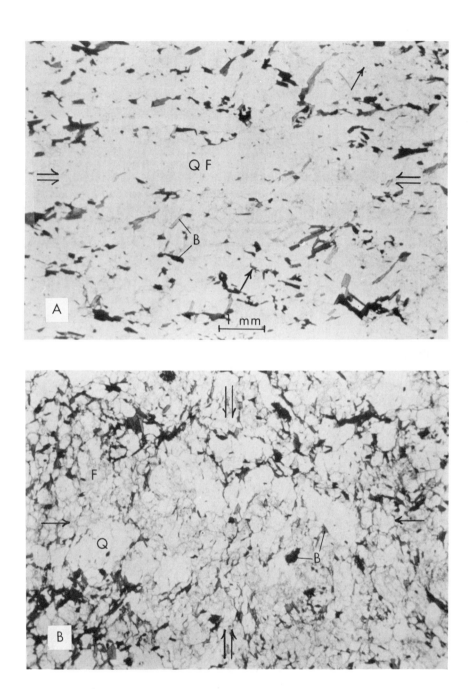

Figure 10. Photomicrographs of foliation in Precambrian adamellite. (A) Sample from east flank of range, at distance from Tertiary pluton, showing biotite (B), quartz and feldspar (QF). Major, nearly horizontal foliation marked by double arrows. Poorly developed second foliation marked by single arrows. (B) Sample collected within 1 m of Tertiary pluton showing biotite (B), quartz (Q), and feldspar (F). Well-developed second (axial-plane) foliation marked by double arrows. Earlier, horizontal foliation (single arrows) is almost obliterated. Second, nearly vertical foliation involves all three minerals. Both views cut perpendicular to earlier, nearly horizontal foliation of Precambrian adamellite; plane-polarized light; scale is same in both views.

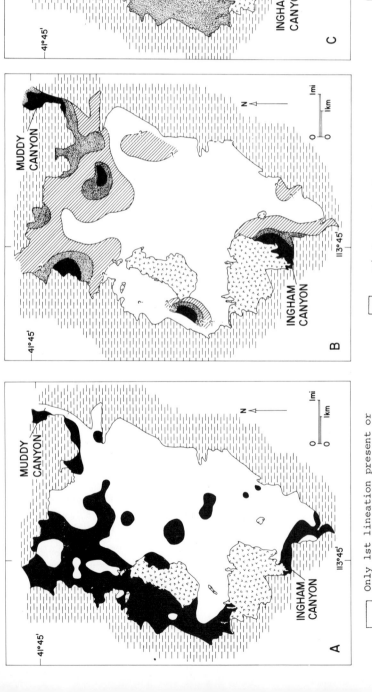
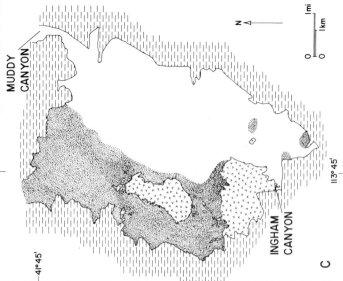

Figure 11. Metasomatic and structural variations in Precambrian adamellite on same base as that of Figure 2. (A) Distribution of first and second lineations at 225 sites in Precambrian adamellite. (B) Distribution of phengitic mica in 80 samples of average Precambrian adamellite, from calculated modes and thin-section estimates. (C) Approximate distribution of outcrops with abundant leucocratic rock in Precambrian adamellite.

TABLE 3. METASOMATIC REACTIONS IN PRECAMBRIAN ADAMELLITE

1) biotite + K-feldspar + H^+ = phengite + chlorite + SiO_2 + ilmenite (+ sphene + rutile + apatite)

2) oligoclase + H^+ + K^+ + Fe^{2+} + Mg^{2+} = phengite + SiO_2 + Na^+ + Ca^{2+}

3) oligoclase + Na^+ + SiO_2 = albite + Ca^{2+} + Al^{3+}

4) oligoclase + K^+ + SiO_2 = K-feldspar + Na^+ + Ca^{2+} + Al^{3+}

5) K-feldspar + H^+ + Fe^{2+} + Mg^{2+} = phengite + quartz + K^+

6) oligoclase + K^+ + H^+ + Fe^{2+} + Mg^{2+} = biotite + SiO_2 + Na^+ + Ca^{2+}

Note: The equations are written under the assumption that no alumina was added to the rocks.

aureole in the Precambrian adamellite in which metasomatic reactions and deformation associated with the second folding have been most intense (Figs. 11B and 11C).

The zone of crushed Precambrian adamellite at the base of the impermeable Elba Quartzite apparently conducted and partly trapped the rising fluids. The existence of relatively high fluid pressure and waning metamorphic temperatures at the time of doming is implied by the preferential development of a retrograde mineral assemblage, principally phengitic mica, in the upper part of the gneiss dome and in gneiss adjacent to the Tertiary stock. Presumably this assemblage developed in a late stage of the northwest-trending (second) folding because biotite, staurolite, kyanite, and sillimanite grew parallel to northwest-trending fold axes in the older than 2.5 b.y. metapelitic inclusions in the Precambrian adamellite of the dome and were in part replaced by retrograde minerals at a later stage. Muscovite, chlorite, and, to some extent, andalusite in these inclusions are preferentially aligned parallel to northwest-trending fold axes.

Metamorphic Reactions in Gneiss Dome

The upward increase in deformation that accompanied doming was matched by increased replacement of feldspar and biotite by the retrograde assemblage phengitic mica–quartz–albite–ilmenite and/or sphene-(chlorite). Figure 11B depicts the distribution of phengite in relation to the upper surface of the gneiss dome. Individual reactions observed in thin section are listed in Table 3, but all mineral phases participated simultaneously in an overall reaction. Growth of new mica in plagioclase was accompanied by the alteration of oligoclase to sodic oligoclase or albite. Biotite was the chief replacement mineral in tonalite, whereas phengitic mica is more abundant in adamellite; this suggests control by bulk composition. Plagioclase in tonalite is less altered than that of adamellite. There was an overall reaction of biotite to phengite as indicated by grains of white mica growing across biotite folia, relict biotite in coarse white mica bands, and the presence of small amounts of white mica, chlorite, and titanium minerals even in large intact biotite grains. Where biotite and phengite occur together in a single plagioclase grain, biotite is subhedral or euhedral, and phengite is ragged and anhedral; this suggests that biotite was earlier. Experimental studies suggest that biotite will react to phengite where fluid pressures are high or where temperatures are falling (Velde, 1965).

Table 4 presents modal and chemical data for igneous and metamorphic rocks of the central Grouse Creek Mountains. Changes in the Precambrian adamellite during formation of the gneiss dome can be followed by comparing data for the first three samples with those of the border zone and with leucoadamellite samples.

Leucocratic Rock. One of the most striking aspects of the gneiss dome is the abundance of leucocratic rock which increases markedly toward the top where it is interlayered on all scales with adamellite, amphibolite, and pelitic schist. Figure 11C shows the approximate distribution of leucocratic rock. Leucoadamellite and leucogranodiorite gneiss and gneissic pegmatite form the bulk of the upper 180 m of the gneiss dome, and textural and field relationships indicate that a substantial part of the leucocratic rock may have originated by metasomatism of Precambrian adamellite. Modal and chemical similarities between leucoadamellite and average Precambrian adamellite can be seen in Table 4. The gneissic leucocratic rocks contain the same accessory minerals, including metamict zircon, as the gneissic Precambrian adamellite and tonalite. Fe, Mg, Ca, and probably Al were lost from leucoadamellite relative to a Precambrian adamellite parent. The average Precambrian adamellite is itself leucocratic, with low total Fe and MgO, and high SiO_2.

The concordant, but irregular, shapes of thin (2 to 15 cm) leucocratic layers and their partly gradational contacts with surrounding adamellite support a replacement origin. Layers thicken and thin along strike, form recumbent isoclinal folds, and can be traced for several metres. Leucocratic rocks also occur as irregular pegmatitic pods and lenses parallel to foliation, stringers of quartzofeldspathic material around 4-cm K-feldspar porphyroblasts, and vague concentrations of quartz and feldspar. Replacement and exsolution microtextures in the leucocratic rocks are extreme, regardless of their vertical position, and locally obliterate the gneissic texture; otherwise, leucoadamellite is petrographically identical to average adamellite gneiss. Opaque oxide grains form pseudomorphs after biotite, which averages only 0.5% in these rocks; the total of other biotite replacement minerals (phengitic mica, chlorite) is about 2%.

Weakly foliated to nonfoliated leucocratic dikes and sills probably resulted directly from Tertiary intrusion. Locally, the inclusion of both Precambrian and Tertiary accessory minerals in leucoadamellite gneiss near the Tertiary pluton suggests some mixing of Tertiary magma with Precambrian adamellite. The origin of the abundant leucocratic rock, whether by injection or replacement or a combination of the two, seems clearly related to Tertiary intrusion.

Border Zone. At the top of the dome, fine-grained muscovite gneiss and schist form a zone 50 m thick that grades downward into less-altered adamellite. Schistose rocks also formed in inclusions of basement older than 2.5 b.y., when these occurred near the top of the dome, and in leucoadamellite. The schistose zone bears the same accessory minerals as deeper-lying adamellite, except that metamict, dark brown zircons have partial overgrowths of clear zircon or are replaced by aggregates of tiny, clear zircons. The uppermost 2- to 3-m thickness of the gneiss dome consists of silvery-gray phengite-quartz-albite schist with an average grain size of 0.75 to 1 mm (Figs. 9 and 12). Segregations of white quartz 15 to 20 cm long and as much as 1 cm thick pinch and swell along foliation or cross it at low angles and locally form isoclinal recumbent folds. Some of these quartz segregations contain flattened, quartz-lined cavities a fraction of a millimetre long.

If an initial composition of average adamellite is assumed, modal compositions of a vertical series of fine-grained gneiss and schist samples suggest that the chief reaction in their formation was K-feldspar + water = phengite + quartz (Table 3). Although most visibly altered, the plagioclase and biotite content remains approximately constant to within 2 to 3 m from the contact with the Elba Quartzite. The importance of K-feldspar in the formation of phengite is also observed in deeper-lying, less-altered rocks where phengitization of biotite occurred only adjacent to large, fresh-looking grains of K-feldspar.

Chemical analyses of schist (Table 4) show it to be largely depleted in CaO and Na_2O and somewhat enriched in FeO and MgO, compared to the adamellite. The depletions reflect disappearance of almost all plagioclase, and the small enrichments suggest removal of other components. The total amount of CaO and Na_2O lost, however, seems insufficient to raise FeO and MgO percentages as much as observed; this suggests loss of an additional component, probably SiO_2.

The coincidence of the zone of maximum recrystallization with the upper shell of Precambrian adamellite suggests that the Elba Quartzite acted as a less plastic and impermeable cover against which adamellite was slowly deformed. The mobility of the adamellite in this zone was probably enhanced by the buoyant rise of water-rich fluids into the upper part, where they became trapped beneath the quartzite. Intense deformation in this zone would have allowed more thorough permeation by fluids and complete reconstitution of adamellite. Trapped fluids may also have provided the buoyancy needed for a surface of detachment to form and for low-angle faulting to occur. Only in this reconstituted shell were Rb- and Sr-isotope ratios homogenized (Compton and others, 1977) and elements lost.

CONCLUSIONS

Mobilization of the gneiss dome of the central Grouse Creek Mountains apparently began during the first (north-trending) folding because the size of mineral rods parallel to the axes of first folds decreases upward in the Precambrian adamellite in concert with the overall decrease in grain size and increase in strain and recrystallization. This rodding appears to have resulted from a combination of crenulation (monoclinic shear), with extension in the direction of fold axes, and recrystallization. The first folds are probably older than late Eocene or early Oligocene (38 ± 2.0 m.y. B.P.; Compton and others, 1977). Development of the dome was closely related to the second folding (northwest-trending axes) and locally to the third folding (east-trending axes) which were still going on in late Oligocene time when the Tertiary pluton was emplaced (24.9 ± 0.5 m.y. B.P.). East-trending metamorphic folds were still forming in Miocene time (20 ± 10 m.y.; 10.2 ± 1.9 m.y.) in the eastern Raft River Mountains (Compton and others, 1977).

Autochthonous and allochthonous cover rocks above the gneiss dome were attenuated by a combination of metamorphic flow and low-angle faulting. Remobilized Precambrian adamellite rose diapirically into the flattening and extending cover rocks during Tertiary intrusion; this indicates that the formation of the dome was controlled by a "hot spot" within a broader region of metamorphism.

The 24.9-m.y.-old Tertiary pluton dates late development of the gneiss dome, retrograde metamorphism, and the later part of the second (and third?) metamorphic folding and associated low-angle faulting. The absence of foliation and lineation in the interior part of the pluton might indicate that deformation had ceased by about 25 m.y. ago, yet other evidence suggests that deformation continued for a considerable time after the pluton solidified. North-northeast–trending late-metamorphic folds refold first and second metamorphic folds in the allochthonous sheets that lay immediately above the pluton. These late-metamorphic folds barely affected the autochthon and are nowhere seen in the Tertiary pluton. Apparently, the locus of deformation had shifted upward to the cover rocks by the time the core of the pluton solidified, possibly in response to gravitational instability created by uplift. Three fission-track ages indicate that the pluton did not complete its crystallization until Miocene time (18.3 ± 1.9 m.y. on zircon; 18.9 ± 6.3 m.y. and 13.7 ± 3.7 m.y. on apatite) (Compton and others, 1977).

Overturn of northwest-trending (second) folds to the northeast in the northern Grouse Creek Mountains (Compton, 1972) and their predominantly southwestward overturn in the central Grouse Creek Mountains suggest flowage off two sides of a rising dome of unknown (northwest?) extent. The development of the incipient, steeply dipping mica foliation associated with axial planes of northwest-

TABLE 4A. CHEMICAL, SPECTROGRAPHIC, NORMATIVE AND MODAL ANALYSES OF ROCKS FROM CENTRAL GROUSE CREEK MOUNTAINS, UTAH

Rock type	1 Adamellite gneiss	2 Adamellite gneiss	3 Adamellite gneiss	4 Muscovitic ada-mellite gneiss (border zone)	5 Quartzo-feldspathic schist (border zone)	6 Phengite-quartz schist (border zone)	7 Phengite-quartz schist (border zone)	8 Tonalite gneiss	9 Granodiorite gneiss	10 Leucogranodiorite gneiss
Protolith	Precambrian adamellite	Precambrian adamellite	Precambrian adamellite	Precambrian adamellite	Precambrian adamellite	Precambrian adamellite	Precambrian adamellite	Precambrian tonalite	Precambrian granodiorite	Precambrian granodiorite
Lat (°N)	41.73	41.69	41.69	41.75	41.75	41.75	41.75	41.67	41.67	41.74
Long (°W)	113.71	113.74	113.74	113.69	113.69	113.69	113.68	113.74	113.74	113.75
Chemical analyses										
SiO_2	71.6	72.8	73.75*	72.7	73.7	72.6	74.6	67.29*	70.4	76.0
Al_2O_3	14.3	14.4	14.06	14.1	14.1	14.1	14.4	17.37	15.5	13.3
Fe_2O_3	0.49	0.23	0.16	0.39	1.4	1.9	0.72	0.48	0.55	0.02
FeO	1.0	0.96	1.35	1.0	0.60	1.0	0.88	2.10	2.3	0.40
MgO	0.30	0.30	0.17	0.40	1.0	0.85	0.56	0.81	0.69	0.20
CaO	1.4	1.4	1.49	0.86	0.06	0.09	0.10	2.73	2.4	0.43
Na_2O	3.2	2.8	4.20	4.0	0.93	0.18	0.49	6.00	4.0	4.7
K_2O	5.3	4.9	4.40	4.1	5.1	5.4	4.8	2.15	1.7	3.0
H_2O+	0.57	0.48	0.14	0.61	1.5	1.9	1.6	0.57	0.68	0.32
H_2O-	0.08	0.00	0.05	0.02	0.06	0.08	0.03	0.05	0.00	0.00
TiO_2	0.18	0.12	0.30	0.26	0.22	0.22	0.24	0.35	0.32	0.02
P_2O_5	0.07	0.11	0.18	0.02	0.02	0.04	0.03	0.16	0.05	0.05
MnO	0.02	0.01	0.01	0.00	0.00	0.00	0.00	0.02	0.02	0.00
CO_2	0.05	0.01	0.00	0.01	0.01	0.07	0.04	0.00	0.08	0.02
Total	99	99	100.26	99	99	99	99	100.08	99	99
Quantitative spectrographic analyses [Ti and Ba in percent; all others in parts per million]										
Ti	0.1	0.07	N.d.	0.15	0.1	0.1	0.15	N.d	0.15	0.02
Mn	150	100		100	50	50	20		200	20
B	N	N		N	20	50	50		N	N
Ba	0.14	0.14		0.13	0.10	0.091	0.085		0.038	0.079
Be	1.5	N		2	N	N	N		N	N
Co	N	N		2	2	3	2		3	2
Cr	1.5	N		3	5	10	7		5	N
Cu	30	3		2	1.5	2	30		15	0.7
La	50	30		2	70	100	50		70	N
Mo	N	N		N	N	N	N		N	N
Nb	N	N		10	N	N	15		10	N
Ni	N	N		N	10	20	15		3	N
Pb	150	100		70	20	N	N		50	30
Sc	N	3		N	N	N	N		7	N
Sn	N	N		N	N	N	N		N	N
Sr	150	150		150	100	30	70		150	70
V	N	N		N	15	15	10		10	N
Y	15	10		N	15	15	10		70	N
Zr	150	100		200	150	150	200		300	30
Ce	100	N		N	100	150	150		150	N
Ga	20	15		20	20	20	20		30	15
Yb	N	N		N	N	1	N		3	N

					Norms					
Q	29.506	34.529	28.632	31.380	47.889	50.510	53.058	16.222	33.765	36.406
C	1.053	2.265	0.110	1.614	7.109	8.093	8.512	0.592	3.061	1.732
Or	31.803	29.390	25.946	24.609	30.553	32.445	28.808	12.701	10.179	18.005
Ab	27.495	24.049	35.465	34.380	7.978	1.549	4.211	50.755	34.296	40.392
An	6.267	6.256	6.203	4.137	0.105		0.048	12.494	11.221	1.706
Wo	··	··	··	··	··	··	··	··	··	··
En	0.759	0.758	0.423	2.525	2.058	1.417		2.017	1.741	0.506
Fs	1.190	1.414	1.866	1.102	··	0.635		2.918	3.321	0.696
Fo	··	··	··	··	··	··	··	··	··	··
Fa	··	··	··	··	··	··	··	··	··	··
Mt	0.721	0.338	0.231	0.574	1.314	2.629	1.060	0.696	0.808	0.029
Hm	··	··	··	··	0.513	0.119	··	··	··	··
Il	0.347	0.231	0.569	0.502	0.424	0.425	0.463	0.665	0.616	0.039
Ap	0.168	0.264	0.425	0.048	0.048	0.096	0.072	0.379	0.120	0.120
CC	0.115	0.023	··	0.023	0.023	0.068	0.092	··	0.184	0.046
Mg	··	··	··	··	··	0.079	··	··	··	··
Total	99.425	99.519	99.870	99.381	98.480	98.070	98.376	99.439	99.314	99.678

[tr, trace; ··, not present; N.d., not determined]

					Modes					
Quartz	25.0	23.0	31.5	30.0	44.0	40.5	N.d.	18.5	31.0	35.5
Plagioclase	34.0	38.5	38.5	30.5	5.5	6.5		61.5	47.5	45.0
K-feldspar	35.0	33.5	23.5	19.5	2.5	0.5		6.0	6.5	19.5
Biotite	5.5	4.5	6.0	4.0	2.0			11.0	9.5	tr
White mica	0.5	0.5	0.5	11.0	45.0	51.5		1.5	3.5	tr
Allanite	tr	tr	tr	··	··			··	tr	··
Chlorite	tr	··	··	0.5	··			0.5	1.5	··
Zircon	tr	··	tr	tr	··			··	tr	··
Apatite	tr	··	tr	tr	tr			tr	tr	··
Epidote	tr	··	··	··	··			tr	tr	··
Sphene	··	tr	··	2.5	tr			tr	0.5	··
Total	100.0	100.0	100.0	99.0	99.0	99.0		99.0	100.0	100.0
An content of plagioclase	An_{20}	$An_{19.6}$	An_{20}	An_{15}	An_{10}	An_8		$An_{20.1}$	An_{20}	An_{15}

Note: Rapid-rock analyses: Analyst, Floyd Brown; analytical method used was a single solution procedure described by Shapiro (1967). Conventional chemical analysis by Alberto Obregón, Instituto de Geología, Universidad Nacional Autónoma de México. Quantitative spectrographic analyses: Analyst, Chris Heropoulos. Results are to be identified with geometric brackets whose boundaries are 1.2, 0.83, 0.56, 0.38, 0.26, 0.18, 0.12, and so forth, but are reported arbitrarily as mid-points of these brackets, 1.0, 0.7, 0.5, 0.3, 0.2, 0.15, 0.1, and so forth. The precision of a reported value is approximately plus or minus one bracket at 68%, or two brackets at 95% confidence. Symbols used are N, not detected at limit of detection; the following elements were looked for but not detected at limit of detection: Ag, As, Au, Bi, Cd, Pd, Pt, Sb, Te, U, W, Zn, Ge, Hf, In, Li, Re, Ta, Th, Tl; N.d., not determined. Analysis of barium by Esma Y. Campbell.

*Conventional chemical analysis.

TABLE 4B. CHEMICAL, SPECTROGRAPHIC, NORMATIVE AND MODAL ANALYSES OF ROCKS FROM CENTRAL GROUSE CREEK MOUNTAINS, UTAH

Rock type	11 Leucoadamellite gneiss	12 Leucoadamellite gneiss	13 Leucoadamellite gneiss	14 Gneissic pegmatite	15 Trondhjemite gneiss	16 Amphibolite	17 Biotite-quartz-epidote schist	18 Biotite-quartz-oligoclase schist	19 Average Tertiary adamellite	20 Tertiary aplite	21 Tertiary aplite
Protolith	Precambrian adamellite?	Precambrian adamellite?	Precambrian adamellite?	. .	Precambrian trondhjemite	Precambrian gabbro?	Precambrian gabbro?	Precambrian feldspathic shale	. .	. .	. .
Lat (°N)	41.67	41.69	41.73	41.71	41.73	41.73	41.73	41.67	41.66	41.67	41.66
Long (°W)	113.74	113.76	113.70	113.72	113.71	113.71	113.73	113.72	113.76	113.73	113.76

Chemical analyses

	11	12	13	14	15	16	17	18	19	20	21
SiO_2	75.7	74.9	75.61*	74.60*	74.7	49.8	50.4	62.8	73.04*	74.09*	74.75*
Al_2O_3	13.9	14.3	14.44	14.24	13.7	17.4	14.7	14.9	13.41	15.38	14.76
Fe_2O_3	0.15	0.02	0.03	0.00	0.24	3.1	3.9	1.8	0.77	0.24	0.01
FeO	0.48	0.03	0.15	0.16	1.5	6.7	10.4	5.3	1.20	0.38	0.15
MgO	0.08	0.05	0.00	0.00	0.80	3.8	5.6	3.7	0.30	0.13	0.00
CaO	1.1	1.0	1.01	0.30	1.6	14.6	3.7	1.9	1.66	0.30	0.42
Na_2O	3.0	3.2	3.95	2.10	3.6	0.42	0.30	2.0	4.30	4.20	4.50
K_2O	4.2	4.5	5.00	8.45	1.4	0.41	5.7	3.3	5.00	4.80	5.35
H_2O+	0.43	0.37	0.20	0.10	0.63	0.77	2.0	1.7	0.69	0.47	0.26
H_2O-	0.00	0.00	0.04	0.03	0.00	0.08	0.15	0.15	0.05	0.06	0.04
TiO_2	0.07	0.03	0.00	0.00	0.20	0.86	1.7	0.74	0.27	0.00	0.00
P_2O_5	0.02	0.03	0.11	0.12	0.09	0.08	0.28	0.13	0.17	0.05	0.05
MnO	0.00	0.00	0.00	0.00	0.00	0.27	0.17	0.09	0.03	0.09	0.00
CO_2	0.07	0.03	0.00	0.00	0.02	0.06	0.02	0.07	0.00	0.00	0.00
Total	99	99	100.39	100.10	99	99	99	99	100.39	100.19	100.29

Quantitative spectrographic analyses
[Ti and Ba in percent; all others in parts per million]

	11	12	13	14	15	16	17	18	19	20	21
Ti	0.05	0.003	N.d.	N.d.	0.15	0.5	1.0	0.5	N.d.	N.d.	N.d.
Mn	15	50			150	1,500	1,500	1,000			
B	N	N			N	N	N	N			
Ba	0.11	0.062			0.066	0.009	0.059	0.055			
Be	N	N			N	3	N	2			
Co	N	N			3	50	50	30			
Cr	3	1.5			3	200	300	200			
Cu	N	N			2	30	1.5	50			
La	N	N			70	N	N	50			
Mo	N	N			N	N	N	10			
Nb	N	N			N	N	N	N			
Ni	N	N			N	100	70	150			
Pb	100	100			30	30	20	30			
Sc	N	N			N	70	70	20			
Sn	N	N			N	15	N	N			
Sr	100	70			100	700	70	150			
V	N	N			10	150	200	150			
Y	N	N			15	70	30	30			
Zr	70	70			300	100	70	150			
Ce	N	N			100	N	N	100			
Ga	15	15			15	20	20	20			
Yb	N	N			1	7	5	3			

Norms

	1	2	3	4	5	6	7	8	9	10	11
Q	39.976	37.594	31.375	29.589	11.267	43.741	5.376	26.653	25.612	30.252	27.153
C	2.650	2.527	0.952	1.380	··	3.671	2.050	5.138	··	2.846	0.920
Or	25.019	27.008	29.399	49.898	2.465	8.401	34.068	19.812	29.300	28.328	31.536
Ab	25.590	27.501	33.257	17.757	3.616	30.932	2.568	17.193	36.082	35.493	37.983
An	4.923	4.647	4.271	0.704	45.163	7.335	16.587	8.264	2.501	1.160	1.753
Wo	··	··	··	··	··	··	··	··	··	··	··
En	0.201	0.126	··	··	11.535	2.023	14.106	9.362	1.906	··	··
Fs	0.647	··	0.249	0.294	9.631	2.260	13.537	7.305	0.741	0.323	0.267
Fo	··	··	··	··	8.980	··	··	··	1.168	0.666	··
Fa	··	··	··	··	··	··	··	··	··	··	··
Mt	0.219	··	0.043	··	4.574	0.353	5.719	2.651	1.107	0.348	0.014
Hm	··	0.010	··	··	··	··	··	··	··	··	··
Il	0.134	0.013	··	··	1.662	0.386	3.266	1.428	0.509	··	··
Ap	0.048	0.058	0.259	0.284	0.193	0.216	0.671	0.313	0.399	0.118	0.118
CC	0.160	0.072	··	··	0.139	0.046	0.046	0.162	··	··	··
Mg	··	0.069	··	··	··	··	··	··	··	··	··
Total	99.567	99.626	99.807	99.906	99.224	99.365	97.994	98.281	99.325	99.534	99.743

Modes

[tr, trace; ··, not present; N.d., not determined]

	1	2	3	4	5	6	7	8	9	10	11
Quartz	34.0	27.5	29.0	N.d.	27.0	18.0	33.0	16.5	27.0	N.d.	N.d.
Plagioclase	37.0	40.0	39.5		52.0	9.0	26.5	1.0	41.0		
K-feldspar	27.5	31.0	30.5		··	··	··	··	26.0		
Biotite	0.5	0.5	tr		5.5	1.0	39.0	69.0	6.5		
White mica	1.0	0.5	0.5		15.0	1.0	1.0	1.0	··		
Chlorite	··	··	··		··	tr	tr	tr	··		
Apatite	··	··	··		··	··	tr	0.5	··		
Epidote	··	··	··		··	0.5	··	··	··		
Ilmenite	··	··	··		··	23.5	0.5	12.0	··		
Sphene	··	··	··		··	2.5	tr	1.0	··		
Zircon	··	··	··		··	1.0	··	··	··		
Allanite	··	··	··		··	tr	tr	··	··		
Hornblende	··	··	··		··	tr	tr	··	··		
Monazite	··	··	··		··	44.5	tr	··	··		
Garnet	··	··	··		··	··	··	··	··		
Total	100.0	99.5	99.5		99.5	99.5	100.0	100.0	100.5		
An content of plagioclase	An$_{20}$	An$_{20}$	An$_{18}$	N.d.	An$_{25}$	Variable; An$_{30}$–An$_{50}$	An$_{30}$–An$_{50}$	An$_{10}$–An$_{30}$	An$_{20}$–An$_{30}$	N.d.	An$_5$–An$_6$

Note: Rapid-rock analyses: Analyst, Floyd Brown; analytical method used was a single solution procedure described by Shapiro (1967). Conventional chemical analysis by Alberto Obregón, Instituto de Geología, Universidad Nacional Autónoma de México. Quantitative spectrographic analyses: Analyst, Chris Heropoulos. Results are to be identified with geometric brackets whose boundaries are 1.2, 0.83, 0.56, 0.38, 0.26, 0.18, 0.12, and so forth, but are reported arbitrarily as mid-points of these brackets: 1.0, 0.7, 0.5, 0.3, 0.2, 0.15, 0.1, and so forth. The precision of a reported value is approximately plus or minus one bracket at 68%, or two brackets at 95% confidence. Symbols used are N, not detected at limit of detection; the following elements were looked for but not detected at limit of detection: Ag, As, Au, Bi, Cd, Pd, Pt, Sb, Te, U, W, Zn, Ge, Hf, In, Li, Re, Ta, Th, Tl; N.d., not determined. Analysis of barium by Esma Y. Campbell.

*Conventional chemical analysis.

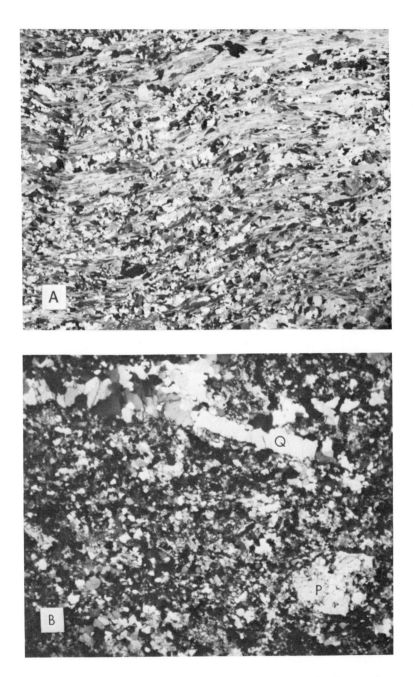

Figure 12 (facing pages). Photomicrographs showing stages in the gradation from gneissose to schistose adamellite. (A) Schist 2.5 m below Elba Quartzite, consisting of phengitic mica (51%), quartz (40.5%), and albite (6.5%). The coarse mica, its grain size probably reflecting abundant water trapped beneath the quartzite, has been polygonized around folds whose axes trend west-northwest. (B) Fine- to medium-grained, strongly lineated gneiss 100 m below projected base of Elba Quartzite. Quartz aggregates (Q) are strongly lenticular, relict plagioclase (P) is recrystallized, and biotite is reduced to small single grains. (C) Medium- to coarse-grained gneiss at base of east side of range. Modified relict K-feldspar (K) and plagioclase (P) phenocrysts occur, but quartz (Q) and biotite phenocrysts have recrystallized.

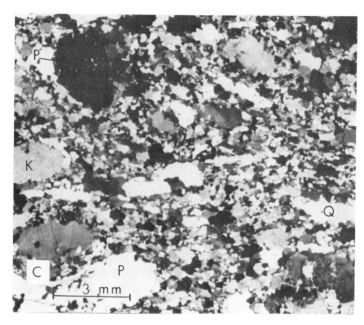

trending folds in the Precambrian adamellite indicates a late-metamorphic overprint of compression on gravitational tectonics. This overprint is also suggested by a change in fold style in the allochthonous sheets from recumbent and strongly overturned to slightly overturned and upright. In the middle allochthon, older-on-younger imbricate thrusts cut the late-metamorphic folds. Locally, these faults themselves are folded in tight, upright, north-trending folds.

About 10 to 15 km to the east, rocks that are lithologically, metamorphically, and structurally similar to those of the middle and upper allochthons lie on Tertiary sedimentary rocks that contain 11-m.y.-old tuff in the upper part (Todd, 1975). This anomalous relationship is repeated in a series of superimposed thin, westward-dipping slices of pre-Tertiary and Tertiary rocks. These structures could be explained by a separate, late Tertiary episode of gravity sliding, except that similar tuffaceous rock occurs under the middle allochthonous sheet in at least two places in the central Grouse Creek Mountains (Fig. 2), where there is no indication that the allochthonous upper Paleozoic rocks have been reshuffled since late-metamorphic folding and imbricate faulting. It is not clear whether these overridden tuffaceous sedimentary rocks were once continuous with the 11-m.y.-old deposits in the Matlin Mountains. Older Tertiary sedimentary rocks do occur in the region; for example, a sample of tuffaceous sandstone from the southern Grouse Creek Mountains yielded K-Ar ages of 32.1 ± 1 m.y. on volcanic biotite and 32.4 ± 1 m.y. on sanidine. The problem remains regardless of the age of the overridden, unmetamorphosed Tertiary rocks because they are overlain by greenschist-facies metamorphic rocks. How did apparently unmetamorphosed Tertiary sedimentary rocks come to lie structurally within the heart of the metamorphic terrane?

The telescoping of metamorphic and postmetamorphic events in the area has made it difficult to decipher its late Tertiary history. Extensive low-angle faulting involving thin sheets of rock probably occurred before, during, and after metamorphism; the resulting structures are concordant and very similar. The large, recumbent north-northeast–trending folds appear to have formed under modest load, presumably after 25 m.y. ago, because small satellitic north-northeast–trending folds refold northwest-trending folds that were still forming at that time. They must also have formed before the end of metamorphism, which from Rb-Sr whole-rock ages for biotite in Precambrian adamellite (11.9 and 8.0 m.y., Compton and others, 1977) was as late as 8 to 12 m.y. ago. These large folds are cut by older-on-younger westward-dipping imbricate thrusts that are late-metamorphic or postmetamorphic. Postmetamorphic events resulted in the eastward emplacement of identical rocks over 11-m.y.-old Tertiary rocks in a series of westward-dipping imbrications. The similarity in these structures and the reasonable fit with the time scale of events suggest a continuum of uplift and basinward shedding of thin sheets in late Tertiary time.

Kehle's model (1970) by which shear flow distributed throughout a zone of decollement results in translation of overlying plates of rock has been adopted by Davis (1975) and Compton and others (1977) to explain extensional tectonic features found in Tertiary metamorphic terranes in Arizona and Utah. In simplest view, the Precambrian adamellite was overlain by upper Precambrian through Triassic sedimentary rocks before and during most of Tertiary metamorphism. The attenuation and elimination of strata and juxtaposition of variably metamorphosed and unmetamorphosed rocks might then be entirely the result of extensional tectonics (gravitational gliding and recumbent folding under load) following intrusion and uplift during regional heating. If, on the other hand, the middle allochthon was emplaced before the onset of Tertiary metamorphism and low-angle faulting, as the hinterland counterpart of movements in the Idaho-Wyoming thrust belt (recently suggested by Compton and Todd, 1979), then the Paleozoic section may already have been considerably thinned before middle Tertiary time. However, this would not substantially alter the above picture of late Tertiary events.

The stratigraphy of the Tertiary sedimentary rocks in the Matlin Mountains is the reverse of that now exposed in the Grouse Creek Mountains; that is, clasts in the lower part of the Tertiary sequence consist largely of unmetamorphosed carbonate, chert, and argillite (uppermost allochthon), and distinctive metamorphic clasts appear higher in the sequence roughly in the order in which allochthonous sheets would have been eroded—black metashale (middle allochthon) followed by micaceous, quartzofeldspathic rocks characteristic of the lower allochthon and autochthon. Yet the source of the sediments was not the present Grouse Creek Mountains, which owe their elevation to basin-range faulting of Pliocene and younger age; their greatly abbreviated stratigraphy was intact until that time. The patches of overridden Tertiary sedimentary rocks now exposed in the central Grouse Creek and Matlin Mountains must represent detritus shed from other, earlier culminations within the region. Because of their position at or near the top of the tectonic pile, the Tertiary sedimentary rocks, like most of the Permian and Triassic rocks, were not metamorphosed or penetratively deformed.

In this view, uplift of the central Grouse Creek Mountains dome did not culminate until after 11 m.y. ago. At that time, parts of the upper and middle sheets with their Tertiary cover moved eastward to become intercalated in a series of tectonic slices. There is confirmation from the radiometric data for late uplift of the Precambrian adamellite and underlying Tertiary pluton: the fission-track ages and two Rb-Sr whole-rock biotite ages cited above indicate that the autochthon of the central Grouse Creek Mountains finally cooled below metamorphic temperatures as late as 8 to 12 m.y. ago.

ACKNOWLEDGMENTS

This research was done for a Ph.D. dissertation at Stanford University, and I would like to thank many people there, in particular Robert R. Compton, my thesis advisor, for unstinting advice and encouragement. Some of the data resulted from work done as a postdoctoral research associate at the U.S. Geological Survey under Max D. Crittenden, Jr., who made many helpful suggestions and whose enthusiastic interest was much appreciated. The field study would not have been possible without the help of the Bureau of Land Management and local ranchers, especially Ed and Diane Mott and their family. My field work was supported by a Geological Society of America Research Grant, and thin sections were financed by the Shell Research Fund at Stanford University. The Instituto de Geología,

Universidad Nacional Autonoma de Mexico, and the U.S. Geological Survey provided chemical analyses. Keith Howard, Kenneth F. Fox, Jr., Peter J. Coney, Max D. Crittenden, Jr., and Robert R. Compton read the manuscript at various stages and made many helpful suggestions.

REFERENCES CITED

Armstrong, R. L., 1968, Mantled gneiss domes in the Albion Range, southern Idaho: Geological Society of America Bulletin, v. 79, p. 1295–1314.

Compton, R. R., 1972, Geologic map of the Yost quadrangle, Box Elder County, Utah, and Cassia County, Idaho: U.S. Geological Survey Miscellaneous Geologic Investigations Map I-672, scale 1:31,680.

——1975, Geologic map of the Park Valley quadrangle, Box Elder County, Utah, and Cassia County, Idaho: U.S. Geological Survey Miscellaneous Geologic Investigations Map I-873, scale 1:31,680.

Compton, R. R., and Todd, V. R., 1979, Oligocene and Miocene metamorphism, folding, and low-angle faulting in northwestern Utah: Reply: Geological Society of America Bulletin, Part I, v. 90, p. 305–309.

Compton, R. R., and others, 1977, Oligocene and Miocene metamorphism, folding, and low-angle faulting in northwestern Utah: Geological Society of America Bulletin, v. 88, p. 1237–1250.

Cook, K. L., and others, 1964, Regional gravity survey of the northern Great Salt Lake Desert and adjacent areas in Utah, Nevada and Idaho: Geological Society of America Bulletin, v. 75, p. 715–740.

Crittenden, M. D., Jr., 1979, Oligocene and Miocene metamorphism, folding, and low-angle faulting in northwestern Utah: Discussion: Geological Society of America Bulletin, Part I, v. 90, p. 305–309.

Davis, G. H., 1975, Gravity-induced folding off a gneiss dome complex, Rincon Mountains, Arizona: Geological Society of America Bulletin, v. 86, p. 979–990.

Fox, K. F., Jr., and others, 1976, Age of emplacement of the Okanogan gneiss dome, north-central Washington: Geological Society of America Bulletin, v. 87, p. 1217–1224.

Higgins, M. W., 1971, Cataclastic rocks. U.S. Geological Survey Professional Paper 687, 97 p.

Kehle, R. O., 1970, Analysis of gravity sliding and orogenic translation: Geological Society of America Bulletin, v. 81, p. 1641–1664.

McMillan, W. J., 1970, West flank, Frenchman's Cap gneiss dome, Shuswap terrane, British Columbia, *in* Wheeler, J. O., ed., Structure of the southern Canadian Cordillera: Geological Association of Canada Special Paper no. 6, p. 99–106.

Reesor, J. E., 1965, Structural evolution and plutonism in Valhalla gneiss complex, British Columbia: Canada Geological Survey Bulletin, v. 129, 128 p.

Reesor, J. E., and Moore, J. M., Jr., 1971, Petrology and structure of Thor-Odin gneiss dome, Shuswap metamorphic complex, British Columbia: Canada Geological Survey Bulletin, v. 195, 149 p.

Shapiro, Leonard, 1967, Rapid analysis of rocks and minerals by a single solution method, *in* Geological Survey research 1967: U.S. Geological Survey Professional Paper 575-B, p. 187–191.

Todd, V. R., 1973, Structure and petrology of metamorphosed rocks in central Grouse Creek Mountains, Box Elder County, Utah [Ph.D. thesis]: Stanford, California, Stanford University, 316 p.

——1975, Late Tertiary low-angle faulting and folding in Matlin Mountains, northwestern Utah: Geological Society of America Abstracts with Programs, v. 7, p. 381–382.

Velde, B., 1965, Phengite micas: Synthesis, stability and natural occurrence: American Journal of Science, v. 263, p. 886–913.

Williams, H., Turner, F. J., and Gilbert, C. M., 1954, Petrography: An introduction to the study of rocks in thin sections: San Francisco, California, W. H. Freeman and Company, 406 p.

Manuscript Received by the Society June 21, 1979

Manuscript Accepted August 7, 1979

Printed in U.S.A.

Fabrics and strains in quartzites of a metamorphic core complex, Raft River Mountains, Utah

ROBERT R. COMPTON
Department of Geology
Stanford University
Stanford, California 94305

ABSTRACT

Solid-state flow during metamorphism was studied in a 300-km^2 exposure where Elba Quartzite forms the upper part of an autochthon and is overlain by two major allochthonous sheets. The quartzite is very locally thrown into north-verging recumbent folds, the largest having a wavelength of 1 km. Fold axes and elongate metamorphic grains trend approximately east, and foliation is subhorizontal.

A vertical sequence of samples was collected from the large fold, and a second sequence was collected from unfolded quartzite 4 km away. Axes of strain ellipsoids measured in five quartzites from the unfolded sequence are close to 0.5:1.1:1.7 (assuming an original sphere of rádius one). Five samples from the fold have ellipsoid axes ranging from 0.4:1.1:2.5 at the top of the sequence to 0.2:1.4:4 near the base. Most quartz grains deformed plastically without recrystallizing and developed strong *c*-axis fabrics. The degree of orientation of both quartz *c* axes and muscovite plates increased with increasing strain. Most of the quartz fabrics have orthorhombic symmetry and cross-girdle patterns like fabrics produced experimentally by J. Tullis and computer-simulated by G. S. Lister and coworkers, for quartzites extended in plane strain.

The results thus indicate a gradual east-west extension and consequent flattening by the force of gravity. Perhaps the fold formed concurrently where material flowing laterally under a broad dome was arrested locally and thus forced to buckle. Dating of nearby granite bodies indicates that the deformation began in Eocene time and continued until Miocene time.

INTRODUCTION

Rocks with subhorizontal foliations and strongly developed lineations characterize most of the metamorphic core complexes exposed in the North American Cordillera west of the thrust belt. One interpretation of the lineations is that they formed by simple shear of more or less equidimensional bodies, which thus became elongated parallel to the direction of shear. In this interpretation, lineations oriented at large angles to the strike of the thrust belt are due to transport of overlying rock bodies toward the overthrusts. A relation contradictory to this interpretation, however, is exact parallelism between the lineations and the axes of folds that are overturned so strongly and consistently as to indicate tectonic transport at right angles to the lineations. This paper describes a case of the latter type, one in which the folds and lineated rocks are well exposed in their relation to overlying allochthonous sheets. The principal subject is an analysis of strains and mineral fabrics in the upper part of the autochthon, these being based on field relations and rock ages that have been described elsewhere in detail (Compton, 1972, 1975; Compton and others, 1977).

The Raft River Mountains make up about one-fourth of a 5,000-km^2 exposure of autochthonous metamorphic rocks and associated allochthonous sheets of metamorphic and sedimentary rocks (Fig. 1). The area is one in which metamorphism, emplacement of granitic plutons, folding, and low-angle (decollement) faulting went on episodically and more or less concurrently over a long period of time. K-Ar ages of metamorphic rocks in the Albion Range suggest that metamorphism and deformation began in the Mesozoic, perhaps in the Jurassic (Armstrong, 1976). Rb-Sr emplacement ages of granitic rocks in the Grouse Creek Mountains show that a first phase of folding ended there before 38 m.y. ago and that a second phase of folding began at about that time and continued, perhaps episodically, at least until 25 m.y. ago, probably until ~20 m.y. ago (Compton and others, 1977). Cooling ages suggest, further, that metamorphism ended about 20 m.y. ago in large parts of the lined area shown in Figure 1, but stratigraphic relations prove that folding and displacements on some low-angle faults took place as recently as 10 m.y. ago in the Raft River Mountains, the Grouse Creek Mountains, and the Matlin Mountains (Compton and others, 1977). The rocks described in this paper are from the eastern half of the Raft River Mountains, where the earliest phase of deformation as well as the latest one had only local, minor effects. The second phase of deformation can thus be interpreted quite firmly, and it is a truly pervasive phase in this part of the area.

GENERAL STRUCTURAL RELATIONS

The Raft River Mountains consist of a broad autochthon overlain by two major allochthonous sheets. All three of these structural units are subparallel and roughly stratiform in themselves, although they are locally folded and cut by subsidiary low-angle faults. The three units were close to horizontal during the deformation described in this paper, and all were arched upward in latest Miocene or in Pliocene time, forming a broad, doubly plunging anticline that defines the mountain range. Erosion has removed most of the allochthonous sheets and has cut downward as much as 800 m into the autochthon, so that the present range exposes an arched autochthonous core overlain locally by remnants of the sheets (Fig. 2). The specific rock formations have been described (Compton, 1972, 1975) and are summarized in Figure 3, which shows, in effect, the structural sequence in the range.

The rocks in the authochthon and the lower allochthonous sheet are metamorphic, and those near the base of the upper sheet are locally metamorphic. In the Grouse Creek Mountains, the grade of metamorphism increases downward from greenschist facies in the highest metamorphosed levels to amphibolite facies in the autochthon. In the Raft River Mountains, however, amphibolite-facies rocks of the lower allochthonous sheet have been displaced over a lower grade part of the autochthon, a part characterized by greenschist facies and, locally, kyanite-chloritoid-muscovite assemblages.

Regardless of these variations in metamorphic grade, deformation of the rocks decreases downward in the autochthon. This relation is seen most clearly in the Precambrian granite (adamellite) that forms the deepest exposed unit in the autochthon. In the Raft River Mountains, this rock has well-preserved igneous textures in the deeper exposures, above which it becomes more and more gneissose and finally

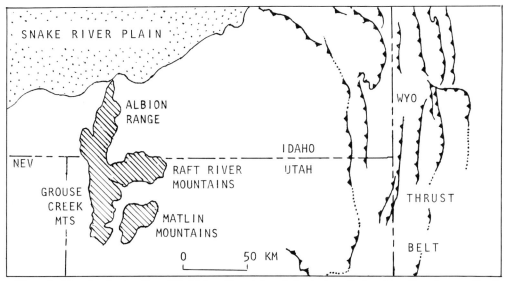

Figure 1. Map showing location of Raft River Mountains relative to thrust belt and to adjacent ranges (lined) exposing similar rocks and structures.

is converted to a mica-quartz schist just beneath the Elba Quartzite. Deformation just above that level, in the quartzite, is the special topic of this paper.

Of the two main allochthonous sheets, the upper sheet was displaced a distance at least as great as the smallest diameter (60 km) of the patterned area shown in Figure 1. The relation implying this distance is the superposition of uppermost Mississippian or younger rocks onto Ordovician or older rocks throughout the exposed extent of the upper decollement in the Raft River Mountains and most of the Grouse Creek Mountains and the Albion Range. High-grade metamorphic variants that were displaced over lower grade parts of the autochthon suggest that the lower allochthonous sheet, or parts of it, were displaced ~30 km during metamorphism or afterward (Compton and others, 1977, p. 1242).

Metamorphism during or after low-angle fault displacements recrystallized crushed materials on or near most fault surfaces. Mylonite formed locally, however, in the vicinity of Indian Creek in the Raft River Mountains (Fig. 2). This porphyroclastic rock differs markedly from deformed quartzites described in this study. It is consistently dark, nearly black, and has a submicroscopic matrix surrounding angular rock fragments.

Directions of displacements on low-angle faults have been determined with only moderate success. Vergence of folds of the first phase is the only indication of this direction during the earliest phase of deformation. These folds are overturned toward the west and the northwest in the western Raft River Mountains and the Grouse Creek Mountains. Northward displacement of the allochthonous sheets during the second phase of deformation is indicated by the vergence of folds and by lineated blastomylonites formed where the sheets were displaced across the tops of two granitic stocks that were themselves lineated during the second phase of folding (Compton and others, 1977, p. 1245, 1247). In addition, a vertical strike-slip fault that curves through 180° along strike, so as to become a local low-angle thrust, suggests northward transport for at least part of the lower allochthonous sheet near the southwestern edge of the Raft River Mountains. Transport along the decollement late in metamorphism and after metamorphism was generally eastward, as shown by (1) eastward-verging folds, (2) two sets of westward-dipping imbricate thrusts, (3) an eastward-striking high-angle strike-slip fault, and (4) eastward displacements of high-grade metamorphic variants of the lower allochthonous sheet, already mentioned.

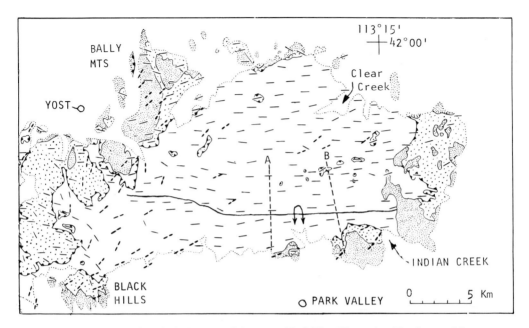

Figure 2. Map showing the principal structural elements of Raft River Mountains. Closely spaced dots = upper allochthonous sheet; widely spaced dots = lower allochthonous sheet; unpatterned = autochthon; dotted boundaries are with Cenozoic deposits; short double dashes = axial lineations of first-phase metamorphic folds; long dashes = axial lineations of second-phase metamorphic folds; single line = crest line of large recumbent anticline. Postmetamorphic folds and high-angle faults not shown. Figure 4 shows vertical sections along lines A and B.

Folds and Folding

Folds are abundant in the lower allochthonous sheet and in the metamorphosed parts of the upper sheet, but they are rather sparse in the autochthon. Second-phase folds are dominant in the autochthon in the eastern Raft River Mountains, where their axial trend is approximately east (Fig. 2). The one large fold of this phase extends for more than 21 km, which is the distance from one end of the autochthon exposure to the other (Fig. 2). This solitary fold expresses a remarkable degree of deformation and was thus one of the principal features sampled in this study (Fig. 4). Note that the lower contact of the Elba Quartzite is unfolded, yet the upper part of the formation at the anticlinal crest shows a relative displacement of 2 km northward since the fold started to form. Figure 4 also indicates moderate variations in the fold along its axial trend; 8 km west of section A it changes further

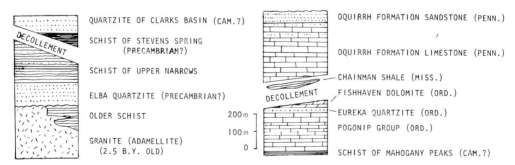

Figure 3. Vertical section of pre-Cenozoic rock units of Raft River Mountains, in typical structural order and at maximum thickness for this area.

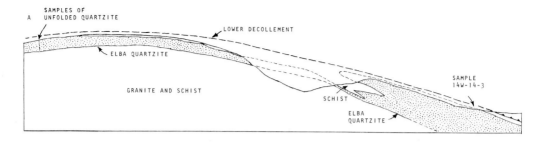
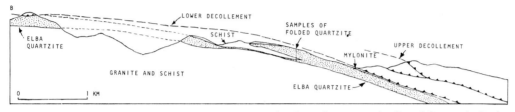

Figure 4. Vertical sections in Raft River Mountains showing the major structural units and the sample sites in the Elba Quartzite. Figure 3 shows locations. Based on Compton (1975).

to a set of three closely spaced anticlines and synclines (Compton, 1975).

Besides this one large fold, folds with wavelengths of 0.1 to 10 m are scarce but can be seen here and there, and folds with wavelengths of a few centimetres are locally numerous. Almost all of these folds are overturned northward and have profiles like the large fold, commonly showing similar evidence of horizontal displacement with respect to underlying unfolded layers (Fig. 5). Some of these folds, however, are much less overturned, and one with a wavelength of 10 m, on the north flank of the range, is overturned toward the south.

Schistosity in metashales and grain foliations in quartzites are parallel to fold axial planes at and near the axial planes, and diverge only slightly from this orientation in fold limbs. Except at crests and troughs of folds, metamorphic foliation is thus parallel or nearly parallel to bedding. Lineations

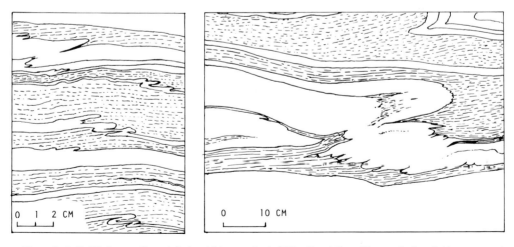

Figure 5. Left, thin layers of quartzite in schistose variant of Elba Quartzite, with second-phase folds overturned toward the north. Right, small folds in quartzite layers that have flowed to the right (north) in a decollement relation to the unfolded thick quartzite layer beneath. Schistosity indicted by dashes.

formed by small folds and elongated grains of quartz, mica, kyanite, and tourmaline are typically parallel to fold axes but locally diverge by as much as 10° from that direction. Mapped lineations based on these small features are parallel to the crest line of the large fold (Fig. 2). The obliqueness between the crest line and the lineations shown in the western part of the map is due to the rise in elevation of the crest in that sector.

Fold wavelengths vary systematically with thicknesses of quartzites that are intercalated with thick schist layers. The ratios of wavelength to quartzite thickness range from approximately 1:12 to 1:18, and the consistency of this relation indicates that the folds were generated by buckling of quartzite layers. The disharmony among folds in a given group of layers also suggests buckling (Fig. 5, left). The nearly recumbent forms of the folds, however, suggest considerable lateral flow and flattening under gravity, as do the subhorizontal schistosity and the association of folds with decollement sheets that are themselves folded locally. Possibly the folds formed where flowing layers were impeded locally and thus buckled, an idea that can be considered more exactly after measurements of the deformation are described.

FABRICS AND STRAINS IN THE ELBA QUARTZITE

The study of fabrics and strains was designed to compare strongly folded quartzites with unfolded quartzites of the same formation. A suite of Elba Quartzite samples was collected from the large fold and another suite from unfolded rocks 4 km north of the fold (Fig. 4). Each suite was collected in a nearly vertical sequence, with the idea of determining the vertical distribution of strain and thus of testing for effects of the allochthonous sheets that lie just above the two exposures (Fig. 4). The Elba Quartzite was selected because it is exposed widely in the area shown on Figure 1 and locally contains pebbly beds that afford a means of estimating strains elsewhere.

The unfolded rocks were collected on the cirque wall west of Bull Lake, and those from the fold were collected on the west wall of Indian Creek Canyon, with an additional sample (14W-14-3) from the south end of section A (Fig. 4). The quartzite layers in the unfolded sequence are typically 1 to 2 m thick and are separated by thinner layers of quartzite and muscovite-quartz schist. Rocks in the lower 80 m of the sequence have moderately developed lineations on most bedding planes, and the lineations trend N80° E to east. The trend of lineations changes to N82° W in the upper 20 m of the formation and differs by as much as 8° between successive layers. In contrast, almost every quartzite layer at the fold locality is distinctly lineated, and trends of the lineations differ by no more than 7° from N85° E, which is the axial trend of the large fold at this locality. The layers are also thinner than those in the unfolded sequence, typically 10 to 20 cm thick, and the thinner layers have a more flaky parting than those in the unfolded sequence. At both localities, schist layers between the quartzite layers form only a small percentage of the total sequence.

Minerals and Their Textural Relations

Fifteen samples from the unfolded sequence and sixteen samples from the fold were studied in thin section. Minerals formed during metamorphism indicate low-grade (greenschist-facies) conditions, except for locally occurring kyanite-chloritoid-muscovite assemblages. Besides grains of recrystallized quartz, muscovite (or possibly phengite) is the principal metamorphic mineral, forming 1% to 6% of the quartzites. Large single muscovite grains are commonly bent or lenticular, or are polygonized into subparallel, slightly curved groups of grains. The larger of these folia lie along quartz grain boundaries and are distinctly elongated parallel to the lineation, especially in the samples from the fold. Crystals of muscovite less than 0.05 mm across are typically simple tablets that are elongated in the direction of the lineation. The tablets lie within quartz grains as well as at their boundaries and in both places are about equally oriented relative to the foliation. Apatite and tourmaline are scarce but widespread metamorphic accessories, and kyanite and chloritoid occur in a few samples from the fold. Elongate grains of all four minerals are generally aligned with the lineation, and the grains may be euhedral or deformed in any one rock. In the samples from the fold, some of these grains are broken and pulled apart in the direction of the lineation.

The most obviously relict sedimentary minerals are microcline and albite, which constitute 5% to 15% of the feldspathic quartzites. Thin (~0.01 mm) shells of muscovite around some of these grains suggest instability of the feldspars and scarcity of water during metamorphism. Texturally, the feldspars indicate that the original sandstones were medium to coarse grained and well sorted to very well sorted. These inferences are supported by fairly numerous detrital grains of zircon that are well sorted by size in a given rock. The feldspars may be broken and pulled apart by extension parallel to the lineation, and in all cases they acted as solid bodies against which quartz grains were deformed plastically (Fig. 6, left).

Quartz grains are mainly relict sand grains that have been deformed into somewhat irregular triaxial ellipsoids with shortest axes perpendicular to foliation and longest axes parallel to fold axes and mineral lineations (Fig. 6, left). Plastic deformation of the grains is indicated by these relations: (1) by far most of the grains are single crystals; (2) the deformed grains have volumes similar to feldspar grains in the same rock; and (3) the small proportion of quartz grains that are undeformed (Fig. 6, right) have c axes oriented perpendicular to the foliation, an orientation unfavorable to either basal or prism slip under a compression acting perpendicular to the foliation.

All samples contain some quartz grains that have partly recrystallized into groups of equidimensional small grains having a great variety of crystallographic orientations. Recrystallization is most common along boundaries of ellipsoidal grains, especially at their thinned ends, but took place locally in the central parts of grains. The amount of recrystallized quartz ranges from nil to 10% in the unfolded rocks and from 10% to 80% in the rocks from the fold.

Most quartz grains also show various indications of unrecovered or only partially recovered strains. Most quartz grains extinguish unevenly under crossed polars, and in a given rock from a few to many grains have polygonized; that is, they have developed elongate subgrains disoriented 1° or less relative to one another. Deformation lamellae are scarce in most samples but are abundant in one in which the lamellae have predominantly sub-basal orientations (Fig. 7). Despite these indications of postcrystalline strain, however, no part of the samples is crushed or granulated to angular or submicroscopic material like that in the mylonite already mentioned.

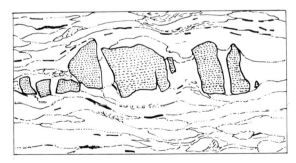 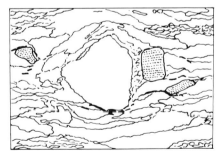

Figure 6. Left, two feldspar grains (dotted) in sample 14W-14-3, pulled apart in a direction parallel to the lineation. Finer lines are quartz grain boundaries and heavy lines mica plates. View is 7 mm high. Right, undeformed single-crystal quartz grain, 3 mm in diameter, surrounded by deformed and locally recrystallized quartz grains and by undeformed feldspar (dotted) in sample IND-16. Muscovite shown by heavy lines. Section perpendicular to lineation.

In summary, all of the metamorphic minerals occur in simple crystallized grains as well as in grains with notable unrecovered strains. Platy and linear grains of quartz, muscovite, kyanite, chloritoid, tourmaline, and apatite are oriented parallel to the subhorizontal foliation and to the lineation, the latter being parallel to the axes of large and small folds. These relations indicate that metamorphism and deformation, including folding, were concurrent at both localities sampled.

Methods of Analysis

Quartz c-axis fabrics were measured in one thin section from each of 15 samples and in an additional section for the 7 samples illustrated in this paper. Where two thin sections were used, one was parallel to the lineation and the other perpendicular to it, both having been approximately perpendicular to the foliation. Grains were selected by moving the thin section 1 mm and measuring the grain at the crosshairs. In cases where this grain was too small to measure, the next grain along the traverse was measured. In analyses of muscovite, the grain with its center nearest the crosshairs was measured. One hundred grains were measured in each thin section and plotted separately, then rotated into single plots that had the same orientation of fabric axes.

Strains were determined by measuring the dimensions of 60 grains in each of the two thin sections used in the fabric analyses. The dimensions measured were those parallel to the three principal directions of the rock fabric. These directions typically coincided with the shortest, intermediate, and longest axes of grains in the more deformed rocks, but the directions typically did not coincide with the three grain axes in the less deformed rocks. The sections were moved 1 mm, and the grain beneath the crosshairs was measured; however, small grains resulting from recrystallization were not measured unless they could be included with the grain from which they were derived. The sets of dimensions were graphed as width:length ratios, and an average ratio was obtained by fitting a straight line to the array of points. The resulting ratios were used to calculate the three axes of the strain ellipsoid by assuming an original sphere with radius = 1.

The accuracy of the method was estimated by a set of measurements on grains making up an ellipsoidal cobble from the base of the Elba Quartzite in Clear Creek Canyon (Fig. 2). The form of the cobble indicates that strain ellipsoid axes are 0.32:0.87:3.6, and the ellipsoid axes obtained by measuring the grains 0.34:0.92:3.1. Possibly the differences are due to the fact that the cobble originally was somewhat elongated parallel to the strained form, although it was collected because it appeared to have an average shape for the outcrop. Another possible cause of the differences is the presence of recrystallized grains that were not recognized as such when they were measured. Such grains would tend to be more equidimensional than grains that did not recrystallize during strain. This possibility is supported by the fact that smaller grains in two measured rocks are not as elongated, on the average, as larger grains in the same rocks. Probably, then, the strains obtained from the grain measurements are somewhat smaller, perhaps 10% smaller, than the actual strains.

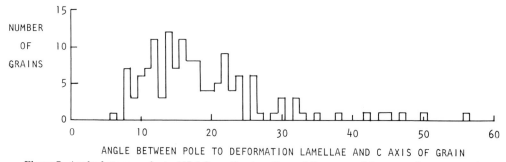

Figure 7. Angles between poles to 157 deformation lamallae and the c axes of the respective quartz grains.

Results of the Analyses

The strain ellipsoid axes measured for samples with illustrated fabrics are included in Figure 8, and the caption gives data on vertical location, texture, and composition of each sample. In addition to the unfolded rocks shown in Figure 8A and 8B, measurements of a conglomerate at the base of the sequence gave ellipsoid axial ratios of 0.55:1.1:1.7, and measurement of a quartzite 8 m above sample 14W-10-8 (Fig. 8B) gave 0.54:1.1:1.7. Quartz c-axis fabrics were measured for four additional rocks located at vertical distances of 34, 115, 120, and 130 m above the base of the formation; these fabrics are similar to the two illustrated, as is the fabric of the quartzite cobble from Clear Creek Canyon already mentioned as having ellipsoid axes of 0.34:0.92:3.1. Finally, the lowest quartzite in the unfolded sequence, which lies just above the basal conglomerate, has a nearly random c-axis fabric and a texture indicating that both quartz and muscovite are largely recrystallized.

Of the rocks from the fold, two strains were measured in addition to those included in Figure 8: (1) sample IND-9, which has ellipsoid axial ratios of 0.20:1.4:3.9, from 47 m above the base of the section, and (2) sample IND-17, which has axial ratios of 0.39:1.1:2.5, from 161 m above the base. Many of the other samples from the fold are too recrystallized to be measured, such as the samples of Figure 8G and 8H, but strains in rocks in the lower 50 m of the sequence are large, with ellipsoid ratios similar to sample IND-4. The overall strain was checked by measuring the thickness of the Elba Quartzite in the limbs of the anticline and comparing it with the thickness measured along the axial plane. The resulting ratio is approximately 1:8, and the ratio of the shortest to the intermediate ellipsoid axes of the more deformed samples is approximately 1:7. The measured extension and flattening of grains thus account for the extension and flattening in the fold—almost exactly so if the measured strains are assumed to be somewhat smaller than the actual strains.

Of the 11 unfigured quartz c-axis fabrics from the folded section, 2 are similar to the fabric shown in Figure 8D and 2 more are somewhat similar. The remaining 7 fabrics have distinct girdles of points around the axis of grain elongation, which has the same orientation as the fold axis. These fabrics are somewhat like the one in Figure 8G but have more complete girdles. Four fabrics also have strong to moderate monoclinic symmetry relative to the foliation and lineation. The remaining 3 of the 7 fabrics have an asymmetry similar to the one shown in Figure 8F, and thus have triclinic symmetry. In all the triclinic fabrics, the geographic sense of the asymmetry is the same.

The measured muscovite fabrics, Figure 8I, 8J, 8K, and 8L, are representative of other rocks in the two sequences. The degree of preferred orientation is stronger in rocks from the folded sequence and tends to increase with depth in that section. In all rocks from both localities, the mica fabrics define a point maximum normal to the foliation with a small to moderate arc around an axis parallel to the lineation. Small circles are drawn on the diagrams to make this relation clear. The arcs extend through more than 90° in fabrics from the unfolded sequence and are shorter in fabrics from the folded sequence. One probable cause of the arcs is that many micas lie on quartz grain boundaries and thus are more nearly parallel in sections cut parallel to the most elongate sections of quartz grains, as suggested by Figure 6. A second cause is a microfolding of the large muscovite folia, the folds occurring in symmetrical trains with wavelengths of 0.5 mm and wave amplitudes of 0.1 mm or less. A tendency for the folds to be straight-limbed is suggested in Figure 8K by the two pole concentrations that do not lie on the perimeter of the plot. The fold axes trend approximately parallel to the lineations in each rock.

The results of the fabric and strain studies may be summarized as follows:

1. The strains in the unfolded rocks are similar throughout that sequence and are less than those in the folded rocks.

2. Strains in the folded sequence increase downward and are distinctly greater in the lower 50 m of the total section.

3. All the strains are close to plane strains (in plane strain the intermediate axis of the strain ellipsoid remains unchanged and would thus be 1 in the axial ratios used here).

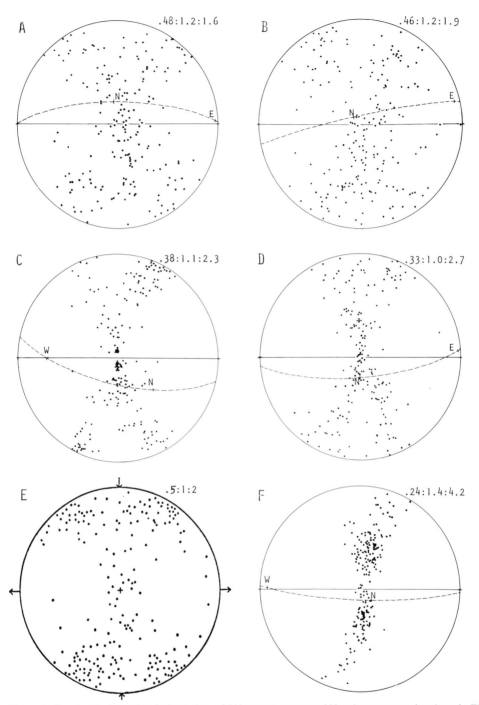

Figure 8. Equal-area, lower hemisphere plots of 200 quartz c axes or 200 poles to muscovite plates in Elba Quartzite (except for E). Lineation and trace of foliation are indicated by horizontal solid line. Dashed arc is trace of horizontal plane, with geographic directions. Numbers by plots are axes of strain ellipsoid, assuming an original sphere of radius 1. In each diagram, longest axis of ellipsoid is parallel to the lineation (horizontal line); shortest axis is perpendicular to this line; and intermediate axis is perpendicular to the page. A. Quartz in sample 14W-10-16, 115 m above base of unfolded section; quartz commonly polygonized; 3% muscovite, 5% feldspar. B. Quartz in sample 14W-10-8, 10 m above base of unfolded section; quartz 10% recrystallized; 3% muscovite, no feldspar. C. Quartz in sample 14W-14-3, 200 m above base of folded sequence (near top of upright upper limb); quartz commonly polygonized; 2% muscovite, 9% feldspar. D. Quartz in sample IND-16, 150 m above base of folded sequence

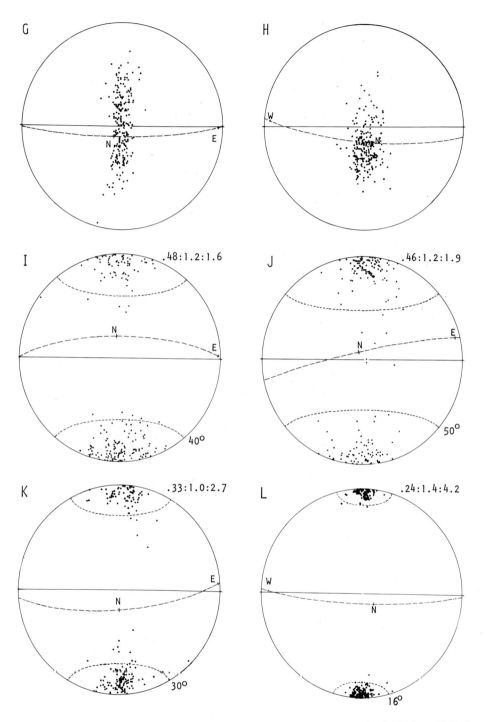

(in lower upright limb); quartz generally polygonized, 10% recrystallized; 3% muscovite, 8% feldspar. E. Preferred orientation of deformed original grains of quartz in quartzite Q-1, deformed in plane strain at 15 kb, 800 °C, and 10^{-6}/s. Reproduced by permission from Tullis (1977, p. 98). F. Quartz in sample IND-4, 10 m above base of folded section; quartz generally polygonized, 20% recrystallized; 5% muscovite, no feldspar. G. Quartz in IND-2, 4 m above base of folded section; quartz polygonized, 30% recrystallized; 3% muscovite, no feldspar. H. Quartz in sample IND-1, 1 m above base of folded section; quartz 80% recrystallized; 3% muscovite, no feldspar. I. Muscovite in sample 14W-10-16 (see A for data). J. Muscovite in sample 14W-10-8 (see B for data). K. Muscovite in sample IND-16 (see D for data). L. Muscovite in sample IND-4 (see F for data).

396   R. R. COMPTON

4. Quartz and muscovite fabrics in the unfolded section and the upper part of the folded section have orthorhombic symmetry, with maxima becoming stronger and pole-free areas becoming larger with increase in strain.

5. Quartz fabrics of many rocks in the lower part of the folded section have monoclinic symmetry coaxial with the large fold.

6. Several rocks in the folded section have similarly oriented triclinic quartz fabrics.

Interpretation of the Strains and Fabrics

Four relations prove that dislocation glide and climb in quartz grains is by far the chief mechanism of deformation in the unfolded rocks and in rocks from the upper part of the folded sequence. (1) The quartz grains have the same volumes as undeformed feldspar grains in the same rocks. (2) Most quartz grains remain unrecrystallized through various degrees of strain. (3) Occasional quartz grains, those with c axes perpendicular to foliation, remain essentially undeformed. (4) Quartz c-axis fabrics and mica fabrics have symmetries identical with those of the strain ellipsoids in the same rocks and show more strongly preferred orientations with increasing strain.

Experimental deformation of quartzites by Tullis (1977) and quartzite deformation models computer-simulated by Lister and others (1978) provide a basis for interpreting the kinetic relations in the sampled rocks. Of the variety of experimental and simulated deformations described by these persons, deformations of quartzite in plane strain are the only ones that produced fabrics similar to those of the unfolded rocks and the rocks in the upper part of the folded sequence.

One of Tullis's fabric diagrams is reproduced here for comparison (Fig. 8E). The arrows indicate the directions of compression and extension, and the resulting ellipsoidal grains lie with their foliation horizontal and their lineation parallel to the axis of extension. The strain ellipsoid is thus oriented exactly as in the other diagrams. The grains in Tullis's experiments deformed by dislocation glide and climb. Strains were similar to several measured in this study, and resulting grain shapes are also similar, even to the occasional grains that remained undeformed.

The computer-simulated fabric most like the ones from the unfolded rocks is based mainly on basal glide, supplemented by rhomb glide systems, a model called "Quartzite I" by Lister and others (1978, Fig. 13B). The importance of basal glide in the Raft River Mountains quartzites is suggested by the sub-basal deformation lamellae and by the undeformed grains with c axes perpendicular to foliation, an orientation in which a compression perpendicular to folation cannot generate a shear stress along basal or prism slip planes.

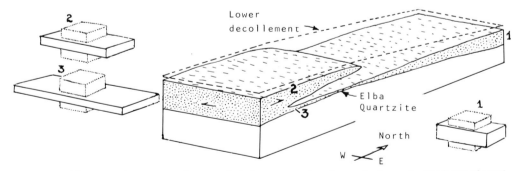

Figure 9. Diagrammatic summary of measured strains, expressed as changes in shape of cubic forms of equal original volume: 1 = average for the unfolded sequence; 2 = top of the folded sequence; and 3 = lower part of the folded sequence. Single-barb arrow on right side of diagram represents northward shear expressed by the fold vergence, and that on the front of diagram represents the local westward shear suggested by fabrics like that of Figure 8F.

In all the rocks from the fold, the direction of extension is parallel, within a few degrees, to the axis of the large fold. Neither experiments nor simulation have produced fabrics like the monoclinic fabrics of the more strongly deformed rocks in the fold (Lister, 1977, personal commun.). The shape of the fold suggests shear of the upper part of the fold northward over the lower part (Fig. 9). Such shear is presumably the cause of the monoclinic symmetry of the fabrics.

The different amounts of extension measured within the folded sequence imply local east-west shear parallel to the quartzite layers (Fig. 9). Possibly this shear caused the triclinic fabrics in some samples, as shown in Figure 8F. Lister (1977) illustrated a simulated fabric based on simple shear of his model Quartzite I, and it is similar to the fabrics measured in this study. The similarity indicates that the overlying rocks in the Raft River Mountains were displaced westward relative to those beneath.

The greater amount of extension of the folded as compared to the unfolded sequence also requires east-west shear between the two localities. The intervening terrane is well exposed, and no obvious shear zones are visible. Probably the shear was accomplished by small increments distributed through the 4 km of intervening rocks.

DISCUSSION

The principal deformation mechanism in the quartzites studied is evidently dislocation glide and climb in quartz. The prominent lineations were produced by east-west extension in nearly plane strain, modified by a moderate component of axial flattening and locally by moderate components of shear. The orthorhombic symmetry of the quartz and muscovite fabrics indicates that the deformation was mainly pure shear rather than simple shear on the foliation planes. The lineations thus do not indicate the direction of displacement of the allochthonous sheets. Moreover, the strains cannot have been produced by overriding of the allochthonous sheets, for the strains do not increase upward (toward the sheets) at either locality studied.

The single major east-west fold implies displacements and presumably shear strains in a north-south direction, but these appear to be minor compared with the flattening normal to the foliation or axial plane of the fold and elongation parallel to its axis. Several relations indicate that the folding was concurrent with the flattening and extension: (1) the major fold axis coincides within a few degrees with the direction of extension; (2) elsewhere in the Raft River Mountains the axes of second-phase folds vary in orientation by as much as 20°, yet mineral lineations are everywhere parallel to the local fold axes (Fig. 2); and (3) flattening and extension are distinctly greater in the folded sequence studied.

Concurrent folding and extension are difficult to explain, however, as we have already noted that the folds appear to have originated by buckling, and buckling requires a compression directed subparallel to the layers, whereas extension and flattening require a compression approximately perpendicular to the layers. A possible explanation is that folding and extension were both generated within a heated and updomed area that was much larger than the area studied. As suggested by the large arrows in Figure 10, the heating and doming might lead to flow of material outward, under gravity. This action

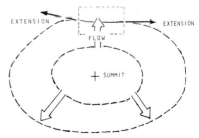

Figure 10. Map view of large hypothetical uplift that might generate outward flow of rocks and thereby extension parallel to its perimeter, as well as local folding.

would cause a gradual stretching (extension) tangential to the contours of the dome. The maximum compressive stress would be approximately vertical in such a model, and the minimum compressive stress would be parallel to the direction of stretching. If the outward flow of the rocks were arrested locally, the orientation of the stresses would be changed at that place and a buckle fold might form. Flow and flattening might then modify the initial buckle into a recumbent form, and local thickening at the site of buckling might be a principal cause of the greater degrees of strain observed in the lower part of the fold. Finally, if the form of the dome were to change substantially, the ongoing phase of deformation would be ended and the next phase would be initiated.

As speculative as this model must be, the idea of gravity-driven deformation in a broad, heated dome is supported by several relations: (1) close correlation between the thermal and deformational histories over a large area, (2) rapid decrease in strain downward in the underlying Precambrian granite, (3) folds of one phase that verge in opposite directions (Compton and others, 1977), (4) subhorizontal foliations and typically recumbent folds, and (5) thinning of all metamorphosed formations within most of the area crosslined in Figure 1.

ACKNOWLEDGMENTS

Field trips and discussions with Victoria Todd and Max Crittenden have contributed to the ideas presented, as have discussions and correspondence with Jan Tullis, Gordon Lister, and Ray Fletcher. Petrofabric work by Stanford University students provided a much appreciated check of the fabric analyses. An early version of the paper was read by Jan Tullis and John Christie, whose suggestions improved it considerably.

REFERENCES CITED

Armstrong, R. L., 1976, The geochronometry of Idaho, Part II: Isochron/West, no. 15, p. 1–33.

Compton, R. R., 1972, Geologic map of the Yost quadrangle, Box Elder County, Utah, and Cassia County, Idaho: U.S. Geological Survey Miscellaneous Geologic Investigations Map I-672.

——1975, Geologic map of the Park Valley quadrangle, Box Elder County, Utah, and Cassia County, Idaho: U.S. Geological Survey Miscellaneous Geologic Investigations Map I-873.

Compton, R. R., and others, 1977, Oligocene and Miocene metamorphism, folding, and low-angle faulting in northwestern Utah: Geological Society of America Bulletin, v. 88, p. 1237–1250.

Lister, G. S., 1977, Discussion: Crossed-girdle c-axis fabrics in quartzites plastically deformed by plane strain and progressive simple shear: Tectonophysics, v. 39, p. 51–54.

Lister, G. S., Paterson, M. S., and Hobbs, B. E., 1978, The simulation of fabric development in plastic deformation and its application to quartzite: The model: Tectonophysics, v. 45, p. 107–158.

Tullis, J., 1977, Preferred orientation of quartz produced by slip during plane strain: Tectonophysics, v. 39, p. 87–102.

MANUSCRIPT RECEIVED BY THE SOCIETY JUNE 21, 1979
MANUSCRIPT ACCEPTED AUGUST 7, 1979

Structural geology of the northern Albion Mountains, south-central Idaho

DAVID M. MILLER*
Department of Earth and Space Sciences
University of California
Los Angeles, California 90024

ABSTRACT

The northernmost of a chain of four or five domes in the Albion Mountains was studied in order to understand its origin and deformational history. The dome is part of a metamorphic terrane that is composed of Precambrian W basement gneiss that is overlain by a Precambrian(?) and Paleozoic metasedimentary rock succession that was metamorphosed to the amphibolite facies in the Albion Mountains.

Four phases of ductile deformation are recognized in minor structures. The first two phases of deformation followed extensive low-angle faulting of younger-on-older type and resulted in the formation of overturned and flattened folds in the weaker rock units and penetrative lineations and foliations in the thick quartzite units. Vergence of minor folds that formed during D_1 and D_2 indicates that shear occurred along planes approximately parallel to bedding such that strata moved to the northwest (D_1) and the northeast (D_2) relative to the basement. Third-phase deformation locally created overturned folds and small bedding-plane faults with geometries that indicate that higher strata moved outward from the apex of the present dome relative to the underlying strata, which suggests that these structures formed during the rise of the dome. The fourth phase of deformation is typified by kink folds whose axes trend N15°E and are horizontal. Movement on a low-angle fault that separates Pennsylvanian sedimentary rocks from underlying metamorphic rocks followed the second phase of deformation and may have been concurrent with late ductile deformation in the more deeply buried rocks.

Timing of the metamorphism and intense deformation in the Albion Mountains is elusive, but it apparently occurred between Late Triassic and Early or middle Cretaceous times; numerous Tertiary K-Ar and Rb-Sr ages from metamorphic rocks indicate that less intense deformation and moderate temperatures continued in the more deeply buried rocks until the Miocene.

The structural history of the Albion metamorphic terrane is thus interpreted as an early to middle

---

*Present address: U.S. Geological Survey, Branch of Western Environmental Geology, 345 Middlefield Road, Menlo Park, California 94025.

Mesozoic period of extensive younger-on-older faulting and ductile shearing on nearly horizontal planes over a little-deformed basement, and a late Mesozoic to early Cenozoic period of less intense deformation and metamorphism. This history is similar to that of other metamorphic terranes in the Great Basin.

INTRODUCTION

The Albion Mountains are a north-trending uplift in the northern Basin and Range province and lie on the southern edge of the Snake River downwarp. The mountain range is part of a metamorphic terrane that is characterized by a complex history of deformation, metamorphism, and plutonism of post-Paleozoic age. Much of the metamorphic terrane is exposed in the Albion, Raft River, and Grouse Creek mountain ranges; it is here termed "the Raft River–Albion metamorphic terrane." The area mapped encompasses Big Bertha dome, the northernmost of four or five domes in the Albion Mountains (Fig. 1).

Pioneering studies of the metamorphic terrane were reviewed by Armstrong (1968b). More recent reconnaissance studies were conducted by Misch and Hazzard (1962) and Armstrong and Hansen (1966). Armstrong and Hills (1967) presented radiometric dates that indicated that the basement gneisses that form the "cores" of the gneiss domes of the Albion Mountains are about 2.5 b.y. old (Precambrian W) and that a major metamorphic event took place during the Mesozoic and possibly the early Cenozoic. Armstrong (1968b) described gneiss domes in the Albion Mountains and outlined a multistage history of penetrative deformation for the area. Contributions by Compton (1972, 1975), Todd (1973), and Compton and others (1977) further clarified the structural history and demonstrated the importance of low-angle faulting in the Raft River and Grouse Creek Mountains. Compton and others (1977) showed that part of the penetrative deformation and metamorphism was as young as Oligocene (25 m.y.). They suggested that the area was part of a regional metamorphic dome, which would explain reversed directions of fold vergence and other indications of gravity tectonics in the southern part of the metamorphic terrane.

This study describes structures formed during low-angle faulting and penetrative deformation in the Mount Harrison region of the northern Albion Mountains. Parts of the metasedimentary succession that overlie the little-deformed basement gneiss complex are extensively deformed, while other parts, such as thick quartzite units, are relatively little deformed. The structures are inferred to have resulted from vertical shortening and shear within weak rocks between stiffer stratigraphic units. Intense deformation and metamorphism are interpreted to be of Mesozoic age, and later moderate temperatures and minor ductile deformation occurred during the early Cenozoic. Doming is demonstrated to be a late, relatively minor phase of deformation.

GEOLOGIC SETTING

The metamorphic terranes of the eastern Great Basin lie in the hinterland of the Cordilleran fold and thrust belt of Idaho, Wyoming, Utah, and Nevada (Armstrong, 1972), and they are part of an enigmatic structural province that has been the subject of much debate (Misch and Hazzard, 1962; Armstrong and Hansen, 1966; Armstrong, 1972; Hose and Danes, 1973; Roberts and Crittenden, 1973). Some of the major questions in the debate over the tectonic interpretation of the eastern Great Basin pertain to the nature and timing of deformation in the hinterland (including the metamorphic terranes) and the relationship of this deformation to the Cordilleran fold and thrust belt.

The hinterland is characterized by a long, but fragmentary, record of high-angle normal faulting

(Loring, 1976), low-angle faulting that generally places younger over older rocks, a moderate amount of plutonism, and the presence of discontinuous metamorphic terranes (Armstrong, 1972). Penetrative deformation and metamorphism apparently occurred throughout much of Mesozoic and Cenozoic times and locally affected a large part of the thick Paleozoic miogeoclinal section of the region. Complex Cenozoic structures generally obscure earlier formed structures; this has resulted in many different interpretations of the Mesozoic tectonic history. Numerous low-angle faults in the hinterland were active during the Mesozoic and Cenozoic (Armstrong, 1972), and movement on some faults recurred as a result of reactivation in changing tectonic settings. The field data that permit a distinction between low-angle faults of regional extent that resulted from large-scale deformation of the crust and those of small extent that were caused by gravity sliding of rising fault blocks or domes are commonly lacking in the isolated mountain ranges of the Great Basin. In addition, the widespread occurrence of Tertiary radiometric dates for metamorphic rocks possibly obscures evidence for a Mesozoic metamorphic event. As a result, while it is evident that Tertiary metamorphism and deformation are

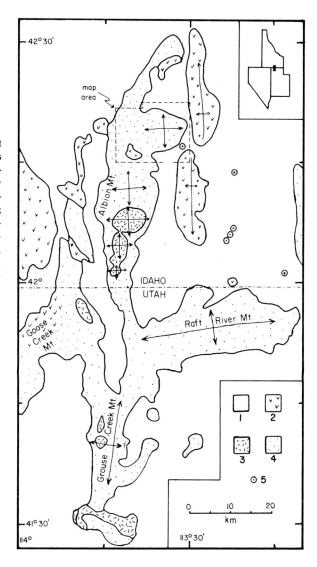

Figure 1. Geologic sketch map of the Raft River–Albion metamorphic terrane. Domes and arches are indicated. 1 = surficial deposits; 2 = Tertiary volcanic rocks; 3 = Tertiary granitic plutons; 4 = Paleozoic and Precambrian meta-morphic and sedimentary rocks; 5 = location of drill hole that penetrates metamorphic rocks. Adapted from R. L. Armstrong (unpub. maps); Compton (1972, 1975); Compton and others (1977); V. R. Todd (unpub. maps); and Williams and others (1975).

important in the metamorphic terranes, the extent to which Mesozoic ductile deformation contributed to the tectonic development of these areas is difficult to document.

The Cordilleran fold and thrust belt (Sevier belt of Armstrong, 1968a) lies east of the hinterland of the Great Basin. It is a north- to northeast-striking belt of folds and low-angle faults consisting of older-on-younger rocks; the structures indicate that east-directed overthrusting took place. Stratigraphic relations indicate that deformation in the belt ranged from Late Jurassic to Paleocene or Eocene (Armstrong and Oriel, 1965; Oriel and Armstrong, 1966), although deformation may have ended somewhat earlier in Utah and Nevada (Armstrong, 1968a, 1974).

High-angle faulting and mafic to felsic volcanism accompanied the formation of the basin-range structures in late Tertiary times. Generally north-trending grabens formed during the late Tertiary throughout much of the Great Basin. Structural downwarping along a northeasterly trend in southern Idaho during approximately the same time span created the eastern Snake River Plain, which is filled to a depth of several kilometres with young volcanic rocks that range in composition from rhyolite to basalt.

ROCK UNITS ON BIG BERTHA DOME

The rocks are divided into three groups: (1) ancient metamorphic-igneous basement complex, (2) unconformably overlying metasedimentary and sedimentary strata of Precambrian(?) and Paleozoic age, and (3) Cenozoic rocks. Most of the rock units were named and described by Armstrong (1968b) and Compton (1972, 1975).

The rocks of the basement complex (Green Creek complex of Armstrong, 1968b) have been assigned a Precambrian W (Archean) age on the basis of Rb-Sr isochrons (Armstrong and Hills, 1967; Compton and others, 1977). The rocks are predominantly weakly foliated quartz monzonite gneisses that contain potassium feldspar megacrysts. Various mica and amphibole schists are also present. Armstrong (1968b) inferred that the quartz monzonite had intruded the schists.

Metasedimentary rock units that unconformably overlie the Green Creek complex maintain a characteristic stratigraphic sequence in an area larger than 4,000 km^2, despite the generally high degree of deformation of the rocks. The sequence, about 4 km thick, is composed of quartzite (about 60%), with lesser amounts of schist and carbonate. The metamorphic rocks are overlain tectonically by Pennsylvanian carbonate and clastic rocks that are unmetamorphosed. Only one group of the metasedimentary rocks is definitely correlated with known sedimentary strata (Ordovician). Underlying and overlying the Ordovician units are structural groups of rocks that are possibly Paleozoic, Precambrian X, and Precambrian Z, respectively (Fig. 2); however, the ages of these rock units are in considerable doubt (Armstrong and Hills, 1967; Armstrong, 1968b, 1976; Compton, 1972, 1975; Compton and others, 1977; Compton and Todd, 1979; Crittenden, 1979; Crittenden and Sorensen, 1980). More complete lithologic descriptions and proposed correlations are presented elsewhere (Miller, 1978).

Cenozoic rocks in the map area include rhyolite ash flows that have yielded 8.5 and 9.2 m.y. ages (Armstrong, 1976; Williams and others, 1975) and Cenozoic glacial deposits, alluvium, colluvium, and landslide deposits.

METAMORPHISM

Metamorphism of the amphibolite facies is typical of the Raft River–Albion metamorphic terrane (Armstrong, 1968b; Compton and others, 1977), although greenschist-facies rocks are also present. In the Mount Harrison area, pelitic rocks show typical mineral assemblages of the quartz-albite-epidote-almandine subfacies of the greenschist facies and staurolite-almandine subfacies of the amphibolite facies (Winkler, 1967). The Pennsylvania Oquirrh Group locally contains metamorphic or hydrothermally produced white mica in otherwise unmetamorphosed rocks.

Highly deformed schists in one area in sheet I contain retrograded staurolite porphyroblasts and unaffected garnets, which indicates that some deformation took place after the temperatures decreased.

The metamorphic rocks are generally quartz rich, and thus occurrences of mineral assemblages that are diagnostic of metamorphic facies are rare. As a result, the positions of isograds are imprecisely defined. All known mineral assemblages of amphibolite facies (chiefly staurolite-bearing pelitic rocks) occur in the western part of the map area; this area also contains the structurally highest rocks, suggesting either that there is an east to west increase in metamorphic grade, or that the structurally higher rocks were metamorphosed to higher grade. The data from this study permit either hypothesis.

The age of metamorphism in the Raft River–Albion metamorphic terrane is not clearly determined, and my results from the northern part appear somewhat discordant with those of Compton and others (1977) from the areas to the south. This subject is dealt with later in this paper.

STRUCTURAL ANALYSIS

The structural history of the study area was investigated by field mapping and measurement of structural features such as foliations, fold axes, and lineations; stereographic plotting of the field measurements; and the laboratory examination of collected rock specimens. Field mapping at 1:24,000 clarified the distribution of the rock units that were mapped by R. L. Armstrong (1977, unpub. data). The orientations and geometries of small-scale structures were recorded, and the relations between sets with distinct geometries were used to establish the sequence of development. Field mapping at 1:8,000 and structural measurements were carried out in two structurally complex areas in order to identify the sequence of superposition of minor structures. The structural data recorded during field studies were divided, on the basis of their geometric characteristics and their superposition, into four phases of penetrative deformation.

The dominant structural feature of the map area (Fig. 2) is Big Bertha dome. The culmination of the dome is near the center of the map as seen by the quaquaversal distribution of dips. A thick stack of metamorphic rocks on the western flank of the dome forms a nearly homoclinal, westward-dipping sequence. Only the lowermost stratigraphic units, the Elba Quartzite and schist of Upper Narrows, are exposed over much of the eastern part of the dome.

Low-angle faults are important in the area. The four major low-angle faults are shown in Figure 2; they define four major allochthonous sheets.

Penetrative lineations and minor folds are present in all the metamorphic rocks and are especially abundant in the schists and carbonate rocks of sheet I. They record a complex history of four separate deformational phases, the structures produced by the first two phases being most extensively developed Despite abundant evidence from the minor structures that large deformations occurred in the map area, the major lithologic units are generally unfolded on the map scale and on the scale of the complex as a whole; if the effect of doming and arching is removed, the units are nearly horizontal (Compton and others, 1977).

High-angle faults are common in the map area, but most have small separations and are not of major importance. They cataclastically deform the metamorphic rocks and therefore postdate metamorphism and ductile deformation. The high-angle faults are described in Miller (1978) and will not be considered further in this paper.

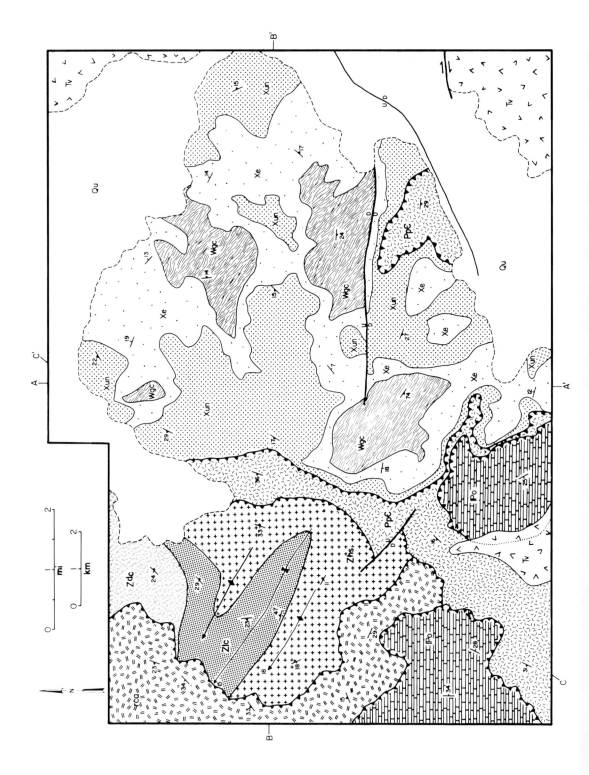

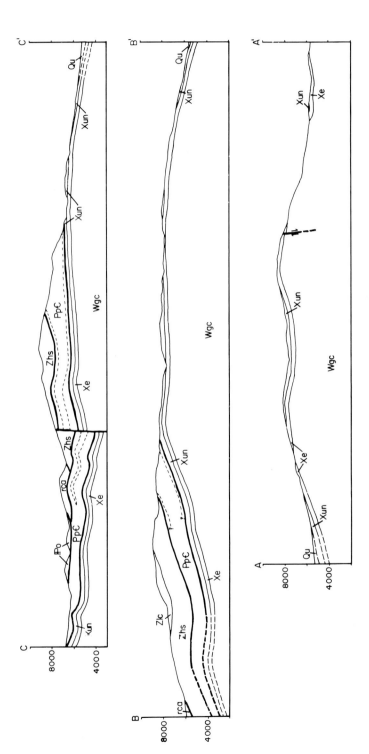

Figure 2 (facing pages). Generalized geologic map and cross-sections. Qu = surficial deposits; Tv = rhyolite flows; Po = Oquirrh Group; rca = Robinson Creek Assemblage; Zhs = Harrison Summit Quartzite; Zlc = Land Creek Formation; Zdc = Dayley Creek Quartzite; PpЄ = Paleozoic and Precambrian(?) metasedimentary rocks; Xun = schist of Upper Narrows; Xe = Elba Quartzite; Wgc = Green Creek complex. Sheet I = PpЄ; sheet II = Zdc, Zlc, and Zhs; sheet III = rca; sheet IV = Po. Dashed line in sheet I in cross-sections represents the base of Ordovician strata.

Low-Angle Faults

The low-angle faults are identified largely by stratigraphic criteria, including (1) truncation, (2) repetition, and (3) elimination of stratigraphic units. Slices of exotic rock types along some boundaries between lithologic units provide useful criteria for recognizing some low-angle faults. Metamorphic discontinuities and tectonic breccias are also useful as criteria for recognizing some faults.

With the exception of fault IV, at the base of the Oquirrh Formation, the low-angle faults shown in Figure 2 have the following characteristics: (1) The faults are subparallel to bedding and to the major foliation. (2) They carry younger rocks over older. (3) No discontinuities in metamorphic grade or in the orientations of minor structures are identifiable across the faults. (4) Mineral assemblages and metamorphic fabrics near and within the fault zones are similar to those observed throughout the map area. (5) Most of the fault boundaries are sharp lithologic breaks, but a few apparently contain intersliced rock units that may resemble metamorphosed interbedded sedimentary rocks.

Four major faults underlie and define allochthonous sheets I to IV (Fig. 2). Fault I is located at the top of the schist of Upper Narrows, and from east to west along its exposure it brings successively higher stratigraphic units in sheet I over the autochthon. A minimum separation of several tens of kilometres along fault I is indicated, but the total amount and direction of displacement are not determinable. Fault II juxtaposes rocks of sheet I with those of sheets II and III. Rocks within sheet II are overturned, as indicated by cross-bedding (Armstrong, 1970), and they probably represent the overturned limb of a major recumbent fold. Schists immediately below one portion of fault II are retrograded as described above, apparently as a result of local fault movement. Fault III juxtaposes the Robinson Creek Assemblage (sheet III) with various rock units of sheet II. The rocks in sheet III resemble those of sheet II, but cross-bedding shows them to be mostly right way up. The similarity of the lithologies in the two sheets makes it very difficult to map the trace of fault III in many areas, since there are no cataclastic textures or distinctive fabrics developed in the fault zone. Fault III may represent minor dislocation along the axial surface of a large recumbent fold whose upper limb is sheet III and whose overturned limb is sheet II. Fault IV lies at the base of the two klippes of Oquirrh Group. The Oquirrh rocks are unmetamorphosed to slightly metamorphosed and lie on rocks with staurolite-garnet assemblages. Bedding in sheet IV is truncated by the fault. A zone of brecciation at the base of the sheet is succeeded upward by pervasive fracturing, and rocks immediately below sheet IV are moderately fractured in the few places where they are exposed. Sheet IV lies on rocks of sheets I and III and truncates stratigraphic and structural boundaries in the metamorphic rocks. Thus, it is evident that sheet IV was emplaced after the close of metamorphism.

Subsidiary low-angle faults are mapped within the autochthon and sheets I, II, and III. Most of these minor faults cut out thin stratigraphic units. Low-angle faults are abundant in sheet I and result in considerable thinning of the stratigraphic section that was described by Compton (1972, 1975) elsewhere in the Raft River–Albion metamorphic terrane. The thick section of Silurian and Devonian carbonates typical of the eastern Great Basin is missing between the Ordovician rocks and rocks of probable Mississippian age. This carbonate section is observed in the southern extremity of the Raft River–Albion metamorphic terrane where the metamorphic grade is low (R. R. Compton, 1977, oral commun.); this indicates that the rocks are not missing due to a major unconformity in the region. The tectonic elimination of these rocks requires large displacement on one or more faults within sheet I, perhaps associated with drastic ductile thinning of the sequence.

One low-angle fault within sheet I brings rocks of the Ordovician Pogonip Group over younger rock units. The stratigraphy of the Pogonip Group is the reverse of that observed elsewhere in the Raft River–Albion metamorphic terrane and indicates that the stratigrapic sequence was overturned. Moreover, the overturned sequence is succeeded upward by a thin slice of Cambrian(?) rocks and by sheet II rocks which are also overturned. Thus, the whole of the overturned sequence may represent the overturned limb of a major recumbent fold including the major low-angle fault (fault II) that separates Precambrian(?) from Cambrian(?) and Ordovician rocks.

Ductile Deformation

After allowing for the effects of late doming, bedding is roughly horizontal and a penetrative foliation (S_1) formed subparallel to the bedding (S_0) during the first phase of deformation (D_1). Figure 3 shows contoured plots of the orientations of these structures. S_1 is defined by the preferred orientation of platy minerals and lies close to the axial planes of first-phase folds (F_1). First-phase fold axes and the parallel mineral elongation lineations (L_1) trend northeast and plunge gently to the northeast and southwest. They are overturned to recumbent, isoclinal or tight folds (Fig. 4) and are systematically overturned to the northwest.

First-phase structures are rarely observed in the autochthon, particularly in the Elba Quartzite and the Green Creek complex. Several occurrences of these structures are known in the rocks of sheet I, but none were found in rocks of sheets II and III. However, all the metamorphic rocks contain a foliation that is folded by D_2 folds; it is thus inferred to be S_1. No major D_1 fold closures have been mapped in the area, but the thick, overturned rock sequences such as sheet II suggest that large-scale recumbent folds were present.

Second-phase (D_2) structures are superposed on D_1 structures, and at least one low-angle fault is folded by D_2 folds. D_2 structures are pervasive in all the metamorphic rocks with the exception of the gneisses of the Green Creek complex, in which D_2 lineations (L_2) are poorly defined. S_2 is best developed in carbonates and schists and is due to parallel micas; it dips steeply to the southwest and northeast (Fig. 5a). It should be noted that many of the measurements of S_2 represented by Figure 5a were taken in the wildly folded marbles of the Pogonip Group; Figure 5a is therefore not representative of the average orientation of F_2 axial planes. S_2 is parallel to the axial planes of second-phase folds (F_2) whose axes trend northwest and plunge at low angles. F_2 folds occur as two distinct geometric types but are consistently oriented and are accompanied by a pervasive mineral lineation (grain elongation) that is parallel to F_2 axes. Most of the folds occur in sheet I, where they are overturned to the northeast and are tight to moderately tight (Fig. 6). The shorter limbs of asymmetric folds in quartzose schists are commonly thickened. Many of the folds are disharmonic, and folds in thick beds have a longer "wavelength" than those in thinner beds (Fig. 6b). F_2 folds in much of the autochthon and sheet II are upright to inclined and have very small amplitudes relative to wavelength. The upright F_2 folds are observed mainly in thick quartzite units and, to a lesser degree, in quartzose schists, suggesting that the geometry of F_2 folds was strongly influenced by differences in ductility.

Biotite, garnet, and staurolite porphyroblasts oriented parallel to S_2 contain internal foliations consisting of parallel chains of inclusions that are oriented differently from S_2 and possibly represent S_1. Some porphyroblasts also show rims in which the inclusions are oriented parallel to the surrounding S_2, indicating that at least the late growth of the minerals took place during the development of S_2. Earlier growth may possibly have occurred during development of S_1 foliations.

Major folds trending N60°W and plunging shallowly to the northwest are mapped in sheet II. These folds have a wavelength of approximately 1.5 km and a 300-m crest to trough amplitude, and their axial planes dip steeply to the northeast. Great circle distributions of S_0 and S_1 about this axis are observed in rocks of sheet II and in the southwestern part of the area of sheet I also, indicating large, northwest-trending folds in these areas. Large folds in the Connor Creek area (southeastern flank of the dome) are complex but also appear to have a northwest trend. Intermediate-scale, tight, overturned F_2 structures in this area are presumed to be larger analogs of the minor F_2 structures, which are strongly overturned to the northeast and are nearly similar in form. The minor F_2 folds of the area are rotated about the major N60°W fold; therefore, the broad folds of F_2 trend developed after the minor,

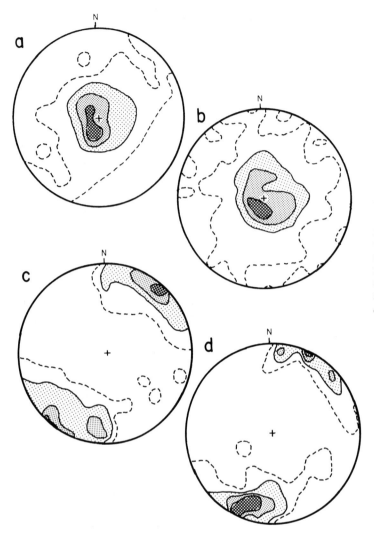

Figure 3. Structural elements of D_1, plotted on lower hemisphere equal-area projections. Contouring done by the Schmidt method. (a) 467 poles to bedding (S_0); contours 2%, 6%, 10% per 1% area. (b) 566 poles to foliation (S_1); contours 2%, 5%, 7% per 1% area. (c) 208 lineations; contours 2%, 5%, 10% per 1% area. (d) 69 fold axes; contours 3%, 6%, 8% per 1% area. Dashed line outlines the pole-free area.

overturned F_2 folds. Because of their trend approximately parallel to minor F_2 folds, their superposition by F_3 folds, and their similar geometry to some minor F_2 folds in quartzite units, the broad, N60°W-trending folds are considered to represent a late part of the second phase of deformation.

Third-phase (D_3) structures are not as widely distributed as those of the two earlier phases. They have been mapped in four areas, and in several other areas are represented only by "fracture" cleavage. The areas of well-developed structures are (1) the northeastern flank of the dome (Elba Quartzite), (2) the eastern flank (Elba Quartzite), (3) the western flank (the upper schist member of the Land Creek Formation), and (4) the Connor Creek area on the southeastern flank of the dome (sheet I).

The structures included in D_3 on the eastern flank of the dome are mullionlike structures and rare, tight folds of several metres wavelength. The "mullion" structures trend approximately north, are ~0.5 m in amplitude, and appear similar to mullions in bedding-plane views. Where observed in cross-section, the "mullions" are seen to be incipient folds or thrusts (Fig. 7a). Vergence in these structures is basinward. Larger folds (F_3) in the same area are also overturned basinward and are commonly related to small, bedding-plane faults (Fig. 7b). The trend of the F_3 structures swings from north on the

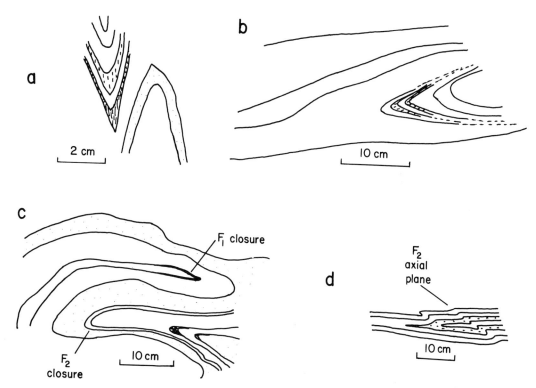

Figure 4. Diagrams of F_1 folds, drawn from collected specimens and photographs. (a) Tight fold in quartzose schist. (b) Tight fold in thin-layered quartzose schist enclosed by thicker quartzite layers. (c) Refolded isoclinal F_1 in layered marble. (d) Refolded F_1 in quartzite.

east flank of the dome to north-northeast in outcrops on the southeastern flank. Probable F_3 folds on the northeastern flank of the dome are identical to the "mullion" structures previously described. Because their axes are parallel to L_2 and F_2 axes and are not observed to be superposed on D_2 structures, they are not definitely established as D_3 structures. Their striking similarity in style to D_3 structures on the eastern flank of the dome is considered to be sufficient evidence that they are D_3 structures. They also verge basinward (to the northeast).

North-trending folds of outcrop scale in the Land Creek Formation on the western flank are observed to be superposed on D_2 structures. The folds are overturned to the west and are nearly similar (Fig. 7c). A crenulation cleavage (S_3) is parallel to axial planes of many of the folds.

Folds that postdate structures of D_2 origin are developed extensively in a quartzite unit in sheet I in the Connor Creek area. These F_3 folds trend S35°W and plunge 20° to 30° to the southwest and are consistently overturned to the southeast. The fold geometry varies consistently with height above the base of the quartzite, which is the locus of a minor thrust fault that truncates the internal stratigraphy of the quartzite unit. Folds are relatively open and overturned, with moderate thickening of the shorter, overturned limb (Fig. 8a, 8b, 8c) in the upper part of the sequence. The folds are progressively tighter downward in the section (Fig. 8b, 8c, 8d, 8e), until they are nearly isoclinal (Fig. 8f) ~10 m above the base. The lowest 10 m consist of platy quartzite with rootless isoclinal folds. Since the folds in the quartzite show progressive development and flattening downward to the basal fault, I consider that they were formed along with movement on the fault during the third phase of deformation (D_3).

The orientations of F_3 folds are shown in Figure 9. The significant feature of the F_3 folds is their consistent vergence away from the apex of the present dome, which indicates that they were probably

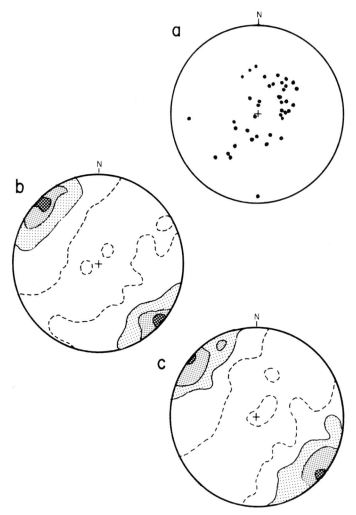

Figure 5. Orientations of D_2 structures. (a) 40 poles to foliation (S_2). (b) 650 lineations (L_2); contours 2%, 5%, 8% per 1% area. (c) 213 fold axes (F_2); contours 2%, 5%, 10% per 1% area.

formed in association with the dome. The growth of metamorphic muscovite during the development of D_3 structures indicates that the rocks near and within the autochthon were still at relatively high temperature.

Kink folds (F_4) that are sporadically distributed throughout the map area are assigned to a fourth phase of deformation on the basis of their superposition upon all earlier structures. The fold attitudes form a single maximum with plunge 10° to N15°E. F_4 folds have amplitudes of about 1 cm and are repeated at 10- to 15-cm intervals in foliation surfaces. F_4 axial planes are nearly vertical, and their sense of vergence is not systematic. These kinks probably represent a late, minor episode of deformation, or possibly several such episodes.

Big Bertha Dome

The geometry of Big Bertha dome is represented in Figure 10 as a structure contour map which was constructed using the top of the Green Creek complex (base of the Elba Quartzite) as a reference surface. The curvature of the dome top is gentle over a considerable area, whereas the flanks of the dome are relatively steeply dipping (25° to 35°). The mean diameter of the dome is approximately 15

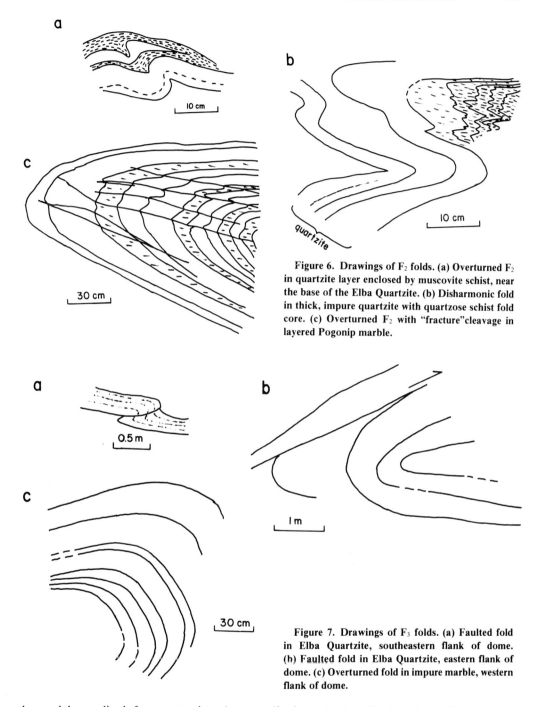

Figure 6. Drawings of F_2 folds. (a) Overturned F_2 in quartzite layer enclosed by muscovite schist, near the base of the Elba Quartzite. (b) Disharmonic fold in thick, impure quartzite with quartzose schist fold core. (c) Overturned F_2 with "fracture" cleavage in layered Pogonip marble.

Figure 7. Drawings of F_3 folds. (a) Faulted fold in Elba Quartzite, southeastern flank of dome. (b) Faulted fold in Elba Quartzite, eastern flank of dome. (c) Overturned fold in impure marble, western flank of dome.

km, and the amplitude from crest to interdome syncline is nearly 1 km. The dome is roughly triangular in shape.

Bedding and D_1 and D_2 structures are warped over the dome, whereas D_3 structures are symmetrically related to the dome; therefore, doming took place following the D_2 phase of deformation and was associated with the D_3 phase of deformation.

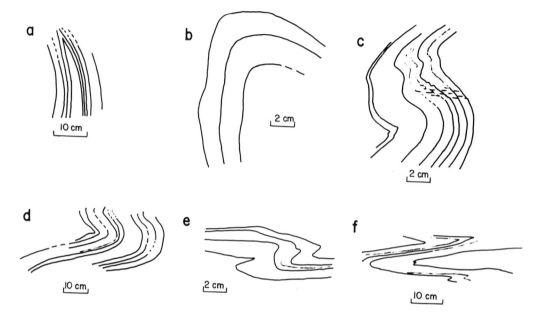

Figure 8. Drawings of F_3 folds in quartzite in sheet I, Connor Creek area. All the folds are viewed down the plunge of the fold axis (to the southwest) except fold (e) which is viewed to the northeast, parallel to its axis. Sequence of fold styles illustrates the progressive tightening of folds downward toward the basal thrust fault. Note the gently refolded isoclinal F_1 fold in (a).

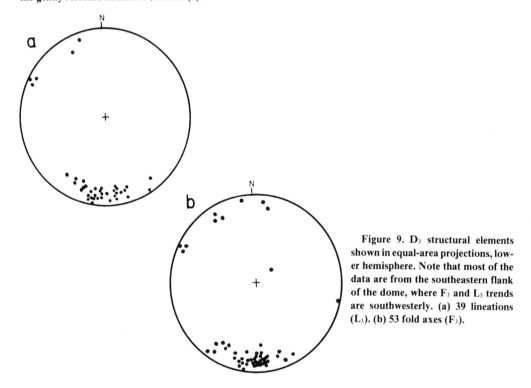

Figure 9. D_3 structural elements shown in equal-area projections, lower hemisphere. Note that most of the data are from the southeastern flank of the dome, where F_3 and L_3 trends are southwesterly. (a) 39 lineations (L_3). (b) 53 fold axes (F_3).

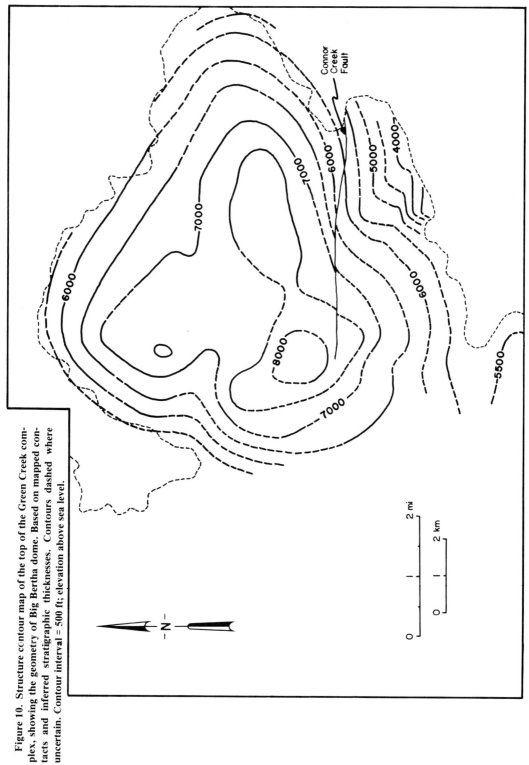

Figure 10. Structure contour map of the top of the Green Creek complex, showing the geometry of Big Bertha dome. Based on mapped contacts and inferred stratigraphic thicknesses. Contours dashed where uncertain. Contour interval = 500 ft; elevation above sea level.

INTERPRETATION

Low-Angle Faults

Low-angle faulting occurred after, during, and probably before the D_1 and D_2 phases of ductile deformation.

The Oquirrh klippes were emplaced in a brittle fashion following high-grade metamorphism, as indicated by the metamorphic discontinuity and the extensive brecciation associated with the base of the klippes. This brittle deformation may have been caused by late, minor movement, such as sliding off the rising Big Bertha dome, that reactivated an earlier formed fault. That the last movement on this fault postdated most of the metamorphism and ductile deformtion (D_1 and D_2) is clear, but its relation to phases D_3 and D_4 has not been determined; the faulting may have been contemporaneous with the latest phases of ductile deformation that took place at a deeper level.

The minor low-angle, bedding-plane faults associated with F_3 folds and retrograde metamorphism that was related to low-angle faulting indicate that at least minor low-angle faulting took place during the late ductile deformational phases at lower metamorphic temperatures.

Most of the rocks from the low-angle fault zones, however, show no textural evidence of extreme deformation. The faults are folded by minor F_2 folds, and thus movement on them predated at least part of the D_2 deformation. Sheet II probably represents part of the overturned limb of a major recumbent fold that was subsequently broken by faulting. The shape of the major fold thus inferred (isoclinal and recumbent) is most similar to the F_1 folds, and some faulting therefore postdated at least part of the D_1 deformation. Structures resulting from fault displacement in rocks at high temperatures are expected to be found throughout a thick zone adjacent to the fault because the rocks are weak, whereas brittle deformation on low-angle faults commonly is confined to a relatively narrow zone. Those rocks in which sedimentary structures are preserved show no evidence of such distributed deformation adjacent to the fault surface, and much of the deformation on the low-angle faults is therefore considered to have predated the D_1 ductile deformation. Less-pronounced faulting on some or all of the faults probably continued during the early phases of ductile deformation as plastic deformation and recrystallization became more important.

Ductile flow alone probably did not produce the discontinuities mapped as low-angle faults. Although the carbonate and schist units may have been considerably more ductile than the intervening quartzite, and the difference in the amount of flow may have resulted in drastic thinning of the weaker rock units, complete elimination of rock units in the large areas observed is unlikely. I believe that earlier low-angle faults were required, but that these fault movements predated most of the metamorphism, so that any cataclastic textures along the fault planes were obliterated by recrystallization.

Most low-angle faults are of younger-on-older geometry, which is most easily explained as a result of "extensional" faulting (extension occurred parallel to the fault plane). Low-angle faults that form as a result of compression parallel to the fault plane (thrusts) are expected to produce younger-on-older geometries only in certain special situations, such as faulting of (1) overturned folds, (2) earlier faults, and (3) unconformities (Armstrong, 1972). The fact that most of the faults in the map area are younger-on-older indicates that the structural relations observed are not special cases of "compressional" faulting, but are due to extension parallel to the fault plane. The two older-on-younger faults mapped can be explained as faulted folds and thus are compatible with "extensional" faulting.

The directions and magnitudes of displacement along the low-angle faults are poorly understood. Evidence for several tens of kilometres of movement of sheet I relative to the autochthon was described above. The faults within sheet I and bounding sheet II appear to be related to and disrupt major D_1 folds so that the northeastward sense of overturning of F_1 folds suggests a similar sense of displacement for at least part of the movement on these faults. Compton and others (1977) cited evidence for an eastward movement of their equivalent of sheet IV, but no such evidence exists in the area mapped in this study.

Ductile Deformation

Information regarding the boundary conditions for ductile deformation of the rocks, the state of strain, and the shape and orientation of minor structures can be combined to indicate some aspects of the type of deformation that the rocks have undergone. The strain varies considerably within individual folds and from one rock unit to another, but these small-scale differences are ignored in the following discussion in which the generalized strains on the scale of the map area are summarized.

The first two phases of deformation occurred pervasively throughout the map area, whereas the last two phases occurred locally; therefore, the first two phases will be discussed separately. Minor structures associated with the first two phases of deformation are abundant in carbonate- and mica-rich strata and are less commonly observed in the quartz-rich rocks, suggesting that the strata had different mechanical properties during deformation. Pebbles in quartzite beds of sheet I are strongly deformed, whereas pebbles in thick quartzite units are much less deformed (Harrison Summit Quartzite) or moderately deformed (Elba Quartzite). In rocks that contain qualitative strain markers such as distorted sedimentary structures and mineral grain shapes, the strain deduced from the markers corresponds directly to the abundance of minor structures. It is thus concluded that several of the strata on Mount Harrison were more extensively deformed than were other strata during the D_1 and D_2 phases of deformation.

Gneisses of the Green Creek complex exhibit a pronounced upward increase in the abundance of minor structures within 50 m of the base of the Elba Quartzite. The lack of pronounced foliations and lineations in the deeper exposures of the gneisses suggests a downward decrease in deformation. The overlying autochthonous strata (Elba Quartzite and schist of Upper Narrows) are moderately deformed. Sheet I rocks are apparently extensively deformed, and sheets II and III are less deformed.

If this interpretation of the deformation experienced by the various rock units is correct, the D_1 and D_2 deformations were strongly controlled by the anisotropic mechanical properties of the strata. The basement apparently behaved much more stiffly than the overlying rocks and thus provided a lower boundary for the overlying deforming mass. Some thick stratigraphic units, particularly the Precambrian Z(?) quartzites, behaved as nearly horizontal, relatively stiff sheets. The extensive deformation observed in certain of the intervening strata may have resulted from the locally increased strain constrained by these stiff layers.

Strain was studied in the Elba Quartzite by measuring distorted pebble shapes and is the subject of a paper in preparation. The maximum extensions measured are parallel to the dominant mineral grain lineations and minor fold axes, and the maximum shortening directions are normal to the foliations. If this relationship between the strain ellipsoid and minor structures is valid for other rock units, we find that the direction of maximum shortening is approximately vertical throughout the map area and that the direction of maximum extension varies, but is generally northwest-southeast oriented, parallel to the dominant L_2 lineation. The L_1 mineral grain lineation is parallel to F_1 fold axes and also may have been a direction of maximum extension during D_1.

The sense of vergence of F_1 and F_2 folds probably indicates the sense of shear of the stiffer strata during the respective deformational phases, when considered in the framework of the layered, mechanically anisotropic model proposed above. Thus, the northwesterly vergence of F_1 folds suggests that overlying strata moved northwest with respect to the basement during D_1, and the northeasterly F_2 fold vergence suggests that the strata moved northeast relative to the basement during D_2.

Deformation during the D_1 and D_2 phases was probably much more complex than shear parallel to stratigraphic boundaries suggested above. Vertical shortening was probably great because the stratigraphic section is drastically thinned, and large parts are missing entirely. I consider that low-angle faulting is necessary to account for much of the tectonic removal of the stratigraphic section, but that ductile thinning during the D_1 and D_2 phases of deformation was also important. The directions of maximum extension indicated by mineral lineations were perpendicular to the directions of shear indicated by fold vergence during the early phases of deformation. This relationship requires complex deformation during the fold development or complex superposition of different deformations. The details of the history of the early phases of deformation are obscure because the data are insufficient.

The type of strain inferred for the early phases of deformation suggests that a model similar to Kehle's (1970) is appropriate, where different shear rates are developed in layers of contrasting mechanical properties, the weaker layers undergoing more shear than the stiffer ones. In addition, vertical shortening and horizontal extension are probably necessary in a model that is applicable to the northern Albion Mountains. The parallelism of the shear plane to the stratigraphic horizons in this model may not result directly from the orientation of the stress field but from the constraints placed on the flow by the mechanical differences between rock units.

Intense deformation during the D_1 phase was primarily localized in sheet I rocks, while during the D_2 phase all of the metamorphic rocks were penetratively deformed. Thus, at least part of the D_1 phase of deformation occurred when the contrast in mechanical properties between the rock units was greater than during the D_2 phase, and the rocks were therefore probably cooler during part of the D_1 phase than during the D_2 phase of deformation. I consider it probable that the D_1 deformation included most of the plastic deformation that succeeded low-angle faulting and preceded the onset of the relatively high-temperature deformation that characterized the D_2 phase.

Large, low-amplitude, upright folds whose axes are parallel to the minor F_2 folds are interpreted as resulting from deformation late in the D_2 phase. The geometry of these folds requires at least a small component of layer-parallel shortening.

Structures of the later phases of deformation, D_3 and D_4, are distinct from the earlier sets of structures (D_1 and D_2) in that they are locally developed and their structural styles suggest that the mechanical properties of the rocks were different. D_3 structures include many folds with associated bedding-plane faults of small extent, which suggests that the rocks behaved in a more brittle fashion. Kink folds of D_4 require pronounced planar heterogeneities in the mechanical properties for slip to take place along discrete layers in the rock. Both sorts of deformation suggest that the rocks were stronger and mechanically more heterogeneous, and therefore probably colder during D_3 and D_4.

Structures assigned to the D_3 phase include (1) folds whose axes trend approximately circumferentially with respect to Big Bertha dome, and (2) bedding-plane faults along which higher rocks are displaced basinward relative to lower rocks. The F_3 folds are overturned toward the basin, which is compatible with the sense of displacement on the bedding-plane faults. The geometry of the D_3 folds and faults indicates that flow during D_3 was away from the apex of the present dome and suggests that D_3 is a result of doming. The fact that broad, open folds whose axes are parallel to that of the interdome syncline occur within the syncline suggests that doming (D_3) took place soon after or in the late part of the D_2 phase. Lobes of an Oligocene intrusive body in the centers of two or three dome structures farther south in the Albion Mountains (R. L. Armstrong, 1977, unpub. map) suggest that doming is the result of intrusion. These relations suggest that a similar but somewhat deeper intrusive body may be present under Big Bertha dome and may have caused its uplift.

F_4 folds are minor kink folds whose axes generally trend N15°E and plunge shallowly. A component of compressive stress must have been present parallel to bedding and normal to the F_4 trend during the development of these structures. D_4 structures are inferred to have formed significantly later than D_1 and D_2 structures, and they probably formed after doming (D_3) as well, since the F_4 fold axes were apparently not reoriented by being warped over the dome.

Structural History

The earliest recognizable deformation was low-angle faulting of younger-over-older type. This is inferred to have resulted in horizontal extension and vertical shortening. The total separation along these faults is unknown, but probably exceeded several tens of kilometres.

Temperatures in the rocks rose and resulted in metamorphism and initiated penetrative plastic deformation (D_1). This deformation probably produced a few large, isoclinal, recumbent folds, and faulting caused minor displacement along the axial planes of the large folds. Ductile deformation was concentrated largely in the weaker rock units and resulted in extensive development of minor structures in these units while the intervening stiffer units remained little deformed. Deformation was constrained by the boundaries imposed by these stiffer rock units, so that most of the strain occurred as shear along planes parallel to stratigraphic boundaries. The strata sheared northwest relative to the basement during D_1 and to the northeast during D_2. Deformation was the most intense during D_2, and it occurred at peak metamorphic temperatures; as a result, all of the metasedimentary rocks and the uppermost part of the Green Creek complex were deformed. Temperatures decreased and large gentle folds formed parallel to the earlier, overturned F_2 folds. Low-angle faulting probably became important in the higher rocks during this period of cooling. This faulting juxtaposed little-metamorphosed sedimentary rocks with underlying metamorphic rocks.

Doming (D_3) during the end of D_2 deformation resulted in locally intense folding and low-angle faulting as movement of rocks occurred outward from the apex of the dome. D_3 was probably related to the intrusion of felsic plutons at depth into the Green Creek complex.

The last ductile deformation was manifested as kinking that was caused by approximately east-west, horizontally directed compression.

DISCUSSION

Age of Deformation and Metamorphism

Misch and Hazzard (1962) outlined the evidence for metamorphism and deformation of late Precambrian and early Paleozoic strata in the Schell Creek Mountains, and they argued that the conformable sequence of Paleozoic through Lower Triassic strata in the area precluded an age older than Early Triassic for a regional orogenic event. In a few areas of eastern Nevada, undeformed Cretaceous sedimentary rocks unconformably overlie deformed Paleozoic rocks (Armstrong and Hansen, 1966), which suggests that regional deformation had ceased by Cretaceous time. Although it is clear that local basins developed during this period, deep deformation and metamorphism may have continued considerably longer.

Postkinematic plutons that have been dated radiometrically are distributed throughout the region. K-Ar ages of these plutons summarized by Armstrong and Hansen (1966) and Armstrong (1972) indicate that metamorphism and deformation ended in several areas of eastern Nevada and western Utah by about Early Cretaceous time (109 to 150 m.y. ago). However, some deformed plutons in the Grouse Creek Mountains have Tertiary Rb-Sr ages (Compton and others, 1977); thus, the end of deformation and metamorphism is not synchronous over the region.

Metamorphic minerals and rocks in the eastern Great Basin yield radiometrically determined dates ranging from 259 to 13 m.y. (for instance, Armstrong and Hansen, 1966; Mauger and others, 1968; Lee and others, 1970; Snoke, 1975; and Armstrong, 1976). The predominance of early Tertiary K-Ar ages, which are about 100 m.y. younger than the age of metamorphism and deformation based on geologic arguments, is interpreted as indicating that the rocks had a protracted high-temperature history and did not finally cool until just before or even during basin-range tectonism (Armstrong and Hansen, 1966; Mauger and others, 1968).

Radiometric dates from the Raft River–Albion metamorphic terrane exhibit many of the features described above. Jurassic K-Ar ages (157 and 162 m.y.) obtained from hornblende schist from the Albion Mountains (Armstrong and Hills, 1967; Armstrong 1977, written commun.) suggest strongly that, at least locally, crystallization began in the Jurassic. Because the protoliths of these rocks may be Precambrian, it is possible that the K-Ar ages are older than the age of metamorphism. Seven Late Cretaceous K-Ar ages ranging from 66 to 81 m.y. and an Early Cretaceous K-Ar age (130 m.y.) are also reported by Armstrong from schists and gneisses in the Albion Mountains. Armstrong (1976) considered that Oligocene Rb-Sr ages from the Middle Mountain injection complex (10 km south of Oakley, Fig. 1) indicated that high-temperature conditions persisted until the middle Tertiary. It thus appears that metamorphic conditions were present in some rocks in the Albion–Middle Mountains area for 150 to 200 m.y.

The styles of structures formed during the two major deformational phases (D_1 and D_2) in the Big Bertha dome area appear to correlate well with structures from similar events in the Raft River and Grouse Creek Mountains, where their timing has been partly resolved by radiometric methods (Compton and others, 1977). The Almo pluton (Albion Mountains), Red Butte stocks, and Immigrant Pass intrusion (Grouse Creek Mountains) are not strongly deformed, except locally, and thus probably postdate much of the deformation; their Rb-Sr ages are 28, 25, and 38 m.y., respectively (Armstrong, 1976; Compton and others, 1977). The border zone of a stock from Red Butte Canyon contains a prominent lineation that is parallel to the F_2 and L_2 structures of the area; on this basis, intrusion and crystallization of the stock 25 m.y. ago were considered by Compton and others (1977) to be contemporaneous with the D_2 deformation in the region. They showed, furthermore, that low-angle faulting took place following the intrusion of this stock, since the upper fault slice in the area showed no evidence of "contact" metamorphism. In contrast, the Immigrant Pass intrusion has an older Rb-Sr age, but it is only slightly deformed. The intrusion cuts folds interpreted by Compton and others (1977) as F_1, and some postintrusion, low-angle faulting took place there as well. It may be concluded that the D_1 deformation occurred before 38 m.y. ago and that at least part of the D_2 deformation and some low-angle faulting persisted until the Oligocene. Armstrong and Hills (1967) obtained 70- to 80-m.y. K-Ar ages on metamorphic minerals developed in association with the early intense D_2 phase, thus dating F_2, at least in part, as somewhat older than Late Cretaceous. I consider that the local nature of the distribution of intensely developed L_2 lineations in the border of the Red Butte stock is consistent with its development in the later part of the D_2 phase, during which moderate temperatures persisted and slight deformation occurred regionally.

Much low-angle faulting is inferred to have predated metamorphism, implying a Late Triassic or Early Jurassic age for the faulting. Some low-angle faulting (sheet IV) occurred after the metamorphic assemblages developed, and therefore is probably Cenozoic in age.

Comparison with the Remainder of the Raft River–Albion Metamorphic Terrane

A general description of the geology of the Albion Mountains by Armstrong (1968b) and a review of the geology of the Raft River and Grouse Creek Mountains area by Compton and others (1977) form the main basis for comparison of the results of the present study with the remainder of the metamorphic terrane.

The rock units defined by Armstrong (1968b) and Compton (1972), with the exception of quartzite sequences of the Middle Mountain and northern Albion Mountains areas, are present throughout most of the terrane. The rocks on Big Bertha dome that are assigned a Precambrian Z age in this report are apparently absent elsewhere in the terrane. The major rock units thicken and thin markedly in the terrane, and locally large sections are eliminated by low-angle faults.

A westward increase in metamorphic facies from greenschist (chlorite zone) to amphibolite (sillimanite–feldspar zone) is observed in the autochthon (Armstrong, 1968b; Compton and others, 1977). The sillimanite assemblages are associated with synkinematic and postkinematic plutons and possibly resulted from locally high temperatures in the vicinity of the intrusions. Compton and others (1977) demonstrated that lateral changes in metamorphic grade within the allochthonous rocks do not coincide with those in the underlying parts of the autochthon, and thus low-angle faulting postdated or accompanied the highest temperatures of metamorphism.

Felsic intrusive bodies range from slightly deformed (Almo pluton and Immigrant Pass intrusion) to moderately deformed (Red Butte stocks) to highly deformed migmatic complexes such as the Middle Mountain "Injection Complex" (Armstrong, 1976) and the Vipont Mountain intrusion (Compton and others, 1977), both of which lie in the westernmost exposures of the autochthon where highest metamorphic facies are attained. The plutons range in composition from granodiorite to quartz monzonite; the synkinematic intrusives contain two mica phases. Two-mica "granite" is also present in the center of the Almo pluton.

The sequence of development of minor and major structures documented in this study corresponds closely to that outlined for the Albion Mountains by Armstrong (1968b) and for the Raft River and Grouse Creek Mountains by Compton and others (1977). Minor and major folds of the D_1 and D_2 phases are similar in style throughout the terrane, but their orientation varies somewhat. F_1 folds trend from east-northeast to north and north-northwest and verge to the north and west, while F_2 folds trend from west to northwest and are overturned to the north and northeast. D_3 folds in the southern portion of the complex generally trend northeast and are overturned to the southeast, but varying styles and trends are present, indicating that the folds resulted from local movements, as is the case for D_3 folds in the Big Bertha dome area. Because the D_3 structures in the Raft River–Albion metamorphic terrane are apparently the result of many local deformations, they are probably not synchronous. Compton and others (1977) showed that in some areas D_3 folds are related to small imbricate "thrust" faults that moved hanging-wall rocks northeastward and eastward relative to the footwall. F_4 kink folds are also observed in the Utah area; their trend is similar to those described in this study, but they consistently verge eastward and are considered by Compton and others (1977) to be associated with eastward movement of low-angle fault sheets. If the F_4 folds in the terrane are synchronous, I consider it probable that they are a result of post-Oligocene unroofing of the area.

The major low-angle faults described by Compton and others (1977) differ from those of this study, but the differences may be caused by uncertainties in interpretation. Because the low-angle faults are exposed over a much smaller area in the northern Albion Mountains than in the Utah portion of the terrane, the designation in this study of some low-angle faults as "major" and others as "minor" is based upon insufficient evidence. For example, the faults at the base of sheets II and III may well have resulted from small lateral displacements. Unfortunately, exposures of the low-angle faults between the Big Bertha area and the Utah portion of the metamorphic terrane are sparse, and therefore direct correlation of the faults is not possible. I consider it probable that the faults at the base of the lower and middle allochthonous sheets of Compton and others (1977) are correlative with those at the base of sheet I and sheet IV, respectively, of this study.

Relationship to the Cordilleran Fold and Thrust Belt

The numerous exposures of metamorphosed Paleozoic rocks in the eastern Great Basin suggest that a broad belt of metamorphic rocks may be present at depth in the region (Misch and Hazzard, 1962;

Armstrong and Hansen, 1966). This belt is here referred to as the Cordilleran metamorphic belt, following Drewes's (1978) assignment of the metamorphic belt and the fold and thrust belt to the east to the Cordilleran orogeny. The areas of high-grade metamorphism have characteristics broadly similar to those of the Raft River–Albion metamorphic terrane: extensive low-angle faults; two pervasive sets of orthogonal, overturned folds, with one set of folds and its accompanying lineation generally very strongly developed; synkinematic and postkinematic felsic intrusions; and truncation of the metamorphic rocks by a major low-angle fault, above which lie late Paleozoic sedimentary rocks.

The Cordilleran fold and thrust belt in the Great Basin region consists mainly of eastward-directed overthrusts that displace Paleozoic and Mesozoic strata. The faulting transgressed eastward from Late Jurassic or Early Cretaceous to Eocene (Armstrong and Oriel, 1965; Armstrong, 1968a), and thus, in my interpretation, postdated much of the intense ductile deformation in the Cordilleran metamorphic belt.

Models that have been proposed for the relations between the Cordilleran metamorphic and fold and thrust belts fall in three categories: (1) gravity gliding, (2) gravity spreading, and (3) hinterland overthrusting. Gravity gliding (sliding of strata eastward off an uplifted hinterland) was proposed by Mudge (1970) and Hose and Danes (1973). Problems with the model, such as lack of evidence for substantial uplift in the hinterland, were discussed by Price (1971) and Armstrong (1972). Gravity spreading invokes uplift in the hinterland that results in lateral "spreading" under the influence of gravitational forces. Coupling of the spreading mass with the strata to the east in some complex fashion results in overthrusting to the east (Price and Mountjoy, 1970; Price, 1971, 1973). Roberts and Crittenden (1973) proposed a model that has characteristics of the gliding and spreading models. Their mechanism for "spreading" in the hinterland is by low-angle faulting (younger-on-older type); the fault sheets slide eastward downslope and create overthrusts. The gravity models assume that the fold and thrust belt resulted from movement of strata from the hinterland due to an eastward surface slope, and they differ in the nature of the tectonics invoked for the hinterland. Although uplift of the hinterland is implicit in the models, the cause of the uplift need not be horizontal shortening of the crust. Simple uplift and gravity gliding or spreading away from the uplifted area should result in a two-sided orogen, but most data indicate that structures of late Mesozoic age are eastward directed. Therefore, more complicated gravity models are necessary or alternatives to gravitational movement away from an uplift must be favored.

Hinterland overthrusting refers to the horizontal shortening of the crust by overthrusting of basement rocks in the hinterland (Armstrong, 1978; Brown, 1978). In this interpretation, west-dipping basement rock overthrusts beneath and immediately east of present exposures of the metamorphic complexes produced eastward translation of basement rocks and pushed strata eastward to form the fold and thrust belt.

The early Mesozoic tectonic history of the Raft River–Albion metamorphic terrane probably predated the overthrusting in the fold and thrust belt and therefore was not directly related to it. The horizontal extension and vertical shortening inferred for the early Mesozoic low-angle faulting and D_1 and D_2 phases of intense ductile deformation in the northern Albion Mountains are compatible with the gravity spreading model, but the timing precludes a genetic relationship between tectonism in the belts. The less intense deformation in the late Mesozoic (late D_2 phase) is more difficult to relate to regional tectonics. The decrease in metamorphic temperatures and intensity of deformation in the metamorphic terrane occurred at about the same time as the early thrusts in the fold and thrust belt, and these events in the two belts therefore may be genetically related. A mechanism for the production of overthrusts in the fold and thrust belt that accounts for reduction in temperatures in the hinterland is thus suggested. According to my interpretation of the structural history of the northern Albion Mountains, there is no evidence for extensive deformation during the late Mesozoic that is compatible with the gravity models. However, the hinterland overthrusting model can account for the decrease in metamorphic temperatures as a result of uplift and eastward translation of the metamorphic rocks, resulting in their emplacement over cooler strata. The model also provides a straightforward cause for tectonism in the fold and thrust belt.

CONCLUSIONS

Structures studied in the northern Albion Mountains are interpreted to be the result of Mesozoic and Cenozoic tectonism. Extensive younger-over-older low-angle faulting was succeeded by ductile deformation characterized by vertical shortening and horizontal extension in the early Mesozoic. In the late Mesozoic, ductile deformation decreased as temperatures declined, but a lesser amount of ductile deformation and low-angle faulting occurred at moderate temperatures until the middle Cenozoic.

This history is compatible with the geology of the remainder of the Raft River–Albion metamorphic terrane that was described by Armstrong (1968b) and Compton and others (1977), although it differs in interpretation from these studies.

Similar metamorphic terranes in the eastern Great Basin suggest that a broad belt of deep-seated metamorphism is present west of the Cordilleran fold and thrust belt. Much of the intense deformation and metamorphism in the metamorphic belt predates, and therefore is probably not related to, tectonism in the fold and thrust belt. Late Mesozoic and Cenozoic waning of temperatures in the metamorphic belt is hypothesized to have resulted from eastward overthrusting of the metamorphic complexes. The explanation for tectonism in the fold and thrust belt by this mechanism is suggested, following similar models developed by Armstrong (1978) and Brown (1978).

ACKNOWLEDGMENTS

I am grateful to the following people for discussions of various aspects of this work: R. L. Armstrong, J. M. Christie, R. R. Compton, M. D. Crittenden, Jr., C. A. Nelson, G. Oertel, S. S. Oriel, A. W. Snoke, and V. R. Todd. J. M. Christie, C. A. Nelson, and G. Oertel commented on my Ph.D. dissertation, which contained many of the ideas presented in this paper, and M. D. Crittenden, Jr., and R. R. Compton suggested several improvements for an earlier version of this paper.

Sources of financial assistance were Sigma Xi, the Scientific Research Organization of North America; Geological Society of America Penrose Bequest; University of California, Los Angeles Research Expense Grant; and National Science Foundation Grant EAR 76-14758 to J. M. Christie and C. A. Nelson.

REFERENCES CITED

Armstrong, F. C., and Oriel, S. S., 1965, Tectonic development of Idaho-Wyoming thrust belt: American Association of Petroleum Geologists Bulletin, v. 49, p. 1847–1866.

Armstrong, R. L., 1968a, Sevier orogenic belt in Nevada and Utah: Geological Society of America Bulletin, v. 79, p. 429–458.

—— 1968b, Mantled gneiss domes in the Albion Range, southern Idaho: Geological Society of America Bulletin, v. 79, p. 1295–1314.

—— 1970, Mantled gneiss domes in the Albion Range, southern Idaho: A revision: Geological Society of America Bulletin, v. 81, p. 909–910.

—— 1972, Low-angle (denudation) faults, hinterland of the Sevier orogenic belt, eastern Nevada and western Utah: Geological Society of America Bulletin, v. 83, p. 1729–1754.

—— 1974, Magmatism, orogenic timing, and orogenic diachronism in the Cordillera from Mexico to Canada: Nature, v. 247, p. 348–351.

—— 1976, The geochronometry of Idaho (Part 2): Isochron/West, no. 15, p. 1–33.

—— 1978, Core complexes, dejected zones, and an orogenic model for the eastern Cordillera: Geological Society of America Abstracts with Programs, v. 10, p. 360–361.

Armstrong, R. L., and Hansen, E., 1966, Cordilleran infrastructure in the eastern Great Basin: American Journal of Science, v. 264, p. 112–127.

Armstrong, R. L., and Hills, F. A., 1967, Rb-Sr and K-Ar geochronologic studies of mantled gneiss domes, Albion Range, southern Idaho, U.S.A.: Earth and Planetary Science Letters, v. 3, p. 114–124.

Brown, R. L., 1978, Structural evolution of the southeast Canadian Cordillera: A new hypothesis: Tectonophysics, v. 48, p. 133–151.

Compton, R. R., 1972, Geologic map of the Yost quadrangle, Box Elder County, Utah, and Cassia County, Idaho: U.S. Geological Survey Miscellaneous Geologic Investigations Map I-672, scale 1:31,680.

—— 1975, Geologic map of the Park Valley quadrangle, Box Elder County, Utah, and Cassia County, Idaho: U.S. Geological Survey Miscellaneous Geologic Investigations Map I-873, scale 1:31,680.

Compton, R. R., and Todd, V. R., 1979, Reply to Discussion on Oligocene and Miocene metamorphism, folding, and low-angle faulting in northwestern Utah: Geological Society of America Bulletin, v. 90, p. 305–309.

Compton, R. R., and others, 1977, Oligocene and Miocene metamorphism, folding, and low-angle faulting in northwestern Utah: Geological Society of America Bulletin, v. 88, p. 1237–1250.

Crittenden, M. D., Jr., 1979, Discussion on Oligocene and Miocene metamorphism, folding, and low-angle faulting in northwestern Utah: Geological Society of America Bulletin, v. 90, p. 305–309.

Crittenden, M. D., Jr., and Sorensen, M. L., 1980, The Facer Formation, a new early Proterozoic unit in northern Utah: U.S. Geological Survey Bulletin 1482F.

Drewes, H., 1978, The Cordilleran orogenic belt between Nevada and Chihuahua: Geological Society of America Bulletin, v. 89, p. 641–657.

Hose, R. K., and Danes, Z. F., 1973, Development of the late Mesozoic to early Cenozoic structures of the eastern Great Basin, in DeJong, K. A., and Scholten, R., eds., Gravity and tectonics: New York, John Wiley & Sons, p. 429–441.

Kehle, R. O., 1970, Analysis of gravity sliding and orogenic translation: Geological Society of America Bulletin, v. 81, p. 1641–1664.

Lee, D. E., and others, 1970, Modification of potassium-argon ages by Tertiary thrusting in the Snake Range, White Pine County, Nevada: U.S. Geological Survey Professional Paper 700-D, p. 92–102.

Loring, A. K., 1976, Distribution in time and space of late Phanerozoic normal faulting in Nevada and Utah: Utah Geology, v. 3, p. 97–109.

Mauger, R. L., Damon, P. E., and Livingston, D. E., 1968, Cenozoic argon ages on metamorphic rocks from the Basin and Range province: American Journal of Science, v. 266, p. 579–589.

Miller, D. M., 1978, Deformation associated with Big Bertha dome, Albion Mountains, Idaho [Ph.D. thesis]: Los Angeles, University of California.

Misch, P., and Hazzard, J. C., 1962, Stratigraphy and metamorphism of late Precambrian rocks in central northwestern Nevada and adjacent Utah: American Association of Petroleum Geologists Bulletin, v. 46, p. 289–343.

Mudge, M. R., 1970, Origin of the disturbed belt in northwestern Montana: Geological Society of America Bulletin, v. 81, p. 377–392.

Oriel, S. S., and Armstrong, F. C., 1966, Times of thrusting in Idaho-Wyoming thrust belt: Reply: American Association of Petroleum Geologists Bulletin, v. 50, p. 2614–2621.

Price, R. A., 1971, Gravitational sliding and the foreland thrust and fold belt of the North American Cordillera: Discussion: Geological Society of America Bulletin, v. 82, p. 1133–1138.
——1973, Large-scale gravitational flow of supracrustal rocks, southern Canadian Rockies, in Dejong, K. A., and Scholten, R., eds., Gravity and tectonics: New York, John Wiley & Sons, p. 291–502.
Price, R. A., and Mountjoy, E. W., 1970, Geologic structure of the Canadian Rocky Mountains between Bow and Athabasca Rivers—Progress report: Geological Association of Canada Special Paper 6, p. 7–25.
Roberts, R. J., and Crittenden, M. D., Jr., 1973, Orogenic mechanisms, Sevier orogenic belt, Nevada and Utah, in Dejong, K. A., and Scholten, R., eds., Gravity and tectonics: New York, John Wiley & Sons, p. 409–428.
Snoke, A. W., 1975, A structural and geochronological puzzle: Secret Creek gorge area, northern Ruby Mountains: Geological Society of America Abstracts with Programs, v. 7, p. 1278–1279.
Todd, V. R., 1973, Structure and petrology of metamorphosed rocks in central Grouse Creek Mountains, Box-Elder County, Utah [Ph.D. thesis]: Stanford University.
Williams, P. L., and others, 1975, Geology and geophysics of the southern Raft River Valley geothermal area, Idaho, U.S.A.: U.S. Geological Survey Open-File Report 75-322.
Winkler, H.G.F., 1967, Petrogenesis of metamorphic rocks (revised second edition): New York, Springer-Verlag, 237 p.

MANUSCRIPT RECEIVED BY THE SOCIETY JUNE 21, 1979
MANUSCRIPT ACCEPTED AUGUST 7, 1979

Printed in U.S.A.

PART 4

THE NORTHWEST

Bitterroot dome–Sapphire tectonic block, an example of a plutonic-core gneiss-dome complex with its detached suprastructure

DONALD W. HYNDMAN
Department of Geology
University of Montana
Missoula, Montana 59812

ABSTRACT

The Bitterroot dome–Sapphire tectonic block appears to be a well-developed example of a plutonic-core gneiss-dome complex or infrastructure separated from the adjacent suprastructure by a gently dipping zone of mylonitic shearing or an "Abscherungszone." The suprastructural Sapphire block, on the order of 15 km thick, 100 km long, and 70 km wide, apparently moved eastward about 60 km, bulldozing rocks of the eastern Flint Creek Range ahead of it. Movement of the block must have occurred about 75 or 80 m.y. ago during late stages of consolidation of the Idaho batholith which, along with sillimanite-zone regional metamorphic rocks, makes up the infrastructure under the mylonitic detachment zone. Timing of movement in the Sapphire block matches that in the Bitterroot dome.

The Bitterroot dome must have risen *after* off-loading of the Sapphire block, because the shear foliation and lineation that formed during movement completely cross the dome; this indicates that the block must have moved eastward across the whole of the area now occupied by the dome rather than radially down the flanks of an existing dome. The shear lineation maintains its eastward trend even at the south end of the dome where the foliation dips southward. The lineation and shear foliation are strongest along the eastern flank of the dome, over which the greatest thickness of the block would have passed.

INTRODUCTION

The Cordilleran infrastructure is exposed intermittently from the Shuswap metamorphic core complex of southeastern British Columbia, south through the Idaho batholith and the metamorphic core complexes of Nevada and Arizona, and to northern Mexico. In general, this high-grade metamorphic infrastructure with its granitic plutons is surrounded and flanked to the east by low-grade suprastructure, which extends eastward in a thick section to the Cordilleran thrust belt and in a thinner section beyond.

The Bitterroot dome (Chase and Talbot, 1973) forms the eastern part of the northern Bitterroot lobe of the Idaho batholith and extends north-south about 100 km. Immediately east lies the 6,000-km^2 Sapphire tectonic block, bounded by thrust faults of the same apparent age as a mylonitic detachment zone on the east side of the dome (Hyndman and others, 1975). Although the Sapphire block is more than twice the size of Rhode Island, it forms an integral block sandwiched between the front of the overthrust belt and the mylonitic detachment zone.

Earlier interpretations of the mylonitic zone flanking the east front of the Bitterroot Range and locally called the frontal zone gneiss (for example, Lindgren, 1904, p. 47–51; Langton, 1935; Wehrenberg, 1972) recognized that the granites of the Bitterroot Range had risen relative to Belt sedimentary rocks of the Sapphire Range. But there was little recognition of the large eastward displacement of the Sapphire Range.

Elsewhere, similar relationships exist between high-grade infrastructure separated by a gently dipping sheared zone (detachment surface, decollement, or Abscherungszone) from low-grade suprastructure.

In most such regions, the infrastructure consists of sillimanite-zone regional metamorphic rocks involved with migmatites and gneissic granite. Structurally, the infrastructure consists of large domes defined by gneissic foliation and, in many cases, recumbent isoclinal folds and a nearly unidirectional penetrative lineation. The flanking suprastructure consists of low-grade rocks with more-open flexural-slip folds and faults. The infrastructure displays flowage deformation developed at high temperatures and considerable depth, whereas the suprastructure consists of competent rocks deformed in a lower-temperature shallower environment. The difference in deformational behavior of infrastructure and suprastructure probably explains development of the intervening detachment zone (compare Haller, 1961; Armstrong and Hansen, 1966).

Examples of such complexes in the Cordillera include domes in the Shuswap metamorphic complex of British Columbia (Reesor, 1965, p. 110–118; Reesor and Moore, 1971; Hyndman, 1968) and northeastern Nevada (Misch and Hazzard, 1962; Armstrong and Hansen, 1966). The domes display a strong, generally east-west mineral streaking and slickenside lineation lying in a penetrative mylonitic foliation. Both lineation and foliation completely cross the dome, are no less prominent at its crest, and in some cases (Reesor, 1965, p. 110) gradually die out westward across the dome. That happens in the Bitterroot dome of the Idaho batholith as well.

In most cases the mylonitic foliation and lineation evidently formed during or shortly before diapiric rise of the dome. The model proposed here on the basis of characteristics of the Bitterroot dome and Sapphire tectonic block appears to fit characteristics of many of the other metamorphic core complexes. It differs from earlier hypotheses in recognizing the importance of (1) a penetrative foliation that crosses the whole dome but gradually dies out on one flank; and (2) a penetrative, largely unidirectional mineral streaking and slickenside lineation lying within the foliation and dying out in the same direction. The lineation is not radial to the dome. (3) The foliation and lineation are strongest on the flank of the dome toward its contact with the suprastructure from which it is separated by the zone of shearing, fracturing, and attentuation of regional metamorphic isograds. It is also strongest toward suprastructural areas containing rootless thrust faults and klippen. (4) In the case of the Sapphire block, the mylonitic surface of the Bitterroot dome is matched in its north-south dimensions to the east by an arc of thrust faults bounding the eastern edges of the detached block. (5) The conclusions concerning timing, direction, and distance of movement during mylonitization of Idaho batholith granites appear to closely match those that can be inferred from entirely independent evidence for formation of thrust faults in the periphery of the Sapphire block. (6) The stratigraphic and metamorphic levels inferred for infrastructural rocks beneath the mylonitic zone agree with those inferred on independent evidence for the base of the Sapphire block.

The following account presents the evidence leading to the above points and to the conclusion that the Sapphire block became detached from the roof of the Idaho batholith during the later stages of its consolidation and moved eastward about 60 km. It indicates that isostatic response to its removal accounts for the uplift of the Bitterroot dome.

DOME AND DETACHMENT ZONE

Shear foliation imposed on granitic rocks of the Idaho batholith and on adjacent high-grade regionally metamorphosed country rocks defines the Bitterroot dome. This foliation extends through an area comprising most of the eastern half of the main mass of the Bitterroot lobe of the Idaho batholith (Fig. 1). The foliated zone trends north-south along the full 100-km length of the Bitterroot front; it lies primarily within rocks of the Idaho batholith but extends some 25 km north into high-grade metasedimentary gneisses apparently derived from the Precambrian Belt Supergroup.

The main body of the Idaho batholith is nearly massive granite (according to the IUGS classification, Streckeisen, 1976) and granodiorite containing biotite and lesser muscovite. Small plutonic sheets generally concentric to the main contact enclose parts of the northeastern border zone (Nold, 1974; Chase, 1977). The early, more-mafic plutons show strong deformation associated with the last phases of regional metamorphism, but the latest plutons beneath the mylonitic zone show only relatively weak shearing that formed during later development of the mylonitic zone (Chase, 1977). Regional metamorphism of the lower units of the Belt Supergroup immediately north of the batholith and just below the mylonitic zone has reached the sillimanite-muscovite to sillimanite-orthoclase zones of the amphibolite facies. The fabric of these rocks includes a strong, gently dipping foliation and at least two sets of postmetamorphic folds, with gentle to steep axes apparently superimposed during emplacement of the Idaho batholith. Sillimanite is moderately common in the pelitic units and kyanite is present locally. Anorthosite plutons described by Berg (1968) and probably part of the pre-Belt basement complex lie in the core of very large recumbent east-verging folds mapped by Chase (1977) near, and in part beneath, the northern contact of the batholith. Structural analysis by Chase shows that these folds developed during diapiric rise and provide evidence of eastward mushrooming of the batholith. Similar mushrooming of the batholith occurred at its southwestern contact (Wiswall, 1979). If pre-Belt basement rocks lie in the cores of such large folds, the pre-Belt rocks were deformed along with the meta-Belt units, rather than separating along a basement decollement.

The mylonitic detachment zone along the straight eastern border of the Bitterroot dome is about 100 km long and averages about 0.5 km thick (Fig. 1). The geometry of the detachment zone and the general aspects of the deformation that created the mylonitic fabric are similar in the granitic rocks of the Idaho batholith and the high-grade metamorphic rocks beyond its northern contact. The mylonitic zone wraps around the southeastern edge of the dome and then apparently dies out to the west. A very strong penetrative slickenside-style lineation (Fig. 2) is prominent throughout the mylonitic zone and maintains a nearly constant east-southeasterly direction even where the mylonitic zone wraps around the southeastern part of the dome. The crest of the dome lies near the drainage divide along the Montana-Idaho border where the shear foliation flattens. The western boundary of the Bitterroot dome, about 30 km west of its crest, lies within granitic rocks of the batholith. That boundary is difficult to locate precisely because the foliation in the granite fades westward. Its position is probably along a ring of Tertiary granite plutons that surrounds the dome. The eastern flank of the Bitterroot dome, which coincides with the mylonitic zone along the spectacular front of the Bitterroot Mountains, lies between 15 and 20 km east of its crest (Figs. 1, 3). Foliation within the upper part of the dome is penetrative on the scale of an outcrop and strongly asymmetric. It is strongly expressed in the mylonitic zone on the east, but fades out both downward (Lindgren, 1904; Wehrenberg, 1972; Chase,

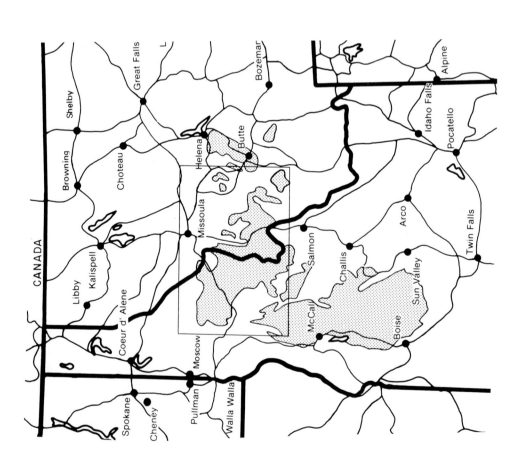

Figure 1 (facing page). Location map of Bitterroot lobe of the Idaho batholith. Enlargement of study area showing the main elements of the Bitterroot dome, mylonitic shear zone, and Sapphire tectonic block on facing page. 1 = Bitterroot Range, 2 = Bass Creek–Lolo Peak area, 3 = Miner's Gulch stock, 4 = Philipsburg batholith, 5 = Race-track Creek pluton, 6 = Philipsburg thrust, 7 = Georgetown thrust.

LEGEND

TQ Tertiary-Quaternary valley fill
Ev Eocene Challis and Lowland Creek Volcanics
    Eocene epizonal granite
Kv Late Cretaceous Elkhorn Mountains Volcanics
    Late Cretaceous granodiorite/granite/granitic rocks undifferentiated
Ms Mesozoic sedimentary rocks
Mv Permian-Triassic volcanic rocks, including Seven Devils volcanics
P   Paleozoic sedimentary rocks
pЄb Precambrian Belt Supergroup sedimentary rocks
pЄm Precambrian metamorphic rocks - meta-Belt and basement metamorphic rocks

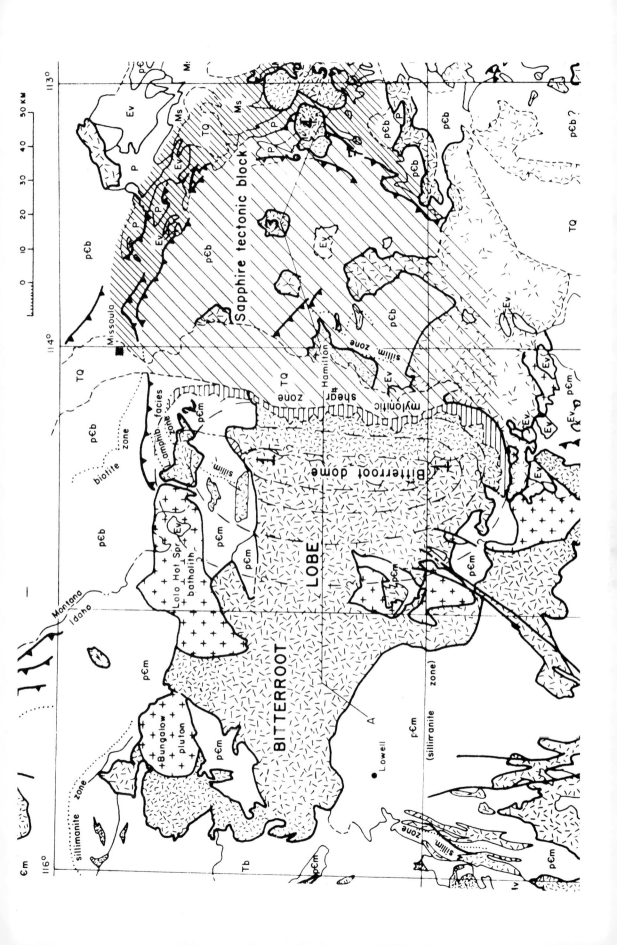

Figure 2. Slickenside lineation in surface of penetrative mylonitic foliation. Mouth of Sweathouse Canyon, Bitterroot Range, Montana. Original rock is pegmatitic granite. Relict coarser white grains are feldspar.

1973) and westward. This asymmetry in intensity of the mylonitic deformation appears to be an original feature which expresses the movements that created it. If mylonitization were related to doming proper, the strong fabric would presumably survive in the granites along its western flank as well as in those along its eastern flank.

In addition to the foliation defined by oriented grains and mineral streaks, the rocks of the mylonitic zone show a slickenside fabric that is also strongest in the eastern mylonitic zone and across the top of the dome. It seems most unlikely that schistosity created by metamorphic growth of parallel grains under stress formed either at the same time or under the same conditions as penetrative development of slickensides, which must have occurred under shallow, brittle conditions. I infer that progressive unloading of the Sapphire tectonic block caused shallowing of the rocks beneath, which led to a change in style of deformation from deep-seated plastic flow to shallow, brittle shearing.

Formation of the mylonitic zone must have followed emplacement of large parts of the Idaho batholith, for which radiometric ages between about 60 and 90 m.y. have been obtained (Hyndman and others, 1975; Chase and others, 1978). It must also have preceded eruption of presumably Eocene volcanic rocks deposited on the southern part of the mylonitic detachment zone. An Eocene age is inferred for the volcanic rocks because they are intimately related to plutons for which 40- to 50-m.y. ages have been obtained (Armstrong, 1974). These limitations bracket the time of mylonitic deformation within the latest stages of consolidation of the batholith. That conclusion is reinforced by numerous late K-feldspar megacrysts widely distributed through the batholith and the mylonitic zone, which show only minor deformation even where they cross individual pronounced shears within the mylonitic zone (Hyndman and others, 1975).

The direction of movement of rocks across the mylonitic zone is recorded in the prominent mica-streak lineations and slickenside striae that plunge uniformly east-southeast throughout the mylonitic zone and in the position of the Sapphire tectonic block apparently detached to the east-southeast of the Bitterroot dome. Some 85% to 90% of these lineations in sheared batholith rocks plunge between 5° and 28°, averaging 18° (Chase, 1977). But the mylonitic zone and its downdip lineations increase in plunge to the east to a nearly constant value of between 20° and 25°. The lineations trend S74°E on the average, 90% of them between due east and S53°E.

The amount of movement of overlying rocks has been inferred from the downslope rotation of steeply plunging axes of small folds in the Belt metasediments within the northeastern part of the Bitterroot dome. Simple shear of individual folds to within 2° of the slip line in the enveloping mylonitic foliation for a zone 0.5 km thick provides a crude estimate of more than 15 km displacement (Chase, 1977). This calculation neglects slip on innumerable individual shears and therefore provides a minimum estimate for the magnitude of displacement (Hyndman and others, 1975).

S. Reynolds (1978, written commun.) has suggested that development of the foliation and rotation of the folds could possibly be due to extreme flattening accompanied by a small component of differential shear, with uplift being related to extreme thinning of cover over the mylonitic shear zone. But both the intensity of the slickenside fabric and its asymmetry across the dome, along with the apparent close spatial relation to the Sapphire tectonic block, strongly suggest that major differential shear occurred. Acceptance of the genetic relationship between the mylonitic zone on the eastern flank of the Bitterroot dome and the Sapphire tectonic block (Hyndman and others, 1975) directly implies large displacement. Csejtey (1963), in his work on a more local level, inferred more than 25 km of eastward movement on the Olson Gulch and related thrust faults of the southeastern Flint Creek Range, which now appear to be at the leading edge of the Sapphire block.

A somewhat more convincing estimate of the amount of movement derives from the origin and dimensions of the Bitterroot dome. If, as suggested below, the still hot and mobile Bitterroot dome rose isostatically in response to tectonic unloading of the Sapphire block, then the structural crest of the dome should be the center of the unloaded area. The domed area appears to be about 60 km wide, double the 30 km distance from the crest of the dome to its western edge. If the domed area approximates the area unloaded, the 60-km distance across the Bitterroot dome should approximate the eastward transport of the Sapphire tectonic block. Asymmetry in original thickness of the block would affect the result to some extent, because the center of the isostatic rebound may not coincide with the center of the unloaded block. Nevertheless, the separation between the center of the Bitterroot dome and the center of the Sapphire tectonic block is about 65 km. The center of gravity of the Sapphire block apparently lies about midway between the eastward-dipping mylonitic detachment zone and the westward-dipping Georgetown thrust in the western part of the Flint Creek Range. The eastern 20 km of the Flint Creek Range is structurally distinct and may be part of either the Sapphire block or, as now seems more likely, material scraped eastward in front of the moving block.

THE SAPPHIRE TECTONIC BLOCK

The Sapphire tectonic block measures about 70 km east-west by 100 km north-south. It consists largely of low-grade late Precambrian Belt sedimentary rocks, which are capped for 20 km to the east of the Georgetown thrust by a nearly complete Paleozoic and Mesozoic sedimentary section. All units, including the Lower Cretaceous Kootenai sandstones and the uppermost Cretaceous Golden Spike conglomerates, have been either cut or overridden by thrusting along the borders of the block. The western edge of the Sapphire block is partially buried under Eocene(?) volcanic rocks in its southern part. The northern edge of the block is defined by a zone of south-dipping, east-northeast–transported thrust faults and asymmetric and overturned folds, some of which are en echelon (compare Desormier, 1975; C. Wallace, 1978, written commun.). That zone extends over a width of 10 to 15 km (Wallace and Klepper, 1976).

The position of the eastern edge of the block is somewhat uncertain. Hyndman and others (1975) defined it as extending to the eastern Flint Creek Range, which is the eastern limit of thrust faulting in the area. In that interpretation, the authors viewed the easternmost 20 km in the Flint Creek Range (east of the Georgetown thrust) as a thin, superficial tongue of the Sapphire block. This arcuate eastern 20-km tongue consists largely of Mesozoic, Paleozoic, and Belt sedimentary rocks tightly crumpled into large folds with steep to overturned axial surfaces and resting on gently dipping thrust faults (Calkins and Emmons, 1915; Csejtey, 1963; McGill, 1965; Hyndman and others, 1976). This apparently thin, eastern tongue contrasts with the thick mass of the Sapphire block proper, which is dominated by Belt sedimentary rocks warped into relatively open folds and broken into steep-sided blocks except near the margins where it developed thrusts and overturned folds. Two alternate interpretations for the distinctive characteristics of the eastern 20-km-wide tongue in the eastern Flint Creek Range are (1) the Sapphire block extends to the eastern part of the Flint Creek Range with the tight folds and gentle thrust faults there being the thin, crumpled toe of the eastward-moving block. The structural style is different at the eastern end of the block, because it consists there of Belt rocks overlain by Mesozoic and Paleozoic sedimentary formations having pronounced stratification and more variable competence between the layers. In this model, Paleozoic and even Mesozoic sedimentary rocks may have been overridden by the main Sapphire Range part of the block (see Fig. 4B). (2) The Sapphire block extends east only to the Georgetown thrust. This interpretation views the tight folds and gentle thrusts of the eastern 20 km in the Flint Creek Range as some of the Belt and Phanerozoic sediments bulldozed to the east from the area of the present Sapphire Range. In this model, Phanerozoic and uppermost Belt units would not have been overridden and therefore do not underlie the Sapphire block (see Fig. 4C).

The second interpretation seems more reasonable because of the abrupt change in structural style east of the Georgetown thrust and the discontinuous and variable nature of thrust faults around the

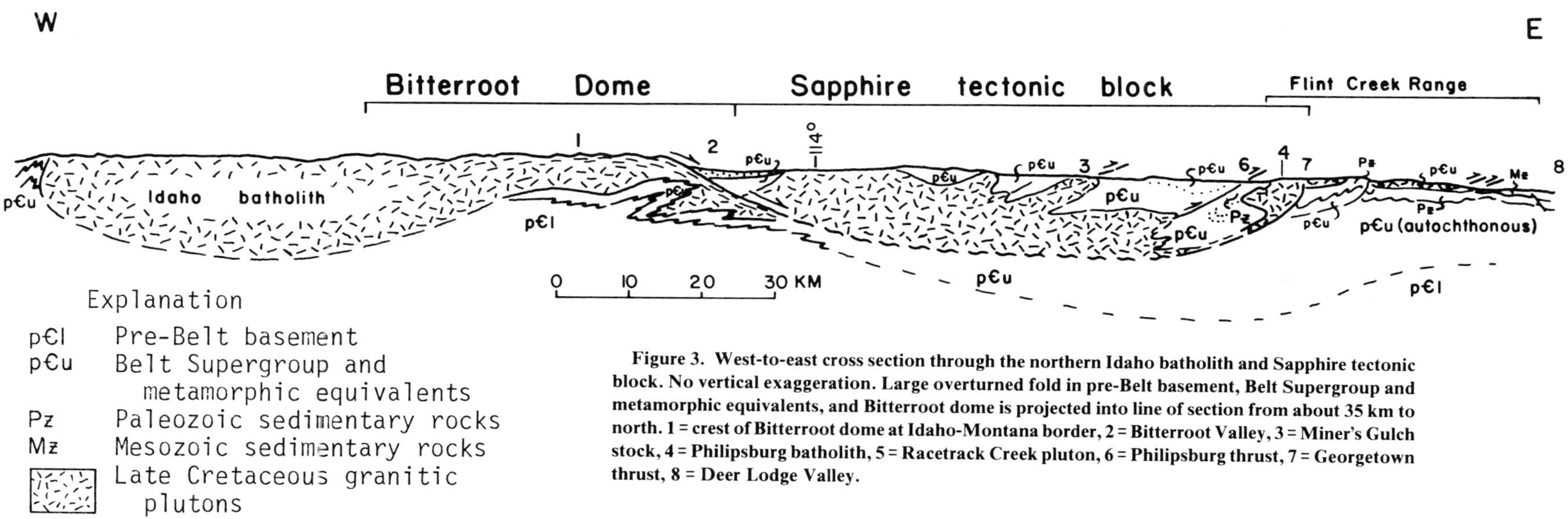

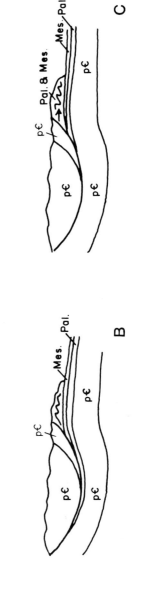

Figure 3. West-to-east cross section through the northern Idaho batholith and Sapphire tectonic block. No vertical exaggeration. Large overturned fold in pre-Belt basement, Belt Supergroup and metamorphic equivalents, and Bitterroot dome is projected into line of section from about 35 km to north. 1 = crest of Bitterroot dome at Idaho-Montana border, 2 = Bitterroot Valley, 3 = Miner's Gulch stock, 4 = Philipsburg batholith, 5 = Racetrack Creek pluton, 6 = Philipsburg thrust, 7 = Georgetown thrust, 8 = Deer Lodge Valley.

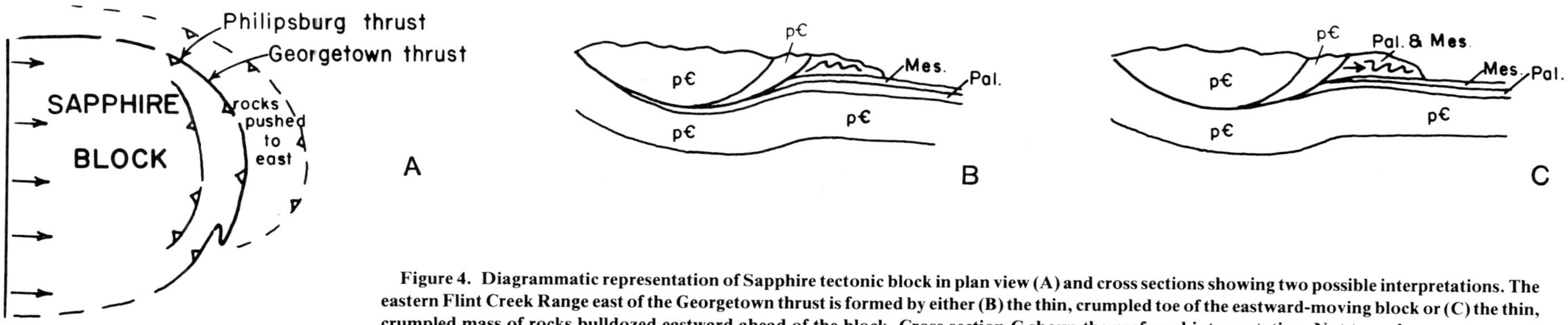

Figure 4. Diagrammatic representation of Sapphire tectonic block in plan view (A) and cross sections showing two possible interpretations. The eastern Flint Creek Range east of the Georgetown thrust is formed by either (B) the thin, crumpled toe of the eastward-moving block or (C) the thin, crumpled mass of rocks bulldozed eastward ahead of the block. Cross section C shows the preferred interpretation. Not to scale.

easternmost fringe of the Flint Creek Range. Although some workers have inferred large local displacements on thrusts (Csejtey, 1963), the lack of significant stratigraphic separation on thrusts elsewhere suggests only limited transport in some areas. Such, for example, appears to be the case between Cretaceous units in the northeastern Flint Creek Range (Gwinn, 1960). Measured stratigraphic sections in the lower Missoula Group (upper Belt Supergroup) and underlying Wallace Formation show abrupt thickening of several thousand feet across the Georgetown thrust to the west and indicate that the rocks on opposite sides of the thrust were originally widely separated (D. Winston, 1978, oral commun.).

Thus, it seems likely that the Sapphire tectonic block extends eastward only to the Georgetown thrust and that crumpled rocks extending for 20 km farther east were bulldozed from the present position of the block. Similar bulldozing of rocks may have occurred also along the north edge of the block, but its limits are not yet clear.

Southwest from the Flint Creek Range, thrust faults and folds mapped by Poulter (1956), Flood (1975), and Wiswall (1977) mark the leading edge of the Sapphire block. Farther west, the main thrusts and folds bounding the block disappear into a poorly known area of large granitic plutons that form the southeastern extension of the Bitterroot lobe of the Idaho batholith. Internal structures of the Sapphire block include many thrust faults (most of which dip inward into the block), open folds, and a number of normal faults that cut the thrusts.

Granitic plutons are scattered through the block (Fig. 3), most densely toward its southern end. Many, such as the Mount Powell and the eastern pluton of the Philipsburg batholith, appear to be emplaced along thrust faults (Hyndman and others, 1976) and are relatively undeformed except for the earliest of the plutons, the Racetrack Creek diorite, which lies in the eastern Flint Creek Range in the eastern edge of the block (Hawley, 1975). It shows locally strong internal foliation and is cut by many thrusts and by the later plutons. Normal faults cut at least the Miner's Gulch stock in the center of the Sapphire block (Hughes, 1975).

Most of the block consists of very low-grade chlorite-zone rocks of the greenschist facies. Metamorphic grade increases to lower, locally upper, amphibolite facies along the southwestern and southern edges of the block (Presley, 1973; LaTour, 1974; Flood, 1975; Wiswall, 1977). Presumably, the regional metamorphic grade increases at depth to the lower amphibolite facies; in the west and closer to the batholith, it increases at depth to the upper amphibolite facies. This increase in grade with depth appears to characterize the lower part of the Belt section on a regional scale.

The latest Cretaceous timing of movement of thrust faults within the Sapphire block (see Fig. 5) is narrowly constrained by the essentially concordant 72.0- and 73.4-m.y. ages on hornblende and biotite, respectively, from a single sample from the eastern undeformed pluton of the Philipsburg batholith (Hyndman and others, 1972). The western, somewhat deformed pluton of the Philipsburg batholith, with essentially concordant ages of 74.0 and 76.7 m.y. on biotite and hornblende, respectively, from a single sample, may have been affected by thrusting, whereas the shallow eastern pluton cuts thrust faults and may have been emplaced eastward on a nearly horizontal thrust fault (Hyndman and others, 1976). A stratigraphic limit on the time of movement is provided by the latest Cretaceous age of the conglomerates of the Golden Spike Formation in the northernmost Flint Creek Range, which were shed from the west (Gwinn and Mutch, 1965) from what is now recognized as the leading edge of the eastward-moving block. These conglomerates contain clasts of and thus are younger than at least the earliest units of the latest Cretaceous Elkhorn Mountains Volcanics, which are dated as 78 m.y. old.

The thickness of the Sapphire block can be estimated by three independent methods, each crude but believed to be reasonable. The first is a summation of the stratigraphic section in the block down to the level exposed in parts of the mylonitic detachment surface. The second is a petrologic estimation of the depth of the mylonitic detachment surface in the Bitterroot dome when it formed. The third is based on growth of apparently primary muscovite with quartz in the granite of the batholith.

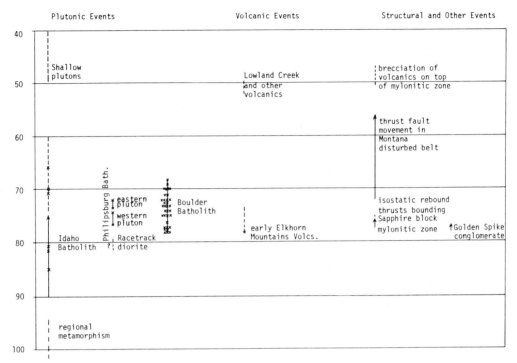

Figure 5. Timing of selected Late Cretaceous and Early Tertiary events (references cited in text).

The mylonitic surface in the Bass Creek–Lolo Peak area of the northeastern Bitterroot dome intersects regionally metamorphosed units most reasonably referred to the Prichard Formation of the lowermost Belt Supergroup and pre-Belt basement (Wehrenberg, 1972; Nold, 1974; Hyndman and Williams, 1977).

The presence of layered anorthosite in fold cores in metamorphosed Belt rocks (Berg, 1968; Chase, 1973; Cheney, 1975) clearly implies the existence of pre-Belt rocks. Such anorthosites elsewhere in the region are known only from lowermost Prichard (Hietanen, 1963; here considered to be pre-Prichard) or pre-Belt basement (Reid, 1957). The regional metamorphic environment in the same area of the northeastern Bitterroot dome was probably imposed immediately before emplacement of the Idaho batholith and thus provides an estimate of the pressure and, therefore, depth of the rocks just before detachment of the Sapphire block. The metamorphosed Prichard Formation contains sillimanite-muscovite-biotite-quartzofeldspathic gneisses grading into sillimanite-orthoclase ± muscovite-bearing gneisses that locally contain kyanite (D. W. Hyndman, 1965, unpub. map; Wehrenberg, 1972; Chase, 1973; Nold, 1974; Cheney, 1975; Hyndman and Williams, 1977).

Summation of the stratigraphic section in the Sapphire block probably should include the rocks between those presently at the surface in the north-central part of the block and those of the lower part of the Prichard Formation inferred to have moved from the mylonitic detachment in the northern part of the Bitterroot dome. Thus, the section should include the Prichard, Ravalli, Wallace or Helena, and much or all of the Missoula Group, various units of which are exposed at the surface in the Sapphire block. Overlying Paleozoic and even Mesozoic formations may have existed, especially in the eastern part. Unless erosion in the past 70 m.y. has removed all 5,000 m or so of Paleozoic-Mesozoic sedimentary rocks in addition to appreciable Missoula Group rocks, these units either never attained appreciable thickness on top of the Sapphire block or were eroded from highland areas over the present Bitterroot dome or Idaho batholith before the block moved. Scholten and Onasch (1977) outlined evidence for erosion and exposure of dominantly Precambrian quartzite in a highland area over at least the southern Idaho batholith during Early Cretaceous time. The Belt section in the Sapphire block, from the Prichard Formation through the Missoula Group, may have been either thickened or thinned by thrusting during movement of the block, so any estimate is tenuous at best. Desormier (1975) estimated that without stacking or thinning by thrust faults about 8.8 km of upper Belt rocks are present in the northern part of the block. This would be added to about 8 km of lower Belt rocks exposed to the northwest, which brings the total to almost 17 km. Chester Wallace (1978, written commun.) estimated 11 km and 10 km, respectively, for the same section, or a total of 21 km. Don Winston (1978, oral commun.) argued, however, that the total Belt section in the area of the Sapphire block can amount to no more than 13 km. Thus, an intermediate value of 17 km is here taken as the thickness of the Belt section and of the Sapphire tectonic block.

The depth of the mylonitic shear surface just before its formation, and, therefore, the thickness of the block removed, can be estimated from the sillimanite-muscovite-quartz ± orthoclase assemblage that locally includes kyanite and occurs with extensive migmatites. Different results may be obtained, depending on the pressure estimate used for sillimanite-kyanite at the appropriate temperatures. A kyanite-sillimanite-andalusite invariant point at 3.5 kb or 13.3 km, as suggested by the experimental data of Richardson and others (1969) and Holdaway (1971), seems most consistent with independent field and petrologic data on metamorphic rocks in general (Hyndman, 1972, p. 313). This would provide a minimum estimate for the pressure on the rocks at the time of metamorphism, because the kyanite-sillimanite isograd exists in the infrastructure just under the mylonitic detachment zone (Wehrenberg, 1972).

Similarly, a few percent or less muscovite with quartz, which commonly occurs in the granite and granodiorite of the Idaho batholith (Chase, 1973), suggests pressures greater than 4 kb or 15 km for a water-saturated granite magma and greater than 6.7 kb or 24 to 25 km for a drier magma with X_{H_2O} = 0.5 (Kerrick, 1972).

Thus, the three rough but independent methods—the stratigraphic section in the Sapphire block, the mineralogy of the metamorphic rocks, and the mineralogy of the granite of the Idaho batholith—all suggest depths to the mylonitic zone before movement of more than 13.3 km, probably more than 15 km, and more likely 17 km.

CONSEQUENCES

The consistent east-southeast direction of the lineation in the mylonitic zone of the Bitterroot dome, even on the southern flank of the dome, dictates that the dome postdates the lineation (see Fig. 6). The strong asymmetry of the shear fabric, which increases in intensity to the east, and the existence of a detached suprastructural block only on the east indicate that unloading was only to the east. That the trailing half of the Sapphire block could not have been dragged up the western side of the existing dome before moving down the eastern side also indicates that the dome must postdate the lineation and movement of the block.

The detachment of more than 15 km of Belt metasedimentary rocks and granite from the eastern flank of the Idaho batholith would give an expected isostatic response of about 0.82 (that is, 2.7:3.3) of the thickness of the block removed or more than 12 km. This assumes an average density of 2.7 g/cm³ for the metasedimentary rocks and granites of the Sapphire block and a density of 3.3 g/cm³ in a perfectly compensating upper mantle. Similarly, the emplacement of a block more than 15 km thick on the crust to the east could cause sinking of more than about 12 km.

The Sapphire block appears to have pushed ahead the Mesozoic and Paleozoic rocks and upper part of the Belt stratigraphic section. The Belt rocks are mostly Missoula Group sedimentary rocks of the

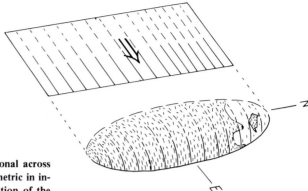

Figure 6. The lineation is one-directional across the Bitterroot dome and strongly asymmetric in intensity. The dome must postdate formation of the lineation.

upper Belt but, locally, may include some Wallace Formation and Ravalli Group of the middle part of the Belt. This suggests that the Belt rocks originally in the present position of the Sapphire block may have been bulldozed to the depth of the middle part of the Belt section. Or, if deformation or erosion had exposed Wallace or Ravalli sedimentary rocks before movement, the block may have bulldozed to a modest depth of perhaps less than 1 km. This latter possibility is more reasonable if, as inferred, the allochthons of the eastern Flint Creek Range are thin and their volume is very small in comparison to that of the Sapphire block. Then, only the thin uppermost sedimentary rocks would have been scraped eastward, and the Sapphire block would have overridden most of the Belt stratigraphic section.

In any case, the Bitterroot dome must have risen isostatically. The Sapphire block presumably sank to a comparable depth unless it implausibly moved into a hole about as deep as the thickness of the block. Rise of more than 12 km in the dome and sinking of a comparable amount to the east would tilt the mylonitic surface more than 25° to 30° to the east. Because the present dip of the surface at the eastern flank of the Bitterroot dome is 20° to 25°, the initial dip of the surface at that position (then near the leading edge of the block) would have had to be more than 5° to the west (Fig. 7). Such a dip in the slippage surface in the toe of a moving block such as in a large landslide or glacier is not unreasonable. Isostatic compensation to form the Bitterroot dome was probably fostered, as noted below, by high temperatures of the batholithic environment. Incomplete sinking of the Sapphire block and therefore a lesser tilt imposed on the detachment surface could reflect lower crustal temperatures east of the Idaho batholith. A lesser tilt could also reflect incomplete compensation of both the Bitterroot dome and the Sapphire block. Before unloading, the Bitterroot Range with a present average ridge elevation of about 2,500 m would have been more than 3,000 m higher or 5,500 m high.

Isostatic compensation of cool crustal blocks characteristically occurs on an order of at least 100 km. The smaller 60 × 100 km dimensions of the Bitterroot dome presumably reflect compensation under the high-temperature and, therefore, low-strength conditions of the Earth's crust at the time of generation, rise, and emplacement of granitic magma of the Idaho batholith.

A high-temperature environment of the batholith at the time of initiation of shearing and isostatic rebound is compatible with the remarkably planar, gently dipping mylonitic zone that is largely within granitic rocks of the Idaho batholith but also extends to the north for 25 km into schistose country rocks. Although the country rocks would still have been hot following metamorphism, shearing or detachment occurred in the nearly massive granite. If the batholith was nearly but incompletely consolidated, its high temperature and interstitial fluids would have minimized its shear strength.

The unloading of the still hot metamorphic aureole also provides an explanation for the presence of andalusite and cordierite (Cheney, 1975) as late-stage phases north of the batholith. Rapid unloading

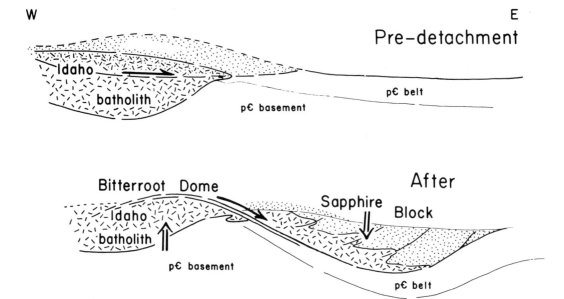

Figure 7. Diagrammatic representation of the area of the Bitterroot dome–Sapphire tectonic block before and after detachment. Not to scale.

of a thick section of the overlying crust would have decreased the pressure while leaving the rocks at high temperature.

Independent evidence for rise of the dome after completion of solidification of the rocks is as follows. Abundant fracture surfaces with steep but otherwise random orientations that consistently show slickensides with downdip lineation have recently been recognized throughout a 30-km section examined across the western half of the Bitterroot dome (D. W. Hyndman, unpub. data). These slickensides must have formed by vertical movement after solidification of the granite. They have not been found in a long section through the Bitterroot lobe west of the dome. They evidently formed during the isostatic rebound that raised the dome.

The initial regional eastward slope required for detachment on the mylonitic surface of low dip would have been provided by either thermal expansion of the crust above the subduction zone west of the Idaho batholith (Hyndman and Talbot, 1976) or diapiric rise of the intruding Idaho batholith in the area of the Bitterroot dome (Chase, 1977). Diapiric rise may well have occurred during intrusion. However, movement of a mass eastward across the whole of the Bitterroot dome, as indicated by the unidirectional slickenside-bearing lineation, implies that the center of high temperature had to lie west of the Bitterroot dome. The regional thermal high represented by sillimanite-zone regional metamorphic rocks, with or without granite, extends for about 95 km west of the crest of the Bitterroot dome, almost to 116° W longitude (see Hyndman and Williams, 1977). Thus, the approximate western edge of the Bitterroot dome lies almost exactly at the center of the thermal high that existed in the continental crust just prior to emplacement of the Idaho batholith. Movement of a thick block eastward from the east flank of this thermal high would create a shear zone and lineation in the position and with the direction preserved in the present Bitterroot dome. That rocks west of the Bitterroot dome but within the thermal high may have been deeper than rocks in the Bitterroot dome is suggested by the abundance of kyanite in metasediments of appropriate composition (Hietanen, 1956, 1963).

Felsic magmas emplaced during Eocene time, 20 m.y. after movement of the Sapphire block, could not have resulted directly from that movement. However, the earlier deformation may well have controlled where they erupted. The felsic rocks include granitic plutons rich in potassium feldspar and full of miarolitic cavities and pegmatites. These were emplaced at shallow depth along with the more extensive Challis–Lowland Creek Volcanics, which extend from the southern Idaho batholith to the area east of the Sapphire block. The Eocene plutons are concentrated in the zone, perhaps a tensional zone at depth, along the western and southern edges of the Bitterroot dome (Hyndman and Williams, 1977). The Lowland Creek Volcanics appear to be concentrated in a moon-shaped arcuate pattern fringing the eastern edge of the Sapphire block, perhaps in a swale created by isostatic sinking of the block. The Eocene volcanics, clearly deposited under surface conditions and lying on the deep-seated rocks of the southern part of the detachment zone, are strongly brecciated. Possibly they collapsed down the detachment surface after formation of the existing dome. Possibly the brecciation is related to that farther north which is in a small patch of probable meta-Wallace Formation that is spatially associated with a normal fault along the base of the Bitterroot front. This fault offsets the mylonitic shear zone by 50 to 100 m.

SUMMARY

Salient aspects of the Bitterroot dome–Sapphire tectonic block are as follows. Rocks now in the dome evolved in a high-pressure–high-temperature environment before the Sapphire block slid off. The fabric of the Bitterroot dome indicates a strong lateral asymmetry, with the shearing and associated lineation being weak on the west side and very strong on the east in the direction of movement of the cover rocks. Presumably, the difference is a consequence of movement of much less rock off the western flank than off the eastern flank. This difference in intensity of the mylonitic fabric would be affected by both the amount of lateral movement and the greater thickness of the cover over the eastern flank. Within the detachment zone itself, the penetrative foliation, schistosity, and mineral alignment and the superimposed brittle-style slickenside deformation in the same rocks appear to reflect the gradual unloading of the cover rocks, with the slickensides appearing only in the later stages when the cover was thin.

The unidirectional rather than radial nature of the lineation, even where the flanks of the dome dip at a high angle to the lineation, indicates that the dome formed after the detachment surface. This important point was unclear at the time of an earlier paper outlining the existence of the Sapphire tectonic block (Hyndman and others, 1975). The Bitterroot dome itself is inferred to be a consequence of the isostatic rebound accompanying unloading of the Sapphire block. Major aspects of the tectonics of the area do not reflect unloading from a preexisting dome because the Bitterroot dome cannot have existed when the block moved. Only the shallow, relatively local brecciation of later Eocene volcanics should be ascribed to movement on the dome after its formation.

Implications of the model presented here may have some relevance for other Cordilleran gneiss domes. For example, the eastern flank of the Shuswap metamorphic complex shows intense shearing, low-angle east-dipping faults, nearby klippen with no apparent source, extensive retrograde metamorphism in flanking gneiss domes, and attenuation of regional metamorphic isograds in the zone between the infrastructure on the west and the suprastructure on the east (Hyndman, 1968). Penetrative, gently east-plunging mineral and slickenside lineation lying in a shear foliation is prominent in the eastern part of the Shuswap complex, and in at least some areas such as the Valhalla dome (Reesor, 1965), both die out gradually to the west across the dome. Read (1977) suggested east-directed structural unroofing of the Shuswap complex during doming. But as suggested for the Bitterroot dome of the Idaho batholith (Hyndman, 1977) and extended here, the evidence in a number of areas indicates that the dome formed after the prominent unidirectional lineation and movement of supracrustal blocks.

In the Basin and Range province of Nevada, Arizona, and northern Mexico, the structural relations of core gneiss domes to units to the east, and in some cases to the west, are controversial and obscured by later basin-range faulting. The asymmetry of fabric and the necessity for strong isostatic response to unloading suggest that the domes themselves may be largely a *consequence* of unloading rather than a *cause*.

ACKNOWLEDGMENTS

I thank Bill Rehrig and Steve Reynolds for discussions during three days in the field and Jim Talbot, Ron Chase, Dave Alt, and other colleagues at the University of Montana for discussions during evolution of my thoughts on the subject. Bob Compton, George Thompson, Chet Wallace, Steve Reynolds, and Dave Alt read the manuscript at various stages and provided constructive suggestions.

REFERENCES CITED

Armstrong, R. L., 1974, Geochronometry of the Eocene volcanic-plutonic episode of Idaho: Northwest Geology, v. 3, p.1-15.

Armstrong, R. L., and Hansen, E., 1966, Cordilleran infrastructure in the eastern Great Basin: American Journal of Science, v. 264, p. 112-127.

Berg, R. B., 1968, Petrology of anorthosites of the Bitterroot Range, Montana, *in* Isachsen, Y. W., ed., Origin of anorthosite and related rocks: New York State Museum and Science Service Memoir 18, p. 387-398.

Calkins, F. C., and Emmons, W. H., 1915, Description of the Philipsburg quadrangle, Montana: U.S. Geological Survey Geologic Atlas of the United States, Folio 196, 26 p.

Chase, R. B., 1973, Petrology of the northeastern border zone of the Idaho batholith, Bitterroot Range, Montana: Montana Bureau of Mines and Geology Memoir 43, 28 p.

——1977, Structural evolution of the Bitterroot dome and zone of cataclasis, *in* Chase, R. B., and Hyndman, D. W., Mylonite detachment zone, eastern flank of Idaho batholith: Geological Society of America, Rocky Mountain Section, 30th Annual Meeting, Field Guide 1, p. 1-24.

Chase, R. B., and Talbot, J. L., 1973, Structural evolution of the northeastern border zone of the Idaho batholith, western Montana: Geological Society of America Abstracts with Programs, v. 5, p. 470-471.

Chase, R. B., Bickford, M. E., and Tripp, S. E., 1978, Rb-Sr and U-Pb isotopic studies of the northeastern Idaho batholith and border zone: Geological Society of America Bulletin, v. 89, p. 1325-1334.

Cheney, J. T., 1975, Kyanite, sillimanite, phlogopite, cordierite layers in the Bass Creek anorthosites, Bitterroot Range, Montana: Northwest Geology, v. 4, p. 77-82.

Csejtey, B., 1963, Geology of the southeast flank of the Flint Creek Range, western Montana [Ph.D. dissert.]: Princeton, N.J., Princeton University, 175 p.

Desormier, W., 1975, A section of the northern boundary of the Sapphire tectonic block, Missoula and Granite Counties, Montana [M.A. thesis]: Missoula, University of Montana, 65 p.

Flood, R. E., 1975, Relationship of igneous emplacement of deformational history in the Anaconda Range, Montana: Northwest Geology, v. 4, p. 9-14.

Gwinn, V. E., 1960, Geology of the Drummond area, central-western Montana: Montana Bureau of Mines and Geology Special Publication 21 (Geol. Map 4), scale 1:125,000.

Gwinn, V. E., and Mutch, T. A., 1965, Intertongued Upper Cretaceous volcanic and nonvolcanic rocks, central-western Montana: Geological Society of America Bulletin, v. 76, p. 1125-1144.

Haller, J., 1961, Account of Caledonian orogeny in Greenland: International Symposium on Arctic Geology, 1st, Calgary, Alberta, 1960, Proceedings, v. 1, p. 155-159.

Hawley, K. T., 1975, The Racetrack pluton—A newly defined Flint Creek pluton: Northwest Geology, v. 4, p. 1-8.

Hietanen, A., 1956, Kyanite, andalusite, and sillimanite in the schist in Boehls Butte quadrangle, Idaho: American Mineralogist, v. 41, p. 1-27.

——1963, Anorthosite and associated rocks in the Boehls Butte quadrangle and vicinity, Idaho: U.S. Geological Survey Professional Paper 344-B, 78 p.

Holdaway, M. J., 1971, Stability of andalusite and the aluminum silicate phase diagram: American Journal of Science, v. 271, p. 97-131.

Hughes, G. J., 1975, Relationship of igneous rocks to structure in the Henderson-Willow Creek igneous belt, Montana: Northwest Geology, v. 4, p. 15-25.

Hyndman, D. W., 1968, Petrology and structure of Nakusp map area, British Columbia: Geological Survey of Canada Bulletin 161, 95 p.
—— 1972, Petrology of igneous and metamorphic rocks: New York, McGraw-Hill Book Co., 533 p.
—— 1977, Mylonitic detachment zone and the Sapphire tectonic block, *in* Chase, R. B., and Hyndman, D. W., Mylonite detachment zone, eastern flank of Idaho batholith: Geological Society of America, Rocky Mountain Section, 30th Annual Meeting, Field Guide 1, p. 25–31.
Hyndman, D. W., and Talbot, J. L., 1976, The Idaho batholith and related subduction complex: Geological Society of America, Cordilleran Section, Field Guide 4, 15 p.
Hyndman, D. W., and Williams, L. D., 1977, The Bitterroot lobe of the Idaho batholith: Northwest Geology, v. 6-1, p. 1–16.
Hyndman, D. W., Obradovich, J. D., and Ehinger, R., 1972, Potassium-argon age determinations of the Philipsburg batholith: Geological Society of America Bulletin, v. 83, p. 473–474.
Hyndman, D. W., Talbot, J. L., and Chase, R. B., 1975, Boulder batholith: A result of emplacement of a block detached from the Idaho batholith infrastructure?: Geology, v. 3, p. 401–404.
Hyndman, D. W., and others, 1976, Petrology of the Philipsburg batholith, western Montana—Satellitic bodies of the Boulder and (?) Idaho batholiths: Montana Bureau of Mines and Geology Open-File Report, 95 p.
Kerrick, D. E., 1972, Experimental determination of muscovite + quartz stability with $P_{H_2O} < P_{total}$: American Journal of Science, v. 272, p. 946–958.
Langton, C. M., 1935, Geology of the northeastern part of the Idaho batholith and adjacent region in Montana: Journal of Geology, v. 43, p. 27–60.
LaTour, T., 1974, An examination of metamorphism and scapolite in the Skalkaho region, southern Sapphire Range, Montana [M.S. thesis]: Missoula, University of Montana, 95 p.
Lindgren, W., 1904, A geological reconnaissance across the Bitterroot Range and Clearwater Mountains in Montana and Idaho: U.S. Geological Survey Professional Paper 27, 123 p.
McGill, G. E., 1965, Tectonics of the northern Flint Creek Range, *in* Geology of the Flint Creek Range, Montana: Billings Geological Society 16th Annual Field Conference Guidebook, p. 127–136.
Misch, P., and Hazzard, J. C., 1962, Stratigraphy and metamorphism of late Precambrian rocks in central northeastern Nevada and adjacent Utah: American Association of Petroleum Geologists Bulletin, v. 46, p. 289–343.
Nold, J. L., 1974, Geology of the northeastern border zone of the Idaho batholith, Montana and Idaho: Northwest Geology, v. 3, p. 47–52.
Poulter, G. J., 1956, Geologic map and sections of Georgetown thrust area, Granite and Deer Lodge Counties, Montana: Montana Bureau of Mines and Geology Map 1.
Presley, M., 1973, Metamorphism in the Sapphire Mountains, Montana: Northwest Geology, v. 2, p. 36–41.
Read, P. B., 1977, Relationship of the Kootenay arc to the Shuswap metamorphic complex, southern British Columbia: Geological Association of Canada Programs with Abstracts, v. 2, p. 43.
Reesor, J. E., 1965, Structural evolution and plutonism in Valhalla gneiss complex, British Columbia: Geological Survey of Canada Bulletin 129, 128 p.
Ressor, J. E., and Moore, J., 1971, Petrology and structure of Thor-Odin gneiss dome, Shuswap metamorphic complex, British Columbia: Geological Survey of Canada Bulletin 195, 149 p.
Reid, R. R., 1957, Bedrock geology of the north end of the Tobacco Root Mountains, Madison County, Montana: Montana Bureau of Mines and Geology Memoir 36, 27 p.
Richardson, S. W., Gilbert, M. C., and Bell, P. M., 1969, Experimental determination of kyanite-andalusite and andalusite-sillimanite equilibria; the aluminum silicate triple point: American Journal of Science, v. 267, p. 259–272.
Scholten, R., and Onasch, C. M., 1977, Genetic relations between the Idaho batholith and its deformed eastern and western margins: Northwest Geology, v. 6-1, p. 25–37.
Streckeisen, A., 1976, To each plutonic rock its proper name: Earth-Science Reviews, v. 12, p. 1–33.
Wallace, C. A., and Klepper, M. R., 1976, Preliminary reconnaissance geologic map of the Cleveland Mountain and north half of the Ravenna quadrangles, western Montana: U.S. Geological Survey Open-File Report 76-527.
Wehrenberg, J. P., 1972, Geology of the Lolo Peak area, northern Bitterroot Range, Montana: Northwest Geology, v. 1, p. 25–32.
Wiswall, C. G., 1977, Structural styles and the sequence of deformation related to the Sapphire tectonic block: Northwest Geology, v. 6-2, p. 51–59.
—— 1979, Field and structural relationships below the Idaho batholith: Northwest Geology, v. 8, p. 18–28.

MANUSCRIPT RECEIVED BY THE SOCIETY JUNE 21, 1979

MANUSCRIPT ACCEPTED AUGUST 7, 1979

Printed in U.S.A.

Structural and metamorphic evolution of northeast flank of Shuswap complex, southern Canoe River area, British Columbia

P. S. SIMONY
E. D. GHENT
D. CRAW
W. MITCHELL
D. B. ROBBINS
Department of Geology
University of Calgary
Calgary, Alberta T2N 1N4, Canada

ABSTRACT

The southern Canoe River area (lat 52°N, long 118°W), situated adjacent to the northeast flank of the Shuswap complex of British Columbia, was mapped in detail in conjunction with mineralogic and petrologic studies. This work is the basis for a section some 50 km long which illustrates the transition from the Rocky Mountain fold and thrust belt on the east to the metamorphic core zone of the Columbian orogen on the west.

A 5-km-thick succession of Proterozoic and Cambrian clastic rocks have been metamorphosed from biotite to sillimanite grade. Three sets of major structures are superimposed in the core zone, although only one major set of buckle folds and thrusts (which formed during the last two deformation phases) is predominant in the Rockies. The Malton Gneiss, a sheet of reworked basement, emerges in the northern part of the area. Mylonite marks the cover-basement decollement that was active early in the orogeny. Southwest-verging isoclinal recumbent nappes sheared off from the Malton Gneiss are refolded by two sets of northeast-verging folds. Isogradic surfaces were imposed late in the second deformation phase with a morphology related to heat flow. These surfaces were subsequently deformed and folded in a third deformation phase. In the core zone, staurolite disappears within the kyanite zone, and trondhjemite migmatite appears concomitantly. In the Rocky Mountains, staurolite persists to the appearance of fibrolite, and no migmatite is found. Geobarometric studies suggest metamorphic pressures in the core zone some 2-kb higher than in the western part of the Rockies and indicate that the vertical motion on the Purcell thrust was about 7 km. Contrasting stratigraphic thicknesses and facies as well as contrasting structural styles and histories across the fault suggest substantial horizontal shortening.

Temperatures and pressures estimated from element fractionation between coexisting minerals are consistent with the structural and isograd pattern.

INTRODUCTION

In southern Canada, two orogenic belts constitute the North American Cordillera—the eastern belt is the Columbian orogen, and the western one is the Pacific orogen (Wheeler and Gabrielse, 1972). The Columbian orogen consists of an extensive foreland fold and thrust belt, the Canadian Rocky Mountains, a metamorphic-plutonic core zone, and a hinterland belt of Paleozoic and Mesozoic volcanic and sedimentary rocks. The core zone, called the Omineca crystalline belt by Wheeler and Gabrielse (1972), is dominated by upper Proterozoic and lower Paleozoic sedimentary rocks within which metamorphic complexes have formed and granitoid plutons have been emplaced. Locally, sheets of Precambrian basement gneisses are tectonically interleaved with the metasedimentary cover.

The Shuswap complex (Fig. 1) is the largest of the metamorphic complexes in the metamorphic-plutonic core zone of the Columbian orogen (Reesor and Moore, 1971). The complex is characterized by upper amphibolite–facies metamorphism and by the presence of migmatite and gneiss. A major anticlinorium (labeled "dome axis," Fig. 1) runs along the east side of the complex, and a series of gneiss domes occupy culminations within the anticlinorium (Reesor, 1965; Reesor and Moore, 1971; Wheeler, 1965). The southern Canoe River area (Fig. 1), which is centered on the big bend of the Columbia River, is particularly well suited for a study of the northeast flank of the metamorphic core zone and its relationship to the Rocky Mountain fold and thrust belt.

Since 1972, we have carried out structural and metamorphic studies in the area, including detailed mapping on the scales of 1:10,000 and 1:24,000. This is the basis for a section, some 50 km long, from the Rocky Mountains into the Shuswap complex. Detailed petrographic, mineralogic, and petrologic studies were combined with the field investigation and enabled us to outline an integrated picture of the metamorphic and structural evolution.

Previous work by Wheeler (1965), Campbell (1968), and Price and Mountjoy (1970) showed that within the southern Canoe River area, the upper Proterozoic (Hadrynian) to Cambrian miogeoclinal succession of the western Rocky Mountains could be traced from the chlorite to the kyanite zone. The Rocky Mountains are separated from the Selkirk and Monashee Ranges of the Columbia Mountains by the Rocky Mountain trench, a valley here occupied by the Columbia and Canoe Rivers. These rivers have been flooded since 1973 when the Mica Dam was completed. West of the trench, the upper Proterozoic (Hadrynian) rocks are complexly folded; their metamorphic grade ranges from upper garnet to muscovite-sillimanite.

In the northwest part of our study area, a mass of granitoid gneiss (the Malton Gneiss) was outlined by Campbell (1968), Giovanella (1967), and Price and Mountjoy (1970). They recognized that the gneiss is present on both sides of the trench and that a normal fault in the trench separates the western, major gneiss segment from the Rocky Mountain segment. They also recognized that the gneiss was thrust over Cambrian rocks on the west flank of the Rocky Mountains. Price and Mountjoy (1970) identified that thrust as the Purcell thrust, which for a strike length of some 300 km forms the leading edge of the imbricated front of the Omineca belt.

The setting of the southern Canoe River area thus permits us to investigate the transition from chlorite-grade rocks along the west flank of the foreland fold and thrust belt to sillimanite-grade rocks in the migmatitic and complexly folded core zone.

STRATIGRAPHY

An understanding of the stratigraphy has played a major role in unraveling structures, and some salient features are summarized here before we discuss the structural and metamorphic evolution.

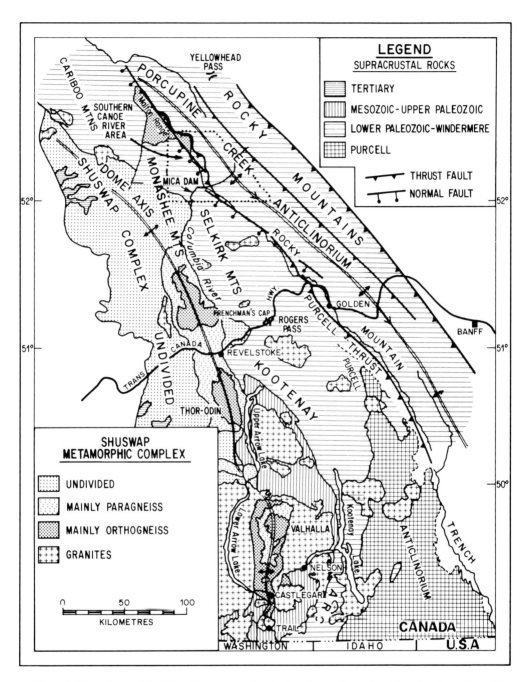

Figure 1. Map of part of the Columbian orogen, showing location and tectonic setting of southern Canoe River area. Map modified from Reesor and Moore (1971).

The area is underlain by metasedimentary rocks of late Proterozoic (Hadrynian) and Cambrian age. The Hadrynian rocks west of the Rocky Mountain trench are in continuity with the Hadrynian succession defined in the northern Purcell Mountains, some 150 km to the southeast of Mica Creek. The subdivisions recognized in the northern Purcell Mountains (Evans, 1933; Wheeler, 1963; Simony and Wind, 1970; Poulton, 1973) also are valid in the Canoe River area. Furthermore, the succession exposed in the western Rockies is in continuity with the Proterozoic and Cambrian succession defined farther to the north by Mountjoy (1962).

A different nomenclature is traditionally used for the strata on either side of the Rocky Mountain trench, but we agree with Young and others (1973) that the lithostratigraphic units can be correlated. Figure 2, together with Tables 1 and 2, summarizes the main features of the stratigraphy. The thicknesses are reliable for the Rocky Mountains but have only semiquantitative significance for the intensely deformed incompetent metamorphic rocks west of the trench.

The succession is well established in the Rockies. Way-up criteria are abundant, and large domains of simple structure can be outlined. Our stratigraphic interpretation is completely in harmony with the stratigraphy etablished farther northwest and southeast along the strike (Fyles, 1960; Wheeler, 1963; Meilliez, 1972; Campbell and others, 1973).

The Lower Cambrian Gog Group is some 200 to 300 m thinner on the west flank of the Rockies than it is in the central Main Ranges (Mountjoy, 1962; Cook, 1970). Furthermore, the Mahto Formation, at the top of the Gog Group, is finer grained and contains more shaley units than farther east. From these observations, we would expect the Lower Cambrian quartzites to thin gradually and give way to shale westward. Unfortunately, Cambrian strata are not preserved above the Hadrynian rocks of the Mica Dam area, but observations to the southwest in the Selkirk Mountains (Wheeler, 1965; Brown and others, 1977, 1978), as well as to the northwest in the Cariboo Mountains (Campbell and others, 1973), indicate that the Lower Cambrian quartzites would be thin or absent in the western part of our area.

West of the Rocky Mountain trench, the structure is complex, metamorphism is intense, and way-up criteria are rare. However, by combining continuous tracing of key markers and observation of the

TABLE 1. SUMMARY OF STRATIGRAPHY EAST OF THE ROCKY MOUNTAIN TRENCH

Unit		Thickness (m)	Description
Chancellor Group	Kinbasket	>1,000 (no top)	Sandy and silty limestone and pelite, with gray marble lenses as much as 200 m thick
	Tsar Creek	200-600	Dark pelitic schist with lenses of sandy carbonate
Gog Group	Mahto	250	Fine-grained pink quartzite with minor pelite and carbonate; local lenses of basal quartz-pebble conglomerate
	Mural	20-150	Pure marble with micaceous partings; dolomitic sandy carbonate and calcareous sandstone
	McNaughton	600	Medium- to coarse-grained white quartzite at the base, overlain by quartzite with pelitic interbeds
Miette Group	Upper clastic	200	Pelitic schist interbedded with fine-grained quartzite; local lenses of sandy carbonate and granule conglomerate
	Carbonate	100-200	Gray marble lenses in interbedded pelite, quartzite, and sandy dolomite
	Slate	800-1,000	Dark, fine-grained pelitic schist and slate with silty laminations and sandstone interbeds; granule conglomerate ("grit") interbeds near the base
	Grit	>1,000	Quartzose and feldspathic granule conglomerate and sandstone ("grit") with pelitic interbeds

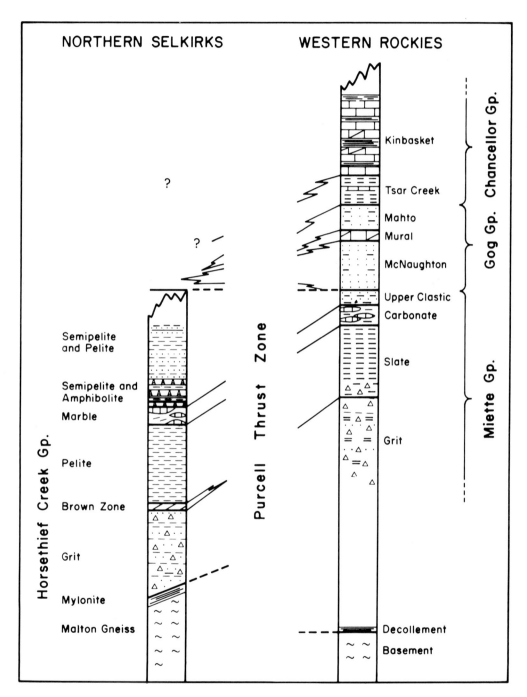

Figure 2. Stratigraphic correlation across Purcell thrust.

structural relationships with the few way-up determinations available, we have been able to establish the succession.

On the ridges west of Canoe Arm (see Fig. 3), an isoclinal synform is outlined by a thin calc-silicate–bearing marker, the "brown zone" (see Table 2). At the drainage divide, south of the Foster Creek inlet, graded bedding and the intersection of the earliest cleavage with bedding indicate that the isoclinal synform is a downward-closing anticline with the grit-bearing unit (which underlies the brown zone) in its core, as shown by cross section C–D in Figure 4. The grit-bearing unit, composed largely of monotonous repetitions of rather fine-grained, gritty feldspathic metasandstone and pelite, resembles closely the lower grit unit of the northern Purcell Mountains (Wheeler, 1963; Poulton, 1973) and the lower grit division of the western Rockies (Craw, 1978). From these observations, the succession given in Table 2 is established. We have some confidence in that sequence because it correlates very well with the succession of units in the Horsethief Creek Group of the northern Purcell Mountains, as well as with the succession outlined in Table 1 for the Miette Group of the western Rockies. The pelite unit, the slate unit (Tables 1 and 2), and the middle slate of the northern Purcells overlie a thick succession of feldspathic grit and underlie a carbonate-bearing unit. They are all about the same thickness and are characterized by a high alumina/alkali ratio and a high iron content. It is probable that all these pelitic units are correlative.

In mapping the western part of our area, we found that the most useful association of markers is the contact between the pelite and the overlying flaggy semipelite-amphibolite unit. Marble and calc-silicate beds occur near this contact, but they are discontinuous and can be found above, at, or below the contact. The semipelite-amphibolite unit is characterized by the interlayering—on a scale of 5 to 500 cm—of amphibolite, dark hornblende-biotite schist with large garnet porphyroblasts, and biotite-muscovite semipelite. Sheets of amphibolite as much as 300 m thick, which probably are metamorphosed gabbroic sills, are virtually restricted to this zone. It is overlain by a great thickness of semipelite with minor amphibolite layers and only rare aluminous pelite horizons.

If the main pelitic unit, with the overlying marble-bearing horizon, is correlated across the Rocky Mountain trench as suggested above, then it is obvious that there is a drastic increase in thickness from the upper clastic unit (which is 200 m thick in the Rockies) to the semipelite-amphibolite unit (which is more than 1,000 m thick in the Selkirks).

The stratigraphy we have outlined can be traced northward to the southern margin of the Malton Gneiss. There, on the ridge north of Windfall Creek, the stratigraphic succession is exposed, as shown in cross section E–F of Figure 4, from the semipelite-amphibolite unit downward to the grit unit. The upper surface of the Malton Gneiss emerges northward from underneath the grit unit. It would appear, then, that the Malton Gneiss is overlain by a right-way-up panel of Proterozoic metasedimentary rocks. Near the contact, the Horsethief Creek strata are not at all migmatized. The clearly metasedimentary schist and granofels stand in sharp contrast to the underlying streaky and banded granitoid Malton orthogneisses. The orthogneiss is mylonitized at the upper contact of the Malton Gneiss, and F_2 folds and metamorphic fabrics (described below) are imposed on the foliated mylonitic rocks. Furthermore, orthogneiss is nowhere seen intruding Hadrynian strata. These relationships show that Malton orthogneiss was already a gneiss before deformation and metamorphism of the Horsethief Creek metasedimentary rocks during the evolution of the Columbian orogen. From these data we infer that the Malton Gneiss represents a piece of Hudsonian (1,800 to 1,600 m.y. old) basement which underlies the southern Canadian Rockies. This inference is strongly supported by a variety of Precambrian radiometric ages obtained in recent years (Campbell, 1973; Chamberlain and others, 1978).

The contact, however, is not a simple unconformity between the Proterozoic cover succession and its gneissic basement. Near its upper surface, the gneiss is highly deformed, streaky, and mylonitic, with a white quartzose blastomylonite zone 20 to 50 m thick at the contact. The contact is approximately parallel to layering in the metasedimentary rocks above, as well as in the gneiss below, but near the head of Windfall Creek (as already recognized by Campbell, 1968), the contact truncates layering in its footwall as well as in its hanging wall. At an early orogenic stage, it could have marked the cover-basement decollement zone such that the cover sequence in its present position is allochthonous on the gneiss.

The stratigraphic relationships depicted in Figure 2 strongly suggest that the successions seen in the northern Selkirks and in the Rocky Mountains immediately to the east represent somewhat different sedimentary facies and that the present juxtaposition of these facies is the result of significant northeastward transport of allochthonous sheets relative to their substratum. The decollement seen at the top of the Malton Gneiss may, in part, account for this transport.

STRUCTURE

The salient features of the large-scale structure are depicted on Figures 3, 4, and 5. From east to west beginning in the western Rocky Mountains, the major structures are (1) the Porcupine Creek anticlinorium, in which one can distinguish a Proterozoic core (which is framed by competent Lower Cambrian beds) and a folded and imbricated west flank; (2) the Purcell thrust zone; (3) the Malton Gneiss; and (4) the Mica Dam antiform, which is developed in a thick package of Proterozoic strata that are tightly folded with axial planes and fold limbs dipping southwest at a moderate angle.

The Purcell thrust clearly divides the area into two separate tectonic domains. It dips to the southwest and brings the western structural edifice against and over the west flank of the Porcupine Creek anticlinorium. The structural effect of the Purcell thrust is clear in the northern part of the area where it brings the Malton Gneiss and Horsethief Creek strata over the Cambrian beds of the Rockies. Farther south, however, this relationship is obscured by a series of en echelon, hinged normal faults with downthrow to the southwest of several hundred metres. The topography of the Rocky Mountain trench closely reflects the traces of these normal faults. The Purcell thrust emerges again from this zone

TABLE 2. SUMMARY OF STRATIGRAPHY OF SOUTHERN CANOE RIVER AREA WEST OF ROCKY MOUNTAIN TRENCH

Unit	Thickness	Description
Semipelite-amphibolite	1,000 m (no top)	Semipelite and pelite with psammite and, locally, amphibolite
		In the lower part, mostly flaggy semipelite with amphibolite and calc-silicate beds, dark garnet-mica schists, and biotite-hornblende schists; some quartzite beds locally
		Marble and quartzite near the base
Marble	0-100 m	Gray and brownish laminated marble, coarse massive lenses, quartzose marble, and calc-silicate beds; semipelite and pelite beds are common
Pelite	600 m	Dark mica schist with widespread aluminum silicates, well laminated and with psammite and semipelite interbeds; locally, calcareous conglomerate near the top; coarse psammite and grit beds near the base
"Brown zone"	5-20 m	Zone that weathers brown and contains calc-silicate, amphibolite, and marble layers in semipelite
Grit	100-600 m (no stratigraphic base)	Semipelite with coarse psammite and grit beds, in regular 30-cm to 2-m-thick beds; local pebbly horizons; much dark mica schist especially near the top
..	..	Contact is a mylonite zone
Malton Gneiss	..	Gray, granitoid hornblende and biotite gneiss with amphibolitic lenses and layers; local zones of mica schist with metasedimentary aspect

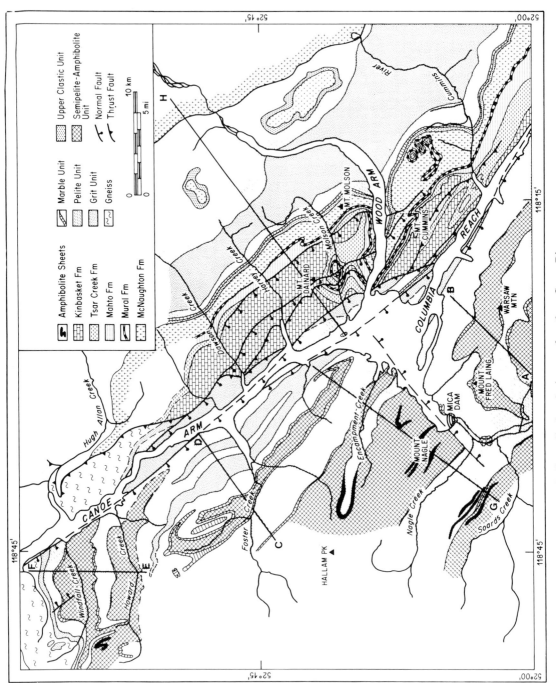

Figure 3. Generalized geologic map of southern Canoe River area.

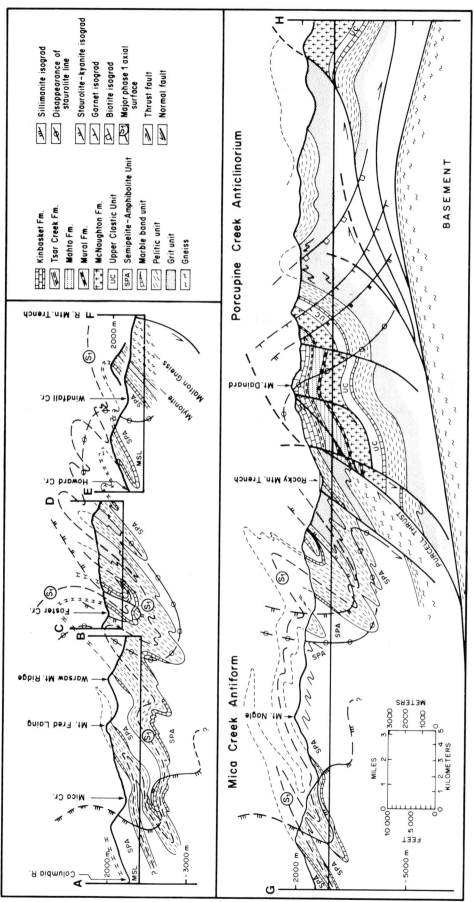

Figure 4. Cross sections illustrating structure of southern Canoe River area. A–B, C–D, and E–F form a composite section to illustrate relationships of structures from Mica Creek to Malton Gneiss. The sections are linked by common structural and stratigraphic elements, and all the geology shown is derived by axial projection from the map in Figure 3, which also shows the locations of the cross sections.

of normal faults, 50 km to the southeast of our area where it brings lower Horsethief Creek strata over Cambrian-Ordovician beds on the floor of the trench. In that area, north of the Trans-Canada Highway, the Purcell thrust fault dips about 35°SW and is the sole fault of a stack of imbricated and folded thrusts which merge with it northward (Wheeler, 1963; Simony and Wind, 1970). The cumulative northeastward displacement on that stack of thrusts is at least 20 km.

In the southern Canoe River area there appear to be no imbrications in the hanging wall, but there are three small thrusts in the footwall on the west flank of the Porcupine Creek anticlinorium. These faults are probably related to the Purcell thrust in that they merge with it.

At least three sets of fold structures—F_1, F_2, and F_3—occur in superposition in the area. Axial-plane foliations S_1, S_2, and S_3 are associated with them. They are not, however, equally developed everywhere, and different fold sets dominate the map pattern in different subareas.

The Porcupine Creek anticlinorium is the dominant structure of the western Rockies. In its core, the grit and slate units of the Miette Group are folded into tight, upright folds with steeply dipping, axial-plane, slaty cleavage. The competent Lower Cambrian beds outline subsidiary folds that constitute the west flank of the anticlinorium. These folds are nearly concentric with minor thickening of incompetent layers in fold hinges. In the Lower Cambrian quartzite beds, sedimentary structures such as pebble bands, cross-bedding, ripple marks, and so forth are well preserved. In incompetent calcareous and pelitic layers, on the other hand, intense deformation is apparent on the mesoscopic and microscopic scale. In particular, a schistosity is present which is nearly parallel to bedding and to axial planes of mesoscopic isoclinal folds. That schistosity is folded by the large-scale upright folds and crenulations, and a strain-slip cleavage congruent with the upright folds is imposed on it. On the west flank of the Porcupine Creek anticlinorium, we thus see two sets of structures; an early schistosity and associated mesoscopic folds and a later set of large folds and associated crenulations. We show below that the large folds are F_3 and the earlier schistosity is probably S_2. The stratigraphic control is sufficiently good to demonstrate that no major recumbent folds are present that are associated with the early schistosity and that predate the upright folds. The metamorphic index minerals lie as porphyroblasts in the early schistosity, but only minor muscovite and chlorite recrystallization is associated with the later phase. This suggests that the main prograde metamorphism overlapped in time with the formation of the early schistosity but that the large upright folds are largely late-metamorphic to postmetamorphic.

The minor thrust faults, which are imbrications related to the Purcell thrust, pass into and are rooted in some of the upright folds. They are therefore linked in time and space with these folds. The quartz c-axis fabric in quartzite adjacent to one of these faults shows (Craw, 1978) that the metamorphic fabric is modified increasingly as the fault is approached. This indicates that the late folds and associated thrusts postdate metamorphism and suggests that the Purcell thrust, to which the minor thrusts appear related, is also postmetamorphic.

The regional plunge of the Porcupine Creek anticlinorium and of the folds on its west flank is about 3°SE from the culmination in the region of Yellowhead Pass to the lower Kicking Horse River region east of Golden, British Columbia, 160 km southeast of Mica Dam. There are local variations, however, and the anticline-syncline pair on the west flank plunges northwestward at 5° to 15° in the vicinity of Wood Arm (see Fig 3). In the core of the anticlinorium, all minor fold hinges are approximately congruent with the major structure. On the west flank, however, tight early folds have plunges that vary within the schistosity plane, locally plunging down the southwest dip.

In the Mica Dam area (Figs. 3 and 4, cross section G–H), a large antiform dominates the structures. It can be traced northwestward to the head of Foster Creek (Fig. 3). The transverse normal fault that passes by the west abutment of Mica Dam drops the fold hinge some 1,500 m to the southeast. This, in conjunction with a 5° to 15° southeasterly axial plunge, exposes higher structural levels in the Warsaw Mountain area southeast of the dam. The axial surface of the antiform can be followed through a

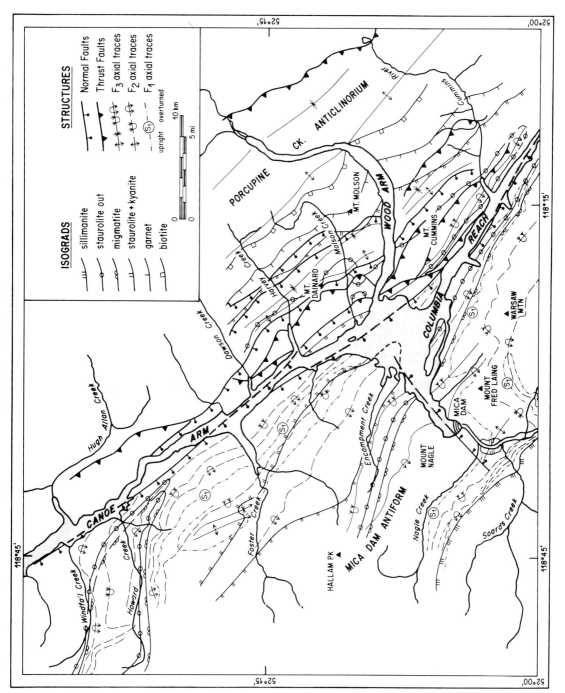

Figure 5. Generalized map of metamorphic isograds and major structures in southern Canoe River area.

structural stack some 5 km thick of earlier isoclinal folds. In the Mount Nagle massif, we observe the broad and complex core of the antiform. Upward, the synform adjacent on the northeast decreases in amplitude, and the fold pair is dissipated into a series of small northeast-verging folds that rumple the southwest-dipping homocline south of Warsaw Mountain. These folds, with wavelengths of 0.5 to 50 m, suggest the presence of a large antiform to the northeast; however, such a large fold is not present, and the small folds may instead be kinematically related to the Purcell thrust.

The Mica Dam fold pair and the associated minor folds are not concentric; in fact, some folds are approximately similar in style. They do, however, have many characteristics suggesting a buckle stage in their evolution; competent layers maintain approximately constant thickness, folds die in zones of disharmony and decollement, and the folds vary greatly in style and attitude as a function of rock type. In particular, the dip of the axial surfaces varies from fold to fold; box folds are not uncommon, and in many folds the axial surface is curved.

The Mica Dam folds and their associated minor structures clearly deform a pre-existing penetrative regional schistosity, S_{1+2}, developed in earlier deformation phases. Crenulation and strain-slip cleavage, S_3, congruent with the Mica Dam folds are superimposed on that earlier schistosity. The metamorphic index minerals—biotite, garnet, staurolite, kyanite, and sillimanite—lie in the schistosity and are themselves bent and kinked in the crenulated zones; these facts demonstrate that the Mica Dam folds postdate the regional development of schistosity and the metamorphic climax. These relationships, as well as the style and vergence, suggest that the Mica Dam folds and the folds of the Porcupine Creek anticlinorium are related.

The folds in the area north of Foster Creek, which have south-trending axes and which are outlined by the marble unit, have a style very similar to that of the Mica Dam folds. They have the same relation to the regional schistosity S_{1+2} as well as to the timing of metamorphism. These folds die out upward and downward with strong disharmony in pelitic horizons. Mesoscopic folds and lineations, with a southeasterly trend congruent with the Mica Dam folds, are superimposed on the Foster Creek folds. The latter therefore began to form a little earlier than the Mica Dam folds.

Mica Dam and Foster Creek folds are superimposed on the long limbs of earlier, nearly isoclinal folds. These earlier folds are largely responsible for the map pattern, and their hinges are exposed north and south of Howard Creek as well as at the mouth of Mica Creek. Outlining these folds on the map does not lead to a simple stratigraphic scheme, and it is clear that yet-earlier folds must be present.

A line that is drawn on the map (Fig. 5) and shown on cross sections A–F and G–H (Fig. 4) represents a surface separating two identical stratigraphic packages facing in opposite directions. That surface lies within the aluminous pelitic unit in much of the area. Beneath it, the sequence downward is pelite to marble to semipelite-amphibolite. Northeastward the surface passes into the grit-bearing core of the downward-closing anticline on the slopes west of Canoe Arm described earlier. Figure 4 illustrates how that represents the axial surface of a major recumbent fold nappe. If our stratigraphic interpretation is correct, then the nappe has older and older rocks in its core to the northeast, and it eventually roots somewhere to the northeast, perhaps on Malton Gneiss now hidden by later eastward thrusting. The core of the overlying recumbent syncline is outlined by the marble unit forming the hook-shaped fold south of Howard Creek.

Once these great recumbent folds are recognized, the stratigraphic relationships are all explained in a simple, consistent manner, and there is no evidence for a yet-earlier set of folds. We therefore conclude that the west-verging recumbent folds are the earliest folds (F_1); these folds are then refolded by the northeast-verging tight folds (F_2) to form a generally southwest-dipping stack of folds on which the third phase of folds (F_3), like the Mica Dam folds, are then imprinted.

F_1 and F_2 folds both have penetrative axial-plane schistosity associated with them. Only locally in some F_2 hinge zones is the F_1 schistosity seen to be folded and crenulated by the F_2 folds. In general, the pervasive schistosity of the area can only be described as an S_{1+2} foliation surface, and it is approximately parallel to the compositional layering. Muscovite, biotite, and flattened quartz grains outline that foliation.

F_1 and F_2 mesoscopic folds as well as associated axial lineations vary considerably in orientation within the S_{1+2} surface. On the mesoscopic scale, F_1 and F_2 lineations are folded by later folds. Statistically, however, the folds of all three phases are approximately coaxial; this is borne out by the map pattern. The long strips formed by formations in fold cores can only be present in an area where polyphase deformation has taken place if the folds are all approximately coaxial and gently plunging.

North of Howard Creek (Fig. 3), the structural "grain," largely composed of F_2 folds, is deflected to the west such as to swing around the south margin of the Malton Gneiss. This swing in strike is a consequence of the steepening southeast plunge on the south flank of the gneiss. It is this steepening in plunge that brings the gneiss to the surface. Because the Purcell thrust obliquely truncates the curving structures, it must postdate F_2 folding and the early development of the structural culmination.

The Yellowhead Pass–Malton culmination is a very large feature (Price and Mountjoy, 1970; Campbell, 1973). Furthermore, we show below that it probably rose late in the structural evolution. We therefore propose that the basement rose into the culmination (cross section G–H, Fig. 4). We draw the basement no higher than 5 km below sea level because no basement outcrops are known in the culmination some 70 km up plunge of cross section G–H. We propose, however, that the basement was not involved in the Rockies in the thrusting which predates the Purcell thrust and that a major decollement surface formed at its upper surface as in the thin-skinned model of Price and Mountjoy (1970). The mylonite zone at the top of the Malton Gneiss may represent that decollement surface.

METAMORPHISM

The lines of first appearance of the index minerals biotite, garnet, staurolite, kyanite, and sillimanite were traced in the field, as were the prograde disappearance of staurolite in the kyanite zone and the first appearance of granitoid leucosomes (Fig. 5). The index minerals first appear as porphyroblasts that are recognizable in the field, and subsequent petrographic work required only minor adjustments of the mapped positions of the isograds. These isograds are, in fact, the traces of isogradic surfaces that outline a series of metamorphic antiforms and synforms (Figs. 4 and 5) on the northeast flank of a very large metamorphic antiform, the Shuswap metamorphic complex.

Metamorphism began during the first deformation phase as indicated by muscovite and biotite which define a schistosity that is axial planar to F_1. The isogradic surfaces cut obliquely across the axial surfaces (S_2) of second phase folds (F_2) without themselves being deflected. This indicates that the metamorphic surfaces were fixed into the rock mass late in F_2 folding or later. Garnet has S_{1+2} muscovite and biotite "wrapped" around it (Spry, 1969), as well as having an internal schistosity (S_i) consisting of biotite, quartz, and ilmenite which locally delineates S trails truncated by an inclusion-free rim. This suggests that garnet crystallized during and a little after the second deformation phase.

The sillimanite isogradic surface south of Mica Creek was traced across 2,000 m of topographic relief; it dips southwest at about 60°, that is, it is overturned. A small inlier of the sillimanite zone crops out in the Columbia Valley north of Mica Creek and coincides exactly with the crest of a small F_3 fold on the southwest flank of the Mica Dam antiform. The sillimanite isogradic surface is thus folded by F_3, as shown in cross sections A–B and G–H of Figure 4. The observation of bent and kinked staurolite, kyanite, and sillimanite in rocks where S_3 is well developed is consistent with the deformation of isogradic surfaces by F_3. Biotite lying in S_{1+2} is strained by F_3 crenulations, but locally it has recrystallized. All these relationships indicate that the higher metamorphic grades were reached late in F_2 but that the metamorphic grade was lower again during F_3.

The isograds, like the structural grain, are deflected westward south of the Malton Gneiss (Fig. 5) and thus show that the culmination cored by the gneiss rose after positioning of the isograds. South of the Malton Gneiss, metamorphic grade decreases from migmatitic upper kyanite grade to staurolite-kyanite grade without migmatite as the gneiss is approached. The gneiss was not emplaced as a "hot mobile tongue" (Price and Mountjoy, 1970) hotter than the surrounding metasedimentary rocks. Metamorphism postdates the establishment of the present relationship between the gneiss and its cover.

The geometry of the isogradic surfaces is only partially explained by F_3 folding. The metamorphic antiforms and synforms were accentuated by postmetamorphic deformation, but they must represent original highs and lows on the isogradic surfaces.

Isobaric surfaces are unlikely to have steep gradients in the core zone of an orogenic belt. Moreover, the stability of most of the index minerals, except kyanite and sillimanite, is more sensitive to temperature than to pressure changes. Therefore, the morphology of the isogradic surfaces must approximately reflect the morphology of isothermal surfaces during the metamorphic maximum. The anisotropy (foliation and lineation) imparted to the rock mass during the second deformation phase appears to have exerted some control on the shape of the isothermal surfaces. They strike approximately parallel to the structural trend and dip southwest, parallel to the southwesterly dip of the foliation.

The surface of first appearance of trondhjemitic leucosome approximately coincides with the disappearance of staurolite within the kyanite zone west of the Purcell thrust. Leucosome appears in the form of blebs, lenses, rods, and layers that are confined within the foliation and the layering. Ten to thirty percent of the rock in the upper kyanite and sillimanite zones is composed of leucosome. Nowhere is the character of the original rock obliterated, and the stratigraphic subdivisions are everywhere recognizable. The leucosome is only weakly foliated and has granitic texture. Pegmatite also appears within the upper kyanite zone in the form of irregular bodies, sheets, and clearly crosscutting dikes. Muscovite, biotite, garnet, and tourmaline are common accessory minerals within the pegmatite. Layers of trondhjemitic leucosome and of pegmatite are folded by F_3; mesoscopic folds, lineations, and boudins congruent with F_3 formed within pegmatite and leucosome masses, which must have formed before F_3. Intensely foliated garnet, biotite, and muscovite pegmatite in the Mount Nagle massif indicates that some pegmatite had begun to form early in F_2 folding. However, most pegmatite and trondhjemite leucosome are only weakly foliated and probably formed during the peak of metamorphism, after the S_{1+2} foliation was well developed in the later stages of F_2 deformation.

PRESSURE–TEMPERATURE ESTIMATES

Metamorphism in the southwestern Canoe River area corresponds to bathozone 5 as defined by Carmichael (1978). Kyanite is the stable aluminum-silicate polymorph at lower grade and sillimanite occurs at higher grade. Staurolite disappears in the kyanite stability field, whereas in the Rockies, staurolite does not break down until near the appearance of fibrolitic sillimanite (Craw, 1978). The lack of stable K-feldspar + kyanite indicates that bathozone 6 conditions (Carmichael, 1978) were not attained during the metamorphism.

Temperature and pressure estimates have been made for staurolite- and kyanite-grade rocks in the Monashee Mountains by Ghent and others (1979) and for sillimanite-grade rocks by Knitter (1978, oral commun.). Temperatures estimated from the distribution of Fe and Mg between garnet and biotite (Thompson, 1976) range from 570 to 640 °C for staurolite- to kyanite-grade rocks. In the Rockies, Craw (1978) obtained temperatures ranging from 480 °C in garnet-grade rocks to 570 °C in kyanite-grade rocks. The spatial distribution of temperature estimates is consistent with estimation of metamorphic grade based on mapping of isograds (Fig. 5). Maximum temperature estimates in muscovite-sillimanite–bearing assemblages west of Mica Dam are near 655 °C (Knitter, 1978, oral commun.).

Pressure estimates have been made from garnet–plagioclase–aluminum-silicate–quartz equilibria (Ghent, 1976; Ghent and others, 1979). If temperatures estimated from garnet-biotite equilibria are used, estimated pressures range from about 6 to 8.5 kb. The highest estimated pressures occurred near the north end of the map area; this is consistent with our interpretation that the deepest structural levels now exposed are at the extreme north end of the map area.

Craw (1978) demonstrated that there was a sharp contrast in the metamorphic load pressures between the rocks exposed in the Rocky Mountains to the west. In the Rocky Mountains, staurolite persists throughout the kyanite zone and disappears only near the first appearance of fibrolite. In the Monashee Mountains and Selkirk Mountains near Mica Dam, on the other hand, the upper kyanite zone is without staurolite and is 5 km thick. Migmatite and pegmatite are absent in the Rockies but are abundant in the thick upper kyanite zone to the west. Craw (1978) concluded that the rocks west of the Purcell thrust were metamorphosed under load pressures 2 to 3 kb higher than those of the adjacent western Rocky Mountains. This corresponds to a difference in depth of burial of 7 to 10 km and suggests a vertical component of postmetamorphic movement on the intervening Purcell thrust on the order of 7 to 10 km.

Important postmetamorphic movement on the Purcell thrust is also suggested by the fact that the fault obliquely truncates the isogradic surfaces in its hanging wall as well as in its footwall and therefore telescopes the metamorphic pattern on the northeast flank of the metamorphic complex.

SUMMARY AND CONCLUSIONS

The cross section G–H through southern Canoe River area (Fig. 4) illustrates the transition from the foreland fold and thrust belt to the metamorphic core zone of the Columbian orogen. The Hadrynian-Cambrian succession can be followed from the biotite to the sillimanite zones of the Barrovian facies series. The structural style changes concomitantly from fairly simple upright folds and associated thrust faults in the northeast to fold nappes that have undergone polyphase deformation in the southwest. The basement of the western Rockies is probably only mildly deformed and may represent a broad bulge corresponding to the Yellowhead Pass culmination. In the core zone, on the other hand, basement rocks occur in sheets deformed and metamorphosed along with their cover.

The change in metamorphism and deformation is not, however, a gradual one. We have emphasized the distinct contrast between the domains east and west of the Purcell thrust. The juxtaposition of different sedimentary facies requires significant horizontal shortening, say, 10 km or more. The 2- to 3-kb difference in pressure requires that the hanging-wall block has risen at least 7 km, and the contrast in structural style is best explained by a combination of uprise and foreshortening.

The tectonic and metamorphic evolution of the area is graphically summarized in Figure 6. The tectonic history began in the west during the Late Jurassic (Price and Mountjoy, 1970) (or much earlier?) with the development of F_1 nappes in the cover sequence which was sliding on the basement. Metamorphic grade gradually rose to that of garnet in the core zone, whereas no deformation or metamorphism was going on in the Rockies at that time. The relationship between the Hadrynian cover and its Malton Gneiss basement was established at that time also and can now be seen west of the Purcell thrust. Neither the cover nor the basement were "hot." The basement was probably significantly less ductile than the schistose cover and developed mylonite near its upper surface.

During the second phase of deformation, the early nappes were refolded. Metamorphic grade gradually rose, and both deformation and metamorphism began to affect the western Rockies. Migmatite formed in the metamorphic core zone during the metamorphic climax late in F_2 deformation. This metamorphism was imprinted on the Malton gneisses as well as on the Hadrynian cover. The emplacement of the gneisses neither caused the metamorphism nor drove the deformation. During F_3

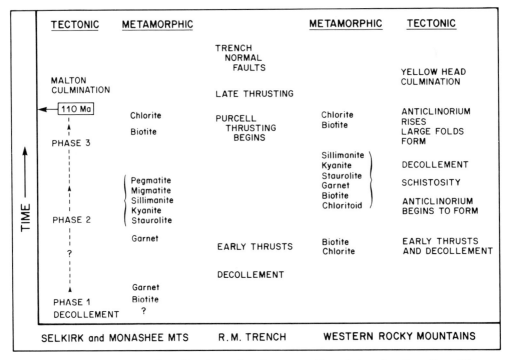

Figure 6. Space-time diagrammatic outline of tectonic and metamorphic evolution of southern Canoe River area.

folding, the Porcupine Creek anticlinorium took its present form, the Yellowhead Pass and Malton culminations rose, and the last major movement on the Purcell thrust took place. The hinged en echelon normal faults of the Rocky Mountain trench only formed after all thrusting had taken place. The eastward swing of the metamorphic isograds into the Rockies (described by Price and Mountjoy, 1970) can probably be explained by the rise of the Yellowhead Pass culmination and its being superimposed on northeast-dipping isogradic surfaces.

Radiometric ages on micas quoted by Price and Mountjoy (1970) for the western Rockies were determined on porphyroblasts that grew late in the metamorphic and deformational history described here. Their Albian age suggests that the evolution of the eastern flank of the core zone outlined here took place in earliest Cretaceous and Jurassic time.

The temperature and pressure gradients determined for the area are consistent with the inferred structure sections and the isograd pattern. That pattern suggests that the isothermal surfaces had significant relief during the metamorphic climax. Pressures of 8 kb require depths of burial on the order of 27 km. The thick stratigraphic succession on the west flank of the Rockies, greatly thickened by deformation, may account for such depths of burial, but we cannot yet document this interpretation.

REFERENCES CITED

Brown, R. L., Perkins, M. J., and Tippett, C. R., 1977, Structure and stratigraphy of the Big Bend area, British Columbia, *in* Report of Activities, Part A: Canada Geological Survey Paper 77-1A, p. 273.

Brown, R. L., Tippet, C. R., and Lane, L. S., 1978, Stratigraphy, facies changes and correlations in the northern Selkirk Mountains, southern Canadian Cordillera: Canadian Journal of Earth Sciences, v. 15, p. 1129–1140.

Campbell, R. B., 1968, Canoe River (83D), British Columbia: Canada Geological Survey Map 15-1967.

—— 1973, Structural cross-section and tectonic model of the southeastern Canadian Cordillera: Canadian Journal of Earth Sciences, v. 10, p. 1607–1620.

Campbell, R. B., Mountjoy, E. W., and Young, F. G., 1973, McBride map area, British Columbia: Canada Geological Survey Paper 72-35, 104 p.

Carmichael, D. M., 1978, Metamorphic bathozones and bathograds: A measure of depth of post-metamorphic uplift and erosion: American Journal of Science, v. 278, p. 769–797.

Chamberlain, V. E., Lambert, R.St.J., and Holland, J. G., 1978, Preliminary subdivisions of the Malton Gneiss Complex, British Columbia, *in* Current research, Part A: Canada Geological Survey Paper 78-1A, p. 491–492.

Cook, D. G., 1970, A Cambrian facies change and its effect on structure, Mt. Stephen–Mt. Dennis area, Alberta, British Columbia: Geological Association of Canada Special Paper 6, p. 27–40.

Craw, D., 1978, Metamorphism, structure and stratigraphy in the southern Park Ranges, British Columbia: Canadian Journal of Earth Sciences, v. 15, p. 86–98.

Evans, C. S., 1933, Brisco-Dogtooth map area, British Columbia: Canada Geological Survey Summary Report, Part A, p. 106–176.

Fyles, J. T., 1960, Geological reconnaissance of the Columbia River between Bluewater Creek and Mica Creek, B.C.: Minister of Mines Annual Report 1959, p. 90–105.

Ghent, E. D., 1976, Plagioclase-garnet-Al_2SiO_5-quartz: A potential geobarometer-geothermometer: American Mineralogist, v. 61, p. 710–714.

Ghent, E. D., Robbins, D. B., and Stout, M. Z., 1979, Geothermometry, geobarometry, and fluid compositions of metamorphosed calc-silicates and pelites, Mica Creek, British Columbia: American Mineralogist, v. 64, p. 874–885.

Giovenella, C. A., 1967, Structural relationships of the metamorphic rocks along the Rocky Mountain trench at Canoe River, British Columbia: Canada Geological Survey Paper 67-1A, p. 60–61.

Meilliez, F., 1972, Structure of the southern Solitude Range, British Columbia [M.Sc. thesis]: Calgary, Alberta, University of Calgary, 113 p.

Mountjoy, E. W., 1962, Mount Robson (southeast) map-area, Rocky Mountains of Alberta and British Columbia: Canada Geological Survey Paper 61-31, 114 p.

Poulton, T. P., 1973, Upper Proterozoic "Limestone Unit," Northern Dogtooth Mountains, British Columbia: Canadian Journal of Earth Sciences, v. 10, p. 292–305.

Price, R. A., and Mountjoy, E. W., 1970, Geological structure of the Canadian Rocky Mountains between Bow and Athabasca Rivers. A progress report: Geological Association of Canada Special Paper 6, p. 7–26.

Reesor, J. E., 1965, Structural evolution and plutonism in Valhalla Gneiss Complex, British Columbia: Canada Geological Survey Bulletin 129, 128 p.

Reesor, J. E., and Moore, J. M., Jr., 1971, Petrology and structure of Thor-Odin gneiss dome, Shuswap metamorphic complex, British Columbia: Canada Geological Survey Bulletin 195, 149 p.

Simony, P. S., and Wind, G., 1970, Structure of the Dogtooth Range and adjacent portions of the Rocky Mountain trench: Geological Association of Canada Special Paper 6, p. 45–51.

Spry, A., 1969, Metamorphic textures: New York, Pergamon Press, 350 p.

Thompson, A. B., 1976, Mineral reactions in pelitic rocks: II. Calculations of some P-T-X (Fe-Mg) relations: American Journal of Science, v. 276, p. 425–454.

Wheeler, J. O., 1963, Rogers Pass map area, British Columbia: Canada Geological Survey Paper 62-32, 34 p.

—— 1965, Big Bend map-area, British Columbia: Canada Geological Survey Paper 64-32, 37 p.

Wheeler, J. O., and Gabrielse, H., 1972, The Cordilleran structural province, *in* Price, R. A., and Douglas, R.J.W., eds., Variations in tectonic styles in Canada: Geological Association of Canada Special Paper 11, p. 1–81.

Young, F. G., Campbell, R. B., and Poulton, T. P., 1973, The Windermere Supergroup of the southeastern Canadian Cordillera, *in* Belt symposium: Moscow, Idaho, University of Idaho and Idaho Bureau of Mines, Symposium Series 1, p. 181–203.

Manuscript Received by the Society June 21, 1979

Manuscript Accepted August 7, 1979

Printed in U.S.A.

Geological Society of America
Memoir 153
1980

Kettle dome and related structures of northeastern Washington

Eric S. Cheney
Department of Geological Sciences
University of Washington
Seattle, Washington 98195

ABSTRACT

The Kettle River Range in Ferry County, Washington, is underlain by sillimanite-grade rocks of the Tenas Mary Creek sequence. Two >800-m-thick sheets of augen gneiss occur above and below feldspathic quartzite, biotitic gneiss, and minor marble. Polyphase deformation (including mylonites and east-trending lineations) and slightly uraniferous aplitic to pegmatitic bodies are common. Cataclasis is common, and rocks of the Tenas Mary Creek sequence appear to be in tectonic contact with overlying upper Paleozoic phyllitic rocks. Fine-grained biotitic metasedimentary rocks occurring locally between the phyllitic rocks and rocks of the Tenas Mary Creek probably are older than the late Paleozoic phyllitic rocks.

Foliation and contacts in the Tenas Mary Creek sequence rarely dip >25° and define the flat-topped Kettle dome (which is >65 km long north-south, 27 km wide, and has about 3 km of structural relief). The Okanogan dome west of the Kettle dome consists primarily of orthogneisses and granitic plutons of Mesozoic(?) age. Rocks in the flat-topped Spokane dome along the Washington-Idaho border are lithologically similar to those of the Tenas Mary Creek and may be pre-Beltian.

The Sanpoil syncline between the Kettle and Okanogan domes and a syncline on the northeastern margin of the Kettle dome contain Eocene sedimentary and volcanic rocks. Because the axes and structural reliefs of the Okanogan dome, the Sanpoil syncline, and the Kettle dome are similar, the present structural relief (as opposed to the internal structure and high-grade metamorphism) of the Kettle dome probably is due to post-Eocene folding. The gently synformal Tertiary Newport fault straddling the Washington-Idaho border may be a related structural feature. Four other low-angle faults, three of which cut Tertiary rocks, occur between the domes.

The low-angle faults commonly are marked by cataclastic zones more than 100 m thick. Cataclasis occurred as the basement of batholiths and pre-Beltian(?) metamorphic rocks became decoupled from overlying Precambrian to Tertiary layered rocks. Whether this decoupling represents one or more zones of Tertiary decollement of regional extent is not yet known. Owing to post-Eocene (possibly late Miocene or younger) folding, the cataclastic zones crop out on the margins of the domes.

INTRODUCTION

Sillimanite-grade rocks in the Kettle River Range serve to identify the Kettle dome of northeastern Washington (Fig. 1). The Kettle dome and the nearby Okanogan and Spokane domes lie geographically between the Shuswap domes, which are thought to be either early Mesozoic gneiss domes (Okulitch and others, 1977) or a Precambrian metamorphic terrane (Duncan, 1978), and the core complexes of the southwestern United States, which are thought to be Tertiary metamorphic rocks cut by mid-Tertiary decollement zones (Davis and Coney, 1979).

The purpose of this paper is to describe the Kettle dome and its significance to Tertiary tectonics of northeastern Washington. The rather novel conclusions of this paper are that (1) the Kettle dome, instead of being a hot diapiric gneiss dome, is the gently upwarped basement of Precambrian(?) metamorphic rocks; (2) the doming occurred during a heretofore unappreciated period of post-Eocene folding; (3) cataclasis prior to the folding is widespread in northeastern Washington; and (4) both cataclasis and thrusting seem to separate a basement largely composed of Precambrian(?) metamorphic rocks and Mesozoic to Tertiary batholiths from Precambrian to Tertiary layered rocks. Much of

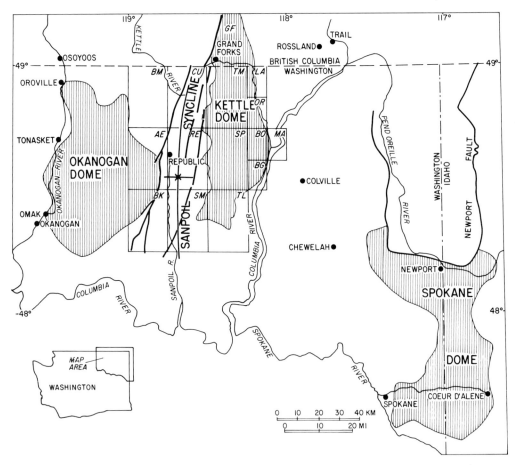

Figure 1. Index map. Quadrangles in Washington are AE = Aeneas, BG = Bangs Mountain, BK = Bald Knob, BM = Bodie Mountain, BO = Boyds, CU = Curlew, MA = Marcus, LA = Laurier, OR = Orient, RE = Republic, SM = Seventeenmile Mountain, SP = Sherman Peak, TL = Twin Lakes, and TM = Togo Mountain. GF = Grand Forks area in British Columbia.

the detailed geology still remains to be done. Therefore, whether the cataclasis in the crystalline rocks and the thrusts in the layered rocks are one or more Tertiary zones of decollement of regional extent is not known (Cheney, 1979).

I began reconnaissance mapping in the Kettle River Range in 1974 to evaluate the numerous radioactive pegmatites discovered during the uranium boom of the 1950s. In 1974 the only published geological mapping (Pardee, 1918; Campbell, 1938, 1946; Bowman, 1950; Lyons, 1967) was along the margins of the range. I soon recognized (Cheney, 1976, 1977) that the metamorphic rocks within the dome matched those described by Parker and Calkins (1964) as the Tenas Mary Creek sequence 20 km west of the dome and by Preto (1970) as the Grand Forks Group in the Canadian part of the dome.

The preliminary nature of this paper must be stressed. I spent only nine weeks from 1974 to 1978 mapping the equivalent of about four 15-min quadrangles in the Kettle dome and a similar period in 1976 and 1977 mapping an area of similar size in the eastern two-thirds of the Okanogan dome. Accordingly, this paper is offered to (1) illustrate the presence of a little-known but fascinating metamorphic terrane, (2) speculate on its regional significance, (3) note problems and areas that need further study, and (4) encourage all investigators to fit their much-needed studies into a regional framework.

ROCK UNITS OF THE KETTLE DOME

Metamorphic Rocks of the Tenas Mary Creek Sequence

General. In the Kettle dome the rocks of Tenas Mary Creek sequence consist of several heterogeneous units of sillimanite-bearing pelitic rocks, marble, and quartzite, a >650-m feldspathic quartzite (Fig. 2), and two sheets of augen gneiss each >800 m thick. More stratigraphic implications may be conveyed by Figure 2 than are intended: (1) because no indicators of tops have been discovered, the metasedimentary rocks are only assumed to be upright; (2) the granitic gneisses are orthogneisses; and (3) the origin of the stratiform amphibolites is unknown.

The rocks of Tenas Mary Creek sequence (including those described by Parker and Calkins, 1964, in the Curlew quadrangle) have common structural characteristics. All are foliated, many are lineated, and many are cataclastically deformed. Lineations generally strike eastward. Outcrops of compositionally layered rocks commonly have recumbent isoclinal folds that deform the foliation but do not have a mesoscopic axial-plane cleavage. These folds are folded by three types of more gentle upright folds (Lyons, 1967; Donnelly, 1978). Boudinage is common in layered rocks and in the slightly uraniferous aplitic to pegmatitic bodies.

Because the petrography of rocks of the Tenas Mary Creek sequence already has been described (Bowman, 1950; Parker and Calkins, 1964; Lyons, 1967; Preto, 1970; Pearson, 1967, 1977; Donnelly, 1978), the field relationships are emphasized below. The units are described from structurally lowest to highest.

Biotite Schist and Gneiss. A heterogeneous unit consisting of biotitic schist and gneiss, marble, and quartzite (B, BM, and BQ, respectively, in Fig. 2) crops out along the crest of the range from Mount Leona southward to Sherman Pass. The thickness of this unit may exceed 700 m at Mount Leona. Where the thick sheets of orthogneiss and the >650-m feldspathic quartzite are absent, the limited criteria in Table 1 are used to distinguish this heterogeneous unit from similar units.

This unit is not present in the type area described by Parker and Calkins (1964) in the Tenas Mary Creek area. Table 2 indicates that if the very thick quartzites in the Kettle dome and Grand Forks area are the same unit, the 700 m of biotitic rock, marble, and quartzite along the crest of the Kettle River Range probably correlates with the basal unit of the Grand Forks Group. This interpretation is compatible with the map pattern of the Tenas Mary Creek rocks north of Sherman Pass.

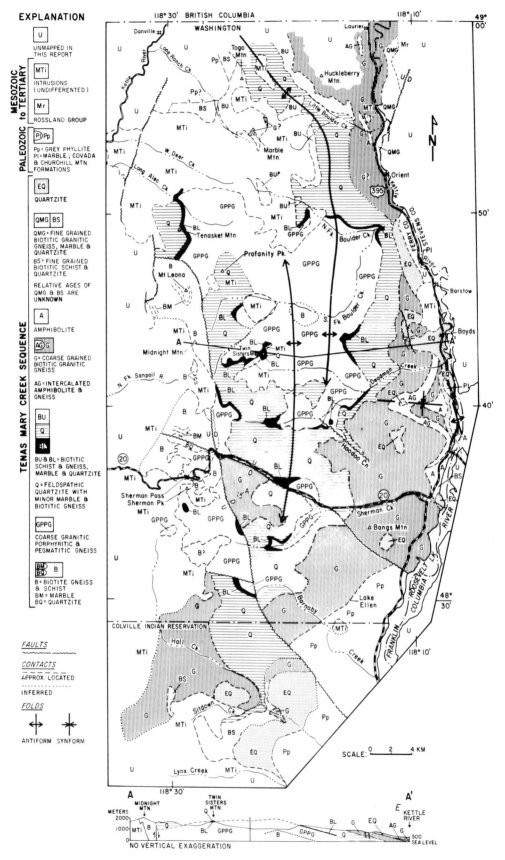

Figure 2. Preliminary geologic map of the Kettle dome, Ferry County, Washington.

TABLE 1. CRITERIA FOR DISTINGUISHING BETWEEN INTERVALS DOMINATED BY BIOTITIC SCHIST AND GNEISS IN THE KETTLE DOME

Map unit*	Approximate thickness (m)	Apparent geographic restriction	Distinctive lithologies and mineralogies	Thickness of marble and quartzite (m)	Amount of pegmatite
BS	200 to 600	Northeast quarter of Togo Mountain and north half of Twin Lakes quadrangle	Fine-grained biotitic schist, calc-silicate schist, minor amphibolite, no sillimanite	No marble <30	None
QMG	>600	Southwest third of Laurier quadrangle	Fine-grained biotitic gneiss, calc-silicate gneiss	<15	None?
BU	0 to 300	North half of Togo Mountain quadrangle	Medium-grained biotitic, sillimanite-bearing schist and gneiss, phlogopitic marble, quartzite	<60	Little
BL	100 to 150	Sherman Peak and south third of Togo Mountain quadrangle	Medium-grained biotitic, sillimanite-bearing schist and gneiss, phlogopitic marble, quartzite	<15	Common
B	800(?)	Mount Leona to Sherman Pass	Medium-grained, biotitic, sillimanite-bearing schist and gneiss, phlogopitic marble, quartzite	<60	Very common

*These map units are shown in Figure 2.

Granitic, Porphyritic, Pegmatitic Gneiss. The lowest widespread unit is a granitic, porphyritic, pegmatitic gneiss (GPPG in Fig. 2, hereafter "pegmatitic gneiss") that commonly has no compositional layering. The megacrysts are potassium feldspar. The pegmatitic portions are dikelets <1 m thick and irregular clots about a metre in diameter that grade outward into nonpegmatitic gneiss. Some of the irregular clots are remnants of dikelets that were disrupted along foliation planes or dismembered by folding. The foliation is sufficiently weak that some previous investigators included the pegmatitic gneiss in adjacent intrusions. This gneiss is one of the thickest (>850 m) and most extensive of all of the Tenas Mary Creek units, extending from the crest of the range on the west to the valley bottoms on the east. Small outcrops of biotitic gneiss and marble along the North Fork of Deadman Creek suggest that the lower contact of the pegmatite gneiss may be exposed. This interpretation is shown in the cross section of Figure 2.

The stratiform map pattern of the thin unit of biotitic schist, gneiss, and rather pure marble in the quartzite-dominated sequence above the pegmatite gneiss led Parker and Calkins (1964) to suggest that the pegmatite gneiss itself is a metasedimentary unit. However, Pearson (1967) was impressed by the lack of compositional layering and suggested that it is an orthogneiss and that the unit of biotitic schist and gneiss and marble was nonconformably deposited upon it.

An alternative interpretation is that the protolith of the pegmatite gneiss originally was a prekinematic or synkinematic pluton intruded into the pelitic rocks. Indeed, near the mouth of the South Fork of Boulder Creek large outcrops of quartzite and of biotitic schist occur within the pegmatitic gneiss. Furthermore, well-foliated, 1- to 2-m-thick concordant bodies of orthogneiss are ubiquitous in the overlying metasedimentary rocks; these orthogneisses are compositionally similar to the pegmatitic gneiss but lack the augen of feldspar and the pegmatitic bodies. Similar concordant bodies occur in the Grand Forks area (Preto, 1970) and in the Curlew area, where they are as much as 12 m thick (Parker and Calkins, 1964). In addition, the biotitic schist and gneiss that overlie the pegmatitic gneiss are so similar to the biotitic schist and gneiss below the pegmatitic gneiss (Table 1) that they may be the same unit. In the Grand Forks area (Table 2), the apparent absence of pegmatitic gneiss plus the presence of 2,000 m of biotitic and associated metasedimentary rocks underlying the quartzite also support the interpretation that the protolith of the pegmatitic gneiss was intrusive into the pelitic rocks.

Preto (1970) correlated Daly's Cascade gneiss (1912) with the pegmatitic gneiss of the Curlew area. However, the Cascade gneiss in the Grand Forks area consists of several bodies only a few square

TABLE 2. CORRELATION AND THICKNESS OF THE TENAS MARY CREEK SEQUENCE AND PALEOZOIC UNITS IN NORTHERN WASHINGTON AND ADJACENT BRITISH COLUMBIA

Kettle Dome (this paper)	Curlew quadrangle (Parker and Calkins, 1964)	Bodie Mountain quadrangle (Pearson, 1967)	Grand Forks, British Columbia (Preto, 1970)	Republic and Aeneas quadrangles (Muessig, 1967)	Bald Knob quadrangle (Staatz, 1964)	Buckhorn Mountain district (McMillen, 1979)
Covada and Churchill Mountain formations	Phyllite assigned to rocks of Tenas Mary Creek	Metamorphic rocks of Buckhorn Mountain; phyllite and marble assigned to the metamorphic rocks in the southwest corner of quadrangle	3	Rocks near Sheep Mountain	Graywacke, phyllite, black shale, and probably, quartzite of Permian or Triassic rocks	Anarchist group
Eastern quartzite, <300 m	Absent	Absent	Absent	Absent	Quartzite within Permian or Triassic rocks?	Absent
Fine-grained biotite schist, 600 m	Schist in rocks of Tenas Mary Creek, 2,750 m	Metamorphic rocks in southwest corner of quadrangle	Part of V(?), >200 m?; IV	Metamorphic rocks near Golden Harvest Creek	Phyllitic quartzite, 1850 m; schist in Permian or Triassic rocks	Goat Ranch metamorphic complex
Amphibolite, 200 m	Hornblende schist in rocks of Tenas Mary Creek?	Amphibolite in rocks of Tenas Mary Creek	IV?, <1100 m; V?	Absent	Absent	Absent
Eastern granite gneiss, >800 m	Quartz-plagioclase gneiss in rocks of Tenas Mary Creek, 500 to 1,000 m	Quartz-feldspar gneiss in rocks of Tenas Mary Creek	IX	Absent	Absent	Absent
Biotitic schist and gneiss with minor quartzite and marble, <300 m	Absent	Metamorphic rocks of Tonata Creek	III, <1200 m	Absent	Absent	Absent
Feldspathic quartzite with minor biotitic schist and marble, >650 m	Quartzite in rocks of Tenas Mary Creek and of St. Peter Creek, 970 m	Quartzite in rocks of Tenas Mary Creek	II, <430 m	Absent	Absent	Absent
Biotitic schist and gneiss with minor quartzite and marble, <150 m	Marble and associated rocks in rocks of Tenas Mary Creek, 3 to 240 m	Marble in rocks of Tenas Mary Creek	I	Absent	Absent	Absent
Granitic, porphyritic, pegmatic gneiss, >850 m	Orthoclase-quartz-oligoclase gneiss in rocks of Tenas Mary Creek, >1,100 m	Granitic gneiss in rocks of Tenas Mary Creek	VII?	Assigned to quartz monzonite east of Sherman fault	Absent	Absent
Biotitic schist and gneiss with minor quartzite and marble, >700 m	Quartz-biotite schist, calc-schist, and quartzite of rocks of St. Peter Creek	Absent	I, 2,000 m	Metamorphic rocks east of Sherman fault	Absent	Absent

kilometres in area. The body about 7 km east of Grand Forks, the only one I have examined, lacks the feldspar megacrysts and pegmatitic patches representative of pegmatitic gneiss. Perhaps the Cascade gneiss and the smaller concordant bodies of gneiss above the pegmatitic gneiss in the Kettle dome were satellitic stocks, sills, and dikes of the pluton from which the pegmatitic gneiss formed.

Quartzite-dominated Sequence. Overlying the pegmatitic gneiss are heterogeneous units (sillimanitic biotitic schists and gneiss with minor quartzite and marble) above and below >650 m of feldspathic quartzite (Table 1). The lower heterogeneous unit is so thin that it was not always encountered during reconnaissance mapping; thus, it is not shown as a continuous unit in Figure 2. It is present in the Tenas Mary Creek and Grand Forks areas (Table 2), and more detailed mapping (Pearson, 1977) shows that it is continuous in the southern part of the Togo Mountain quadrangle.

Rusty- to white-weathering quartzite >650 m thick overlies the lower heterogeneous unit. The quartzite has 5% to 10% white-weathering feldspar 1 to 5 mm in diameter. The feldspar is more commonly orthoclase than plagioclase (Parker and Calkins, 1964; Preto, 1970). The quartzite generally is nonmicaceous, but does contain intercalated biotitic schists and gneisses >20 m thick. On the southern side of Profanity Peak, coarse white marble 30 to 60 m thick occurs within the quartzite. The heterogeneous unit above the quartzite appears to be restricted to the northern part of the Togo Mountain quadrangle and the Grand Forks area.

Eastern Granitic Gneiss. Above the quartzite-dominated sequence on the eastern limb of the dome are >800 m of coarse-grained, very well foliated, unlayered to indistinctly layered, granodioritic orthogneiss. The gneiss generally is not pegmatitic but does have some plagioclase megacrysts as much as 1 cm long. The basal part of this gneiss commonly is more leucocratic and has inclusions of feldspathic quartzite similar to those in the underlying rocks. Amphibolites and thin quartzites along U.S. Route 395 in the Laurier quadrangle are of uncertain origin; they may have been either xenoliths or intercalated sediments. In the Curlew quadrangle (Parker and Calkins, 1964) and in the Boyds quadrangle on the eastern limb of the Kettle dome, the upper part of the gneiss contains stratiform amphibolites. The regionally discordant contacts of this gneiss shown in Figure 2 may indicate the intrusive origin of its protolith.

This unit has caused considerable confusion. Pardee (1918) and Campbell (1938, 1946) included it within the Colville batholith. Still more confusing is the similarity of this unit to the pegmatitic gneiss, especially in the few places where the latter is not particularly pegmatitic or the eastern gneiss has 1-cm plagioclase megacrysts. In these places the >650-m quartzite above or below a gneissic unit is diagnostic. Judging from the description of Parker and Calkins (1964) and Pearson (1967), the pegmatitic gneiss contains more potassium feldspar and less hornblende than the eastern gneiss. Bowman (1950), Campbell and Thorsen (1966), Lyons (1967), and Pearson (1977) lumped the two gneisses as a single unit. Bowman (1950) correlated the eastern orthogneiss at Laurier on the Canadian border with the Cascade gneiss. However, the Cascade gneiss 7 km east of Grand Forks is neither as hornblende-rich nor as well foliated as the eastern gneiss.

The remarkable areal extents of the eastern granitic gneiss and the pegmatitic gneiss deserve comment. Both gneisses occur throughout the Kettle dome (Fig. 2). Both also are present as far northwest as the bend in the Kettle River in the Curlew quadrangle (Parker and Calkins, 1964), and both probably occur in the Grand Forks area of British Columbia mapped by Preto (1970). Thus the pegmatitic gneiss has a minimum distance of outcrop of 50 km and the eastern gneiss a minimum of 70 km. Because both gneisses are overlain or intruded by other units, their total length could be greater. How much of their present form was caused by attenuation of the original plutons during metamorphism is not known.

Amphibolite. On the eastern limb of the dome, a 200-m thick, black amphibolite overlies the eastern granitic gneiss; similar stratiform amphibolites occur within the gneiss. On the basis of elemental ratios, Preto (1970) concluded that similar amphibolites of the Grand Forks area originally were mafic intrusions, but Donnelly (1978) noted that metasomatism may make these results inconclusive. The presence of blue-green hornblende accompanied by oligoclase or andesine suggests that the amphibolite is not of sillimanite grade (Donnelly, 1978).

Fine-Grained Biotitic Rocks. Fine-grained, biotitic, granitic gneiss with intercalated calc-silicate units, quartzite, and amphibolite overlie the eastern granitic gneiss in the Laurier quadrangle; these are labeled QMG in Figure 2. Compositionally similar rocks (labeled BS) in the northwestern and southeastern margins of the dome are schistose to almost phyllitic instead of gneissic (Table 1). Correlation of QMG with BS cannot be demonstrated because the two belts of outcrop cannot be traced into each other. Because neither unit crops out near Boyds, their position relative to the 200-m-thick amphibolite is not known.

The schistose rocks are similar to the fine-grained biotitic schist and biotitic quartzite with intercalated fine-grained amphibolite and calc-silicate schist mapped by Staatz (1964), Muessig (1967), and McMillen (1979) in the Okanogan dome (Table 2). The fine-grained schist does not seem to be sillimanite-bearing (Parker and Calkins, 1964; Staatz, 1964; Muessig, 1967; Preto, 1970; McMillen, 1979), except adjacent to plutons in the Okanogan dome.

Eastern Quartzite. A slabby, slightly rusty weathering, fine-grained quartzite with micaceous partings occurs along the southeastern margin of the Kettle dome. Muscovite and minor biotite on the partings and the presence of isoclinal recumbent folds (outlined by the micaceous partings) suggest that metamorphism of the rocks in the Kettle dome is younger than this quartzite. Sillimanite has yet to be found in this quartzite, even in the schistose partings (Donnelly, 1978). Because the quartzite overlies the 200-m amphibolite, the eastern gneiss, and the fine-grained biotite schists (Fig. 2), the basal contact of the quartzite may be a major unconformity.

Because this quartzite occurs on both sides of the Columbia River near the bridge at Kettle Falls, it must be >200 m thick. Along U.S. Route 395 in the unmapped area between Boyds and Orient, where calc-silicate gneisses of unknown affinity appear to overlie the quartzite, the quartzite appears to be about 300 m thick.

I have suggested (Cheney, 1977) that the eastern quartzite might be equivalent to the 580 to 910 m of thin platy quartzite with sericitic partings near the top of the basal Cambrian Gypsy quartzite described by Park and Cannon (1943) in northeasternmost Washington. However, the eastern quartzite, which seems to have a higher metamorphic grade than all known examples of the Gypsy, could equally well not be Gypsy.

Because the >650-m quartzite within the dome contains intercalated units of biotitic gneiss and marble as much as 60 m thick but the eastern quartzite does not, these two quartzites probably are not correlative. All known examples of Beltian Revett Quartzite in Stevens County are gray and occur in lower grade rocks, so that this correlation may not be likely. Because Donnelly (1978) observed that the eastern quartzite has the same recumbent and other folds as the underlying amphibolite and the other rocks of Tenas Mary Creek described by Lyons (1967), the eastern quartzite is provisionally included within the Tenas Mary Creek sequence.

Paleozoic and Younger Rocks

Black argillite, gray phyllite, dark limestone, and white marble of late Paleozoic age, together with greenstone of reputed Triassic age, overlie the high-grade metamorphic rocks of the dome. In the Orient district, Bowman (1950) called these the Churchill formation; in the Colville Indian Reservation, Pardee (1918) used the name Covada group. Table 2 and Figure 3 indicate tht similar rocks have been described elsewhere. These formations may not be strictly correlative, but all are inferred to be Pennsylvanian to Triassic in age. The next youngest unit is the Jurassic Rossland Group of volcanic rocks. The Tertiary rocks are described later.

Although the fine-grained biotitic schist is compositionally similar to the late Paleozoic phyllitic rocks, the two probably are not the same unit subjected to different grades of regional metamorphism. For example, the late Paleozoic phyllitic rocks contain thick pods of quite pure limestone and marble, whereas fine-grained biotitic schists have only thin calc-silicate schists. Furthermore, the eastern quartzite occurs between the fine-grained schists and the phyllites on the southeastern limb of the dome, and the fine-grained biotitic schist is locally absent along margins of the dome that are overlain by phyllite. Parker and Calkins (1964) suggested that in the Curlew quadrangle an unconformity exists between fine-grained biotitic schist and the overlying phyllites. Such an unconformity would explain the relationships seen in the Kettle dome.

Age and Correlation of the Rocks of the Tenas Mary Creek Sequence

Bowman (1950) applied the name Boulder Creek Formation to the metamorphic rocks in the small part of the Kettle dome that he mapped. However, the sequence is better exposed and described as several mappable units in the area of the Tenas Mary Creek. Hence, the name Tenas Mary Creek sequence proposed by Parker and Calkins (1964) is preferred. However, the basal heterogeneous unit of biotitic gneiss, marble, and quartzite does not crop out in the type area. Furthermore, Parker and Calkins included phyllite in the top part of the Tenas Mary Creek sequence. Because the phyllite has a much lower metamorphic grade (greenschist) and may be unconformable on the rocks of the Tenas Mary Creek, I believe it should be excluded. Until more extensive stratigraphic and petrographic studies are available, the fine-grained biotitic schists, amphibolite, and the eastern quartzite are provisionally included in the Tenas Mary Creek sequence.

The age and regional correlation of the rocks of Tenas Mary Creek are poorly known. Engels and others (1976) listed K-Ar dates on individual minerals of 50 and 67 m.y. for amphibolites. R. L. Armstrong (1977, personal commun.) has obtained preliminary whole-rock Rb-Sr dates of 600 to 1,200 m.y. B.P. on the pegmatitic gneiss and the eastern gneiss of the Kettle dome. He regards the dates as typical of the Shuswap terrane.

Parker and Calkins (1964), Pearson (1967), Preto (1970), and Donnelly (1978) correlated the rocks of Tenas Mary Creek with the Shuswap terrane of southern British Columbia. The mantling metasedimentary rocks in the Shuswap have been regarded as probably mostly Precambrian and Paleozoic but with some as young as Triassic-Jurassic (Wanless and Ressor, 1975; Okulitch and others, 1977). The thick quartzites in the Shuswap terrane were regarded as possibly Lower Cambrian (Okulitch and others, 1977); thus, by analogy, the $>$650-m quartzite in the Tenas Mary Creek or the eastern quartzite in the Kettle dome might be Lower Cambrian. However, the Shuswap now appears to include metasedimentary rocks as much as 3,000 m.y. old that were intruded by 1,960-m.y.-old granitic rocks and then metamorphosed 935 m.y. ago (Duncan, 1978).

If the rocks of Tenas Mary Creek are Precambrian, they could be equivalent to Windermere, Beltian, or pre-Beltian rocks. The Beltian rocks closest to the Kettle dome are in southeastern Stevens County and have been described by Miller and Clark (1975). These and other known examples of Beltian rocks in Washington and adjacent Idaho are of much lower metamorphic grade (the pelitic rocks are still black argillites). Furthermore, the 950-m Beltian Revett Quartzite weathers gray and is not as feldspathic as quartzite of the Tenas Mary Creek, and marbles of the Tenas Mary Creek rocks do not resemble the carbonate rocks of the Belt. These same arguments could be applied to the Windermere–Deer Trail rocks which, in addition, have thick greenstones that have no analogues, except possibly the amphibolites, in the rocks of the Tenas Mary Creek.

Lithologically and structurally, the most probable correlatives of the Tenas Mary Creek sequence are sillimanite-grade rocks in the Spokane area (Fig. 3). These rocks include the cataclastic Newman

Lake orthogneiss with 5-cm potassium feldspar megacrysts (Miller, 1974d; Weissenborn and Weis, 1976) and a feldspathic quartzite near Freeman (Weis, 1968) that appears to be at least as thick as the quartzite of Tenas Mary Creek. Furthermore, Griggs (1973) showed that these rocks define a flat-topped dome, herein called the Spokane dome. The northeasternmost rocks of this terrane in Idaho also are domal and yield ages of about 1,500 m.y. (Clark, 1973). The Pb-α ages of 1,150 m.y. for the Hauser Gneiss (Weis, 1968) are, of course, suspect but are suggestive of a correlation with the Tenas Mary Creek.

Most workers (Griggs, 1973; Clark, 1973; Miller, 1974b; Miller and Clark, 1975; Weissenborn and Weis, 1976) have suggested that the rocks of the Spokane dome are high-grade portions of the Belt Supergroup. However, Armstrong (1975) suggested that paragneiss, quartzite, marble, schist, amphibolite, and orthogneiss in central Idaho and the Spokane dome are part of a pre-Beltian metamorphic terrane. Because Griggs (1973) and Miller and Clark (1975) showed that on a regional scale the Spokane dome is conformably surrounded on the southern and western sides by Beltian strata, the abrupt change in metamorphic grade could be due to an unrecognized, gently domed sub-Beltian unconformity or low-angle fault. I prefer this alternative and believe that the high-grade rocks of the Spokane dome probably are pre-Beltian.

Initial strontium isotopic ratios of Mesozoic plutons in northern Washington also suggest that the rocks of the Tenas Mary Creek may be Precambrian. Plutons as far west as long. 121°W have initial ratios >0.704, which implies that the magmas were contaminated by radiogenic strontium from a Precambrian basement (Armstrong and others, 1977). Thus, the rocks of the Tenas Mary Creek could be part of such a basement. If so, the lithologic similarity of the rocks of the Tenas Mary Creek to those in the Spokane dome would support the suggestion of Armstrong and others (1977) that on the basis of these initial strontium isotopic ratios, the Precambrian basement of pre-Mesozoic North America extended westward into northern Washington.

In summary, meager stratigraphic and radiometric evidence, including a comparison with the Shuswap rocks, favor but do not prove a Precambrian age for the rocks of the Tenas Mary Creek in the Kettle dome. Regional structural interpretations, in turn, favor a pre-Beltian age for the high-grade metamorphic rocks in the Spokane dome and, by analogy, the rocks of the Tenas Mary Creek in the Kettle dome. However, because the Spokane dome is east of the Kootenay arc and Kettle dome is west of it, such a correlation is, admittedly, unconventional.

Plutons

A number of Mesozoic and Tertiary granitic plutons intrude the Tenas Mary Creek, Paleozoic, and Mesozoic rocks. The plutons vary from biotite dominated to hornblende dominated, from fine and medium grained to coarse grained, and from foliated to unfoliated phases. At present it is not known how many discrete plutons there are. All are shown as a single unit in Figure 2.

STRUCTURE OF KETTLE DOME

Kettle Dome

The antiformal nature of the rocks underlying the Kettle River Range was first recognized by Campbell (1946). He noted that near and south of State Route 20 the foliation forms a dome "12 miles wide" elongated to the northeast. He also realized that the cataclasis of the rocks is similar to that of the Okanogan dome.

The antiformal pattern of the Kettle dome is best shown by the map pattern of the >650-m quartzite

(Fig. 2) and by the antiformal dips of this quartzite. Rocks of the Tenas Mary Creek sequence on the eastern and southeastern limbs of the dome form prominent dip slopes near Orient, Kettle Falls, and Lake Ellen. The best-preserved dip slope on rocks of the Tenas Mary Creek sequence defining the western limb of the dome is Tenasket Mountain.

The dome is >65 km long north-south and 27 km wide. The structural relief is only about 3 km (Fig. 2). The northern end of the dome is a northwest-trending antiform, defined by opposing dips in the >650-m quartzite and biotitic rocks on Huckleberry Mountain and on Togo and Marble Mountains.

The dips of bedding and of foliation within the Tenas Mary Creek units crudely define two en echelon north-trending, gently antiformal axes within the dome. Dips generally are <25°, and locally foliation and bedding are nearly horizontal, forming flat-topped ridges in the center of the dome. However, contacts are unusually steep to vertical along the northwesternmost margin of the dome and near Profanity Peak along the north-trending fault on the western side of the dome.

The simple domal structure could be part of a larger, more complex structure. For example, detailed mapping might show that the domal structure is the upper limb of a large recumbent fold similar to the small folds commonly seen in outcrops. Furthermore, if the eastern quartzite could be shown to be correlative with, or older than, the >650-m quartzite within the dome, the structure is more complex than shown in Figure 2. When the reconnaissance nature of the mapping is considered, such possibilities should not be ignored.

Structures within the Dome

Folds larger than those in outcrops but with map patterns smaller than several kilometres are difficult to recognize in reconnaissance mapping. A gently eastward-plunging synform may exist in the eastern gneiss and amphibolites south of Deadman Creek in the Boyds quadrangle. The amphibolite north of the mouth of Sherman Creek in the Bangs Mountain quadrangle may mark a similar synform. An east-trending antiform brings the lowest units of Tenas Mary Creek to the surface along the Kettle River in Canada (Preto, 1970), and this fold could account for the east-trending salient of the eastern gneiss at the Canadian border near Laurier in Figure 2. Gentle northwest-trending folds in the eastern quartzite along the eastern margin of the dome may be minor folds associated with formation of the dome.

A north-trending fault in the western part of the Sherman Peak and Togo Mountain quadrangles juxtaposes structurally higher rocks on the east against structurally lower rocks on the west. The ≥2-km vertical separation on this fault shown on the cross section in Figure 2 could produce the 10 km of apparent left-lateral separation shown on the map in Figure 2. The fault does not seem to offset the hornblende quartz dioritic pluton in the valley of the North Fork of Sherman Creek.

A west-northwest–trending fault with the northeast side up occurs in Hoodoo Canyon. A similar fault in the Sherman Peak and Bangs Mountain quadrangles may offset the north-trending fault mentioned above. Such a northwest-trending fault would explain why marble and quartzite in the biotitic unit north of Sherman Pass dip toward each other. It would also explain the apparent juxtaposition of the eastern gneiss and pegmatitic gneiss along State Route 20 northeast of Sherman Pass. The same fault would account for the northward termination of the upper Paleozoic phyllites along the southeastern edge of the dome, as well as the straight courses of Sherman Creek and of Donaldson Draw on Bangs Mountain. Smaller faults within Tenas Mary Creek rocks probably are more numerous than reconnaissance mapping can resolve. Small northwest-trending faults do cut the eastern quartzite and eastern gneiss in the Boyds quadrangle.

The maps of Parker and Calkins (1964) and Muessig (1967) indicate that neither of the large faults offset the faults that bound Tertiary rocks to the west of the dome. Thus, the large faults within the

Kettle dome probably are pre-Tertiary. On the eastern side of the Kettle River in the Orient quadrangle, pyroxenites and other mafic rocks mark the contact between the rocks of the Tenas Mary Creek and upper Paleozoic phyllites and Tertiary rocks (Bowman, 1950). Perhaps this is a postdome fault (like the serpentinite-bearing Sherman fault west of the dome, shown in Fig. 3).

The contact between the rocks of the Tenas Mary Creek sequence and the overlying low-grade rocks along the eastern margin of the dome appears to be tectonic. Campbell (1938, 1946) described cataclasis in the eastern gneiss on the southeastern limb of the dome. Locally, chloritic fractures and brecciation are well developed along the northeastern margin of the dome (Bowman, 1950; Lyons, 1967; Donnelly, 1978), especially in the small plutons near Orient (Bowman, 1950). Cataclasis (microshears and microbrecciation) in the metasedimentary rocks is parallel to but later than the foliation that outlines the recumbent folds (Lyons, 1967; Donnelly, 1978). Furthermore, unmetamorphosed nonrecrystallized, but brecciated limestone (presumably of late Paleozoic age) overlies rocks of Tenas Mary Creek in three places: just west of the confluence of the Kettle and Columbia Rivers, on the Kettle River 3.3 km northwest of Barstow, and on U.S. Route 395 2 km northwest of Barstow. The granitic gneisses of the Tenas Mary Creek below the limestone of the Kettle River locality are extensively chloritized. It seems likely that detailed mapping would show that the limestone in the Orient and Boyds quadrangles overlies a gently eastward-dipping tectonic zone.

REGIONAL GEOLOGY

Terranes Equivalent to the Kettle Dome

The cataclastic and domal nature of the gneisses between the Republic area and the Okanogan River have been described by Waters and Krauskopf (1941), Snook (1965), and Fox and others (1976, 1977). Although this dome is structurally similar to the Kettle dome, it is predominantly composed of Mesozoic(?) orthogneiss and granitic plutons.

The dioritic gneisses in the western part of the Okanogan dome were regarded as paragneisses by Snook (1965). Fox and others (1976) proposed that Snook's name of Tonasket Gneiss be applied to all such rocks (Fig. 3). My preliminary mapping suggests that virtually all of the Tonasket Gneiss in the central part of the dome is derived from a pluton grading inward from diorite to quartz diorite to porphyritic granodiorite.

The age of the Tonasket Gneiss is not too well known. On the southeastern margin of the dome, upper Paleozoic hornfelsic phyllite occurs adjacent to an orthogneiss that is similar to the interior porphyritic granodioritic phase of the Tonasket Gneiss. Fox and others (1976) reported U-Pb ages of 87 and 100 m.y. and a Th-Pb age of 94 m.y. from a euhedral zircon from hornblende-rich Tonasket Gneiss.

The eastern part of the Okanogan dome is dominated by biotitic quartz monzonitic to granodioritic plutons that intrude the Tonasket Gneiss and the late Paleozoic phyllitic rocks. Portions of these plutons have been described by Waters and Krauskopf (1941), Parker and Calkins (1964), Staatz (1964), Muessig (1967), and Pearson (1967). In general, the plutons are texturally zoned, becoming coarser grained and more porphyritic inward, weakly to moderately foliated, and locally cataclastic. The western contacts of the westernmost plutons that I have mapped in the dome commonly dip ≤25° eastward. Pardee (1918) named various crystalline rocks, including such plutons at the southern ends of both the Kettle and Okanogan domes, the Colville batholith. The term "Colville batholith" probably should be reserved for these variably foliated, leucocratic quartz monzonitic to granodioritic Mesozoic plutons as Staatz (1964) suggested.

Studies of the Okanogan dome have led to three theories of origin that might be applicable to the other domes as well. Waters and Krauskopf (1941) considered the cataclasis of the Tonasket Gneiss to be the protoclastic border, or carapace, of the Colville batholith. Snook (1965) demonstrated the metamorphic nature of the Tonasket Gneiss and pointed out that cataclasis postdates the mylonitization that cuts the foliation within the Tonasket Gneiss. He concluded that the increasingly cataclastic nature of the gneiss adjacent to the border of the dome could be attributed to later folding of an originally flat thrust in the gneisses, rather than to batholithic emplacement. The presumed paragneissic origin of the Tonasket Gneiss and 66- to 46-m.y. ages determined by K-Ar and fission-track measurments led Fox and others (1976, 1977) to suggest that the rocks were emplaced as an Upper Cretaceous gneiss dome that cooled through the Eocene.

As noted above, rocks similar to those in the Kettle dome occur in the Spokane dome. At present, cataclasis (Fig. 3) has been reported only in the coarse Newman Lake Orthogneiss and the Mesozoic Loon Lake batholith (Weissenborn and Weis, 1976; Miller, 1974d). Mylonitic rocks are known at two localities on the eastern edge of the dome between Coeur d'Alene and lat 48°N (Miller and Engels, 1975, p. 524).

Regional Extent of Tertiary Formations and Folding

An understanding of the regional geology (Fig. 3) is helpful in determining the origin and the age of the Kettle dome. The maps of Parker and Calkins (1964), Muessig (1967), and Staatz (1964) demonstrate that a syncline occurs west of the Kettle dome. This fold was recognized by Wright (1949) and named the "Sanpoil syncline" by Muessig. In the center of the fold is the Eocene Klondike Mountain Formation; successively outward (down the section) are the Eocene Sanpoil volcanic rocks, the Eocene O'Brien Creek Formation, and the upper Paleozoic to Triassic rocks. The synclinal map pattern is discernible on Figure 3.

Another north-trending synclinal inlier of the same three Eocene formations occurs near Orient on the northeastern margin of the dome (Fig. 3). Dips as great as 50° occur in the lower part of the Klondike Mountain Formation (Pearson and Obradovich, 1977).

Discordant K-Ar dates similar to those reported by Fox and others (1976, 1977) in the Okanogan dome are common in northeastern Washington and adjacent British Columbia (Miller and Engels, 1975; Armstrong and others, 1977). An alternative explanation to a cooling gneiss dome is that these dates were caused by Eocene volcanism and plutonism (Armstrong and others, 1977). As noted below, the Eocene volcanic rocks (Sanpoil Volcanics and Klondike Mountain Formation) were of regional extent, and Eocene plutons are common; the quartz monzonite of Long Alec Creek (K-Ar age of 51.7 ± 1.6 m.y., according to Engels and others, 1976) in the northern end of the Kettle Dome (Figs. 2, 3) even has batholithic dimensions.

Because the Kettle dome is bounded on the west and the northeast by Tertiary synclines, its present antiformal structure also is most likely Tertiary (Cheney, 1976, 1977). Furthermore, the length, trend, and structural relief of the Sanpoil syncline are similar to those of the dome. The axis of the Kettle dome is not parallel to the axis of the Sanpoil syncline, but this difference may be due to the combined effect of Tertiary folding and older structures within the rocks of the Tenas Mary Creek sequence. The high-grade metamorphism and related folding within the Tenas Mary Creek probably is pre-Tertiary (and probably pre-Beltian), and much of the uplift of the Tenas Mary Creek from the depths at which sillimanite forms probably was pre-Tertiary.

The regional extent of Tertiary folding is best appreciated after recognition of the regional extent of the Tertiary formations. Pearson and Obradovich (1977) have shown that the Eocene O'Brien Creek Formation, the dacites of the Sanpoil Volcanics, and the volcanic and volcaniclastic rocks of the Klondike Mountain Formation in the Republic area (Meussig, 1967) extend across northeastern Washington. The regional presence of the same three unconformity-bounded Tertiary formations suggests that they were not deposited in local basins as most authors—including Parker and Calkins

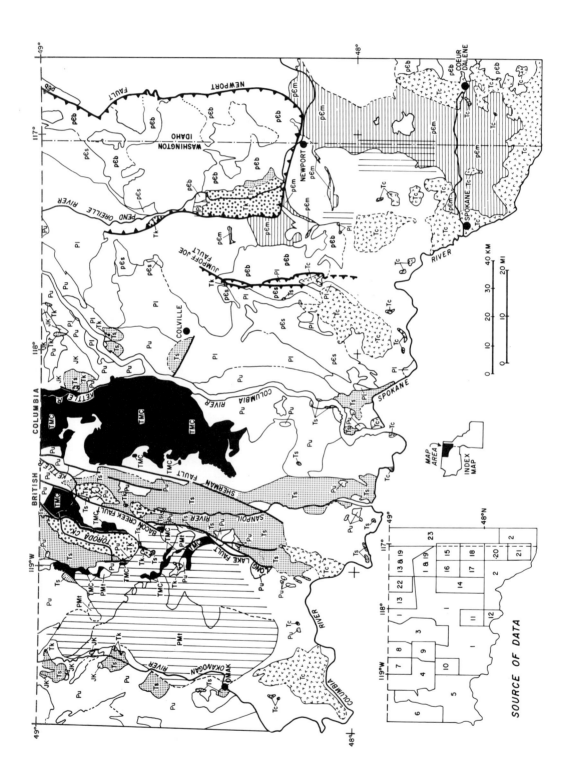

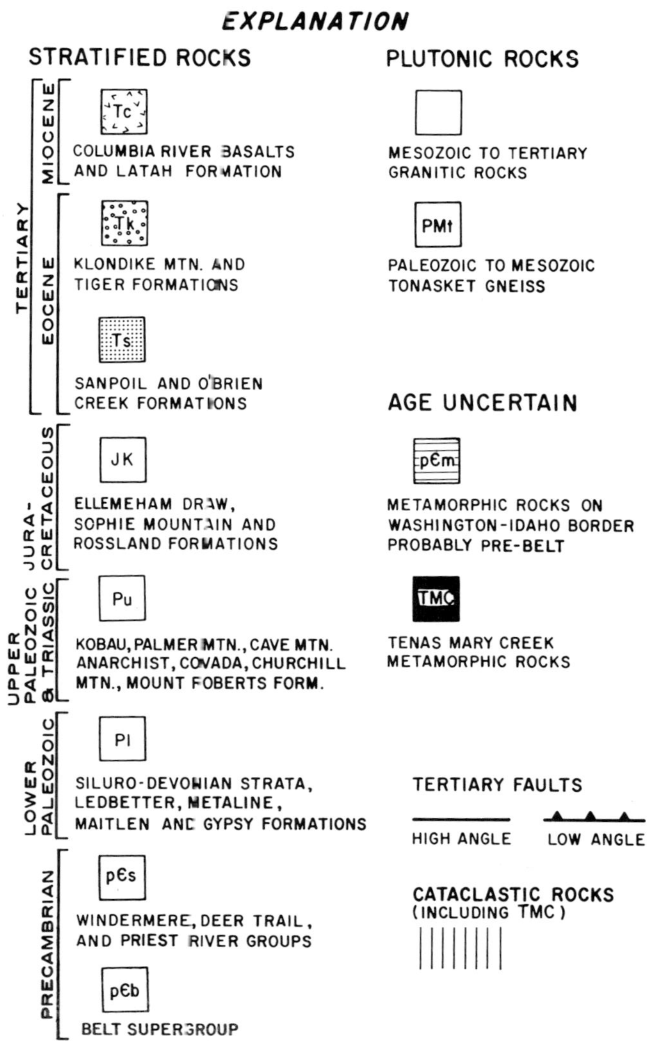

EXPLANATION

STRATIFIED ROCKS

TERTIARY
- MIOCENE — **Tc** COLUMBIA RIVER BASALTS AND LATAH FORMATION
- EOCENE — **Tk** KLONDIKE MTN. AND TIGER FORMATIONS
- EOCENE — **Ts** SANPOIL AND O'BRIEN CREEK FORMATIONS

UPPER JURA- & TRIASSIC CRETACEOUS
- **JK** ELLEMEHAM DRAW, SOPHIE MOUNTAIN AND ROSSLAND FORMATIONS

PALEOZOIC
- **Pu** KOBAU, PALMER MTN, CAVE MTN. ANARCHIST, COVADA, CHURCHILL MTN., MOUNT FOBERTS FORM.

LOWER PALEOZOIC
- **Pl** SILURO-DEVONIAN STRATA, LEDBETTER, METALINE, MAITLEN AND GYPSY FORMATIONS

PRECAMBRIAN
- **pЄs** WINDERMERE, DEER TRAIL, AND PRIEST RIVER GROUPS
- **pЄb** BELT SUPERGROUP

PLUTONIC ROCKS
- MESOZOIC TO TERTIARY GRANITIC ROCKS
- **PMt** PALEOZOIC TO MESOZOIC TONASKET GNEISS

AGE UNCERTAIN
- **pЄm** METAMORPHIC ROCKS ON WASHINGTON-IDAHO BORDER PROBABLY PRE-BELT
- **TMC** TENAS MARY CREEK METAMORPHIC ROCKS

TERTIARY FAULTS
- HIGH ANGLE
- LOW ANGLE

CATACLASTIC ROCKS (INCLUDING TMC)

Figure 3. Geologic map of northeastern Washington and adjacent Idaho. For cartographic clarity, small stocks and the letter designations of Tertiary and Mesozoic plutons have been omitted from the map. Data sources are (1) Hunttting and others (1961); (2) Griggs (1973); (3) Cheney (this paper); (4) Cheney (unpub. mapping); (5) Fox and others (1977); (6) Rinehart and Fox (1972); (7) Pearson (1967); (8) Parker and Calkins (1964); (9) Muessig (1967); (10) Staatz (1964); (11) Campbell and Raup (1964); (12) Becraft and Weis (1963); (13) Yates (1971); (14) Miller and Clark (1975); (15) Miller (1974a); (16) Miller (1974b); (17) Miller (1974c); (18) Miller (1974d); (19) Miller and Engels (1975); (20) Weissenborn and Weis (1976); (21) Weis (1968); (22) Yates (1964); (23) Bond (1978).

(1964), Muessig (1967), and Pearson and Obradovich (1977)—suppose.

The Klondike Mountain Formation as shown in Figure 3 is more extensive than shown by Pearson and Obradovich (1977). The map of Fox (1970) suggests that the Klondike Mountain Formation may exist in the Okanogan Valley. East of the Columbia River the mafic, olivine-bearing flows that locally lie above the Sanpoil immediately east of long. 118°W (Yates, 1971) might correlate with Muessig's (1967) basaltic upper member of the Klondike Mountain Formation. Pearson and Obradovich (1977) gave the following minimum ages: O'Brien Creek, 53 m.y.; Sanpoil, 50 m.y.; and Klondike Mountain, 41 m.y.

For simplicity, the conglomerates and sandstones of the Tiger Formation that unconformably overlie Sanpoil lavas in the Pend Oreille Valley (Pearson and Obradovich, 1977) are shown in the same pattern in Figure 3 as the Klondike Mountain Formation. However, no evidence presently exists as to whether these formations are correlative or not. Indeed, because the Tiger Formation varies greatly in provenance and appearance, it may have been deposited in more than one epoch of the Tertiary (Miller, 1971, 1974b). Additionally, although some parts of the Tiger appear to dip westward into the Newport fault, Miller (1971) pointed out that other parts of the Tiger appear to overlie the fault and no part of the formation is known to show the effects of proximity to such a major fault as the Newport fault. Thus, at least part of the Tiger may be correlative with at least one of the two unconformity-bounded epiclastic units described by Muessig (1967) and Pearson and Obradovich (1977) in the lower part of the Klondike Mountain Formation.

The Sanpoil syncline and the syncline near Orient already have been noted. The regional map of Rinehart and Fox (1972) shows two synclinal remnants of the Eocene formations along the western border of the Okanogan dome near Tonasket. The inliers of Eocene rocks just east of the Columbia River are partly synclinal and partly fault bounded (Yates, 1971; Pearson and Obradovich, 1977) and are aligned along a north-northeast trend. Perhaps the inliers west of the Okanogan dome and east of the Columbia River are remnants of formerly more extensive north-northeast–trending synclinal belts of Tertiary rocks similar to the Sanpoil syncline.

The inlier of Eocene formations on the Canadian border northwest of the Sanpoil syncline has been named the "Toroda Creek graben" by Pearson and Obradovich (1977). This inlier may also be synclinal, but, admittedly, the number of westward-dipping flow structures in the eastern edge of the Klondike Mountain Formation are few (Pearson, 1967), unconformities obscure a synclinal map pattern in the Tertiary rocks, and the eastern edge of the Klondike Mountain Formation is faulted (Pearson, 1967).

Tertiary Faults and Cataclasis

The synformal Newport fault zone in northeastern Washington and northwestern Idaho (Fig. 3) may cut the Tertiary Tiger Formation (Miller, 1974b) and does cut a 45- to 51-m.y.-old pluton (Miller and Engels, 1975). The fault separates structurally lower muscovite-biotite schist, micaceous quartzite, gneiss, and batholithic rocks from Tertiary rocks and only mildly metamorphosed Paleozoic and Beltian rocks (Miller, 1971, 1974b, 1974c, 1974d). The fault is a gently northward-plunging, synformal, cataclastic zone 300 m wide. Figure 3 shows areas of cataclasis beyond the fault described by Miller (1974b, 1974c, 1974d); detailed petrographic studies might enlarge these areas. K-Ar dates of plutons peripheral to the fault are typically 45 to 51 m.y. B.P. (Miller and Engels, 1975). Miller and Engels suggested that the preservation of much older K-Ar dates (typically 93 to 101 m.y. B.P.) in the plutonic rocks 8 to 25 km from the fault and in the upper plate of the fault indicates lateral displacement of 70 to 100 km.

A smaller Newport-type fault may bound the belt of Tertiary rocks of the so-called Toroda Creek graben. Along the northeastern margin of this belt, between the Canadian border and the Kettle River,

Parker and Calkins (1964) described a fault with a 400-m-wide zone of sheared breccia; this fault separates rocks of the Tenas Mary Creek in their type area from the Tertiary rocks to the west. Along this contact southwest of the Kettle River, Pearson (1967) described westward-dipping sheets of breccia as much as 30 m (locally 300 m) thick below and within the Klondike Mountain Formation. Although he suggested that these breccias were debris flows, Pearson also interpreted the eastern contact of the Tertiary rocks as a fault dipping 20° to 30° westward. Pearson and Obradovich (1977) extended this fault southward toward Granite Creek in the Aeneas quadrangle. In the valley of Granite Creek, a very poorly sorted and poorly stratified breccia consisting mostly of granitic fragments in an arkosic matrix occurs beneath the Klondike Mountain volcanic rocks. Muessig (1967) and Pearson and Obradovich (1977) regarded these breccias as sedimentary, but my mapping indicates that (1) locally some of the clasts are "smeared out" in a well-foliated matrix; (2) matrix-filled fractures down to hairline width extend into a few clasts, and (3) the underlying granites have mylonitic seams. Thus, the breccia may be tectonic. If, like the Newport fault, a western limb of this fault does exist, it may explain the juxtaposition of low-grade upper Paleozoic strata against garnet-staurolite mica schist at Wauconda Summit in the northwestern corner of the Aeneas quadrangle. This fault would be on strike with the fault that Pearson (1967) mapped just to the north and would separate the Tertiary volcanic rocks to the east from schist, phyllite, amphibolite, and marble to the west. Because additional mapping is necessary to determine whether these faults are segments of a single system analogous to the Newport fault, a single fault is not shown on Figure 3.

Although the faults on the western side of the Sanpoil syncline (Fig. 3) clearly are regarded as the western boundary faults of the Republic graben (Parker and Calkins, 1964; Muessig, 1967; Staatz, 1964), a number of anomalies exist (Cheney, 1979). Firstly, the traces of these faults are more sinuous than can be shown on Figure 3. Secondly, Wright (1949) concluded that most of the epithermal gold ore in the Sanpoil Volcanics in the Republic district adjacent to the Bacon Creek fault is in thrust faults that dip 55° to 65° eastward. He illustrated (1949, Figs. 3, 4b, 7) the Bacon Creek fault as a major break along which an anticline involving the Sanpoil and Klondike Mountain units was thrust westward over Colville granitic rocks. Furthermore, highly sheared and veined phyllite with concordant rhombic tectonic clasts of limestone dips 20° eastward in an adit in sec. 32, T. 37 N., R. 32 E., where Muessig (1967) interpreted the junction of the Bacon Creek and Scatter Creek faults; the location is virtually on strike with Wright's (1949) cross section showing the Bacon Creek fault.

On Figure 3, the Scatter Creek fault is the unlabeled segment between the Bacon Creek and Long Lake faults. Muessig noted (1967) that in one adit the Scatter Creek fault is horizontal. Two of the three western boundary faults in the Bald Knob quadrangle to the southwest dip gently eastward (Staatz, 1964). In Figure 3, the King Creek and Nespelem River faults are the first and second faults, respectively, east of the Long Lake fault. Staatz showed the King Creek fault as a thrust and only assumed normal movement on the Long Lake fault. He also showed the high-angle Nespelem River fault as up on the eastern side (not the western side as one might expect for a western-bounding fault of a graben).

Another thrust exists at least locally on the eastern limb of the Sanpoil syncline. Muessig (1967) mapped a "major thrust fault," the Lambert Creek thrust, cutting Sanpoil flows and the younger quartz monzonite of Herron Creek. Parker and Calkins (1964) did not recognize such a fault in the neighboring Curlew quadrangle, but the sinuous St. Peter fault, which is cut by the Sherman fault, is a likely candidate. The Sherman fault east of the Sanpoil syncline does appear to be a high-angle fault (Staatz, 1964; Meussig, 1967).

In summary, the so-called Toroda Creek and Republic grabens may be synclinely folded allochthons rather than grabens. Alternatively, if they are bounded only on one side by thrusts, they are only half-grabens. In any case they are not grabens in which the Tertiary rocks were deposited.

Available mapping (Campbell, 1938, 1946; Waters and Krauskopf, 1941; Snook, 1965; Petro, 1970;

Weissenborn and Weis, 1976) indicates that cataclasis within the crystalline rocks of the three domes increases in intensity toward the margins of each dome. Furthermore, the intensely cataclastic marginal zones, including the previously discussed northeastern margin of the Kettle dome, have sinuous traces suggestive of low dips. Snook (1965) has already suggested that the cataclastic zone along the western margin of the Okanogan dome is due to folding of an originally flat thrust and that erosion has removed the cataclastic zone from the crest of the dome.

Although cataclasis is easier to detect in coarse-grained crystalline rocks, the greatest shearing probably occurred in the incompetent rocks (such as the upper Paleozoic argillites) above the crystalline rocks. The Osoyoos and Whiskey Mountain plutons within upper Paleozoic strata peripheral to the northwestern margin of the Okanogan dome do become more cataclastic toward the dome (Fox and others, 1976; Rinehart and Fox, 1972). An intensely shattered and hydrothermally altered pluton occurs in phyllitic rocks south of Lake Ellen on the southeastern margin of the Kettle dome, and Campbell (1938) described cataclastic sills and quartzite in the Paleozoic phyllites in this area. In fact, Campbell (1938) probably was the first to suggest that intense shearing in the phyllitic rocks and the intense cataclasis in the adjacent gneiss were similar to the effects of major thrust faults, but he discarded this idea in favor of a protoclastic border of what he inferred was the Colville batholith.

If these cataclastic zones are antiformal analogues of the synformal Newport fault, a westward-dipping fault zone should occur between the Newport fault and the Kettle dome. A possible candidate is the gently westward-dipping Jumpoff Joe fault in the Chewelah area. Miller and Clark (1975) suggested that thrusting on the Jumpoff Joe fault might be extensive enough to explain the structural and stratigraphic contrasts between the Deer Trail group west of the fault and the Belt rocks to the east. Where the fault cuts 100-m.y.-old plutons, Miller and Clark reported that it forms a cataclastic zone as much as 150 m wide; south of Chewelah, upper Miocene Columbia River Basalt overlies the fault (Miller and Clark, 1975). Miller and Clark also suggested that northeast-striking faults that pass a few kilometres northwest of Chewelah might be the major structures in the area.

If the Jumpoff Joe fault and the imbricate zone beneath it that involves lower Paleozoic strata are equivalent to the Newport fault, the lower Paleozoic and the Precambrian Deer Trail–Windermere–Priest River strata are restricted to the upper plate. Units of the Belt Supergroup in the Chewelah area (Miller and Clark, 1975) would be in the lower plate, but east of the Pend Oreille River (Miller, 1974a), such units are in the upper plate of the Newport fault.

The Newport fault, the low-angle faulting in the Toroda Creek area, the faults bordering the western limb of the Sanpoil syncline, and the Lambert Creek fault may be similar in age. All cut Eocene rocks. The Newport fault cuts a 45- to 51-m.y.-old pluton (Miller and Engels, 1975). The Lambert Creek thrust cuts the quartz monzonite of Herron Creek, which is similar to the Long Alec Creek batholith that has been dated at 53 m.y. B.P. (Pearson and Obradovich, 1977). Furthermore, the 48- to 49-m.y.-old Swimptkin Creek and Coyote Creek plutons in the Okanogan dome are slightly cataclastic (Fox and others, 1977). Whether these faults are portions of a single regional fault, a series of related faults, or merely local zones of decoupling is not yet known.

TIMING OF STRUCTURAL EVENTS

Mylonites and brecciated rocks have been described in the crystalline rocks of each of the three domes. Snook (1965) stressed that, although both commonly occur in the same rocks on the western margin of the Okanogan dome, the directionless microbreccias formed later than the schistose mylonites and that in most mylonites the biotite did not change to chlorite, whereas, chlorite, epidote, and zeolites are prominent in the microbreccias. The same relationships occur on the northern margin of the Okanogan dome in the contact metamorphic aureole of the Mount Bonaparte pluton (one of the Colville plutons) and as discrete sericitic phyllonite zones within the Cretaceous(?) Buckhorn Mountain pluton a few kilometres north of the dome (McMillen, 1979). As McMillen (1979) has stresed, the petrographic descriptions of Campbell (1938), Parker and Calkins (1964), Lyons (1964), and Donnelly (1978) suggest that mylonitization and later brecciation accompanied by retrograde metamorphism also are common on the margins of the Kettle dome.

Thus, although the mylonites and the cataclastic zones characterized by brecciated rocks commonly are coincident, they differ in age. Mylonitization is Cretaceous(?) or younger (McMillen, 1979) but has not been observed in Eocene rocks; whereas, cataclasis is Eocene or younger. Thus, mylonitization in the crystalline rocks of the domes is not related to the Tertiary faults and cataclasis described above. If the Jumpoff Joe fault near Chewelah, which is overlain by Columbia River basalt, is related to the other low-angle faults marked by cataclastic zones, these faults are pre-late Miocene. A study of that part of the Tiger Formation that appears to overlie the Newport fault might provide a better age for the faulting.

The antiformal nature of the cataclastic zones around the margins of the domes indicates that the cataclastic zones have been folded. The age of this folding is not well known. The map and cross section C-C' of Weissenborn and Weis (1976) suggest that the erosion surface beneath the Columbia River Basalt and the interlayered Latah formation dips southwesterly off the Spokane dome; thus, at least part of the doming may be older than the basalt. However, because the greater structural relief of the larger north-trending Cascade arch to the west is younger than the Columbia River Basalt (McKee, 1972), it is tempting to speculate that the present structural relief of the north-trending Spokane, Kettle, and Okanogan domes also may be due to folding younger than the basalt.

CONCLUSIONS

In summary, northeastern Washington is characterized by north-northeast-trending Tertiary folds tens of kilometres long but with amplitudes of only a few kilometres. The synclines are marked by remnants of Tertiary strata that once were regionally extensive. Instead of being diapiric gneiss domes, I believe that the Okanogan dome, the Kettle dome, the Paleozoic and Precambrian rocks near Chewelah between the Pend Oreille and Columbia Rivers, and the Spokane dome are the anticlines. The cores of the Kettle and Spokane domes consist of high-grade metamorphic rocks that probably are pre-Beltian in age. The high-grade rocks near Chewelah probably are pre-Beltian also. Mylonites within the domes probably are Cretaceous in age.

At present, any relationship between the Newport, the Jumpoff Joe, and other low-angle faults must be regarded as speculative, and no unequivocal physical evidence exists for significant displacement along any of them or on the cataclastic zones rimming the domes. Until physical evidence of significant displacement is available and until the ages of most of the cataclastic zones are known, the cataclastic zones should be regarded as local zones of Tertiary decoupling between the crystalline batholithic and metamorphic basement and the stratified cover rocks. Only additional investigations can determine whether the low-angle faults and cataclastic zones are a major folded Tertiary thrust, a series of thrusts, or purely local phenomena. The folding that caused the present distribution of faults and the present structural relief of the domes may be Miocene or younger.

ACKNOWLEDGMENTS

I thank Urangesellshaft, Chevron Resources, and Wold Nuclear for supporting the field work that made this paper possible. I am particularly grateful to D. S. Cowan, S. J. Reynolds, and A. V. Okulitch for very critical reviews of an earlier version of the manuscript and for attempting to moderate some of its more controversial aspects.

REFERENCES CITED

Armstrong, R. L., 1975, Precambrian (1500 m.y. old) rocks of central Idaho; The Salmon River arch and its role in Cordilleran sedimentation and tectonics: American Journal of Science, v. 275-A, p. 437–467.

Armstrong, R. L., Taubeneck, W. H., and Hales, P. O., 1977, Rb-Sr and K-Ar geochronometry of Mesozoic granitic rocks and their Sr isotope composition, Oregon, Washington, and Idaho: Geological Society of America Bulletin, v. 88, p. 397–411.

Becraft, G. E., and Weis, P. L., 1963, Geology and mineral deposits of the Turtle Lake quadrangle, Washington: U.S. Geological Survey Bulletin 1131, 73 p.

Bond, J. G., compiler, 1978, Geologic map of Idaho: Idaho Bureau of Mines and Geology, scale 1:500,000.

Bowman, E. C., 1950, Stratigraphy and structure of the Orient area, Washington [Ph.D. dissert.]: Cambridge, Mass., Harvard University, 149 p.

Campbell, A. B., and Raup, O. B., 1964, Preliminary geological map of the Hunters quadrangle, Stevens and Ferry Counties, Washington: U.S. Geological Survey Map MF-276, scale 1:48,000.

Campbell, C. D., 1938, An unusually wide zone of crushing in the rocks near Kettle Falls, Washington: Northwest Science, v. 12, p. 92–94.

—— 1946, Structure in the east border of the Colville batholith, Washington [abs.]: Geological Society of America Bulletin, v. 57, p. 1184–1185.

Campbell, C. D., and Thorsen, G. W., 1966, Compilation of geological mapping from 1935 to 1966 in the Sherman Peak and Kettle Falls quadrangles: Washington Division of Geology and Earth Resources Open-File Maps, scale 1:62,500.

Cheney, E. S., 1976, Kettle Dome, Okanogan Highlands, Ferry County, Washington: Geological Society of America Abstracts with Programs, v. 8, p. 360.

—— 1977, The Kettle dome: The southern extension of the Shuswap terrane into Washington: Geological Society of America Abstracts with Programs, v. 9, p. 926.

—— 1979, Tertiary decollement in northeastern Washington?: Geological Society of America Abstracts with Programs, v. 11, p. 72.

Clark, S.H.B., 1973, Interpretation of a high-grade Precambrian terrane in northern Idaho: Geological Society of America Bulletin, v. 84, p. 1999–2004.

Daly, R. A., 1912, Geology of the North American Cordillera at the forty-ninth parallel: Canadian Geological Survey Memoir 38, Part 1, 546 p.

Davis, G. H., and Coney, P. J., 1979, Geologic development of the Cordilleran metamorphic core complexes: Geology, v. 7, p. 120–124.

Donnelly, B. J., 1978, Structural geology of the Nancy Creek area, east flank of the Kettle dome, Ferry County, Washington [M.S. thesis]: Pullman, Washington State University, 251 p.

Duncan, I. J., 1978, Rb/Sr whole rock evidence for three Precambrian events in the Shuswap complex, southeast British Columbia: Geological Society of America Abstracts with Programs, v. 10, p. 392–393.

Engels, J. C., and others, 1976, Summary of K-Ar, Rb-Sr, U-Pb, Pb-α, and fission-track ages of rocks from Washington State prior to 1975 (exclusive of Columbia Plateau basalts): U.S. Geological Survey Miscellaneous Field Studies Map MF-70 (two sheets).

Fox, K. F., Jr., 1970, Geologic map of the Oroville quadrangle, Okanogan County, Washington: U.S. Geological Survey Open-File Map, scale 1:48,000.

Fox, K. F., Rinehart, C. D., and Engels, J. C., 1977, Plutonism and orogeny in north-central Washington—Timing and regional context: U.S. Geological Survey Professional Paper 989, 27 p.

Fox, K. F., Jr., and others, 1976, Age of emplacement of the Okanogan gneiss dome, north-central Washington: Geological Society of America Bulletin, v. 87, p. 1217–1224.

Griggs, A. B., 1973, Geologic map of the Spokane quadrangle, Washington, Idaho, and Montana: U.S. Geological Survey Map I-768, scale 1:250,000.

Huntting, M. T., and others, 1961, Geologic map of Washington: Washington Division of Mines and Geology, scale 1:500,000.

Lyons, D. J., 1967, Structural geology of the Boulder Creek metamorphic terrane, Ferry County, Washington [Ph.D. dissert.]: Pullman, Washington State University, 115 p.

McKee, B., 1972, Cascadia: The geologic evolution of the Pacific Northwest: New York, McGraw-Hill Book Company, 394 p.

McMillen, D. D., 1979, The structure and economic geology of Buckhorn Mountain, Okanogan County, Washington [M.S. thesis]: Seattle, University of Washington, 68 p.

Miller, F. K., 1971, The Newport fault and associated mylonites, northeastern Washington: U.S. Geological Survey Professional Paper 750-D, p. D77–D79.

—— 1974a, Preliminary geologic map of the Newport Number 1 quadrangle, Pend Oreille County, Washington, and Bonner County, Idaho: Washington Division of Geology and Earth Resources Map GM-7, scale 1:62,500.

—— 1974b, Preliminary geologic map of the Newport Number 2 quadrangle, Pend Oreille and Stevens

Counties, Washington: Washington Division of Geology and Earth Resources Map GM-8, scale 1:62,500.

——1974c, Preliminary geologic map of the Newport Number 3 quadrangle, Pend Oreille, Stevens, and Spokane Counties, Washington: Washington Division of Geology and Earth Resources Map GM-9, scale 1:62,500.

——1974d, Preliminary geologic map of the Newport Number 4 quadrangle, Spokane and Pend Oreille Counties, Washington, and Bonner County, Idaho: Washington Division of Geology and Earth Resources Map GM-10, scale 1:62,500.

Miller, F. K., and Clark, L. D., 1975, Geology of the Chewelah-Loon Lake area, Stevens and Spokane Counties, Washington: U.S. Geological Survey Professional Paper 806, 74 p.

Miller, F. K., and Engels, J. C., 1975, Distribution and trends of discordant ages of the plutonic rocks of northeastern Washington and northern Idaho: Geological Society of America Bulletin, v. 86, p. 517-528.

Muessig, S., 1967, Geology of the Republic quadrangle and a part of the Aeneas quadrangle, Ferry County, Washington: U.S. Geological Survey Bulletin 1216, 135 p.

Okulitch, A. V., Price, R. A., and Richards, T. A., editors, 1977, Geology of the southern Canadian Cordillera—Calgary to Vancouver: Geological Association of Canada Guidebook to Field Trip 8, 135 p.

Pardee, J. T., 1918, Geology and mineral deposits of the Colville Indian Reservation, Washington: U.S. Geological Survey Bulletin 677, 186 p.

Park, C. F., Jr., and Cannon, R. S., Jr., 1943, Geology and ore deposits of the Metalline quadrangle, Washington: U.S. Geological Survey Professional Paper 202, 81 p.

Parker, R. L., and Calkins, J. A., 1964, Geology of the Curlew quadrangle, Ferry County, Washington: U.S. Geological Survey Bulletin 1169, 95 p.

Pearson, R. C., 1967, Geologic map of the Bodie Mountain quadrangle, Ferry and Okanogan Counties, Washington: U.S. Geological Survey Map GQ-636, scale 1:62,500.

——1977, Preliminary geological map of the Togo Mountain quadrangle, Ferry County, Washington: U.S. Geological Survey Open-File Report 77 371, scale 1:62,500.

Pearson, R. C., and Obradovich, J. D., 1977, Eocene rocks in northeast Washington—Radiometric ages and correlation: U.S. Geological Survey Bulletin 1433, 41 p.

Preto, V. A., 1970, Structure and petrology of the Grand Forks group, British Columbia: Geological Survey of Canada Paper 69-22, 80 p.

Rinehart, C. D., and Fox, K. F., Jr., 1972, Geology and mineral deposits of the Loomis quadrangle, Okanogan County, Washington: Washington Division of Mines and Geology Bulletin 64, 124 p.

Snook, J. R., 1965, Metamorphic and structural history of the "Colville batholith" gneisses, north-central jWashington: Geological Society of America Bulletin, v. 76, p. 759-776.

Staatz, M. H., 1964, Geology of the Bald Knob quadrangle, Ferry and Okanogan Counties, Washington: U.S. Geological Survey Bulletin 1161-F, 79 p.

Wanless, R. K., and Reesor, J. E., 1975, Precambrian zircon age of orthogneiss in the Shuswap metamorphic complex, British Columbia: Canadian Journal of Earth Science, v. 12, p. 326–334.

Waters, A. C., and Krauskopf, K., 1941, Protoclastic border of the Colville batholith: Geological Society of America Bulletin, v. 52, p. 1355–1417.

Weis, P. L., 1968, Geologic map of the Greenacres quadrangle, Washington and Idaho: U.S. Geological Survey Map GQ-734, scale 1:62,500.

Weissenborn, A. E., and Weis, P. L., 1976, Geologic map of the Mount Spokane quadrangle, Spokane County, Washington, and Kootenai and Bonner Counties, Idaho: U.S. Geological Survey Map GQ-1336, scale 1:62,500.

Wright, L. B., 1949, Geologic relations and new ore bodies of the Republic district, Washington: American Institute of Mining and Metallurgical Engineers Transactions, v. 178, p. 264–282.

Yates, R. G., 1964, Geologic map and sections of the Deep Creek area, Stevens and Pend Oreille Counties, Washington: U.S. Geological Survey Map I-412, scale 1:31,680.

——1971, Geologic map of the Northport quadrangle, Washington: U.S. Geological Survey Map I-603, scale 1:31,680.

MANUSCRIPT RECEIVED BY THE SOCIETY JUNE 21, 1979

MANUSCRIPT ACCEPTED AUGUST 7, 1979

Printed in U.S.A.

Geological Society of America
Memoir 153
1980

Metamorphic core complexes of the North American Cordillera: Summary

MAX D. CRITTENDEN, JR.
U.S. Geological Survey
345 Middlefield Road
Menlo Park, California 94025

To attempt to summarize the series of papers that constitute this volume in a few pages, much less a few paragraphs, would require temerity indeed. Nevertheless, I will take this opportunity to point out a few of the geologic relations described that appear to be of unusual interest for someone attempting to unravel the origin of these fascinating geologic terranes.

SALIENT CONCLUSIONS

Clearly the most important conclusion regarding these complexes, particularly those south of the Snake River Plain to which the bulk of this volume is devoted, is that they are startlingly young and that they show consistent evidence of horizontal extension and vertical attenuation. It is primarily the latter processes that have operated to create the array of features by which these structures are characterized (Davis, this volume). In the core, ductile deformation took place at depths and temperatures sufficient to cause quartz to behave plastically producing extensive foliated and lineated mylonitic fabrics. In the cover, horizontal extension and vertical attenuation was accommodated by low-angle, younger-over-older faults and by listric normal faults that commonly affected rocks of Tertiary age. In some areas, however, evidence suggests that these two processes operated together or sequentially over a considerable period of time, in places beginning in the Mesozoic.

TECTONIC SETTING

As Coney (this volume) and many others have pointed out, most of the characteristic metamorphic core complexes lie in the midst of the Cordilleran miogeocline between the zone of imbricate thrusts that define the leading edge of the Cordilleran thrust and fold belt and the inferred edge of the North American Precambrian craton. The Okanogan dome in northeastern Washington (Fox and others, 1977) is a possible exception, as it involves Paleozoic eugeosynclinal assemblages deposited on possible oceanic crust in an arc or back-arc setting. In contrast, the complexes in Arizona and southeastern California developed within the craton itself, only thinly blanketed by Paleozoic platform sediments.

PROTOLITHS

Although most of the examples described in this volume involve Precambrian crystalline basement that has been more or less remobilized, many also involve one or more episodes of intrusive activity ranging in age from Jurassic to Tertiary. Sorting out the identity, correlation, and age of these now-gneissic rocks has long been the major impediment to understanding their origin. In the Catalina-Rincon area, the volume provides a wealth of data on both geochemistry and geochronology (Banks, this volume; Keith and others, this volume). Keith and others have used Rb/Sr ratios to fingerprint the geochemical character of the several sets of plutons. This technique is most successful for the compositional end members; late products of remobilization are characterized by a widely overlapping spread of data. They also used these data to distinguish highly metamorphosed septa of Precambrian Pinal Schist from similar-appearing rocks that were produced by cataclasis and recrystallization of entirely different protoliths. The validity of the result may be compromised, however, by the limited sampling in the lithologically diverse Pinal Schist. Their paper presents for the first time a comprehensive and coherent interpretation of the rocks in the core zone of the entire Catalina-Rincon-Tortolita complex.

ORIGIN OF GNEISSIC CORE ROCKS

In Arizona, areas of nonfoliated, nonlineated Precambrian basement are widely exposed, but in most places the nature of the transition from these normal crystalline rocks to those of the complexes is not decipherable. Davis and others (this volume) appear to have recognized exactly this feature in the Whipple Mountains, where they describe a remarkably abrupt "mylonitic front" across which this change in fabric occurs. This relationship appears also in the South Mountain complex (Reynolds and Rehrig, this volume) and a similar, though more gradual transition has been recorded in northwestern Utah (Todd, this volume). Nonfoliated and nonlineated normal-appearing Archean adamellite is exposed in Clear Creek Canyon at the east end of the Raft River area but can be followed through increasing degrees of remobilization toward the adjoining Grouse Creek Mountains, where it has a strongly developed gneissic texture and has locally invaded its sedimentary cover. A final example is available from the Snake Range in easternmost Nevada (Misch, 1960). The well-known and extensively developed detachment surface in this area cuts across two bodies of igneous rock. Where it transects the Jurassic pluton extensively studied by Lee and others (1970), the surface is a discontinuity overlain by 1 to 15 m of marble tectonite and a thin layer of gouge and breccia; penetrative foliation and lineation in the underlying pluton are apparently absent. North of Sacramento Pass, however, the detachment surface lies on a younger pluton, as yet poorly dated, from which Hose and Blake (1976, p. 27) reported a hornblende K-Ar age of about 56 m.y. In striking contrast to the Jurassic rock to the south, this body is gneissic wherever it is exposed and exhibits the characteristic mylonitic foliation and streaky lineation resulting from ductile flow. These relations suggest that the nature of the response to strain depended on the elevated temperatures that resulted from the emplacement of the northern pluton; this brought bodies of rock adjoining the plane of detachment into the temperature regime where the plutonic rocks responded by penetrative ductile flow rather than by brittle fracture. These and similar examples elsewhere suggest that regionally consistent stress fields existed for long periods of time, but they resulted in sufficient strain to produce lineated mylonitic fabrics only where rocks were brought to the appropriate pressure-temperature environment.

NATURE OF PENETRATIVE STRAIN

Compton in the Raft River Range (this volume) and Miller in the Albion Range (1978 and this volume) have examined quartz fabrics and deformed pebbles to ascertain the internal mechanisms by which penetrative fabrics developed. The rocks studied were quartzite immediately overlying gneissic more-or-less remobilized Archean basement (adamellite). Both Compton and Miller conclude that the gross deformation involved flattening normal to foliation combined with horizontal extension parallel with the dominant lineation and with rare large fold axes. The principal deformation mechanisms in quartz are dislocation glide and climb. Dominantly orthorhombic fabric symmetry suggests to Compton that deformation was mainly by pure shear. Compton reports that axes of the strain ellipsoid in the most severely strained rocks were 0.2:1.4:3.9; Miller reports maxima in the range of 0.7:0.3:1.6. The ratio of shortest to intermediate axes ranges from 1:7 to 1:4. Miller believes he can distinguish early regional strain associated with fold axes and lineation from later localized strain related to the formation of the dome.

DEPTH OF DUCTILE DEFORMATION

Attempts to quantify the pressure-temperature regimes in which the characteristic penetratively lineated mylonite fabrics developed has long proved difficult. The metamorphic grade in the core zone commonly ranges from greenschist to upper amphibolite facies, but in many areas pressure-sensitive mineral assemblages or rocks of the appropriate composition are absent. Moreover, the almost ubiquitous tectonic thinning severely limits the usefullness of estimates based on direct measurements of the overlying nonmetamorphic cover. Evidence available to date is in some degree contradictory and suggests deep burial in some examples and shallow cover in others. Simony and others (this volume) estimate that kyanite-sillimanite-bearing rocks of the Shuswap terrane were developed during the climax of regional metamorphism at depths in excess of 25 km. Subsequently, both the host rocks and the metamorphic isograds were folded and finally cut by undeformed plutons aout 110 m.y. ago. Regional relations suggest that such depths are appropriate for development of observed metamorphic assemblages but not necessarily for the later mylonitic fabrics. In southeastern California, Davis and others (this volume) believe that mylonitic fabrics formed at pressures of 3 to 3.5 kb, corresponding to depths of 10 to 13 km. In the Ruby Mountains, Snoke (this volume) notes that sillimanite-bearing gneiss exhibits lineation that apparently formed during development of the high-grade mineral assemblage. By contrast, Reynolds and Rehrig (this volume) show that penetrative mylonite fabrics in the South Mountain area near Phoenix, Arizona, have been developed on plutons as young as 23 m.y. whose hypabyssal textures and close resemblance to nearby volcanic rocks clearly point to shallow depths of emplacement, perhaps a kilometre or less.

REGIONAL GEOMETRY OF EXTENSION

One of the most remarkable discoveries to emerge from the current episode of detailed study is the great areal extent and apparent continuity of unidirectionally lineated terrane in the various contiguous sets of core complexes. The most remarkable examples are those in western Arizona and southeastern California (Davis, this volume; Davis and others, this volume). Because the rocks involved exhibit both ductile flow and shallow brittle fracture, their development appears to require long-persistent unidirectional extension. This regularity stands in particular contrast to the observations of Compton and others (1977), Todd (this volume), and Miller (this volume) in the Raft River–Albion area of northwestern Utah and southern Idaho, where patterns of lineations—also shown by fabric studies to be the results of vertical shortening and horizontal extension—are not rectilinear but broadly arcuate, two successive sets crossing each other locally at moderate to high angles. As would be expected, accompanying folds, where present, show a vergence and direction of flow that may be 90° apart. In spite of the similarity of the resulting textures and fabrics, it seems possible that the deformation in these two sets of complexes were produced by distinctly different tectonic mechanisms: one (Arizona) by regional unidirectional extension, the other (Raft River) by broad intrusive culminations or diapirs.

TIMING

The clearest impression of participants in the Penrose Conference in Tucson in 1977—where the Canadian complexes were more fully discussed by R. B. Campbell, J. E. Reesor, P. S. Simony, and others—was that the Canadian examples are significantly older than the bulk of those in the United States and that the discontinuity between the two groups lies at the Snake River Plain.

This diachroneity on a regional scale is immediately evident from Table 1, as is the temporal hiatus across the Snake River Plain. Of equal interest, however, is the fact that the patterns of strain that gave rise to the characteristic mylonite fabrics are also diachronous and long persistent within individual complexes. This is perhaps best displayed in the data from the Catalina-Rincon-Tortilita complex in Arizona. Keith and others (this volume) note that the Leatherwood quartz diorite (75 to 60 m.y. old) was subjected to movement that produced lineation and foliation and that xenoliths of these foliated rocks were incorporated in nonfoliated granitic plutons dated at 51 to 44 m.y. B.P. Other parts of those same plutons were extensively foliated and cut by still younger units dated at about 26 m.y. B.P., which themselves locally carry the same unidirectional lineation. Finally, nonfoliated dikes were intruded 24 to 21 m.y. ago and cut all of the previously developed fabrics. A most remarkable feature of this sequence is that for much of this time interval of nearly 40 m.y. the strain orientation apparently remained essentially constant. As Keith and others point out, however, it is difficult to determine whether strain was actually continuous during this interval and is merely recorded periodically by the available intrusive events or whether the strain itself, as well as the intrusive activity, was intermittent.

The prize for precision of temporal data clearly goes to the South Mountain area near Phoenix, Arizona (Reynolds and Rehrig, this volume), where the waning stage of extension that produced characteristic mylonitic foliation is bracketed by a set of andesitic dikes ranging in age from 25 to 20 m.y.

UNROOFING

In a few areas it is possible to date the final erosional exposure of the core of the metamorphic complex by the appearance of foliated and lineated clasts in adjoining alluvial basins. In the Okanogan area of northeastern Washington such clasts were identified by Rinehart (Fox and others, 1977) in rocks of the Republic graben underlying the Sanpoil Volcanics dated at about 50 m.y. B.P. In striking contrast, in the Raft River-Grouse Creek area of northwestern Utah, similar clasts together with large-scale monolithologic breccias are intercalated with tuffs dated at 12 m.y. B.P. (Todd, this volume; Compton and others, 1977). In the Rincon area east of Tucson, masses of nonfoliated granodiorite slid from the rising complex into playa deposits of the adjoining Pantano Formation inferred to be 20 to 30 m.y. old. In the Whipple Mountains lineated clasts first appear in the Gene Canyon Formation, believed to be late Oligocene to early Miocene in age (Davis and others, this volume), but tilting and extension along the underlying detachment surface continued until perhaps 13 m.y. ago.

These data shed some light on the perennial question as to the relation between the intrusive and metamorphic processes internal to the complexes and the uplift that generated their present topographic form. In the case of the Raft River Range, for example, it is clear that some 12 to 13 m.y. intervened between the intrusion of the youngest weakly foliated pluton (Red Butte stock) and the

TABLE 1. AGE, IN MILLIONS OF YEARS, OF SALIENT EVENTS IN SOME CORDILLERAN METAMORPHIC CORE COMPLEXES

	Shuswap (1)	Okanogan (2)	Kettle (3)	Raft River-Albion (4)	W. Ariz. SE Calif. (5)	South Mountain (6)	Catalina-Rincon (7)	Sonora (8)
Sediments with lineated clasts (unroofing)	65?	50	..	12	19-21	..	20-25 (Pantano Formation)	?
Nonfoliated post orogenic unit	70-110	48	40-50	..	..	19	24-21	?
Late foliated unit	100	70	60	25	27	..	26	?
Medial foliated unit	..	.	..	[Snake River Plain]	..	25	51-44	?
Early foliated unit	..	87	90	..	..	..	75-60	75-55
Climax of regional metamorphism	160-175	..	..	38	..	..	..	..

Note: Sources: (1) Simony and others (this volume); (2) Fox and others (1977); (3) Cheney (this volume); (4) Compton and others (1977); (5) Davis and others (this volume); (6) Reynolds and Rehrig (this volume); (7) Keith and others (this volume); (8) Anderson and others (this volume).

exposure of the core of the complex, as recorded by the earliest appearance of lineated clasts and associated monolithologic breccias in the adjoining sedimentary basin. Both the sediments and the enclosed breccia sheets are warped upward around the borders of the present range and show that the present range is less than 10 m.y. old. On a larger scale, Compton and others (1977) concluded that the patterns of metamorphic flow during early to middle Tertiary time point to culminations that are both larger and noncoincident relative to the present topographic highs.

FUTURE DIRECTIONS

Even where protoliths have been clearly identified, the origin and underlying controls leading to the development of gneissic rocks in the core zones have long been enigmatic. One of the earliest and most fundamental questions about the core complexes of the Cordillera is whether they represent samples of metamorphic infrastructure that is essentially continuous at depth or whether they represent local "hot spots." I believe the evidence now available from the Raft River and Arizona examples tilts strongly in the direction of the latter.

Viewed broadly, these relations support Coney's conclusions (this volume) that the core complexes are products of a widespread, but nonuniform thermal event. He has emphasized its expression during early to middle Tertiary time, but I believe the evidence also suggests that though it culminated in the Tertiary, it may have begun during the Jurassic in British Columbia, and locally was expressed during the Cretaceous in other areas, including parts of Arizona. In any event, both the physical relations and the timing indicate that at least so far as the complexes south of the Snake River Plain are concerned, the structures described do not represent a continuous inner gneissic core of the Cordilleran orogen in the classic sense.

An obvious enigma, particularly with respect to the complexes in Arizona and southeastern California, is the underlying tectonic control that gave rise to long persistent unidirectional extension. A general concensus appears to have arisen in recent years that the entire Basin and Range province is a product of such extension. This has led to suggestions that the core complexes are products of similar but earlier and differently oriented episodes of extensional deformation. If so, what are their relations to the basin-range structures of today and at what level are the present flat-lying detachment surfaces?

Only in the Rawhide Mountains (Davis and others, this volume), so far as I know, is there evidence that the basal detachment surfaces have "stepped downward" in a way as to suggest a transition from an early level of extension now brought to the surface to a younger and deeper one. In most areas both the timing and the physical relations suggest that the two episodes of extension operated sequentially and separately—the one associated with core complexes climaxed in early to middle Tertiary time, the basin-range extension mainly in late Tertiary and Quaternary time—but the details of this interaction are poorly understood.

More fundamental and far-reaching questions involve the underlying sources of heat and possible relations to deep-seated or shallow subduction along the Pacific continental margin. Several authors in this volume offer provocative suggestions along these lines and abundant theories for further testing.

REFERENCES CITED

Banks, N. G., 1980, Geology of a zone of metamorphic core complexes in southeastern Arizona: Geological Society of America Memoir 153 (this volume).

Compton, R. R., 1980, Fabrics and strains in quartzites of a metamorphic core complex, Raft River Mountains, Utah: Geological Society of America Memoir 153 (this volume).

Compton, R. R., and others, 1977, Oligocene and Miocene metamorphism, folding, and low-angle faulting in northwestern Utah: Geological Society of America Bulletin, v. 88, p. 1237–1250.

Coney, P. J., 1980, Cordilleran metamorphic core complexes: An overview: Geological Society of America Memoir 153 (this volume).

Davis, G. H., 1980, Structural characteristics of metamorphic core complexes, southern Arizona: Geological Society of America Memoir 153 (this volume).

Davis, G. A., and others, 1980, Mylonitization and detachment faulting in the Whipple-Buckskin-Rawhide Mountains terrane, southeastern California and western Arizona: Geological Society of America Memoir 153 (this volume).

Fox, K. F., Jr., Rinehart, C. D., and Engels, J. C., 1977, Plutonism and orogeny in north-central Washington; timing and regional context: U.S. Geologial Survey Professional Paper 989, 27 p.

Hose, R. K., and Blake, M. C., Jr., 1976, Geology and mineral resources of White Pine County, Nevada; Part 1, Geology: Nevada Bureau of Mines and Geology Bulletin 85, p. 1–25.

Keith, S. B., and others, 1980, Evidence for multiple intrusion and deformation within the Santa Catalina–Rincon–Tortolita crystalline complex, southeastern Arizona: Geological Society of America Memoir 153 (this volume).

Lee, D. E., and others, 1970, Modification of potassium-argon ages by Tertiary thrusting in the Snake Range, Nevada: U.S. Geological Survey Professional Paper 700-D, p. D-92–D-102.

Miller, D. M., 1978, Deformation associated with Big Bertha dome, Albion Mountains, Idaho [Ph.D. thesis]: Los Angeles, University of California.

——1980, Structural geology of the northern Albion Mountains, south-central Idaho: Geological Society of America Memoir 153 (this volume).

Misch, P., 1960, Regional structural reconnaissance in central-northeast Nevada and some adjacent areas: Observations and interpretations, in Intermountain Association of Petroleum Geologists Guidebook for 11th Annual Field Conference: p. 17–42.

Reynolds, S. J., and Rehrig, W. A., 1980, Mid-Tertiary plutonism and mylonitization, South Mountains, central Arizona: Geological Society of Ameriuca Memoir 153 (this volume).

Simony, P. S., and others, 1980, Structural and metamorphic evolution of the northeast flank of the Shuswap complex, southern Canoe River area, British Columbia: Geological Society of America Memoir 153 (this volume).

Snoke, A. W., 1980, Transition from infrastructure to suprastructure in the northern Ruby Mountains, Nevada: Geological Society of America Memoir 153 (this volume).

MANUSCRIPT RECEIVED BY THE SOCIETY JUNE 21, 1979
MANUSCRIPT ACCEPTED AUGUST 7, 1979

WITHDRAWN